Clinical Trial Biostatistics and Biopharmaceutical Applications

T0225329

Clinical Trial Biostatistics and Biopharmaceutical Applications

Edited by
Walter R. Young

Ding-Geng (Din) Chen

CRC Press
Taylor & Francis Group
Boca Raton London New York

CRC Press is an imprint of the
Taylor & Francis Group, an **informa** business

A CHAPMAN & HALL BOOK

CRC Press
Taylor & Francis Group
6000 Broken Sound Parkway NW, Suite 300
Boca Raton, FL 33487-2742

First issued in paperback 2020

© 2015 by Taylor & Francis Group, LLC
CRC Press is an imprint of Taylor & Francis Group, an Informa business

No claim to original U.S. Government works

ISBN 13: 978-0-367-57603-5 (pbk)
ISBN 13: 978-1-4822-1218-1 (hbk)

Visit the Taylor & Francis Web site at
http://www.taylorandfrancis.com

and the CRC Press Web site at
http://www.crcpress.com

To my wife, Lolita Young, whose love of Atlantic City has made me continue my chairmanship of the conference for the past 29 years. I also thank my sons, Walter and Peter, and my daughter, Katharine, for their occasional help with the conference.

Walter Young

To my parents and parents-in-law who value higher education and hard work, and to my wife, Ke, my son, John D. Chen, and my daughter, Jenny K. Chen, for their love and support. This book is also dedicated to my passion for the Deming Conference!

Ding-Geng (Din) Chen

Contents

Section I   Emerging Issues in Clinical Trial Design and Analysis

Section II   Adaptive Clinical Trials

Section III   Clinical Trials in Oncology

Section IV   Multiple Comparisons in Clinical Trials

Section V Clinical Trials in a Genomic Era

List of Figures

List of Tables

Preface

The Annual Deming Conference on Applied Statistics (http://www.demingconference.com/) has long been deemed as an important event in the statistics profession since 1945. This book is to honor the contributions of the Deming conference toward clinical trials and the biopharmaceutical industries for the last decades by inviting prominent Deming conference speakers to contribute chapters on the cutting-edge developments in biostatistical methodologies in clinical trials and biopharmaceutical applications. The novel methodology development and up-to-date biopharmaceutical application validate the importance of this book.

The book starts with two historical notes: one from Stuart Hunter from Princeton University and the other from Walter Young, Deming conference chair, which provide a historical overview of the Deming conference. Following this, the book is divided into five sections containing the contributions from prominent Deming conference speakers.

Section I, Emerging Issues in Clinical Trial Design and Analysis, consists of five chapters. Chapter 1 covers topics ranging from emerging challenges of clinical trial methodologies in regulatory applications to active controlled trial design, adaptive design, multiple comparison considerations, roles of modeling and simulation, as well as multiregional clinical trials. Chapter 2 is a review of randomization methods in clinical trials and discusses the pros and cons of some randomization procedures. Chapter 3 discusses the importance of selecting a wide range of doses in designing Phase II dose-ranging trials. Many practitioners misunderstand this critical issue and tend to add more study doses in a narrow dose range. The chapter clarifies this point. Chapter 4 is about thorough QT/QTc clinical trials. A thorough QT clinical study is required for any nonantiarrhythmic drugs in early drug development to assess the potential of drug-induced prolongation of QT interval and proarrhythmic potential since 2005. This chapter discusses the approaches and issues in the design and analysis of a thorough QT clinical trial. Chapter 5 discusses controversial (unresolved) issues in noninferiority trials, especially on assay sensitivity and the constancy assumption.

Section II, Adaptive Clinical Trials, consists of three chapters. Chapter 6 discusses the roles of adaptive designs in drug development. This chapter gives a brief overview of some of the major advances in adaptive clinical trial designs and presents several scenarios where adaptive design is worthwhile. Chapter 7 is on optimal design of group-sequential survival trials including adaptability. This chapter focuses on the survival trials that are notorious for surprises, such as far fewer events, or smaller treatment effects, than anticipated. Particular focus is on the consequences of group-sequential and adaptive designs. Chapter 8 presents the group sequential design in R.

This chapter demonstrates the `gsDesign` R package using examples of trials planned with a single futility analysis.

Section III, Clinical Trials in Oncology, contains four chapters. Chapter 9 gives an overview on issues in the design and analysis of oncology clinical trials, such as choice of treatment arms, primary endpoint, randomization scheme, trial size, use of historical information, model assumptions, and testing strategy, as well as issues such as those related to missing data, evaluability, subset analysis, competing risks, outcome-by-outcome analysis, and surrogate endpoints, which may be critical to interpretation. Chapter 10 is aimed at competing risks and their application in oncology clinical trials. Competing risks are frequently encountered in clinical cancer research, where individuals under study may experience one of two or more different types (causes) of failures, and the time to the first failure of any type is typically used for efficacy evaluation. This chapter then pays special attention to competing risks since the occurrence of first failure either prevents the occurrence of other types of failures or the disease natural history is altered due to the subsequent first-line treatment. Chapter 11 presents the escalation with overdose control (EWOC) dose finding in cancer clinical trials, which is a Bayesian adaptive dose-finding design that produces consistent sequences of doses while controlling the probability that patients are overdosed. This chapter presents the original model and several extensions, such as the incorporation of individual patient characteristics by continuous or categorical covariates and introduction of intermediate grade toxicities in addition to dose-limiting toxicities. Chapter 12 is on interval-censored time-to-event data and their applications in cancer research. Interval-censored time-to-event or failure time data occur in many areas especially in oncology and biopharmaceutics. This chapter provides a review of the recent developments in the field with a focus on the applications of existing methodologies and available software for biopharmaceutical applications.

Section IV, Multiple Comparisons in Clinical Trials, contains four chapters. Chapter 13 introduces multiple test problems with applications to adaptive designs. This chapter focuses on the general concepts behind multiplicity and some of the basic techniques for correcting for multiplicity with applications to adaptive trial design. Chapter 14 focuses on the graphical approaches to multiple testing that can be applied to common multiple test problems, such as comparing several treatments with a control, assessing the benefit of a new treatment for more than one endpoint, and combined noninferiority and superiority testing along with the graphical user interface from the `gMCP` package in R. Chapter 15 presents pairwise comparisons with binary responses and focuses on estimating simultaneous confidence intervals for multiple comparisons to a control or all pairwise comparisons, using the risk difference as the effect measure. Chapter 16 discusses the weighted parametric multiple testing methods for correlated multiple endpoints. To incorporate the correlations among the test statistics, several weighted parametric multiple testing methods have been proposed, such as Huque and Alosh's flexible fixed-sequence (FFS) testing method, Li and Mehrotra's adaptive α allocation approach (4A), Xie's weighted multiple testing correction (WMTC), Bretz et al.'s graphical approaches, and Millen and Dmitrienko's chain procedures. This chapter provides the relationship between these weighted parametric multiple testing methods and discusses the effect of misspecified correlations in these methods. Tentative guidelines to help choose an appropriate method are provided.

Section V, Clinical Trials in a Genomic Era, contains three chapters. Chapter 17 is for the statistical analysis of biomarkers from *-omics* technologies. This chapter presents an overview of classification, validation, and survival prediction methodologies, and real data examples for analyzing *-omics* data. As new data types evolve, there is ample opportunity for statisticians and bioinformaticians to develop novel methodologies for solving these big data challenges. Chapter 18 presents understanding of therapeutic pathways via biomarkers and other uses of biomarkers in clinical studies. Specifically, this chapter gives an overview of biomarker strategies applicable to clinical development, covering phases I–III, with an emphasis on early development studies, which include examples of successes and challenges associated with these approaches; study designs and analysis appropriate for different uses of biomarkers; practical considerations related to acquiring samples and developing or selecting a suitable assay; basic models useful for a quantitative approach to pharmacology, including the Hill and Gaddum models; regulatory considerations and recent FDA guidance; and concepts around biomarker validation. Chapter 19 is on statistical evaluation of surrogate endpoints in clinical studies, which emphasizes the meta-analytic approach and its information-theoretic versions. The methods are illustrated with the help of case studies.

Each chapter is self-contained with references provided at the end of every chapter, for the readers' convenience.

We would like to express our gratitude to many individuals. First, thanks to David Grubbs, from Taylor & Francis Group, for his interest in the book, and Shashi Kumar, from ITC, for assistance in LaTeX. Special thanks to the authors of the chapters, to the Deming Conference Committee, and to Xinyan (Abby) Zhang, a PhD candidate from the Department of Biostatistics, University of Alabama at Birmingham, and to Jenny K. Chen from Pittsford Mendon High School, for assistance in LaTeX and formatting, which helped to speed up the production of this book.

We welcome any comments and suggestions on typos, errors, and future improvements to this book. Please contact Ding-Geng (Din) Chen at DG.Chen@gmail.com.

Walter Young
Wayne, Pennsylvania

Ding-Geng (Din) Chen
Rochester, New York

Early History of the Deming Conference

J. Stuart Hunter
Princeton University

Today's Deming Conference has its origin at the War Production Board's first Princeton Applied Statistics Conference held on December 4, 1945, and conducted annually thereafter. To help recall 1945, remember that Franklin Roosevelt died and Harry Truman became president on April 12; Germany surrendered on May 8; and Japan on September 2, all in 1945.

The conference organizer was Princeton professor Sam. S. Wilks (1906–1964). Sam Wilks was then a member of the National Defense Research Applied Mathematics panel and the organizer of the Statistical Techniques Research Group (STRG) at both Princeton and Columbia universities. Wilks was widely recognized as a leader in the adaptations of statistics to industrial problems. His 1941 paper "Determination of Sample Sizes for Setting Tolerance Limits" (*Ann. Math. Stat.* 12, 1, 91–96) and his close association with Walter Shewhart as founding editors of the John Wiley & Sons' "Statistical Series" serve as examples of his commitment. The Princeton Applied Statistics seminars became an annual event from 1946. Recall, too, that the American Society for Quality Control (ASQC) was established from 1946, and the Shewhart Medal in 1949. The Princeton conference continued to attract academics and research workers throughout its early years.

A significant change of conference environment occurred in the early 1950s with the appearance of Box and Wilson's 1951 paper "Response Surface Methodology" (*J. Royal Stat. Soc.*, B 13, 1–45). This paper stimulated the invitation to George Box to become a visiting professor at North Carolina State University in 1953 and ultimately led to his appointment as director of a newly established STRG at Princeton in 1955. Sam Wilks' influence throughout is obvious.

Interest in experimental design applied to industrial problems soon exploded. The Metropolitan Section of ASQC held its first annual conference as part of the Princeton Conference in 1954. In 1956, the Chemical Division of the ASQC was established to offer myriad short courses on statistical applications. In 1957, George Box's paper on evolutionary operation (*Appl. Stat.* 6, 2–23) appeared, adding renewed emphasis on the importance of statistics in the industrial environment. And in 1959 the journal *Technometrics* appeared.

As an example of a conference, consider the 18th Princeton Conference cosponsored by the Metropolitan Section ASQC and held December 6–7,

1962. The meeting announcement states "The Princeton Conference has been the traditional meeting place for practitioners of statistical methods and those concerned with their theoretical development." Conference Committee: J. S. Hunter Chr. A. Bloomberg, E. Sector, and L. Pasteelnick. Registration: (luncheon, beer, and cider party) $6.00. Registration only: $1.50. The program: Acheson Duncan on CuSum charts, Cuthbert Daniel on Biased Observations, Frank Anscomb on Outliers, and Barrie Wetherill on Quantal Response Curves. Almost 100 attended.

The conference became too large to be held at the Princeton campus, and responsibility for the conference gradually changed. In 1967, the Bio-Pharmaceutical Section of the American Statistical Association (ASA) was formed and it soon joined the Metropolitan Section of ASQ in cosponsorship. The two national societies, the ASQC and the ASA were thus cosponsoring the conference. Walter Young began his service as conference chairman during this period and the conference was named the Deming Conference in 1994.

Some Nonstatistical Reminiscences of My 44 Years of Chairing the Deming Conference

Walter R. Young
Deming Conference Chair

In 1964, I was group leader, Process Analysis, at Lederle Laboratories: American Cyanamid in Pearl River, New York. I worked closely with Charlie Dunnett, whose department owned the computer and mechanical calculators my group had to use. In December, he was speaking at the Princeton Conference and he invited me to go along. The conference started at 2:00 p.m. on a Friday and ran through lunch on Saturday. It included a beer and cider party on Friday evening and a Saturday lunch for a total registration fee of $6. The topics ran the gamut of the highly theoretical from the Princeton Statistics Department to extremely simple QC talks. I enjoyed the atmosphere and camaraderie so I went every year for the next six years. I usually stayed in a cheap motel on Route 1 but occasionally splurged for the Nassau Inn.

I had joined the ASQC in 1962 and by 1969 was quite active in the Metropolitan Section's executive committee and thus was acquainted with the members of the conference's organizing committee. Harry Howard was the official conference chair but the de facto chair was Art Bobis, who headed the statistics group in Cyanamid's Bound Brook, New Jersey facility. At that point I had to go to Bound Brook every few weeks to use their analog computer to model chemical reactions. After the conference, I stayed at my usual motel and went to Bound Brook on Monday morning via public transportation. I ranted with Art about the conference on such points that most attendees were grandfathered out of paying the $10.50 registration, the Friday afternoon to Saturday schedule was weird, and there was no unified theme to the talks. After listening to me grumble for about five minutes, he told me that he was leaving Cyanamid to become vice president of his family's folding chair business and how would I like to be chair. It sounded like fun, so I accepted.

I enlisted Khushroo Shaikh, Charlie Dunnett, and Don Behnkin from Cyanamid and my old professor from New York University, John Kao, on the committee and retained Art Bloomberg who did a tremendous job on arrangements and registration until his death 25 years later. John stayed on the committee for the next 25 years and in 2012 at the age 96, we gave him a distinguished service award (Figure 1). (John died on June 8, 2014 at the age of 97.)

FIGURE 1
Young and Kao, 2013.

Stu Hunter was also extremely helpful with arrangements for the next nine years until the conference left Princeton. I got together with him once or twice a year for a luncheon at the faculty club where I gained a reputation for eating only the desserts. In 2013, we gave Stu the distinguished service award despite the fact that he was only a youngster of 90 (Figure 2).

Immediately upon becoming chairman, I changed the format to all day Friday with four 3-hour sessions of three 1-hour related talks, while retaining the luncheon, beer, and cider party. (I claim credit for enforcing the term *related*, as prior to this anyone who wanted to talk could. In a similar vein, I added the word *Applied* to the title of the conference in 1972.) We were forced to raise the registration fee to $15 but we also raised attendance to about 100 and tried to let absolutely no one in for free. The Metropolitan Section was pleasantly surprised that the conference actually made money, and with the exception of an experimental outlier year of 2001 when statistics was radically deemphasized, the conference never lost money again.

In 1971, the ASA's Biopharmaceutical Subsection assumed cosponsorship. In 1973, Bill Wooding became their first official representative and the Section has been continually active in the conference. Ivan Chan, who joined the committee in 1996, has been their longest serving agent. In 1979, the Statistics Division of the ASQC assumed cosponsorship. Bill Strawderman joined the committee in 1980 for 14 years and Frank Alt in 1983 for 11 years as their official representatives. They did an excellent job in evenly balancing

FIGURE 2
Hunter, Laroia, and Young, 2013.

biopharmaceutical topics with QC and applied topics in other industries. Once they left the committee, the conference gradually shifted to being completely biopharmaceutical and the Statistics Division became a sponsor in name only. Biopharmaceutical made sense due to the heavy preponderance of pharmaceutical firms in New Jersey, although in recent years more than 10% of our participants have been international. The conference steadily grew in attendance and length. Book sales and tutorials based on recently published texts started in 1976. The conference had a peak attendance of 476 in 1977 when it overflowed the Woodrow Wilson School. For the first time we decided to check badges. The registration fee had at this point been raised to $40 but this included meals. Literally dozens of people hadn't paid. One individual threatened to sue as he said he'd been coming for 5 years without paying and it was illegal for us to check badges without advance notice. An author of a highly acclaimed text who flew in from California also said that she had never paid in the past. Two individuals from a major drug company in NYC were stopped at the door of the meeting room. When they were next caught trying to slip through a back door, they paid, but came back 15 minutes later to ask for a refund. Their boss told them that if they couldn't sneak in, they couldn't attend. A member of the program committee, Kris Arora,

FIGURE 3
Young and Deming, 1974.

owned an excellent local Indian restaurant and he catered a very successful Indian buffet in the basement of the Statistics Building for a number of years. During the meal, four students were caught exiting from a closet where they had earlier secreted themselves.

In 1979, the conference established the W. Edwards Deming Medal in Statistical Excellence. I was undeservedly awarded the first silver medal and for the next 16 years the medal was presented at the conference to distinguished statisticians. It is now awarded by the national ASQ to QC practitioners. Ed (in Figure 3) was the keynote speaker every year and presented the medal until his death in 1993, shortly after the conference. The following year, with his daughter's permission, we renamed the conference in his honor. I was privileged to get to know Ed during this period and was a guest in both his Washington home and Manhattan apartment on a number of occasions. Ed was one of the participants in the first conference and always had a soft spot in his heart for it. He spoke at the conference several times prior to the medal's establishment. One amusing incident occurred in 1974. Ed was scheduled to speak for two hours followed by a famous pollster headquartered in Princeton. The pollster still had not shown up 30 minutes prior to his talk. I called his office, and heard in the background, "Tell them I'm not here!" Ed graciously agreed to continue to talk and did not stop until well after three hours.

The conference switched to its current format of 3 days of tutorials and two 2-day courses in 1980. One course was made the responsibility of the ASA's Biopharmaceutical Section and their agents have done an excellent job of organizing this for the last 33 years. The other one was given to the ASQC's Statistics Division but after they terminated their active sponsorship in 1999, I moderated at our calamitous year after which Fred Balch assumed control and has done a fine job for the last 13 years.

I believe that my major claim to fame was coming up with the idea of the 3-hour tutorial (preferably based on recently published texts) as opposed to three 1-hour talks on related subjects hosted by a moderator. Tutorials made their appearance in 1975 and the last 1-hour talk was essentially given in 1982. Coincident with the assumption of the current format, the conference moved to the Holiday Inn at Newark Airport in 1980, where it remained for the next five years. Until my wife compelled me to move in 1990, I was a lifelong Manhattan resident who twice failed the driving test. This obscure point is relevant, because I insisted on including information on how to get to the Holiday Inn by public transportation. My instructions in the program were to take a nonstop bus to Newark Airport, walk about a quarter of a mile along a busy exit ramp, try to safely cross it, and then go through a hole in a chain link fence. When I got there the day before the conference, I found that the hole had been repaired. I had to take a cab to a Newark hardware store, buy a pair of wire cutters, and regenerate the hole.

After the Holiday Inn, we moved to Atlantic City in 1985, where with two not too successful exceptions and one disaster, we have remained ever since. I got married in 1984 and my three kids virtually grew up in Atlantic City. My two-month-old son got thrown out of the Osmond Family production of *Fiddler on the Roof* at the Claridge for crying. I've never been paid for my tenure at the conference but in 1986 I won a drawing (attended by Mickey Mantle no less) at the Claridge for meeting planners where the grand prize was a trip on the QE2 to England and back on the Concord 11 days later. We had a great time but my wife suffered some odd complications such as physically losing her green card, being allowed on the ship without a British visa and thus having a rather difficult time getting off, and losing her brown Philippine passport two hours prior to our scheduled return, which caused her to somewhat believe in being cursed. The passport was the same color as our hotel room carpet and our son had apparently grabbed it from his crib and dumped it. These incidents somewhat assuaged my guilt for not rebooking the Claridge for 1986.

Besides attending the conference in their formative years, my oldest son redesigned our program into its current format, my youngest son (who is currently in the Air Force in Qatar) served as registrar for a year, and my daughter was awarded the conference's Walter Young scholarship.

In 1984, Karl Peace joined the program committee and for 11 years he did a superlative job in getting speakers and successful short courses. In 2010, Karl graciously accepted our invitation to present a tutorial on his latest text and we recognized him for his contributions to our conference. His coauthor, and my coeditor, Din Chen has done an equally outstanding job in replacing him for the past three years (Figure 4).

FIGURE 4
Young and Chen, 2013.

After five quite successful and enjoyable years at the Sands, we were banished as the Sands replaced their meeting rooms with a gambling floor. We had a multiple-year contract and the Sands negotiated a very favorable contract at Resorts as part of their settlement. Resorts has a very strange meeting facility on the 13th floor consisting of a large auditorium and two very long and narrow meeting rooms. There is a parapet surrounding these rooms and one can walk along the outside of the building with a rather spectacular view of Atlantic City. However, considering the December weather, not many registrants elected to exercise this option. We remained there successfully for seven years, mainly because the contract included a cocktail hour with excellent food and an open bar. This proved to be a mixed blessing as one of our moderators had to stay in the local hospital with an inebriated registrant for a number of hours until a family member came and picked him up.

In 2000, the Metropolitan Section fired me as chairman as they felt the conference should be devoted to Deming's QC principles despite the fact that he was a statistician for the greater part of his career. I did not contest this as I felt a 31-year tenure was adequate. A few weeks later they rehired me as cochair as they couldn't find anyone else to handle the statistics portion of their revamped program. All programs for the past 15 years can be found at www.demingconference.com, and the Spring 2001 program was radically different from anything seen before or since. It was held in the spring at the Newark Airport Holiday Inn because it could not attract sufficient registrants in December. It still lost quite a bit of money in the spring but as so much money had been invested in the conference (e.g., paying a professional design firm to compose the program), it was decided not to cancel. The statistics tracks did marginally better than the QC tracks but both did poorly. The meeting included meals but there were more speakers and committee members at the meals than participants. The statistics course that I moderated came close to making money but the QC course was held with a single registrant and the instructor was paid appreciably more than the registration fee. The one positive change we made that year was that Ed Warner joined the committee as transactions chair and all registrants have been given a bound copy and a CD of all of the speaker's slides ever since. I regained my sole hold on the chair and Satish Laroia (in Figure 2) was ensconced as arrangements chair and has done an outstanding job in improving and strengthening the conference's amenities.

Registration has always been a problem. In our record year at the Woodrow Wilson Center, I personally typed the 476 punched cards and wrote the listing program in Fortran. We had a number of registrars but they did not stay that long due to the hassle of opening the mail and interpreting the poor handwriting of the registration forms. Fred Balch, our currently longest serving committee member joined in 1993 as registrar and served for six years before transferring to the program committee. Our biggest technical innovation came when Wenjin Wang joined the committee in 2005 as Bibliolater. In 2006, he replaced Kalyan Ghosh (who only used the web to publicize our program) as Webmaster and designed a website that allowed for online registration, removing the need for the registrar to type all of the demographic information. We still allow for mail registration but have not received a form for several years.

For the past 11 years we've been at the Tropicana with a very high rate of recidivism as our attendees like the Havana Tower. From past experience, we're reluctant to change what has proved to be a relatively successful format. However, the conference is still evolving. Since 2008, we've awarded a $4,000 college scholarship to the offspring of a participant. For the past five years, Nandita Biswas has run a program where we support the attendance of two to four local biostatistics PhD students who present their work in a poster. In 2012, Manoj Patel, our current registrar, initiated a program where we encourage our attendees to also present posters. We've been lucky in that we've never lost a speaker due to weather. Obviously, I can't continue forever but I've already confirmed Din Chen as my speaker for 2014, my 45th year as chair.

Contributors

Ariel Alonso Abad
Department of Methodology and
  Statistics
Maastricht University
Maastricht, the Netherlands

Keaven M. Anderson
Late Development Statistics
Merck Research Laboratories
North Wales, Pennsylvania

Vance W. Berger
Biometry Research Group
National Cancer Institute
and
University of Maryland
Rockville, Maryland

Frank Bretz
Novartis
Basel, Switzerland

Tomasz Burzykowski
Interuniversity Institute for
  Biostatistics and Statistical
  Bioinformatics (I-BioStat)
Hasselt University
Hasselt, Belgium

and

International Drug Development
  Institute
Louvain-la-Neuve, Belgium

Marc Buyse
International Drug Development
  Institute
Louvain-la-Neuve, Belgium

and

Interuniversity Institute for
  Biostatistics and Statistical
  Bioinformatics (I-BioStat)
Hasselt University
Hasselt, Belgium

Ding-Geng (Din) Chen
Center for Research
and
School of Nursing
and
Department of Biostatistics and
  Computational Biology
School of Medicine
University of Rochester
Rochester, New York

James J. Dignam
Department of Health Studies
The University of Chicago
Chicago, Illinois

Yanqin Feng
School of Mathematics and Statistics
Wuhan University
Wuhan, People's Republic of China

William C. Grant
Department of Economics
James Madison University
Harrisonburg, Virginia

Stephanie Green
Pfizer, Inc.
Groton, Connecticut

Michael D. Hale
Amgen, Inc.
Thousand Oaks, California

Chen Hu
Statistics and Data Management
    Center
NRG Oncology
American College of Radiology
Philadelphia, Pennsylvania

H.M. James Hung
Division of Biometrics I
Office of Biostatistics
Office of Translational Sciences
Center for Drug Evaluation and
    Research
U.S. Food and Drug Administration
Silver Spring, Maryland

J. Stuart Hunter
Department of Civil and
    Environmental Engineering
Princeton University
Princeton, New Jersey

Bernhard Klingenberg
Williams College
Williamstown, Massachusetts

Edward Lakatos
BiostatHaven, Inc.
Croton-on-Hudson, New York

Xuewen Lu
Department of Mathematics and
    Statistics
University of Calgary
Calgary, Alberta, Canada

Ling Ma
Department of Statistics
University of Missouri
Columbia, Missouri

Jeff Maca
Center for Statistics in Drug
    Development
Quintiles, Inc.
Morrisville, North Carolina

Willi Maurer
Novartis
Basel, Switzerland

Geert Molenberghs
Interuniversity Institute for
    Biostatistics and Statistical
    Bioinformatics (I-BioStat)
Hasselt University
Hasselt, Belgium

and

University of Leuven
Leuven, Belgium

Herbert Pang
Department of Biostatistics and
    Bioinformatics
School of Medicine
Duke University
Durham, North Carolina

and

School of Public Health
The University of Hong Kong
Hong Kong, People's Republic
    of China

Scott D. Patterson
Amgen, Inc.
Thousand Oaks, California

Faraz Rahman
Williams College
Williamstown, Massachusetts

André Rogatko
Biostatistics and Bioinformatics
  Research Center
Cedars-Sinai Medical Center
Samuel Oschin Comprehensive
  Cancer Institute
Los Angeles, California

Jianguo Sun
Department of Statistics
University of Missouri
Columbia, Missouri

Mourad Tighiouart
Biostatistics and Bioinformatics
  Research Center
Cedars-Sinai Medical Center
Samuel Oschin Comprehensive
  Cancer Institute
Los Angeles, California

Naitee Ting
Boehringer-Ingelheim
  Pharmaceuticals, Inc.
Ridgefield, Connecticut

Yi Tsong
Office of Biostatistics
Office of Translational Sciences
Center for Drug Evaluation and
  Research
U.S. Food and Drug Administration
Silver Spring, Maryland

Wim Van der Elst
Interuniversity Institute for
  Biostatistics and Statistical
  Bioinformatics (I-BioStat)
Hasselt University
Hasselt, Belgium

Sue-Jane Wang
Office of Biostatistics
Office of Translational Sciences
Center for Drug Evaluation and
  Research
U.S. Food and Drug Administration
Silver Spring, Maryland

Brian L. Wiens
Portola Pharmaceuticals
San Francisco, California

Changchun Xie
Division of Epidemiology and
  Biostatistics
Department of Environmental
  Health
University of Cincinnati
Cincinnati, Ohio

Walter R. Young (Retired)
Pfizer, Inc.
Wayne, Pennsylvania

Guojun Yuan
Cubist Pharmaceuticals, Inc.
Lexington, Massachusetts

Hongyu Zhao
Department of Biostatistics
Yale School of Public Health
New Haven, Connecticut

Section I

Emerging Issues in Clinical Trial Design and Analysis

1

Emerging Challenges of Clinical Trial Methodologies in Regulatory Applications

H.M. James Hung and Sue-Jane Wang

CONTENTS

In the last two decades, the clinical trial methodology considered or employed in regulatory applications is increasingly complex. As the potential ethical issues with use of placebo draw more awareness, use of an active control as a comparative treatment arm is increasingly stipulated for assessing the effect of a test treatment. Noninferiority trial designs have therefore been revisited. On another front, even for the placebo-controlled trial, the conventional trial design has been facing a great many of challenges that arise from possible design adaptations, following an increasing use of group sequential designs. Evidentiary standard may require consideration of more innovative trial designs and inferential frameworks for studying more than one objective or endpoint in regulatory submissions. Globalization of drug development adds a great deal of complexity to trial designs and generates issues with interpretability and applicability of a global estimate of treatment effect for each geographical region. Literature has been growing rapidly on these topics. In this chapter, we shall focus primarily on a number of emerging major challenges of the clinical trial methodology on the topics: active controlled trial designs, adaptive designs, multiple comparison considerations, and multiregional clinical trials. Because of this specific focus and multiple topics, only the references very closely related to the emerging challenges under discussion are selected for inclusion in this chapter. The readers are encouraged to read the articles cited in these selected references.

1.1 Active Controlled Trial Designs

In many disease areas, active control trial designs are employed whereby a carefully selected active control treatment arm is included in the trial with or without a placebo arm. When a placebo arm is used in the trial, evaluation of a test treatment is usually based upon the direct comparison between the test treatment and the placebo. In some disease areas such as psychiatric disorders, the active control arm is used mainly to assess whether the trial has assay sensitivity for detecting a treatment difference if in truth the treatments under comparison have a difference in therapeutic effect. In some other disease areas, the active control arm is employed to compare the treatment effects between the test arm and the active control arm. In these cases, the trial objective may be to show that the test treatment is superior to or not much worse than the active control. The concept of *not much worse* is later evolved into the concept of *noninferiority*.

In recent decades, the effect of a test treatment is assessed via a direct comparison with an active control treatment in a trial without a placebo arm because of ethical reasons by which the patients in the trial cannot be given placebo. The trial design that fosters such assessment is often referred to as a *noninferiority* trial design. This type of trial design has a long history in medical research. For regulatory applications, the U.S. Food and Drug Administration (FDA) issued a draft guidance document about noninferiority trial designs without a placebo arm to discuss extensively the critical design features and statistical analysis approaches that require careful consideration. This document contains many references from the literature closely related to this topic on such designs. Also highly relevant to regulatory considerations is CHMP (2006).

For ease in discussion, without loss of generality, consider that the endpoint of interest is a major adverse clinical event (MACE) and that the goal of a medical treatment is to reduce the risk of MACE. To simplify mathematical notation, we use P, C, and T to label the placebo, active control, and test treatment, respectively, and also to denote the risks associated with these treatment arms interchangeably. As an example, T/P denotes the ratio of the risk associated with T relative to the risk associated with P.

1.1.1 Noninferiority Trial with Placebo Arm

An active control trial design that contains a placebo arm is often referred to as *gold standard design*; see Hauschke and Pigeot (2005). As mentioned earlier, when the active control trial includes a placebo arm, the direct comparison between the test treatment T and the placebo P is the basis for assessing the effect of T, that is, to reject the null hypothesis $H_{0(TP)} : T/P \geq 1$ in favor of the alternative hypothesis $H_{1(TP)} : T/P < 1$. Once the effect of T is established, another hypothesis that may be at issue is whether T is superior to or not much inferior to C. The corresponding hypothesis to support is $H_{1(TC)} : T/C < M$, where M is a margin to select prior to commencement of the trial. The corresponding null hypothesis $H_{0(TC)} : T/C \geq M$. If $M = 1$, then $H_{1(TC)}$ is the superiority hypothesis for T over C. If $M > 1$, then $H_{1(TC)}$ is that T is *noninferior* to C with M being the so-called noninferiority margin.

In most common practice, the margin M is a prespecified fixed constant. However, under such a three-arm trial design, use of a fixed constant margin M is not necessarily the best way to support a *noninferiority* assertion as arguably the degree of acceptable inferiority should depend on how effective the active control is. As such, M can be derived by requiring that a large portion of the effect of the active control be retained by the new therapy. A large portion of the control's effect can be stipulated in two ways (Hung et al. 2005). One way is to require retaining $100\lambda\%$ of the effect of the control on the risk ratio by the test treatment T if the effect of a treatment is quantified to be the reduction of the risk ratio of the treatment versus the placebo; that is,

$$\left(1 - \frac{T}{P}\right) > \lambda \left(1 - \frac{C}{P}\right).$$

By simple algebraic manipulations, we can derive that this inequality is equivalent to

$$P - T > \lambda(P - C),$$

that is, retaining $100\lambda\%$ of the effect of the control on the risk difference by the test treatment T. Consequently, the noninferiority margin for the risk ratio T/C is given by $(1 - \lambda)(P/C) + \lambda$, which is the weighted average of (P/C) and the constant 1 (which corresponds to no effect) with the weights $(1 - \lambda)$ and λ, respectively. This actually corresponds to the noninferiority margin, $(1 - \lambda)(P - C) + \lambda \dot{0}$, for the risk difference $(T - C)$. The margin is exactly the same kind of weighted average of $(P - C)$ and the constant 0 (i.e., no effect) with the same weights $(1 - \lambda)$ and λ, respectively. This is a desirable feature of retention on the risk ratio scale or the risk difference scale.

Retention on the risk ratio scale, however, has an undesirable feature in interpretation. For instance, if the active control decreases the incidence rate of MACE from 50% to 25% relative to the placebo, then the relative risk $C/P = 0.25/0.5 = 0.5$; that is, the risk reduction by administration of C compared with P is 50% $[=(1-0.5) \times 100\%]$. Decreasing the rate from 50% to 25% by C as compared with P means that use of P increases the rate from 25% to 50% as compared with use of C. Yet, the resulting risk increase by P relative to C is $(P/C - 1) = (0.5/0.25 - 1) = 1$, that is, the risk increase by P is 100%. This undesirable feature motivates constructing the margin in terms of the amount of retention required on the log risk ratio scale since $\ln(A/B) = -\ln(B/A)$, where $\ln(\cdot)$ is the natural logarithm function; that is, the magnitude of log ratio is unchanged, as the retention on the log risk ratio scale is also recommended in the FDA noninferiority draft guidance. Retention of $100\gamma\%$ of the control's effect on ln risk ratio scale yields

$$\ln(P) - \ln(T) > \gamma \left\{\ln(P) - \ln(C)\right\},$$

which results in

$$M = \left(\frac{P}{C}\right)^{\gamma} = 1^{1-\gamma}\left(\frac{P}{C}\right)^{\gamma},$$

a weighted geometric mean of 1 and (P/C). This M does not have a direct translation to the margin defined on the risk difference scale and the risk ratio scale. However, there is a one-to-one correspondence between γ and

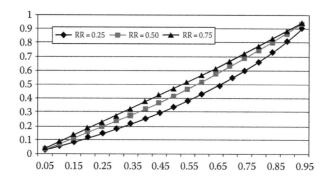

FIGURE 1.1
Relationship between γ and λ as a function of $RR = C/P$. (Excerpted from Hung, H.M.J. et al., *Biometric. J.*, 47, 28, 2005.)

λ, as given by Hung et al. (2005) as seen from Figure 1.1; that is,

$$\gamma = \frac{\log\left(1 - \lambda\left(1 - \frac{C}{P}\right)\right)}{\log(C) - \log(P)}, \; \lambda = \frac{1 - \left(\frac{C}{P}\right)^{\gamma}}{1 - \frac{C}{P}}$$

In an active control trial with a placebo arm, the ratio C/P is estimable from the trial. Thus, there is no need to fix the margin at a constant; instead, use of percent retention may better construct a noninferiority margin. The ratio T/P is also estimable. Thus, under such a three-arm trial, the two hypotheses, $H_{0(TP)}$ and $H_{0(TC)}$, can be tested with valid control of the type I error probability for each hypothesis. The overall type I error can be strongly controlled by testing $H_{0(TP)}$ first and followed by testing $H_{0(TC)}$ with the condition that testing $H_{0(TC)}$ is permitted only after $H_{0(TP)}$ is rejected.

If M is not a fixed margin, for example, $M = (P/C)^{\gamma} = 1^{1-\gamma}(P/C)^{\gamma}$, then the statistical test for $H_{0(TC)}$ can be derived from the estimators $\ln(\hat{T}/\hat{C})$ and $\ln(\hat{C}/\hat{P})$, where $(\hat{T}/\hat{C})$ is the estimator of the risk ratio (T/C). That is, the test is given by

$$Z = \frac{\ln(\hat{T}/\hat{C}) + (1 - \gamma)\ln(\hat{C}/\hat{P})}{\sqrt{\text{var(numerator)}}},$$

where var(numerator) is the estimator of the variance of the numerator. The test statistic Z is asymptotically distributed as a standard normal under the null hypothesis $H_{0(TC)}$: $\ln(P) - \ln(T) > \gamma\{\ln(P) - \ln(C)\}$, equivalently, $\ln(T) - \ln(C) > -(1 - \gamma)\{\ln(C) - \ln(P)\}$. The rejection region of $H_{0(TC)}$ can be easily constructed.

1.1.2 Noninferiority Trial without Placebo Arm

If use of a placebo is not permissible (due to ethical reasons) in the active control trial, the effect of T can only be assessed via the direct comparison between T and C. In regulatory applications, for efficacy or effectiveness assessment, the objectives of such a two-arm trial are (1) to assert that T is effective or efficacious in the sense that had a placebo been studied the test arm would have been superior to the placebo and (2) to assert that T is not much inferior to C. However, the absence of P makes statistical inference extremely difficult under this noninferiority trial setting. In one aspect, determination of the noninferiority margin is extremely difficult because the noninferiority trial does not provide data that can be used to estimate C/P. Relevant historical placebo-controlled trials may need to be identified to provide guidance for projecting the magnitude of C/P under the noninferiority trial setting. This margin selection process involves integration of the identified historical trials using meta-analysis methods or alike for estimating the effect of the selected active control under several relevant historical trial settings. However, most often in practice, planning prospectively such a meta-analysis is rarely feasible since the results of these historical trials are mostly known already, at least, from the literature. The inability to do the necessary prospective planning for the meta-analysis will have an adverse impact on selection of a noninferiority margin because the selection involves subjective judgment based on the knowledge of some results of the trials that may potentially be either selected or not selected. Furthermore, in many situations, the relevant historical data available for the comparison of the active control to placebo are limited or inconsistent and the between-trial variability in the estimated effects of the active control may not be estimable. Regardless of how the identification is made, which involves usually subjective judgment, the historical trial settings and the noninferiority trial setting are rarely comparable with respect to the important observable and unobservable covariates, such as the background medications used, and study endpoints. Such incomparability poses many constraints to the design and analysis of the noninferiority trial. In the design aspect, for margin selection, the estimate of the active control effect derived from the selected historical trials would require some kind of discounting in projecting the effect of the active control under the noninferiority trial setting because of the two unverifiable critical assumptions the noninferiority inference relies on. The two assumptions are (1) the assay sensitivity assumption that the noninferiority trial is able to detect a treatment difference if in truth the two treatments under comparison differ in their effects, and (2) the constancy assumption that the effect of the active control remains virtually unchanged from the historical trial settings to the current noninferiority trial setting. Because of all these difficulties, the FDA noninferiority draft guidance document lays out a two-step process for margin selection. The process begins with selection of M_1, the effect of the active control in the noninferiority trial, followed by the selection of M_2, the proportion of M_1 to retain, and the noninferiority margin M is then set to M_2. In a two-arm active control trial without a placebo, the FDA guidance document recommends that M be a fixed constant constructed according to the two-step process and needs to be prespecified prior to commencement of the trial.

After M is selected, the noninferiority hypothesis $H_{1(TC)} : T/C < M$ is defined and tested using a 95 or larger percent confidence interval about the ratio T/C in order to rule out M. The sample size or the amount of statistical information (e.g., number of events) for the noninferiority trial can then be planned to rule out M with sufficient statistical power.

1.1.3 Margin Determination in Noninferiority Trial without Placebo

The literature provides a number of viable approaches for selecting M_1. One approach is use of the worst limit of 95 or greater percent prediction, confidence or Bayesian credible interval of the active control effect estimated from the historical trials as the estimate of the effect of the active control in the noninferiority trial. In some situations, for example, when the test treatment and the active control are in the same drug class, use of a 90 or smaller percent interval can be considered. Hung et al. (2007) explored the potential utility of *across-trial* type I error probability as guidance for selection. The essential condition behind this approach is to attain a small but not fixed level of the across-trial type I error probability. The across-trial type I error probability depends on the amount of statistical information (e.g., sample size or total number of events) of the selected historical trials as a whole and the amount of statistical information of the planned noninferiority trial. Since the amount of statistical information that needs to be planned for the noninferiority trial is not yet available, it seems not reasonable to require that this across-trial error be controlled at a fixed level. Wang and Blum (2011) explored use of a likelihood measure as guidance for selection. Hung and Wang (2013) present a Bayesian-based approach that treats the active control effect as a random parameter to account for statistical uncertainty. Regardless of which approach is used to guide selection of M_1, the resulting estimate may or may not be conservative because of many aforementioned challenges to projection of the effect of the active control under the noninferiority trial setting from the frequently limited and inconsistent historical data. Further discounting may need to be considered in selection of M_1. Certainly, subjective judgment is rendered in the selection.

Selection of M_2 from M_1 is, in most practices, based on subjective judgment that incorporates the consideration of how much loss of the active-control effect may be acceptable and the feasibility of the noninferiority trial. However, arguably the acceptable extent of loss of the active-control effect by use of the test treatment should not be compromised by the feasibility of conducting the noninferiority trial.

A so-called *biocreep* problem that is often of great concern in considering use of noninferiority trial designs is described as follows. Suppose that a test treatment A is shown effective by showing noninferiority to an active control C but in truth A is a little worse than C. Suppose that a new treatment B is shown effective by showing noninferiority to A but in truth B is a little worse than A. Such a chain of noninferiority testing will eventually lead to approve an ineffective new treatment by showing noninferiority to the treatment just prior to this new treatment. That is, had a placebo been in that noninferiority trial, the new treatment would have been inferior to the placebo. Now, instead, suppose that treatment A has been shown to be effective by showing superiority over the control C. To assess the effectiveness or efficacy of the new treatment B, treatment A may be a better active comparator than C. The use of A as the active control for assessing B may mitigate the concern of *biocreep*. Hung and Wang (2013) stipulate a viable approach to derive a reasonable M_1 that may be larger than the margin derived if C is the comparator. However, selection of M_2 based on the former M_1 may in fact be smaller if the acceptable loss of the clinical benefit (e.g., survival) is smaller and consequently the final noninferiority margin M may still be smaller.

1.1.4 Intent-to-Treat versus Per-Protocol versus On-Treatment Analysis

For showing a difference between treatments, statistical testing is commonly based on the intent-to-treat (ITT) principle by analyzing all patients according to their randomized treatment assignment. It is well known, however, that the ITT-based analysis may yield a bias in favor of falsely showing noninferiority simply because of sloppiness in trial conduct such as patient's noncompliance in taking the assigned treatment or dropping out of the trial. Wang and Dunson (2011) point out that misclassification at baseline may increase the probability of falsely showing noninferiority. As an alternative, the per-protocol (PP) analysis that uses only the patients who are in compliance with the study protocol was often argued favorably in some literature to replace the ITT analysis as the primary analysis for testing noninferiority. However, by excluding some patients, the PP analysis does not adhere to the randomization principle; consequently, the PP analysis is subject to selection bias, which is extremely difficult to quantify and severely affects the interpretability of noninferiority test results.

The primary analysis for noninferiority testing remains a difficult open question, though this topic has been researched in the literature, such as Wiens and Rosenkranz (2013) and the references cited therein. For time to event endpoints, the on-treatment (OT) or as-treated (AT) approach has been proposed in regulatory applications. This approach captures all randomized patients but censors the events that occur beyond a time window after discontinuation of their assigned treatment arms. As an example, in a number of investigational new drug applications using warfarin as the active control in noninferiority trials, the OT analysis was initially defined using 7 days as a time window; that is, the events occurring 7 days or later after patient's permanent discontinuation of the assigned study medication in each treatment arm were censored from the analysis and not counted against the treatment arm assigned to that patient. While the idea behind such censoring in the OT analysis seems sensible, difficulty lies in how to select the time window. In addition, the events censored may not be equally relevant or irrelevant to each treatment arm; that is, the relevant time window for censoring may not be the same for each treatment arm. Moreover, who should select the time window? How can it be certain that those censored could never be the clinical sequelae associated with the assigned treatment arms? For analysis, because of the uncertainty about time window, there may be a tendency to use a maximum time window as an initial selection and then performing multiple analyses based on multiple time windows to look for internal consistency. Depending on the noninferiority decision tree, there may be a multiplicity issue. For example, the decision tree that if at least h of K analyses rule out their respective predefined margins then noninferiority indexed by the margins is concluded will definitely cause a multiplicity issue.

1.1.5 Testing for Superiority and Noninferiority

There is a close relationship between the noninferiority hypothesis and the superiority hypothesis. The noninferiority hypothesis $H_{1(TC)} : T/C < M$ is indexed by the margin M, which is greater than one. Thus, for a particular clinical endpoint of interest, the parameter space for the superiority hypothesis is a subspace of that of any noninferiority hypothesis defined by the margin M. If the same confidence interval can be used to rule out M or the constant 1, then testing superiority and noninferiority can be performed in any order without multiplicity adjustment; see Morikawa and Yoshida (1995). In regulatory applications, the topic of testing noninferiority and superiority in a noninferiority clinical trial is often complicated because there may be multiple endpoints being tested. One reason is that when the test treatment is shown to be *noninferior* but not superior to the active control on the primary endpoint, it is desirable that the test treatment is superior to the active control on at least one important secondary endpoint. Thus, for regulatory applications, it is desirable to consider in trial planning the possibility of superiority testing after a clinical endpoint shows noninferiority. Then there needs to be a plan to control the familywise type I error probability associated with testing superiority and noninferiority for multiple endpoints. As the endpoints are often tested following a hierarchical scheme in a prespecified order only after the preceding endpoints achieve statistical significance, the strategy for testing superiority and noninferiority for each endpoint can be very challenging in terms of possibly illogical interpretation of the trial results and a combination of alpha allocation and alpha propagation, as articulated in

Hung and Wang (2010). An example of an alpha allocation strategy for testing superiority and noninferiority on one primary endpoint is given in Figure 1 of Hung and Wang (2010).

Interim analyses that may lead to termination of the noninferiority trial based on a hard clinical endpoint, such as mortality, will create additional challenges in controlling type I error, as Hung and Wang (2013) point out. For instance, suppose a nonmortality primary endpoint will be tested for noninferiority and then for superiority after noninferiority is concluded. Mortality as a secondary endpoint will be tested after the primary endpoint achieves noninferiority. However, mortality will be monitored by formal interim analyses with a prespecified alpha spending function during the course of the trial. The mortality may lead to early termination of the trial to declare the effectiveness of the test treatment. If mortality does not lead to early termination, then the trial will continue to the end and follow the planned hierarchical testing for the primary endpoint and then for mortality. As Hung and Wang (2013) point out, additional consideration for alpha management is necessary to ensure that the overall type I error probability is properly controlled.

1.2 Adaptive Design

A commonly used design, often referred to as *fixed design*, for an evidence-setting or *confirmatory* clinical trial, relies on prespecification of major design elements that are not altered during the course of the trial. Modification of such design elements may have an undesirable impact on interpretability of the trial results. Therefore, when the trial design is modified or adapted during the trial, use of proper statistical methods for testing and estimation is required to ensure statistical validity.

Two classes of adaptive designs increasingly seen in regulatory applications are adaptive statistical information designs and adaptive selection designs. Following the seminal article of Bauer and Köhne (1994), a great deal of advancement in statistical methodology has been made, and the literature is still fast growing for such adaptive designs. A sample of articles relevant to this topic, published within the last 10 years, are Chen et al. (2004), Bauer and König (2006), Branson et al. (2005), Hung et al. (2006a,b, 2011), Chow and Chang (2006), Wang et al. (2007, 2009, 2010, 2012, 2013), König et al. (2008), Liu and Andersen (2008), Gao et al. (2008), Brannath et al. (2009), Mehta et al. (2009), Posch et al. (2010), Mehta and Pocock (2011), Hung and Wang (2011, 2013), and Wang and Hung (2013), reflection paper by EMA 2007, and the articles cited therein. In 2010, the FDA issued a draft guidance document on adaptive designs to highlight some critical issues that require careful attention when adaptive designs are employed for regulatory applications (FDA 2010).

1.2.1 Adaptation of Statistical Information

Sample size or in broader sense statistical information (e.g., number of events for time-to-event analysis) is one of the basic design elements that are critical to success or failure of a clinical trial under the statistical testing paradigm. However, planning the amount of statistical information needed for a trial to be successful is not a simple task and requires many assumptions, such as the postulated treatment effect to detect and the variance of the response variable of interest. Benefit–risk assessment, which most often can be made only after the trials are completed, may be needed, together with a profit margin, in selecting the magnitude of treatment effect to detect. Thus, sample size re-estimation at some point of an ongoing evidence-setting trial may be needed. Moreover, in regulatory applications, a pediatric *written-request* trial can give the sponsor 6 months patent exclusivity for their medical products if the trial provides *interpretable* information in pediatric patients regardless of whether the trial achieves statistical significance or not. To meet this objective, regulatory agencies may require that the trial have sufficient statistical power to detect a treatment effect so that a trial that does not achieve statistical significance is still interpretable; thus, sample size re-estimation may be demanded. It is well known, however, that sample size adjustment based on a treatment difference that is observed during the course of the trial can substantially compromise the type I error probability. In what follows, this type I error problem and the statistical methods for dealing with this problem will be discussed.

Let T denote the test treatment, C the control comparator, and N the sample size per arm for evaluating the parameter of mean treatment difference, $\Delta = \mu_T - \mu_C$, of a normally distributed response variable Y, where μ_g is the mean change at a targeted postrandomization time point from baseline for treatment arm g ($g = T$ or C), assuming for simplicity that the common standard deviation $\sigma = 1$ for Y. The main goal of the trial is to test the null hypothesis $H_0 : \Delta = 0$ against the alternative hypothesis $H_1 : \Delta \neq 0$, where a positive Δ indicates that T is more beneficial than C. The postulated magnitude of the treatment effect or difference between T and C is denoted by $\delta > 0$, which is to be detected at the one-sided level of statistical significance α (traditionally, one-sided $\alpha = 0.025$ or less, say) and power $1 - \beta$, using the test statistic

$$Z = \sqrt{\frac{N}{2}} \hat{\Delta}, \tag{1.1}$$

where $\hat{\Delta}$ is the estimator of Δ based on all $2N$ patients. For notational convenience, we shall use the terms *estimator* and *estimate* interchangeably.

During the course of the trial, when the first n patients in each arm have contributed data for Y, the treatment effect Δ can be estimated. Denote the estimate by Δ_t, where $t = n/N$, the information time, and $n < N$. Depending on how hopeful the interim estimate Δ_t is as compared to δ, one may elect to increase the per-arm sample size N to N^* such that the statistical power for detecting a treatment effect in the vicinity of Δ_t or the conditional power evaluated based on a future projection of Δ is sufficiently large. Under this adaptive statistical information design, N^* is a random variable depending on Δ_t, and in practice N^* is capped by a limit N_{max}, which is most often fixed. Note that under the fixed design, the conventional Z test statistic given in (1.1) can be expressed as

$$Z = \sqrt{t}Z_t + \sqrt{1 - t}Z_{1-t},$$
(1.2)

where

$$Z_t = \sqrt{\frac{n}{2}}\Delta_t, \quad Z_{1-t} = \sqrt{\frac{(N - n)}{2}}\Delta_{1-t}$$

and Δ_{1-t} is the estimate of Δ based on the Y data from the $(N - n)$ patients in each treatment arm after the information time t. With possible sample size adaptation, the conventional Z test statistic becomes $Z^* = \sqrt{N^*/2}\hat{\Delta}^*$, where $\hat{\Delta}^*$ is the estimate based on the sample size of N^* per arm, which can be expressed as

$$Z^* = \sqrt{\frac{n}{N^*}}Z_t + \sqrt{\frac{1 - n}{N^*}}Z^*_{\gamma - t},$$
(1.3)

where

$$\gamma = \frac{N^*}{N}, \quad Z^*_{\gamma - t} = \sqrt{\frac{(N^* - n)}{2}}\Delta_{\gamma - t}.$$

It is well known that Z^* is not a standard normal random variable under H_0 and use of the critical value z_α, the αth upper percentile of the standard normal distribution, can substantially increase the type I error probability; see Proschan and Hunsberger (1995). The inflation depends on the time t, sample size increase algorithm denoted by D_t, final sample size N^*, and the fixed maximum sample size N_{max} for N^*.

A natural approach to statistical adjustment is by using a constant critical value c_α larger than z_α for Z^* in (1.3) to control this type I error probability at level α; that is, solve for c_α by setting

$$P\left(Z^* > c_\alpha | \Delta = 0\right) = \int P\left(Z^*_{\gamma-t} > \frac{c_\alpha - \sqrt{n/N^*}Z_t}{\sqrt{1 - n/N^*}} \middle| Z_t = z_t : \Delta = 0\right) \phi(z_t)\, dz_t$$

$$= \int \Phi\left(-\frac{c_\alpha - \sqrt{n/N^*}Z_t}{\sqrt{1 - n/N^*}}\right) \phi(z_t)\, dz_t$$

to α. It is worth noting that this constant critical value c_α depends on the design specification $(t, D_t, N^*, N_{\max})$; that is, it is likely to change as the design specification is changed. An alternative approach of statistical adjustment is by fixing weights applied to the data before and after the sample size re-estimation, for example, in the general class of weighted Z test statistics by Lehmacher and Wassmer (1999) and Cui et al. (1999), that is,

$$Z_w = \sqrt{w}Z_t + \sqrt{1 - w}Z^*_{\gamma-t}$$

where w is a fixed constant. The Cui–Hung–Wang (CHW) test statistic considered in Cui et al. (1999) is the weighted Z test statistic using $w = t = n/N$, the original weight in the conventional Z test statistic in (1.2). Under H_0, Z_w follows the standard normal distribution and its α-level critical value is z_α. A potential controversy with the use of Z_w is that unlike Z^* the test Z_w does not have the desirable *one patient one vote* or *all patients are equal* property because the final sample size N^* is a random variable, though $N^* \le N_{\max}$. This controversy should not be overemphasized because even under the fixed design several test statistics commonly used under generalized linear modeling do not possess this desirable property.

Muller and Schafer (2001) introduce the viable concept of controlling conditional type I error probability to control the (unconditional) type I error probability that is usually an issue. Chen et al. (2004) propose a concept of adapting sample size or the amount of statistical information only under some condition, such as when the conditional power at an information time is larger than 50%. Also using the concept of Muller and Schafer (2001), Gao et al. (2008), and Mehta and Pocock (2011) propose a *promising zone* approach whereby sample size could be increased only when Δ_t falls into a promising zone. The promising zone approach employs the critical value $c_\alpha(Z_t)$, which is a function of the statistic Z_t based on the data from the first n patients per arm, by equating the conditional type I error probability of Z^* to that of the CHW test. As Hung et al. (2014) point out, the conversion from the weighted Z test statistic to the conventional test statistic Z^* does not seem fruitful because it may create unnecessary confusion in application. In one aspect, the weighted Z test is always valid and flexible, irrespective of any sample size adaptation rule, but Z^* is valid only if there is full compliance with critical value function $c_\alpha(Z_t)$ and the sample size adaptation rule. Moreover, the required full compliance cannot be verified in practice. Thus, when the compliance is in doubt, for example, when the final sample size is short of the planned sample size for some reason, application of the weighted Z test may be needed before using the Z^* test to ensure that statistical validity is achieved. In another aspect, the weighted Z test provides a straightforward way to compute the p-value and confidence interval, whereas the Z^* test would need to rely on numerical integration and numerical differentiation for finding the solutions to the relevant estimating equations for the estimators, unless Z^* is converted back to the weighted Z test statistic. Thus, the conversion back to the weighted Z statistic may be best to obtain the p-value and confidence interval for Δ (Lawrence and Hung 2003). Such compliance issues deserve attention in regulatory applications.

1.2.2 Adaptive Selection

Adaptive selection employs the data accumulated up to some point in the trial to make a selection, change of dose(s) or patient subgroup(s) to target in the remainder of an ongoing trial. The selection can be based on data on the primary efficacy endpoint at stake or on data of some biomarkers thought to be predictive of the effect of the test treatment. If the trial is to provide intended conclusive evidence for the selected dose or patient population, the overall type I error control and statistical interpretability of the trial results are at issue.

An important question relevant to regulatory decisions is whether the conventional intention-to-treat analysis that might be impacted by changing the initially targeted dose or patient population remains statistically valid. Another question is whether adaptive selection based completely or partially on marker data will add more statistical uncertainty than the adaptive selection based on purely the primary efficacy endpoint data.

A simple viable statistical framework based on a two-stage design can be constructed to foster discussion. Consider that the data collected at stage 1 (at information time t) on the primary efficacy endpoint and/or the markers of interest are used to select either the original intention-to-treat patient cohort or a prespecified patient subgroup for testing the test treatment on the primary efficacy endpoint in the end of the trial. Let G_0 be the initially prescribed overall intention-to-treat population, G_1 the prespecified patient subgroup, G_2 the complementary subgroup, Δ_i the mean effect of the test treatment relative to the control comparator in patient cohort $i(i = 0, 1, 2)$ on the primary efficacy endpoint, Z the test statistic for detecting this mean effect parameter Δ, and $\mathbf{W}$ a vector of statistics for the markers. At time t, a decision will be made to either enrich to the G_1 population subgroup or stay with G_0 for collecting data at the second stage using a prespecified binary decision criterion $D_t \equiv h(Z_t, \mathbf{W_t})$, where $(Z_t, \mathbf{W_t})$ is the vector of statistics on the primary efficacy endpoint and biomarkers for patient cohort i, and h is a prespecified measurable function. $D_t = 1$ if only the G_1 patients will be enrolled at the second stage and 0 otherwise.

If the total sample size per treatment arm is fixed and unchanged regardless of the selection at the time t, the intention-to-treat analysis of all $2N$ patients yields the overall Z test procedure characterized by either

$$Z_0 = \sqrt{t}Z_{0t} + \sqrt{1-t}Z_{0,1-t},$$

if the interim decision is to stay with G_0, or

$$Z_1^* = \sqrt{t}Z_{0t} + \sqrt{1-t}Z_{1,1-t},$$

if the interim decision is to study G_1 only, where Z_{0t} is the test statistic on the G_0 data up to time t, $Z_{0,1-t}$ the test statistic on the G_0 data after the time t when $D_t = 0$, and $Z_{1,1-t}$ the test statistic on G_1 after the time t when $D_t = 1$. Hung et al. (2012, 2014) point out that this overall Z test procedure using z_α as the critical value is statistically valid in terms of type I error control under the global null hypothesis $H_0 : \Delta_0 = 0$ and $\Delta_1 = 0$. However, the type I error probability associated with testing the treatment effect Δ_1 for G_1 by use of

$$Z_1^{**} = \sqrt{t_1}Z_{1t} + \sqrt{1-t_1}Z_{1,1-t},$$

where t_1 is the information fraction of the G_1 subgroup by the adaptive selection, is

$$
\begin{aligned}
T1ER(G_1) &= P\{Z_1^{**} > z_\alpha | \Delta_1 = 0\} \\
&= P\left(Z_{1,1-t} > \frac{z_\alpha - \sqrt{t_1}Z_{1t}}{\sqrt{1-t_1}} \,\middle|\, \Delta_1 = 0 \right) \\
&= E\left[E\left\{ \Phi\left(-\frac{z_\alpha - \sqrt{t_1}Z_{1t}}{\sqrt{1-t_1}} \,\middle|\, Z_t, W_t, \Delta_t = 0 \right) \right\} \right].
\end{aligned}
$$

This type I error probability is most likely not equal to α and generally not evaluable unless the joint distribution of (Z_t, W_t) is known. The same issue applies to the overall test procedure given by either Z_0, if $D_t = 0$, or Z_1^{**}, if $D_t = 1$. That is,

$$
\begin{aligned}
T1ER &= P\{Z_0 > z_\alpha, D_t = 0\} + P\{Z_1^{**} > z_\alpha, D_t = 1\} \\
&= P\left(Z_{0,1-t} > \frac{z_\alpha - \sqrt{t}Z_{0t}}{\sqrt{1-t}}, D_t = 0 \right) + P\left(Z_{1,1-t} > \frac{z_\alpha - \sqrt{t_1}Z_{1t}}{\sqrt{1-t_1}}, D_t = 1 \right) \\
&= P\left(Z_{0,1-t} > \frac{z_\alpha - \sqrt{t}Z_{0t}}{\sqrt{1-t}}, D_t = 0 \right) + P\left(Z_{0,1-t} > \frac{z_\alpha - \sqrt{t_1}Z_{1t}}{\sqrt{1-t_1}}, D_t = 1 \right),
\end{aligned}
$$

which may or may not be exactly α and may not be evaluable. The type I error problem can be worse if t_1 depends on $\mathbf{W_t}$.

For estimation, adaptive selection can result in an overestimation bias to the naive estimator of Δ and its variance. Let f be the prevalence of the G_1 patients in G_0. The treatment effect estimate that is ultimately observed for the potentially modified overall population is based on the naive estimator given by

$$\hat{\Delta}_0^* = t\hat{\Delta}_{0t} + (1-t)\left\{\hat{\Delta}_{0,1-t}I[D_t = 0] + \tilde{\Delta}_{1,1-t}I[D_t = 1]\right\},$$

where

$$\hat{\Delta}_{0,1-t} = f\hat{\Delta}_{1,1-t} + (1-f)\hat{\Delta}_{2,1-t},$$

$(\hat{\Delta}_{1,1-t}, \hat{\Delta}_{2,1-t})$ are the estimators based on the sample sizes $N(1-t)f$ and $N(1-t)(1-f)$ for G_1 and G_2, respectively, $\tilde{\Delta}_{1,1-t}$ is based on the sample size $N(1-t)$ for G_1, and $I(E) = 1$ if event E and 0 otherwise. By the same arguments as given in Hung et al. (2012, 2013), the adaptive selection based on the value of the interim statistics $(Z_t, \mathbf{W_t})$ yields the relative bias given by

$$\frac{b\left(\hat{\Delta}_0^*\right)}{\Delta_0} = \frac{\left\{E\left(\hat{\Delta}_0^*\right) - \Delta_0\right\}}{\Delta_0}$$

$$= (1-t)P(D_t = 1)\left(\frac{\Delta_1}{\Delta_0} - 1\right)$$

$$= \frac{(1-t)P(D_t = 1)(1-f)(\Delta_1 - \Delta_2)}{\Delta_0}.$$

Hence, if $\Delta_1 = \Delta_2 = \Delta_0$, the naive estimator $\hat{\Delta}_0^*$ of Δ_0 is still unbiased for Δ_0 on the initially planned overall population regardless of how the prevalence is at the time t. When $\Delta_1 > \Delta_0$, it is not surprising that the treatment effect on the initially intended overall population is overestimated. The bias of the overall effect estimator $\hat{\Delta}_0^*$ increases as the adaptive selection is performed earlier, the chance of selecting the G_1 patients at the second stage is larger, or the prevalence of the G_1 patients is smaller. Given the values of Δ_1 and Δ_0, the probability of selecting the G_1 patients will likely be an increasing function of the prevalence rate f. Thus, given the decision rule D_t at the given time t, the factor $P(D_t = 1)(1-f)$ in the bias formula will likely be bounded away from one. The bias would largely be driven by the difference between Δ_1 and Δ_2.

The variance of the naive estimator is

$$\text{Var}\left(\hat{\Delta}_0^*\right) = \frac{\sigma^2}{N} + (1-t)^2(\Delta_1 - \Delta_0)^2 P[D_t = 0]P[D_t = 1]$$
$$+ 2t(1-t)(\Delta_1 - \Delta_0)E\left\{\left(\hat{\Delta}_{0t} - \Delta_0\right)I[D_t = 1]\right\}.$$

Therefore, when $\Delta_1 > \Delta_0$, the variance may also be inflated, which is often the case for adapting to G_1 patients. If $\Delta_1 = \Delta_2 = \Delta_0$, there is no inflation of the variance of the naive estimator $\hat{\Delta}_0^*$.

The belief that adaptive selection based on the marker data does not add more uncertainty to statistical testing is unfounded. On the contrary, adaptive selection with biomarker data can generally lead to a substantial inflation of the type I error probability and the degree of inflation might not be evaluable, unlike adaptive selection based purely on the primary efficacy endpoint. Adaptive selection can also bias the estimation of the treatment effect and its variance.

1.2.3 Trial Conduct and Logistics

The more critical issues with adaptive designs concern whether the trial conduct is compromised by an unblinded examination of the trial data for adaptation at the interim analysis and any logistics with planning and conducting the trial. Observing the interim data path may generate anxiety that can be an issue with classical group-sequential designs and certainly even more with adaptive designs. Adaptive designs open rooms for potential influences on the trial conduct that may lead to operational biases that are not quantifiable. The FDA draft guidance document on adaptive designs pinpoints the criticality of all these issues and tremendous difficulties in handling them properly in practice so that the trial results are interpretable. On the one hand, the interim trial data would need to be handled by a party that is independent of the sponsor, such as Data Monitoring Committee (DMC) employed for handling group sequential trials or safety monitoring. On the other hand, it is unclear whether a DMC in a current form has a proper role in recommending the prespecified adaptation that always involves much more complex trial conduct issues. Other big issues include the sponsor's role and the communication between the parties that handle the interim data for adaptation and the sponsor. A standard operation procedure (SOP) must be in place to address the issues of *who see what?*, what knowledge of the internal trial needs to be protected from investigators or patients and the sponsor management, how to minimize the possible impact of adaptation on the investigator and patient behaviors, the ability of the SOP to prevent information leaking, etc. Measures would need to be developed to assess the ability to comply with the SOP, check and monitor quality of compliance, ensure that confidentiality is protected from leaking the internal data or the adaptation process to internal constituents and external communities. Ultimately, when an adaptive design trial is submitted to the regulatory agencies, whether the trial results are reviewable is definitely a big issue, as the regulatory agencies may need to know every detail of how the trial proceeded and hence the paper trail for trial conduct regarding adaptation would be needed. All these issues will be related to the infrastructures and software that are necessary to be able to capture the internal trial data path and every detail of how the trial proceeded, have simulation capability to evaluate design efficiency, and detect and model the impact of the trial misconduct or noncompliance.

1.2.4 Issues of Statistical Efficiency

The statistical efficiency of adaptive designs, particularly adaptive statistical information designs, has been challenged in the literature such as Emerson et al. (2011) and the relevant articles cited therein. The key message is that there are well-planned group sequential designs that can have better statistical efficiency than adaptive statistical information designs. This topic seems to remain controversial in the sense that the traditional notion of statistical efficiency is not the only criterion for design comparison. While statisticians often tend to rely on *statistical efficiency* as guidance for design comparison, the reality is that not only individual trial by itself is at stake. In this sense, depending on the extent of complexity in the practical scenario, a well-considered adaptive design can remain to be a good choice.

1.3 Multiple Comparison Considerations

As clinical trial designs are increasingly more complex, it can be expected that multiple testing or multiple comparison problems are more difficult to handle properly. This indeed has occurred in regulatory applications. Multiple doses, multiple endpoints, together with multiple times of analyses during the course of a trial are increasingly seen in study designs. The convention that is often adopted for type I error control applies to a trial per se as an experiment; that is, the studywise or overall type I error probability is an important issue with interpretation of the results of the confirmatory trials in regulatory applications.

In some cases such as cardiovascular large outcome trials, often at most two doses of a test treatment are studied and not expected to yield very different treatment effects on the clinical outcome endpoints. Pooling doses is, therefore, entertained to mitigate the complexity of the multiple comparison. However, as Hung et al. (2013) point out, pooling the doses may not be desirable as it takes away the opportunity of drawing a definitive dose conclusion for each individual dose, when the effects of the doses are substantially different. A good example is RE-LY trial (Connolly et al. 2009) that studied the effects of dabigatran at 150 and 110 mg twice daily on stroke or systemic embolism in patients with nonvavular atrial fibrillation. Multiplicity issues with the efficacy endpoints generally can be more complex than those with doses. This is because potentially additional efficacy claims may be involved to gain market advantages. Depending on the type of the endpoint, it can be tested using repeated statistical significance testing over time during the course of the trial, that is, under a group sequential design. Recent advances for possible adaptation of the trial design would definitely add complexity to the type I error control problem.

1.3.1 Primary Endpoint versus Secondary Endpoint

Categorization of endpoint into *primary* versus *secondary* depends on many factors, such as whether the endpoint can capture meaningful clinical benefits or risks, whether the postulated treatment effect can be detected with sufficient power, etc. There is possibly another category—tertiary endpoints, which mostly are only for exploration purposes. In practice, usually there is only one primary endpoint that is selected based on the primary clinical hypothesis under study. This does not mean that the secondary endpoints are always less important than the primary endpoints. For instance, in most cardiovascular large outcome trials, a composite of death or some other irreversible clinical outcomes may be a primary endpoint while the mortality endpoint as a secondary endpoint is in fact more important to patients. In practice, there is usually not a clear line to draw between *primary* and *secondary*. Nonetheless, some secondary endpoints should be used only for descriptive purposes or for providing information to support the important endpoints that yields claims. Given the vagueness described earlier, controlling experimentwise or studywise type I error probability seems to be a good scientific principle for analysis of multiple endpoints. Those secondary endpoints that are not claim based should be excluded from the formal statistical testing.

The FDA draft guidance document on multiple endpoints in clinical trials makes an attempt to set clear distinction between primary endpoints and secondary endpoints, based on a hierarchical order of importance. The primary endpoint(s) is that which will be the essential basis for supporting a claimed efficacy effect, whereas the secondary endpoints are those that may provide useful additional, supportive information about a drug's effect but are either not sufficient or not required to demonstrate efficacy. Therefore, there may be multiple primary endpoints, just as there may be multiple secondary endpoints. Under this hierarchically structured framework, the necessity of controlling the conventional studywise type I error probability, that is, including the primary endpoint family and the secondary family, may be questionable, if the secondary endpoints by themselves cannot lead to an additional claim. However, even with this distinction, where a serious clinical outcome endpoint such as mortality should be placed often remains unclear. On the one hand, mortality should belong to the primary endpoint family because it clearly signifies perhaps the most important clinical benefit. On the other hand, if the mortality rate is dismal, then placing it in the primary endpoint family may demand more statistical adjustments to the alpha level because the mortality is always evaluated for safety consideration even though the mortality rate is too low to have a chance to detect a potential efficacy. Therefore, it seems desirable to place all the endpoints that can generate claims, respectively, by themselves in the primary endpoint family. Once the primary endpoint family and the secondary endpoint family are clearly defined following this hierarchical ordering structure, it can then be stipulated that the most important familywise type I error probability is pertinent only to the primary endpoint family. Each primary endpoint would then be allocated an alpha level for testing and the level of alpha allocated can depend on how likely the endpoint would succeed for generating a claim. For instance, if the mortality rate is very small, then the alpha allocated for testing the mortality effect can be very small, say, 0.001. The familywise type I error probability for the secondary endpoint family may also need to be controlled, depending on the relationship between the secondary endpoints and the primary endpoints.

1.3.1.1 Testing Multiple Endpoints in Fixed Design Trial

Statistical testing can be pursued, following the same hierarchical structure. At the minimum, the type I error probability for the primary endpoint family needs to be strongly controlled. Alpha allocation and hierarchical management of alpha can be combined for use to control the familywise type I error probability. The graphical procedures, such as Bretz et al. (2009) and Burman et al. (2009), can be very helpful to illustrate how each null hypothesis is tested and whether it is rejected or not, how the allocated alpha for each hypothesis is propagated for possible reuse, etc.

In the cases where there are multiple doses for testing and dose-specific assertions for the primary and secondary endpoints are necessary, the families would have to be expanded to incorporate doses. As Hung and Wang (2010) articulate, the hypotheses for doses and the hypotheses for endpoints may have different clinical contexts. If in a trial the major objective is to select a dose that has a favorable margin of clinical benefits that are characterized by a single primary endpoint, the hypotheses corresponding to the doses for that endpoint may be the main focus and hence the hierarchical structure may be set up to reflect the potentially multiple layers of hypotheses. As an example, for this primary endpoint, a single-branch hierarchical testing procedure may order the testing structure such that a high dose is tested first at level α, which is then followed by testing a low dose at level α after the high dose achieves statistical significance. With addition of another primary endpoint, the single-branch hierarchical testing procedure may have to force the placement of a dose-endpoint pair hypothesis before or after another dose-dependent pair with different doses, for example, both doses must achieve statistical significance for the first endpoint in order to test the second endpoint. This single-branch hierarchical testing procedure is likely to confuse the decision tree and should be avoided. If the ultimate clinical decision tree may entertain the dose–endpoint pair hypotheses for different doses and endpoints, then the family of all dose-endpoint pairs may become important and the maximum type I error rate associated with this family may need to be controlled. Nevertheless, it is most desirable to think through the ultimate clinical decision tree at the stage of trial planning, in order to construct a suitable alpha propagation rule in using the graphical procedure by Bretz et al. and Burman et al. and possibly bypass some of the unnecessary routes in the statistical significance decision tree, as proposed by Liu and Shen (2009).

1.3.1.2 Testing Multiple Endpoints in Group Sequential Design Trial

Under a typical group-sequential design, certain endpoints may be analyzed through repeated statistical significance testing during the course of the trial. Repeated significance testing adds tremendous challenges to handling the test problems for multiple endpoints. Hung et al. (2007) expose a commonly seen problem that a primary endpoint is tested possibly multiple times using a prespecified alpha-spending function (Lan and DeMets, 1983) but another key endpoint will be tested only after the primary endpoint achieves statistical significance and the trial is stopped. For instance, let the total alpha for the primary endpoint E1 be α and there is only one interim analysis for testing E1. If the key endpoint E2 is tested at a nominal level α whenever E1 achieves statistical significance and the trial is stopped, then the probability of type I error for testing E2 will be larger than the desired level α, though E2 is tested at most once. In fact, this type I error probability is a function of the size of the treatment effect on E1 as a nuisance parameter. The global type I error probability (i.e., under the state that there is no treatment difference on both E1 and E2), however, is controlled at α. As a solution to this problem, Glimm et al. (2010) and Tamhane et al. (2010) recommend an optimal alpha-spending function for testing E2. It is worthwhile to note that in practice, if E2 is mortality, the alpha-spending function that is recommended for testing E2 often deviates from the optimal alpha-spending function that is suggested in these articles because the mortality endpoint usually should be tested conservatively at interim analyses.

1.3.1.3 Testing Coprimary Endpoints

In contrast to testing multiple endpoints to seek a positive finding from at least one endpoint, testing coprimary endpoints requires that all primary endpoints show positive findings. The most critical issue concerns the level of statistical significance for each individual endpoint. The theory of statistical inference for joint statistical significance of multiple tests is well known in the statistical literature, for example, Laska and Meisner (1989), and the relevant articles cited therein. For simplicity, consider the case of two coprimary endpoints. Denote by (Δ_1, Δ_2) the effect sizes of the test treatment compared to the placebo on the two coprimary endpoints, where a positive Δ_i means that the test treatment is favorable on endpoint i and thus the hypothesis to show is $H_1: \Delta_1 > 0$ and $\Delta_2 > 0$ by rejecting the null hypothesis $H_0: \Delta_1 \leq 0$ or $\Delta_2 \leq 0$. The respective test statistics are denoted by (T_1, T_2), assuming that sample size is sufficiently large to justify use of asymptotic normality; that is, $(T_1, T_2) \sim N((n/2)^{1/2}(\Delta_1, \Delta_2), [1, 1, \rho])$, where ρ is the asymptotic correlation of the two tests and n the sample size per treatment arm. The rejection region of H_0 is $[T_1 > c_1, T_2 > c_2]$, where c_1 and c_2 are the critical values to be determined.

It can be shown that the type I error probability associated with this rejection region is an increasing function of (Δ_1, Δ_2) and in fact an increasing function of $|\Delta_1 - \Delta_2|$, which is a nuisance parameter. Thus, without any restriction on the parameter space of (Δ_1, Δ_2) under H_0, it is necessary to evaluate the type I error probability at $(n/2)^{1/2}(\Delta_1, \Delta_2) = (0, \pm\infty)$ or $(\pm\infty, 0)$, equivalently, at $(n/2)^{1/2}|\Delta_1 - \Delta_2| = \infty$. Consequently, the maximum type I error probability is $\max[\Phi(-c_1), \Phi(-c_2)]$, where Φ is the cumulative distribution function of the standard normal distribution. In fact, this maximum is almost achieved at $(n/2)^{1/2}|\Delta_1 - \Delta_2| = 3.5$. Hence, the α-level critical value of (T_1, T_2) is $c_1 = c_2 = z_\alpha$, where z_α is the upper αth percentile of the standard normal distribution. Thus, the only way to control all possible type I error probability associated with such joint significance testing at one-sided 2.5% level requires that each endpoint be significant at one-sided 2.5% level, which is consistent with the convention that the statistical significance criterion for an individual endpoint in drug labeling is that its two-sided p-value is no greater than 5%.

A subtle question that is often raised in practical applications is whether it is appropriate to control only the pointwise type I error probability conditional on the value of the nuisance parameter $|\Delta_1 - \Delta_2|$. It is rarely possible to know the true value of this nuisance parameter; therefore, controlling this pointwise type I error probability at each possible value of the nuisance parameter is necessary, which is equivalent to controlling the maximum type I error probability. Controlling the pointwise type I error probability at the specifically assumed values for the nuisance parameter is likely to generate fallacy. For instance, the type I error probability associated with the joint testing each at one-sided 2.5% level has the lowest level of $(0.025)^2 = 0.000625$ at

$(\Delta_1, \Delta_2) = (0, 0)$ and $\rho = 0$ in the domain of $\rho \geq 0$. The fallacy with controlling type I error only at $\Delta_1 = \Delta_2 = 0$ is that one can have sufficiently many endpoints each with a nominal p-value close to 0.50 (no signal at all) but that the type I error probability associated with the joint significance testing ≤ 0.025 at all $\Delta_k = 0$, that is, $P(T_k \geq 0 \text{ for } k = 1, \ldots, K \mid \text{all } \Delta_k = 0) \leq 0.025$ for a sufficiently large K. Furthermore, use of a critical value $< z_{0.025}$ (i.e., two-sided $p > 0.05$) for asserting that the test treatment has a treatment benefit requires justification why a two-sided p-value substantially larger than 0.05 is statistically significant, which does not conform to the convention for drug labeling.

An important caveat with requiring each endpoint to be significant at level α is that the pointwise type I error probability in the true and yet unknown state of practical scenario can be very conservative and the test procedure demands a big sample size in most applications. The first approach to this problem is to limit the number of endpoints that are required to show statistical significance; this requires careful consideration of clinical significance. After the number of endpoints is fixed, a group sequential design can be considered to make the trial efficient with careful selection of the rejection boundary for each endpoint, as stipulated by Hung and Wang (2009).

1.3.2 Composite Endpoint

Mortality and serious morbidities are frequently studied in a form of a composite of adverse clinical outcomes, such as a composite of death, myocardial infarction or stroke in large outcome clinical trials for studying acute coronary syndrome. The effect of a test treatment on such a composite endpoint is usually evaluated using the time to first occurrence of a clinical event. A statistically significant treatment effect on the composite endpoint paves a way to understand where the test treatment is likely to have an effect in terms of adverse clinical outcomes and the potential benefits the patients may obtain. Two subsequent analyses are usually performed. One analysis examines the decomposition of the first composite events. For instance, in RENAAL trial (Brenner et al. 2001), as shown in the time to first composite endpoint of Table 1.1, the first composite events that occurred in 43.5% of the patients in the losartan group comprised 21.6% double serum creatinine, 8.5% end-stage renal disease, or 13.4% death. In the placebo group, the first composite events that occurred in 47.1% of the patients comprised 26.0% double serum creatinine, 8.5% end-stage renal disease, or 12.6% death. This analysis is descriptive to capture how the component events make up of the first composite events.

Analysis of the time to the first occurrence of individual component endpoint is also performed routinely. In RENAAL, the time to first occurrence of double serum creatinine, the time to end-stage renal disease, and the time to death were all analyzed, as shown in Table 1.1. The time to end-stage renal disease or death, whichever occurred first, was also analyzed. These analyses

TABLE 1.1

Incidence of Adjudicated Events in RENAAL Trial

	Losartan (N = 751)	Placebo (N = 762)	Hazard Ratio (95% CI)	p-Value[a]
Time to first occurrence of composite endpoint				
Doubling SC, ESRD or death	327 (43.5%)	359 (47.1%)	0.84 (0.72, 0.97)	0.022
Doubling SC	162 (21.6%)	198 (26.0%)		
ESRD	64 (8.5%)	65 (8.5%)		
Death	101 (13.4%)	96 (12.6%)		
Time to first occurrence of component endpoint				
Doubling SC	162 (21.6%)	198 (26.2%)	0.75 (0.61, 0.92)	
ESRD	147 (19.5%)	194 (25.5%)	0.71 (0.57, 0.89)	
Death	158 (21.0%)	155 (20.3%)	1.02 (0.81, 1.27)	
ESRD or death	255 (34.0%)	300 (39.4%)	0.80 (0.68, 0.95)	

Source: Excerpted from FDA/CDER/DBI Statistical Review, April 16, 2002.

[a] Nominal p-value.

SC; serum creatinine; ESRD; end-stage renal disease.

as a whole are often thought of as a better descriptor for the treatment effect. However, these component endpoint analyses are not directly related to the time to the first composite events. Thus, for such component endpoint analyses to yield an adequate description of where the treatment effect on the time to first composite event lies, some kind of multiple comparison adjustments arguably may be needed. Potentially dependent competing risks among the component endpoints impose additional difficulties to the interpretation of the treatment effect on each individual endpoint. On the one hand, it seems to make sense to put the component endpoints in one cluster and the resulting composite endpoint seems to be a natural representative of this cluster in the family of hypotheses for strong control. On the other hand, the time to first occurrence of a component endpoint may be better represented by something that incorporates other component events. The more relevant view of the treatment effect on each component event may need to be based on analysis of all component events that occur during the entire course of the trial. For instance, if a patient had a stroke before the first experience of myocardial infarction, the analysis of the time to the first occurrence of myocardial infarction should capture the occurrence of stroke in this patient and its potential influence on the risk of having a myocardial infarction. In addition, if a patient did not experience a myocardial infarction before the first occurrence of stroke, censoring at the time of the stroke for this patient may not capture the risk of myocardial infarction correctly. In this sense, strong control of type I error probability for the component endpoints as a family seems to be needed for statistical interpretation, particularly when a labeling claim is generated from a specific component endpoint. Lubsen and Kirwan (2002) expose additional problems with interpretation of the risk ratio in the composite endpoint analysis.

Given the problem of dependent competing risk among the component events articulated earlier, the composite endpoint seems to provide a preliminary global assessment of treatment effect on the adverse clinical outcomes of interest and avoids the problem of multiple testing. However, if the test treatment demonstrates a statistically significantly favorable effect on the composite endpoint, it is necessary to describe the treatment effect on each component. Consequently, as a common practice, some type of statistical analysis of the time to each component is performed, and the nominal *p*-value, the effect estimate and confidence interval are reported. These statistics will directly or indirectly have inferential implications. Thus, in some sense, the component endpoints form a cluster and the analyses on them are subsequent to the analysis of the composite endpoints. If strong control of the type I error probability associated with testing the components together with the composite endpoint is necessary, the hierarchical test strategy that first tests the composite endpoint as the primary endpoint and then tests the component endpoints using a strong control multiple testing method may be quite appealing. However, difficulties arise as to how to expand this test strategy in order to incorporate another claim-based endpoint in the hierarchical testing chain. If this claimed-based endpoint is tested after the cluster of the component endpoints, the chain will require that all the components meet statistical significance. Thus, to avoid confronting this restriction, the components must be placed at the end of the chain. This is awkward. Why does testing of the component endpoints have to be conditional on a positive finding on that claim-based endpoint? A more fundamental question is which family this cluster of component endpoints should be placed in. This needs further thought.

1.4 Multiregional Clinical Trials

International Conference on Harmonization Guidance E5 document (1998) has pinpointed that the treatment effect of a medical product may differ across intrinsic factors such as gender and genetic factors or across extrinsic factors such as standard of medical care and environmental factors. For some time, the concerns with such potential heterogeneity were addressed using the so-called *bridging study strategy*. With this strategy, for a medical product to be approved for marketing in a new geographical region, the results of existing clinical trials from foreign region(s) may need to be borrowed for planning and analyzing a study (called bridging study) in that new region. The major utility of this strategy stems from the spirit of extrapolation. The bridging study may be a pharmacokinetic trial, a pharmacodynamic trial, or a clinical trial. The main philosophy underlying the bridging study strategy is to reduce duplication of large Phase III clinical trials for the new region. A number of early stipulated concepts, statistical methods, and sample size ramifications for bridging can be found in several statistical articles such as Shih (2001), Chow et al. (2002), and Liu et al. (2002). The recent book by Liu et al. (2013) contains many relevant articles on this topic. It is commonly known that extrapolation relies on consistency assessment based on across-trial comparisons that are highly controversial and difficult because of the inherited differences in many aspects between the foreign trials and the new-region trial. As such, a bridging trial often places a greater demand on sample size; see Hung et al. (2013). Such inherited heterogeneity between the trials can be reduced by having all regions participate in a trial under the same study protocol. This is the essential concept behind the so-called multiregional or global trial strategy that incorporates internal bridging.

1.4.1 Values of Multiregional Clinical Trial Strategy

Randomization and averaging are two essential statistical methods for generating scientific evidence from clinical trials. The multiregional clinical trial strategy employs these methods to estimate the effect of a test medical product for the patients in geographical regions under the same study protocol. All the regions contribute their respective estimates to produce the global estimate of the treatment effect and in return the global effect is expected to be applicable to each region.

As an example, suppose a multiregional clinical trial involves four geographical regions for evaluating a blood pressure lowering effect of a test drug. Table 1.2 presents four scenarios for the effect of the test drug relative to the placebo in terms of blood pressure reduction from baseline in each local region. In all four scenarios, the global average leads to an identical effect

TABLE 1.2

Treatment Effect (mmHg) of a Test Drug Relative to Placebo Based on Mean Change from Baseline in Blood Pressure at a Targeted Clinical Visit

	Region I	Region II	Region III	Region IV	Global Average
Scenario 1	4.0	4.0	4.0	4.0	4.0
Scenario 2	3.4	3.8	4.6	3.0	4.0
Scenario 3	0	4.0	6.0	3.0	4.0
Scenario 4	−2.0	2.0	8.0	3.0	4.0

Note that a positive value is in favor of the test drug.

of 4.0 mmHg blood pressure reduction for the drug. This average has a big advantage, as compared to each regional estimate, in terms of the precision; that is, the standard error of the global effect is much smaller than that of each regional estimate. This is a power of averaging.

The accuracy of an effect estimate is a different aspect. Scenario 1 shows perfect homogeneity of the drug effect across the regions, which is the best scenario we hope for. Scenario 2 shows some degree of heterogeneity in drug effect among the four regions, but it appears reasonable to use the global effect of 4.0 mmHg to approximate the drug effect for each region. However, for Region I, the drug appears to have no effect in Scenario 3 and a harmful effect in Scenario 4. The global estimate of 4.0 mmHg may not be applicable to Region I in Scenarios 3 and 4. This is a pitfall of averaging. Therefore, in practice, while averaging is powerful to handle the precision of an estimate, it must be noted that the global average may or may not always be interpretable. When the global average is interpretable, evaluation of the prespecified regions nested in multiregional trials through informative internal bridging may be possible (Wang 2009) but extra care is still needed, like Scenario 2. Challenges in practice stem from the fact that at best we can only deal with estimates that are subject to statistical uncertainty and perturbation. In fact, if all the numbers in Table 1.2 are estimates, then the probability that Scenario I occurs is zero.

As articulated earlier, a multiregional clinical trial essentially serves two interrelated purposes: (1) assess the global treatment effect and (2) use the trial results to bridge from global to local or between local regions. The potential regional differences in treatment effect can still exist because of intrinsic factors or extrinsic factors. These regional differences are of scientific interest so that understanding them may help to best apply the approved medicine to patients, such as how to give a right dose to a patient. Another source of regional heterogeneity in treatment effect may be caused by the trial conduct or the data quality problem but the regional differences due to this problem are rarely of scientific interest and should be minimized. The values of the multiregional clinical trial strategy are

1. It can yield a global effect estimate with best precision
2. The global estimate may be best for bridging if the effect estimates are not very dissimilar between regions
3. It offers opportunities to study the regional differences of scientific interest
4. It can stimulate collaborative clinical research among regions for improving worldwide public health
5. It raises awareness of the concept of *quality* and can help to enhance the trial quality for all local regions
6. It can harness global harmonization on trial standard

7. It can nurture clinical trial leadership with a global view
8. It can help to raise awareness of cost effectiveness, ethical standard, regulatory standard, data or trial quality assurance

1.4.2 Challenges of Multiregional Clinical Trial Strategy

There are a great many of challenges with the multiregional clinical trial strategy (Wang 2010). Regional differences of treatment effect estimate indeed appear in many multiregional clinical trials, such as MERIT-HF (1999), RENAAL (Brenner et al. 2001), IDNT (Lewis et al. 2001), PLATO (Walletin et al. 2009), and Gomes et al. (2009). When such heterogeneity appears, its causes are often unknown and the interpretation of the global effect is very difficult. It is unclear how to tease out the real differences of scientific interest from the observed regional differences. It is also unclear how to use the regional differences for planning a future multiregional clinical trial. Above all, how to best inform the consumers regarding the regional differences of the drug effect and which estimate should be taken as a guided value for a patient are unknown. In the MERIT-HF trial, metaprolol appears to yield little effect on all-cause mortality in the United States while the mortality effect demonstrated in that trial seems entirely driven by outside the United States. In patients with diabetic nephropathy, losartan and ibersartan seem to have a consistently large effect on reduction of the risk of doubling serum creatinine, end-stage renal disease or death in Asia but consistently little effect in North America, based on the results of RENAAL and IDNT. The fact of the matter is that when regional differences in a drug effect appear in trials, statistical analyses that are necessary to search for causes and explanation are post hoc and hence the credibility issue with such post hoc analyses is often too great to allow the analyses to draw a definitive conclusion.

Another challenge stems from the reality that among the geographical regions there is large disparity in the concept of quality, trial or data monitoring at the local region, regulatory enforcement and difficulty in trial or data inspection due to translation, cultural aspects, resources, etc. The quality of data or the trial logistics can accentuate or attenuate the regional estimate of the drug effect and the regional differences in the drug effect. Let Y_h denote the estimate of the drug effect in region $h, h = 1, \ldots, K$, where K is the number of regions considered. When the regional estimate can be affected by the intrinsic/extrinsic factors (jointly denoted by X) and the quality factor Q, the expectation of Y_h may be expressed in a simple form as

$$E(Y_h) = \Delta + \beta X_h + \lambda Q_h + \gamma X_h Q_h,$$

where
Δ is the common component of treatment effect for all regions
β measures the joint effects of the intrinsic/extrinsic factors on the regional effect

λ measures the impacts of the quality factor
γ quantifies the joint effects of the intrinsic/extrinsic factors and the quality
factor.

When the value of λ is nonzero, the expectation of the regional estimate may
differ from region to region, even if the effects of intrinsic or extrinsic factors
influence the expectation of the regional estimate uniformly by the coefficient β. Moreover, a nonzero value of γ will affect the impacts of intrinsic
or extrinsic factors on the expectation of regional estimate differentially from
region to region. Consequently, the value of γ will confound the relationship
between the intrinsic or extrinsic factors and the clinical outcome of interest
such that the trial result may not provide accurate guidance for incorporating
the regional differences in the intrinsic or extrinsic factors to best recommend
how to use the approved medicine based on the global effect and the regional
estimates. Clearly, we hope to conduct the multiregional clinical trial in the
best way to minimize the influence of the quality factor; that is, ideally, we
should ensure that $\lambda = \gamma = 0$.

1.4.3 Design Considerations

The challenges articulated earlier present great difficulty to the design of a
multiregional clinical trial. At first, attention needs to be paid to the question of whether the multiregional clinical trial strategy fits the scenario of the
application in hand. If the primary clinical endpoint is sensitive to regional
cultures, such as that relying on quality of life instruments, the multiregional
trial strategy might not be suitable, unless the instruments are validated
across the local regions so that the necessary calibration can be made to properly adjust the regional estimates. Once the strategy is deemed a reasonable
method, it is still often extremely difficult and complex to define regions.
Geography might not be the most relevant determinant. If the question is how
to apply a medicine to patients according to the intrinsic or extrinsic factors,
then regions should be more sensibly defined using these factors. To achieve
this goal, research is definitely needed to explore the relationship between
intrinsic/extrinsic factors and the clinical outcomes of interest and to search
for possible effect modifiers that are related to intrinsic or extrinsic factors.
Such research work relies on a lot of learning tasks prior to considering how
to define regions in any multiregional clinical trials.

Regional differences in treatment effect estimate may have substantial
ramifications on sample size planning. Hung et al. (2010) characterized the
relationship between the sample size inflation factor and the ratio of inter-
region variance to intraregion variance. For instance, if the inter-region
standard deviation can be 50% of the intraregion standard deviation, then
the sample size planned by ignoring the interregion variability can severely
be underestimated and the consequence is under powering the multiregional
trial for showing a treatment effect. If the interregion standard deviation

exceeds 30% of the postulated size of the global effect, the multiregional clinical trial can also be severely underpowered.

Assessment of consistency or inconsistency in regional estimates is necessary in the analysis stage after the multiregional clinical trial demonstrates a treatment effect. Naturally, it is desirable to consider such assessment in trial planning because it also has implications on sample size planning. Japan MHLW issued a points-to-consider document Basic Principles on Global Clinical Trials in 2007 to stipulate the idea of considering consistency in the results between the Japanese population and the patients outside Japan in sample size planning. Essentially, there are two methods: (1) ensure the probability that the effect estimate in Japanese patients is at least 50%, say, of the global estimate is at least 80%, say, and (2) ensure the probability that all regions show a positive treatment effect, given the global effect is conclusively positive, is at least 80%, say. Kawai et al. (2008) and Quan et al. (2009) computed the minimum proportion of total sample size for Japan in order to meet the respective methods. To meet such a consistency assessment criterion may have large impacts on the sample size distribution to the regions involved in the trial (Uesaka 2009). As shown in Tsou et al. (2012), if the true state of nature is that all regions share the same treatment effect, ensuring all regions to meet such a consistency criterion can double total sample size of the multiregional trial or even substantially more, depending on the sample size distribution to the regions. Another undesirable impact of overly using consistency criteria for designing a multiregional trial is that each local region may tend to value the selected regional estimates more than the global estimate. Consequently, the local regions may develop their own consistency criteria, which may generate counterproductive competitions on sample size among the regions.

From the design standpoint, consideration of consistency assessment entails accounting for another element. This element is not a total surprise. Each regional estimate is accompanied with substantial statistical uncertainty, partly because each region has only a portion of total sample size and the regional estimate has less precision than the global estimate. Consequently, a play of chance can be a source of the apparent heterogeneity seen in the regional estimates. Hung et al. (2010) investigated the probability of play-of-chance heterogeneity. Suppose that a multiregional clinical trial of four regions is planned to detect a postulated effect size δ at 5% level of statistical significance and 90% power and the clinical response variable has a standard deviation that is uniform in all the regions. Based on the mathematical formula given in Hung et al. (2010), the probability that m of the four regions show a nonpositive treatment effect where in truth all four regions share a common effect $\Delta > 0$ can be plotted as a function of (Δ/δ), when the sample size distribution to the regions is (20%, 10%, 30%, 40%), as illustrated in Figure 1.2.

If the global effect is positive at borderline statistical significance, that is, p-value for the global effect is approximately 0.05, then there can be more than

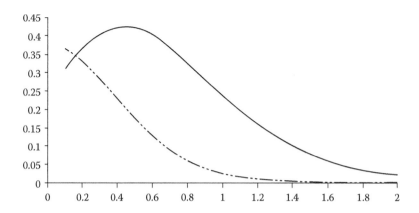

FIGURE 1.2
Probability that m of four regions show a nonpositive treatment effect where sample size allocation for the four regions: (0.20, 0.10, 0.30, 0.40). Solid curve: $m = 1$, dashed curve: $m = 2$. (Excerpted from Figure 7.4 of Hung, H.M.J. et al., Issues of sample size in bridging trials and global clinical trials, in: Liw, J.P. et al. (eds.), *Design and Analysis of Bridging Studies*, CRC Press, Boca Raton, FL, 2013.).

35% probability that one of the four regions shows a nonpositive effect, even though in truth all the regions share the same effect (i.e., no regional heterogeneity). Even in the case that the global effect is highly significant, $p = 0.001$, there can still be more than 20% probability that one of the four regions shows a nonpositive effect. Such a probability of a play-of-chance regional heterogeneity may increase as the number of regions increases or sample size allocation to the regions is more unbalanced, as also noted in Li et al. (2007). In MERIT-HF (1999), there were 14 countries. Assume the global average is bridgeable to each country. We calculated the probability of observing at least 1 out of the 14 countries showing a numerical effect reversal (hazard ratio greater than 1) assuming that the global observed 34% hazard reduction (95% confidence interval: 19%–47%) is the true hazard reduction. This probability (e.g., at least US showed an effect reversal) ranges from 0.22, 0.45 to 0.84 corresponding to the upper, the point and the lower estimates, respectively. These probabilities are substantial. In addition, the smaller the sample size fraction for a region, the larger the probability of showing an effect reversal by chance in this region will be; see Figure 7.5 of Hung et al. (2013). Thus, it makes sense to strive for balance in sample size allocation to regions and the apparent regional heterogeneity due to a play of chance should not be ignored in designing a multiregional clinical trial.

The essential idea behind the multiregional clinical trial strategy is that the global effect is expected to represent all the regions. If this is very much doubtful in the design stage, applicability of this strategy to a practical scenario needs to be challenged. Without careful consideration, the multiregional clinical trial as planned by ignoring the critical applicability issues may present extreme difficulty in interpretation of the trial results in the end. Once we are committed to this strategy, we should recognize the global estimate is superior to regional estimates in terms of better precision and statistical stability. As discussed earlier, we should recognize that an effect reversal from one of many regions may appear entirely or partially due to chance. We can reduce the chance of apparent heterogeneity, for instance, by planning the trial sample size more conservatively and minimizing the imbalance of sample size allocation to the regions.

Another critical aspect of trial design is to minimize the differences in trial conduct and data quality among the regions. A trial monitoring process needs to be developed to embrace the regional differences in many aspects, such as culture, political system, and inspection of trial logistics and data. Clearly, there is a need of global harmonization on clinical trial standard (e.g., good clinical practices, data or site managements), clinical trial leadership, regulatory standard and guidance, clinical trial infrastructure, and operational aspects such as regional versus global issues.

Acknowledgment

Thanks are due to Dr. Lisa LaVange for the editorial comments that helped improve the presentation of the materials.

Disclaimer

This article reflects the views of the authors and should not be construed to represent the views or policies of the US Food and Drug Administration.

References

Active Control Trial Designs

Committee for Medicinal Products for Human Use (CHMP). (2006) Guideline on the choice of the noninferiority margin. *Statistics in Medicine*; 25:1628–1638.

Hauschke, D. and Pigeot, I. (2005) Establishing efficacy of a new experimental treatment in the 'Gold Standard' design. *Biometrical Journal*; 47:782–786. DOI: 10.1002/bimj.200510169.

Hung, H. M. J. and Wang, S. J. (2010) Challenges to multiple testing in clinical trials. *Biometrical Journal*; 52:747–756.

Hung, H. M. J. and Wang, S. J. (2013) Multiple comparisons in complex clinical trial designs. *Biometrical Journal*; 55:420–429. DOI: 10.1002/bimj.201200048747-756.

Hung, H. M. J. and Wang, S. J. (2013) Statistical considerations for non-inferiority trial designs without placebo. *Statistics in Biopharmaceutical Research* 5:239–247.

Hung, H. M. J., Wang, S. J., and O'Neill, R. (2005) A regulatory perspective on choice of margin and statistical inference issue in non-inferiority trials. *Biometrical Journal*; 47:28–36.

Hung, H. M. J., Wang, S. J., and O'Neill, R. (2007) Issues with statistical risks for testing methods in noninferiority trial without a placebo arm. *Journal of Biopharmaceutical Statistics*; 17:28–36.

Morikawa, T. and Yoshida, M. (1995) A useful testing strategy in phase III trials: Combined test of superiority and test of equivalence. *Journal of Biopharmaceutical Statistics*; 5:297–306.

U.S. Food and Drug Administration Guidance for Industry: Non-Inferiority Clinical Trials. 2010. http://www.fda.gov/downloads/Drugs/.../Guidances/UCM202140.pdf.

Wang, S. J. and Blum, J. D. (2011) An evidential approach to non-inferiority clinical trials. *Pharmaceutical Statistics*; 10:440–447.

Wang, S. J., Hung, H. M. J., and O'Neill, R. (2011) Genomic classifier for patient enrichment: Misclassification and type I error issues in pharmacogenomics non-inferiority trial. *Statistics in Biopharmaceutical Research*; 3:310–319.

Wiens, B. and Rosenkranz, G. K. (2013) Missing data in non-inferiority trials. *Statistics in Biopharmaceutical Research* 5:383–393.

Adaptive Designs

Bauer, P. and Köhne, K. (1994) Evaluations of experiments with adaptive interim analyses. *Biometrics*; 50:1029–1041.

Bauer, P. and König, F. (2006) The reassessment of trial perspectives from interim data—A critical view. *Statistics in Medicine*; 25:23–36

Brannath, W., Zuber, E., Branson, M., Bretz, F., Gallo, P., Posch, M., and Racine-Poon, A. (2009) Confirmatory adaptive designs with Bayesian decision tools for a targeted therapy in oncology. *Statistics in Medicine*; 28:1445–1463.

Branson, M., Brannah, W., Dunger-Baldauf, C., and Bauer, P. (2005) Testing and estimation in flexible group sequential design with adaptive treatment selection. *Statistics in Medicine*; 24:3697–3714.

Chen, Y. H., DeMets, D. L., and Lan, K. K. (2004) Increasing the sample size when the unblinded interim result is promising. *Statistics in Medicine*; 23:1023–1038.

Chow, S. C. and Chang, M. (2006) *Adaptive Design Methods in Clinical Trials*. Chapman and Hall/CRC Press, Taylor & Francis: New York.

Cui, L., Hung, H. M. J., and Wang, S. J. (1999) Modification of sample size in group sequential clinical trials. *Biometrics*; 55:321–324.

Emerson, S. S., Levin, G. P., and Emerson, S. C. (2011) Comments on 'Adaptive increase in sample size when interim results are promising: A practical guide with examples'. *Statistics in Medicine*; 30:3285–3301.

FDA. (2010) FDA Draft Guidance for Industry: Adaptive Design Clinical Trials for Drugs and Biologics. www.fda.gov/downloads/Drugs/GuidanceCompliance RegulatoryInformation/Guidances/UCM201790.pdf. Released on February 25, 2010 for public comments.

Gao, P., Ware, J. H., and Mehta, C. R. (2008) Sample size re-estimation for adaptive sequential design in clinical trials. *Journal of Biopharmaceutical Statistics*; 18:1184–1196.

Hung, H. M. J., O'Neill, R., Wang, S. J., and Lawrence, J. (2006a) A regulatory view on adaptive/flexible clinical trial design (with rejoinder). *Biometrical Journal*; 48:565–573, 613–615.

Hung, H. M. J. and Wang, S. J. (2012) Sample size adaptation in fixed-dose combination drug trial. *Journal of Biopharmaceutical Statistics*; 22:679–686.

Hung, H. M. J., Wang, S. J., and O'Neill, R. (2006b) Methodological issues with adaptation of clinical trial design. *Pharmaceutical Statistics*; 5:99–107.

Hung, H. M. J., Wang, S. J., and O'Neill, R. (2011) Flexible design clinical trial methodology in regulatory applications. *Statistics in Medicine*; 30: 1519–1527.

Hung, H. M. J., Wang, S. J., and Yang, P. (2014) Some challenges with statistical inference in adaptive designs. *Journal of Biopharmaceutical Statistics* (to appear).

König, F., Brannah, W., Bretz, F., and Posch, M. (2008) Adaptive Dunnett tests for treatment selection. *Statistics in Medicine*; 27:1612–1625.

Lawrence, J. and Hung, H. M. J. (2003) Estimation and confidence intervals after adjusting the maximum information. *Biometrical Journal*, 45:143–152.

Lehmacher, W. and Wassmer, G. (1999) Adaptive sample size calculation in group sequential trials. *Biometrics*; 55:1286–1290.

Liu, Q. and Andersen, K. M. (2008) On adaptive extensions of group sequential trials for clinical investigations. *Journal of the American Statistical Association*; 103: 3267–3284.

Mehta, C. R., Gao, P., Bhatt, D. L., Harrington, R. A., Skerjanec, S., and Ware, J. H. (2009) Optimizing trial design: Sequential, adaptive, and enrichment strategies. *Circulation*; 119:597–605.

Mehta, C. R. and Pocock, S. J. (2011) Adaptive increase in sample size when interim results are promising: A practical guide with examples. *Statistics in Medicine*; 30:3267–3284.

Müller, H. H. and Shäfer, H. (2001) Adaptive group sequential designs for clinical trials: Combining the advantages of adaptive and of classical group sequential approaches. *Biometrics*; 57:886–891.

Posch, M., Maurer, W., and Bretz, F. (2010) Type I error rate control in adaptive designs for confirmatory clinical trials with treatment selection at interim. *Pharmaceutical Statistics*; 10:96–104.

Proschan, M. A. and Hunsberger, S. A. (1995) Designed extension of studies based on conditional power. *Biometrics*; 51:1315–1324.

Reflection paper on methodological issues in confirmatory clinical trials planned with an adaptive design. European Medicines Agency EMEA CHMP/EWP/2459/02 October 2007.

Wang, S. J., Brannath, W., Brückner, M., Hung, H. M. J., and Koch, A. (2013) Unblinded adaptive information design based on clinical endpoint or biomarker. *Statistics in Biopharmaceutical Research* 5:293–310.

Wang, S. J. and Hung, H. M. J. (2013) Adaptive enrichment with subpopulation selection at interim: Methodologies, applications and design considerations. *Contemporary Clinical Trials* 36:673–681.

Wang, S. J., Hung, H. M. J., and O'Neill, R. T. (2009) Adaptive patient enrichment designs in therapeutic trials. *Biometrical Journal*; 51:358–374.

Wang, S. J., Hung, H. M. J., and O'Neill, R. T. (2010) Impacts of type I error rate with inappropriate use of learn for confirm in adaptive designs. *Biometrical Journal*; 52:798–810.

Wang, S. J., Hung, H. M. J., and O'Neill, R. T. (2012) Paradigms for adaptive statistical information designs: Practical experiences and strategies. *Statistics in Medicine*; 31:3011–3023.

Wang, S. J., O'Neill, R. T., and Hung, H. M. J. (2007) Approaches to evaluation of treatment effect in randomized clinical trials with genomic subset. *Pharmaceutical Statistics*; 6:227–244.

Multiple Comparison Considerations

Brenner, B. M., Cooper, M. E., de Zeeuw, D., Keane, W. F., Mitch, W. E., Parving, H. H., Remuzzi, G., Snapinn, S. M., Zhang, Z., and Shahinfar, S. (2001) RENAAL study investigators. *New England Journal of Medicine*; 345:861–869.

Bretz, F., Maurer, W., Brannah, W., and Posch, M. (2009) A graphical approach to sequentially rejective multiple test procedures. *Statistics in Medicine*; 28: 586–604.

Burman, C. F., Sonesson, C., and Guilbaud, O. (2009) A recycling framework for the construction of Bonferroni-based multiple tests. *Statistics in Medicine*; 28:739–761.

Connolly, S. J., Ezekowitz, M. D., Yusuf, S., Eikelboom, J., Oldgren, J., Parekh, A., Pogue, J. et al. (2009) RE-LY Steering Committee and Investigators. Dabigatran versus warfarin in patients with atrial fibrillation. *New England Journal of Medicine*; 361:1139–1151.

FDA Draft Guidance for Industry: Multiple Endpoints. In review.

Glimm, E., Maurer, W., and Bretz, F. (2010) Hierarchical testing of multiple endpoints in group sequential trials. *Statistics in Medicine*; 29:219–228.

Hung, H. M. J. and Wang, S. J. (2009) Some controversial multiple testing problems in regulatory applications. *Journal of Biopharmaceutical Statistics*; 19:1–11.

Hung, H. M. J. and Wang, S. J. (2010) Challenges to multiple testing in clinical trials. *Biometrical Journal*; 52:747–756.

Hung H. M. J. and Wang S. J. (2013) Multiple comparisons in complex clinical trial designs. *Biometrical Journal*; 55:420–429. DOI: 10.1002/bimj.201200048747-756.

Hung, H. M. J., Wang, S. J., and O'Neill, R. (2007) Statistical considerations for testing multiple endpoints in group sequential or adaptive clinical trials. *Journal of Biopharmaceutical Statistics*; 17:1201–1210.

Lan, K. K. G. and DeMets, D. L. (1983) Discrete sequential boundaries for clinical trials. *Biometrika*; 70:659–663.

Laska, E. M. and Meisner, M. J. (1989) Testing whether an identified treatment is best. *Biometrics*; 45:1139–1151.

Liu, Y. and Hsu, J. (2009) Testing for efficacy in primary and secondary endpoints by partitioning decision paths. *Journal of the American Statistical Association*, 104:1661–1770.

Lubsen, J. and Kirwan, B. A. (2002) Composite endpoints: Can we use them? *Statistics in Medicine*; 21:2959–2971.

Tamhane, A. C., Mehta, C. R., and Liu, L. (2010) Testing a primary endpoint and a secondary endpoint in a group sequential design. *Biometrics*; 66:1174–1184.

Multiregional Clinical Trials

Brenner, B. M., Cooper, M. E., Zeeuw, D. D., Keane, W. F., Mitcg, W. E., Parving, H. H., Remuzzi, G., Snapinn, S. M., Zhang, Z., and Shahinfar, S. (2001) Effects of losartan on renal and cardiovascular outcomes in patients with type 2 diabetes and nephropathy. *New England Journal of Medicine*; 345:861–869.

Chow, S. C., Shao, J., and Hu, O. Y. P. (2002) Assessing sensitivity and similarity in bridging studies. *Journal of Biopharmaceutical Statistics*; 12:385–400.

Gomes, M. F., Faiz, M. A., Gyapong, J. O., Warsame, M., Agbenyega, T., Babiker, A., Baiden, F. et al., and for the Study 13 Research Group. (2009) Pre-referral rectal artesunate to prevent death and disability in severe malaria: A placebo-controlled trial. *Lancet*; 373:557–566.

Hung, H. M. J., Wang, S. J., and O'Neill, R. T. (2010) Challenges in design and analysis of multi-regional clinical trials. *Pharmaceutical Statistics*; 9:173–178. DOI: 10.1002/pst.440.

Hung, H. M. J., Wang, S. J., and O'Neill, R. T. (2013) Issues of sample size in bridging trials and global clinical trials. In: Liu, J. P., Chow, S. C., and Hsiao, C. F. (eds.), *Design and Analysis of Bridging Studies*, CRC Press, Boca Raton, FL, Chapter 8.

ICH International Conference on Harmonization Tripartite Guidance E5. (1998) Ethnic factor in the acceptability of foreign data. *The US Federal Register*; 83:31790–31796.

Kawai, N., Chuang-Stein, C., Komiyama, O., and Li, Y. (2008) An approach to rationalize partitioning sample size into individual regions in a multiregional trial. *Drug Information Journal*; 42:139–147.

Lewis, E. J., Hunsicker, L. G., Clarke, W. R., Berl, T., Pohl, M. A., Lewis, J. B., Ritz, E. et al. (2001) Renoprotective effect of the angiotensin-receptor antagonist irbesartan in patients with nephropathy due to type 2 diabetes. *New England Journal of Medicine*; 345:851–860.

Li, Z., Chuang-Stein, C., and Hoseyal, C. (2007) The probability of observing negative subgroup results when the treatment effect is positive and homogeneous across all subgroups. *Drug Information Journal*; 41:47–56.

Liu, J. P., Chow, S. C., and Hsiao, C. F. (2013) *Design and Analysis of Bridging Studies*. CRC Press, Boca Raton, FL.

Liu, J. P., Hsueh, H. M., and Chen, J. J. (2002) Sample size requirement for evaluation of bridging evidence. *Biometrical Journal*; 44:969–981.

Ministry of Health, Labour and Welfare of Japan. *Basic Principles on Global Clinical Trials*, September 28, 2007.

Quan, H., Zhao, P. L., Zhang, J., Roessner, M., and Aizawa, K. (2009) Sample size considerations for Japanese patients in a multi-regional trial based on MHLW guidance. *Pharmaceutical Statistics*; 8:1–14. DOI: 10.1002/pst.380.

Shih, W. J. (2001) Clinical trials for drug registrations in Asian-Pacific countries: Proposal for a new paradigm from a statistical perspective. *Controlled Clinical Trials*; 22:357–366.

The MERIT-HF Study Group. (1999) Effect of metropolol CR/XL in chronic heart failure: Metropolol CR/XL randomised intervention trial in congestive heart failure (MERIT-HF). *Lancet*; 353:2001–2007

Tsou, H. H., Hung, H. M. J., Chen, Y. M., Huang, W. S., Chang, W. J., and Hsiao, C. F. (2012) Establishing consistency across all regions in a multi-regional clinical trial. *Pharmaceutical Statistics*; 11(4):295–299. DOI: 10.1002/pst.1512.

Uesaka, H. (2009) Sample size allocation to regions in a multiregional trial. *Journal of Biopharmaceutical Statistics*; 19:580–594. DOI: 10.1080/10543400902963185.

Wang, S. J. (2009) Bridging study versus prespecified regions nested in global trials. Special Issue on "Partnership in harmonization for global drug development in Asia/pacific". *Drug Information Journal*; 43:27–34.

Wang, S. J. Editorial. (2010) Special issue on multi-regional clinical trials—What are the challenges? *Pharmaceutical Statistics*; 9(3):171–172.

Wallentin, L., Becker, R. C., Budaj, A., Cannon, C. P., Emanuelsson, H., Held, C., Horrow, J. et al. A. (2009) Ticagrelor versus clopidogrel in patients with acute coronary syndromes. *New England Journal of Medicine* 361:1045–1057.

2

Review of Randomization Methods in Clinical Trials

Vance W. Berger and William C. Grant

CONTENTS

2.1 Introduction

To many, the name W. Edwards Deming is synonymous with quality. What would this man, who personified the very essence of quality, have to say about the bewildering array of randomization procedures that are available today, and the manner in which one of these gets chosen for use in any given trial? Would he grant carte blanche and follow the Nike mantra of *Just do it* (without regard to how you just do it)? Would he use more modern jargon and go with *It's all good*? Or would he instead concern himself with the properties of various randomization procedures, recognize that some are distinctively better than others, and then rail in favor of those better procedures? We would like to believe that he would take this matter far more seriously than the typical researcher does, and everything in the record supports this assertion of his diligence.

Anyone who cares to apply a similar level of diligence to the problem of how to randomize will quickly become aware of a bewildering array of randomization procedures in modern use. By itself, this is neither good nor bad. A variety of solutions to what seems at first glance to be one unique problem may represent (1) variations in the precise objectives and formulations that lead to these various solutions; (2) a situation in which it just really does not matter which of these solutions is chosen; or (3) a Wild West situation in which everyone is free to do his or her own thing, based on precedent or any other reason, whether or not the chosen solution is reasonable. We reject the second possibility out of hand; as we shall demonstrate, it certainly does matter how one randomizes. Regarding the first possibility, we note that in fact there are variations in the precise objectives and formulations that should lead to the various solutions.

However, different choices of randomization procedures do not simply reflect different objectives. In fact, the third situation, anarchy, is the one that governs this decision in practice. The subtle differences that should guide the choice of how to randomize are generally ignored completely, so any linkage from the precise formulation of the problem to the solution of that problem is severed. This leads, paradoxically, to simultaneously too much and too little conformity. There are both inertia, the resistance to change even if that change represents improvement, and a maverick spirit that spurs researchers on to try novel randomization procedures, even if they are not appropriate to the situation at hand. Hence, we cannot advocate for either more or less conformity across the board. Rather, we must content ourselves with advocating for an infusion of reason and strategic thinking into the selection of randomization procedures, as this will, in one fell swoop, address the dual problems of excessive conformity and insufficient conformity and will also improve practice.

2.2 What Do We Expect from a Randomization Procedure?

What is it that makes one randomization procedure better than another? Do we compare them on the basis of the number of letters in their respective names? Or is there some more appropriate basis for comparing them? Clearly we expect something, as evidenced by the fact that so many randomized trials are criticized after the fact for not being properly randomized. For example, Chapter 3 of Berger (2005) lists 30 actual trials that were suspected of selection bias from prediction of future allocations, and more recently, Fayers and King (2008) describe another one. We shall have more to say about this matter in further sections, but for now, suffice it to say that randomization can fail, even when the trial actually is randomized. This is a separate issue from studies that are falsely labeled as randomized when in fact they were not randomized (see Berger and Bears, 2003). How does randomization fail, and how can this failure be prevented?

There are all sorts of statistical adjustment techniques for unbalanced covariates, but as with human health, so too is it the case in this arena that an ounce of prevention is worth several pounds of cure. Many statistical adjustment techniques, such as the analysis of covariance, are based on normality and other assumptions that cannot possibly be true. These assumptions must necessarily open up the results to criticism, since the plausibility of the results cannot exceed the plausibility of the assumptions upon which they are based. Other statistical adjustment techniques are not based on models, and, therefore, would seem more robust.

In fact, this appearance is true; they are more robust than their parametric counterparts, yet they may nevertheless not be as robust as they are perceived to be. For example, everyone knows that a baseline imbalance in a key covariate cannot affect treatment comparisons made within the levels of this covariate. In other words, even if there are 90% males in one group and 90% females in the other group, this is OK as long as we compare males to males and females to females. But consider an unmasked trial comparing one diet to another for weight loss, with group meetings to train subjects in meal preparation and also to discuss other health-enhancing behaviors.

In the group that is predominantly female, the female subjects impress upon the male minority the need for step aerobics, whereas in the group that is predominantly male, the male subjects impress upon the female minority the need for weight lifting. This means not only that the males in one group differ systematically (other than in the study treatment) from those in the other group, and that the females in one group differ systematically (other than in the study treatment) from those in the other group, but also that these differences are caused precisely by the baseline imbalance in gender. In other words, the within-gender treatment comparisons are still confounded, and this is precisely because of the overall gender imbalance across the treatment (diet) groups.

Suffice it to say, then, that statistical adjustment after the fact, though often helpful or even necessary, cannot in general offer the same unambiguous inference as comparison groups that are comparable to begin with any more than removing the bullets can return the shooting victim to the state he was in prior to being shot. These problems are prevented with proper randomization that is intended to produce balanced or comparable comparison groups. On the surface, this sounds rather simple, but it becomes complicated by the multiplicity of baseline covariates, which serve as predictors of response to the study treatments. So we want to balance group sizes as well as the key covariates that are known to be predictive, other covariates that may or may not relate to the disease, and even observable but unrecorded patient characteristics.

2.3 Critical Survey of Randomization Procedures

In this section, we shall enumerate several randomization procedures, covering both those that are used frequently in practice and those that are not (but should be). Our treatment shall be critical, rather than following the frequently self-congratulatory evaluations that in essence state that any methods must be good or they would not be used. After all, these are professionals. They know what they are doing. This inverse logic is invoked rather more than we care to see and is used to justify methods based on the credentials of those who use them. We prefer direct logic. Evaluate the methods first, and these methods then validate the researcher using them, or not, as the case may be. A randomized clinical trial is not a game. Patients depend on researchers to use valid methods so that the results are valid and reflect the reality of the situation. Just as an athlete might claim that his body is a temple, so too, then, can a trial be considered the holy of holies, and only the valid research methods may pass. If only.

In this section, we shall discuss stratification, alternation (which is not actually randomization), unrestricted randomization, the random allocation procedure, permuted blocks, variable block designs, minimization, mirror image randomization, the big stick procedure (Soares and Wu, 1983), Chen's procedure (Chen, 1999), and the maximal procedure (Berger et al., 2003). First, we note that these are not all competitors. Stratification can be used with any or all of the other methods listed, and mirror image randomization, which was used in a pivotal trial of etanercept for juvenile rheumatoid arthritis (Berger, 1999; Lovell et al., 2000), can be used only when stratification is also used. There would need to be paired strata to employ this method, and it, too, can be used in conjunction with other methods. In the etanercept trial, it was used in conjunction with permuted blocks of size two.

The overlap between some of these methods goes beyond the fact that some can be used in conjunction with others. Beyond that, some also reduce to others in special cases. For example, if the total sample size of the trial is n, then one variation of the random allocation rule can be described as using permuted blocks with block size n (one big block). Unrestricted randomization can be affected by using the big stick procedure with the maximally tolerated imbalance (MTI) set equal to n. If the MTI is set at one, then any of the big stick procedure, Chen's procedure, and the maximal procedure will reduce to permuted blocks with fixed block size two, as we shall explain.

Unrestricted randomization consists of essentially tossing a fair (for 1:1 allocation) coin for each allocation. As the name indicates, there are no restrictions, so it would be possible (even if highly unlikely) to end up with all allocations to the same treatment group. Far more likely would be a situation in which not all but at least a strong majority of allocations are to the same group, thereby resulting in grossly unequal group sizes. This is something that trial planners typically wish to avoid, so they have introduced restrictions on the randomization to ensure reasonably balanced group sizes. There are many types of restriction that can be imposed on the randomization, but one of the simplest is the random allocation rule, which specifies only that the group sizes must be equal at the end of the trial. Beyond that, there are no other restrictions. That is, the random allocation rule takes the total number of patients and randomly divides them into two equal subgroups, or randomly selects half of the accession numbers to serve as one of the treatment groups, and the rest serve as the other treatment group. Just as the flaw in unrestricted randomization is best illustrated with the extreme (and highly unlikely) scenario in which all patients end up in the same treatment group, so too is it the case here that an extreme scenario best exposes the flaws of the random allocation rule.

It is not possible for all patients to end up in the same treatment group, but it is possible for the first half of the patients to all receive A and the next half to all receive B. Far more likely would be the situation in which many more early patients are randomized to A and many more late patients are randomized to B. Why is this a problem? In fact, by itself this is not a problem. But when we overlay the possibility of time trends in the patient characteristics, then we can see the potential for trouble. Suppose, for example, that there is a study conducted from August of 1 year to March of the next year and that patients are followed for a few weeks. The patients recruited after the New Year may be more motivated to get fit and/or lose weight due to New Year's resolutions or just wanting to get ready for the summer months. So the later patients may tend to be healthier than the early patients, and if many more early patients are allocated to A and many more late patients are allocated to B, then we see the potential for confounding not only time with baseline characteristics but also treatments with baseline characteristics. This is chronological bias (Matts and McHugh, 1983; Berger, 2005).

Clearly, the control of chronological bias involves additional restrictions, above and beyond the one restriction of the random allocation procedure (equal group sizes at the end). Specifically, we would want to ensure also that the group sizes never get too far apart from each other. This ongoing balance in group sizes is achieved by the permuted blocks procedure, which is probably the most commonly used method today. One would hope that this frequency of use is based on rational consideration of the merits of the various procedures, so that one would be able to argue that permuted blocks are used so often because they should be used so often. Alas, this is not the case.

The way this procedure controls chronological bias is by randomizing within blocks. Block sizes may vary, but typically there is a fixed block size, such as four, in which case groups of four consecutive patients constitute blocks. Randomization is used to determine which two of these are to receive A and which two are to receive B. So with a block size of four, there would be six different permutations: AABB, ABAB, ABBA, BABA, BAAB, BBAA. Consider an unmasked study with permuted blocks. In this case, at the time a new patient is screened, all prior allocations will be known. If this current patient is the last patient in a block, then the treatment to be allocated next will also be known, as it is easily deduced from the knowledge of the prior allocations. Selection bias occurs when this advance knowledge is exploited so that patients are preferentially selected by the researcher based on the treatment to be allocated next. That is, healthier patients may be selected if A will be allocated and sicker patients may be selected if B will be allocated, thereby subverting the randomization. In fact, as mentioned, this is not a hypothetical concern, as Berger (2005) listed 30 trials suspected of this type of selection bias and Fayers and King (2008) described another one. Generally, the smaller the block size, the more predictable the upcoming allocations are, but larger sized blocks also allow for predictable allocations. Permuted blocks of any size should never be used, given the availability of better procedures, but this is especially true for blocks of size two.

The next method we consider is alternation, which is actually not randomization, yet is often misrepresented as such. This procedure simply alternates the treatments assigned, ABABABAB, ... Clearly such a procedure, in the presence of any unmasking at all (even if only one allocation is unmasked), precludes the possibility of allocation concealment, since knowledge of even one allocation serves as de facto knowledge of all future allocations. This remains true even if the usual steps toward allocation concealment (central randomization, hidden allocation sequence, opaque envelopes) are used, since the upcoming allocations can be predicted even if they cannot be directly observed. Beyond this, alternation is also inferior, relative even to other deterministic procedures (let alone truly random ones), for reasons that are not generally appreciated.

For randomization to succeed, it must avoid confounding the treatment allocations with time. And yet this is exactly what alternation does. If we consider blocks of two patients each, then the first patient recruited in each matched set (block) always receives Treatment A, and the second always receives Treatment B; this order is never varied, randomly or otherwise. With as few as 6 matched pairs, or 12 patients total, we would find a statistically significant level of confounding of time and treatments at the customary 0.05 level, with $p = 1/32$ by Fisher's exact test (two-sided). So alternation is even worse than permuted blocks of size two and is also worse than other deterministic schemes. Hence, alternation should never be used in practice, and a study that has used it should never be called randomized. We are quite certain that Deming would agree with the need for higher standards than this.

Stratification is not a randomization technique per se but rather is used in conjunction with other randomization techniques. Just as one can split in black jack and still use whatever betting strategy (e.g., hold at 17, hit at 16) one wants to, so too can one stratify the randomization (akin to splitting in black jack) to create strata and then create separate allocation sequences within each stratum. The purpose of stratification is to avoid baseline imbalances in key covariates, such as gender. Without stratification, it would be possible to have many more males in one treatment group and many more females in the other, and this would lead to confounding. Stratification handles this by creating one list for males and another separate list for females, with allocation ratios (e.g., 1:1, as is most commonly used) to match the overall allocation ratio.

Minimization is somewhat of a competitor to stratification in that it attempts to do essentially the same thing but takes a different approach to balancing key predictors. Specifically, it defines an imbalance function that measures how unbalanced the treatment groups are relative to a few specified covariates and then allocates patients so as to minimize this imbalance function, either deterministically or probabilistically. This procedure can look at more factors compared to standard stratification.

Mirror image randomization is not reported very often, but it was used in one study of etanercept for children with juvenile rheumatoid arthritis (Lovell et al., 2000), and probably in other studies as well (we note that it is a grave failure of medical journals as a whole that they do not insist on clear descriptions of randomization procedure; therefore we have, no way of knowing how often this procedure is used in practice). This study used permuted blocks of size two. Small block sizes invite selection bias because they make it easy to determine upcoming allocations. In this study, the two blocks were from one center and in two strata, but they were also mirror images of each other, meaning that if one block was AB, then its matching block was necessarily BA. This makes it especially easy to determine upcoming allocations and should not be used.

We have noted that the more restrictive the randomization procedure, the more predictable its allocations can be, but less restrictive procedures allow for chronological bias. So there is a trade-off, and it is impossible (without demonstrable perfect masking) to simultaneously eliminate both selection bias and chronological bias (Berger, 2005). However, there are newer and better methods that do a better job at controlling selection bias than permuted blocks do, even when matched for chronological bias.

Three particular methods—the big stick procedure (Soares and Wu, 1983), Chen's procedure (Chen, 1999), and the maximal procedure (Berger et al., 2003)—share some especially useful features. Each of these methods (1) specifies a certain MTI, (2) forces the sequence back toward (but not all the way to) balance when the MTI is reached, and (3) assigns equal probabilities to each treatment when there is perfect balance. The only difference among these three procedures occurs when there is some imbalance but not yet reaching the MTI, in which case the big stick procedure uses equal allocation probabilities to the treatment groups, Chen's procedure uses a fixed biasing probability that needs to be specified in advance by the user, and the maximal procedure uses a more complex procedure that results in more extreme biasing probabilities as the imbalance increases. That is, if the MTI is three, then with two treatment groups and 1:1 allocation, all three procedures will use equal allocation (50% to each group) when the imbalance is zero (meaning that to this point the same number of patients have been allocated to each group), and extreme probabilities (100% and 0%) when the MTI has been reached and one group has three more patients allocated than the other. The big stick procedure will use 50% also when the imbalance is one or two. Chen's procedure will use the specified biasing probabilities, possibly 75% to the less represented group and 25% to the more represented group, when the imbalance is either one or two, and the maximal procedure will have a more extreme biasing probability when the imbalance is two than it will when the imbalance is one.

All three procedures can be envisioned as follows. Enumerate all possible strings of 0s and 1s of the appropriate length (the sample size), and for each one of these potential allocation sequences, keep a running tab of the current imbalance (as in, the absolute difference in the numbers of patients allocated to the two groups). Compute the maximum imbalance for each sequence, and eliminate those sequences whose maximum imbalance exceeds the MTI. The maximal procedure will then select one of the remaining (admissible) sequences at random. The other two procedures will use the exact same set of permissible allocation sequences but, as mentioned, will use different allocation probabilities when there is some imbalance that has not reached the MTI.

We note in passing that the permuted blocks procedure can also be cast as an MTI procedure, but with one critical difference. Imagine if your ideal indoor temperature is 70°F and that your tolerance is 5° in either direction. So you can set your thermostat to allow free range within the limited range of 65–75, but 75 is too hot and 65 is too cold, so at these extremes, it will need to kick in. We could specify that when the temperature falls to 65, then the heater kicks in and remains on until the temperature is increased to 67, and that when the temperature rises to 75, then the air conditioner kicks in and remains on until the temperature is dropped to 73. This strategy would correspond to the three MTI procedures that force returns toward balance while not requiring that the balance be restored all the way. The permuted blocks procedure, in contrast, would essentially say that any time the MTI is reached, any time the temperature gets to 65 or 75, it must be forced all the way back to perfect balance (70°). Intuitively, this is excessive, and practically, it leads to excessive prediction and therefore also to more selection bias. As an example, consider a block size of four (MTI = 2) with a given initial sequence of AAB and MTI = 2.

The imbalance is one, so the MTI has not been reached. Chen's procedure will assign its specified unequal probabilities to each treatment group, possibly 60% to B and 40% to A. The maximal procedure would also make B more likely, in that it would encourage but not force a return to perfect balance. The big stick would not even encourage this return, as A and B would be equally likely. But permuted blocks would force the return to perfect balance, AABB. It would not allow for AABA, as the other three procedures would. This is the essence of its inferiority relative to the three MTI procedures. Moreover, there is no compensating factor to render the permuted blocks procedure superior. Hence, the MTI procedures should be used exclusively.

2.4 Factors to Consider in Selecting a Randomization Procedure and Sensitivity Designs

We already noted the weaknesses of the permuted blocks procedure in Section 2.3. These weaknesses alone do not render the procedure obsolete; after all, one who is parching in the desert does not discard his tap water unless or until filtered water comes along. But the filtered water has come along, in the form of the maximal, Chen, and big stick procedures. These uniformly better procedures render the permuted blocks procedure obsolete. This discussion may suggest that we are endorsing one single randomization procedure for all researchers to follow in all trials. In fact, we are not.

Recall that one of our biggest concerns in randomization is the ability of the investigators to predict upcoming allocations. We are dealing with an intelligent adversary, so it behooves us to take whatever steps we can to avoid becoming predictable. This means, among other things, that the last thing we want to do is to use the same randomization procedure all the time, even if we have found one that we consider to be uniquely best. Clearly, the permuted blocks procedure is not uniquely best, as discussed in Section 2.3, but even if it were, we would still be hard pressed to find any justification in using it as a standard, as has become the case in practice. If another procedure were merely as good as it, let alone better, then just for the sake of variation, we would use the newer procedure. So much more so when the newer methods also happen to be substantially better, as is the case here.

But these same considerations apply when choosing among the three MTI procedures. Variation is the key (Berger et al., 2010), and there are important advantages to mixing the procedures across sites within a given multisite trial. By generating a treatment sequence at one site with one procedure, and in the same trial generating a treatment sequence at a different site with a different procedure, predictability is reduced because each investigator would be unaware of which procedure applied to their particular site. In addition, varying the procedure across sites may provide evidence concerning what kinds of bias affected outcomes. Suppose, for example, that the random allocation rule is used for some sites in some particular trial, and permuted blocks of size four are used for other sites in that trial. If the former sites show a stronger treatment effect, then this suggests chronological bias, whereas if the latter sites show a stronger treatment effect, then this suggests selection bias. The variation in procedures across sites allows the kinds of comparisons necessary for these kinds of deductions (Berger et al., 2010). If the same randomization procedure is employed at every site, then potential sources of bias are harder to distinguish. This is the case partly because all sites will be more likely to have similar vulnerabilities. Varying randomization procedures across sites within a given trial is not without difficulties, however. Frequently, the exact number of sites is not known with certainty at the outset and may change during the course of the trial. New sites may join the study, or sites may drop out or be forced out, which would of course alter the mix of procedures generating the final data.

Aside from the need to vary the randomization procedures, there are also certain other factors to consider when selecting a randomization procedure. For example, how well can we identify the key predictors and how many levels are there in their cross product? This consideration determines if we stratify or use minimization. For example, if there is but one key covariate, and it is binary (gender, perhaps), then it can be handled with stratification. But if several key covariates each has three or four levels (ordered or unordered categories), then stratification may not be feasible.

In trials with multiple sites, an important consideration is whether to stratify by site or not. Frequently this is done, but this allows investigators to keep track of the numbers allocated to each group and, therefore, to predict upcoming allocations. Unless there is very strong reason to suspect that site affects the dependent variable, we suggest that randomization might not be stratified by site, or at least not by all the sites. It would be possible to isolate certain sites that would each have their own individual randomization lists and to still combine other sites together with one comprehensive randomization list. Numerous researchers have observed that investigators' information in multi site trials is imperfectly correlated (Brown et al., 2005; Barbachano et al., 2008; Berger et al., 2010). It is commonly presumed that investigators learn treatment histories at their own sites but are unaware of treatment histories at other sites. This fact is strategically useful for any statistician seeking to minimize treatment predictability, as long as randomization does not have to be stratified by site.

There will of course be some trials where perfect treatment balance at each and every site is required. For instance, intersite variation in investigator expertise can sometimes affect the dependent variable, which could necessitate perfect treatment balance at each site. These circumstances appear to be the exception rather than the rule. McEntegart (2010), for instance, examined a sample of multisite trials that required perfect treatment balance for the overall trial and found that only 12% of such trials required treatment balance at each site. Hills et al. (2009) reported that randomization is stratified by site in less than 40% of intentionally unmasked clinical trials published in the *British Medical Journal*, the *Lancet*, and the *New England Journal of Medicine*.

The way patients are enrolled in a study can also be a factor in the choice of a randomization procedure. The ideal, but highly unusual, situation is having all patients present and enrolled at the same time, because when this is feasible it practically eliminates both chronological bias and selection bias. The more typical situation involves patient enrollment over time, meaning that it is especially crucial to randomize properly (by using one of the three MTI procedures). Perfect masking would eliminate selection bias, but the catch is that this perfect masking would need to be demonstrable and knowable ahead of time to be exploited. If we could know ahead of time that masking would be perfect, then we would be satisfied that selection bias is not a problem, and we would then focus on chronological bias, meaning that we would use a small MTI. However, the more typical situation is that masking is claimed, but never demonstrated, and that we cannot know even after the fact, let alone before the trial, how successful masking will be. In these cases, we need a larger MTI to control selection bias.

Another consideration is the size of the trial. If we expect to enroll thousands of patients, then chronological bias may be negligible even with a larger MTI such as eight. A smaller trial, of course, may necessitate a smaller MTI as well, but this would entail more restrictions, more prediction, and more selection bias. But these considerations are all finer points. The primary focus should be on using more appropriate methodologies (the MTI procedures) in general, rather than continuing to use the permuted blocks procedure just because it has become the (rather misguided) precedent.

2.5 Summary and Conclusions

The spirit of W. Edwards Deming lives on in many circles that concern themselves with ongoing improvement and insistence on quality. A careful look at the literature will make clear that, sadly, medical studies in general, and trials in particular, would not constitute one of these circles (Berger and Ioannidis, 2004; Berger and Matthews, 2005; Berger et al., 2008). Permuted block randomization, particularly blocks of two, blocks of four, or mixed blocks of two and four, should not be used (Berger, 2006a–c); yet these seem to be the most common methods. The advantages of block designs can be obtained by better procedures without the predictability. Predictable randomization reduces a so-called randomized study to an essentially observational study. In fact, a predictable randomized study is qualitatively worse than an observational study because it charades as statistically valid in ways that it is not. The issues of predictability and selection bias are complicated and unlikely to be well understood by participants in clinical trials. This makes false claims of statistical validity more ethically objectionable because it ties funding and participation in randomized clinical trials to false promises. At the very least, there should be standards requiring that both the randomization method and the particular randomization sequence generated by a given trial must be disclosed in an easily accessible fashion. Examination of clinical trial publications reveals that this is yet to be the case.

Any trial seeking to limit predictability while still keeping desirable constraints on treatment imbalance should consider the use of the better designs discussed in Section 2.3. For a single-site trial to randomize in an unpredictable fashion that constrains treatment imbalance, the best designs may be the big stick rule (Soares and Wu, 1983), the modified big stick rule (Chen, 1999), the maximal procedure (Berger et al., 2003), or a Nash-equilibrium-type of procedure (Grant and Anstrom, 2008). For a multisite trial, avoiding predictability is easier as long as the randomization procedure exploits the fact that an investigator's prediction is likely based on the treatment history at his own site but not at treatment histories at other sites. What is revealed in publications of clinical trial results is a troubling inconsistency regarding the role of site-specific factors. When randomization is stratified by site, which makes treatments more predictable than if randomization was not stratified by site, the study results should include site as an explanatory variable. If study leaders claim that there is no need to include site as an explanatory variable, which is commonly the case, then there is no excuse for creating excess predictability through site-stratified randomization. In short, more consistency and more transparency regarding randomization will lead to less biased, more efficient, more ethical, and more impactful clinical trials.

References

Barbachano Y, Coad DS, Robinson, DR (2008) Predictability of designs which adjust for imbalances in prognostic factors. *Journal of Statistical Planning and Inference* 138: 756–767.

Berger VW (1999) FDA product approval information—Licensing action: Statistical review. http://www.fda.gov/cder/biologics/review/etanimm052799r2.pdf (accessed February 2, 2009).

Berger VW (2005) *Selection Bias and Covariate Imbalances in Randomized Clinical Trials.* Chichester, U.K.: John Wiley & Sons.

Berger VW (2006a) Do not use blocked randomization. *Headache* 46(2): 343.

Berger VW (2006b) Misguided precedent is not a reason to use permuted blocks. *Headache* 46(7): 1210–1212.

Berger VW (2006c) Varying block sizes does not conceal the allocation. *Journal of Critical Care* 21(2): 229.

Berger VW, Bears J (2003) When can a clinical trial be called 'randomized'? *Vaccine* 21: 468–472.

Berger VW, Grant WC, Vazquez LF (2010) Sensitivity designs for preventing bias replication in randomized clinical trials. *Statistical Methods in Medical Research* 19(4): 415–424.

Berger VW, Ioannidis JPA (2004) The Decameron of poor research. *British Medical Journal* 329: 1436–1440.

Berger VW, Ivanova A, Deloria-Knoll M (2003) Minimizing predictability while retaining balance through the use of less restrictive randomization procedures. *Statistics in Medicine* 22(19): 3017–3028.

Berger VW, Matthews JR (2005) Conducting today's trials by tomorrow's standards. *Pharmaceutical Statistics* 4: 155–159.

Berger VW, Matthews JR, Grosch EN (2008) On improving research methodology in medical studies. *Statistical Methods in Medical Research* 17: 231–242.

Brown S, Thorpe H, Hawkins K, Brown J (2005) Minimization: Reducing predictability for multi-center trials whilst retaining balance within centre. *Statistics in Medicine* 24: 3715–3727.

Chen YP (1999) Biased coin design with imbalance tolerance. *Communications in Statistics* 15: 953–975.

Fayers PM, King M (2008) A highly significant difference in baseline characteristics: The play of chance of evidence of a more selective game? *Quality of Life Research* 17: 1121–1123.

Grant WC, Anstrom KJ (2008) Minimizing selection bias in clinical trials: A Nash equilibrium approach to optimal randomization. *Journal of Economic Behavior & Organization* 66: 606–624.

Hills R, Gray R, Wheatley K (2009) Balancing treatment allocations by clinician or center in randomized trials allows unacceptable levels of treatment prediction. *Journal of Evidence Based Medicine* 2: 196–204.

Lovell DJ, Giannini EH, Reiff A et al. (2000) Etanercept in children with polyarticular juvenile rheumatoid arthritis. *New England Journal of Medicine* 342: 763–769.

Matts JP, McHugh RB (1983) Conditional Markov chain designs for accrual clinical trials. *Biometrical Journal* 25: 563–577.

McEntegart D (2010) In response to minimization, by its nature, precludes allocation concealment and invites selection bias. *Controlled Clinical Trials* 31: 507.

Soares JF, Wu CFJ (1983) Some restricted randomization rules in sequential designs. *Communications in Statistics: Theory and Methods* 12: 2017–2034.

Zhang J (2009) Moxifloxacin and placebo can be given in a crossover fashion during a parallel-designed thorough QT study. In *DIA Cardiovascular Safety, QT, and Arrhythmia in Drug Development Conference*, April 30–May 1, 2009, Bethesda, MD.

3

First Dose Ranging Clinical Trial Design: More Doses? Or a Wider Range?

Guojun Yuan and Naitee Ting

CONTENTS

3.1 Background

In the drug development process, a candidate compound needs to go through a stringent series of testing for toxicity, pharmacokinetics, efficacy, and safety before it can be released to the market for patient use. The process involves different phases of nonclinical and clinical studies on animals, healthy volunteers, and the patient population with the target disease (Ting 2008, 2011). In clinical development stage, the process is usually categorized into four phases (Phase I–IV, see ICH E8 General Considerations for Clinical Trials). Phases I and II are generally considered exploratory and learning phases. The goal of Phase III studies is mainly for therapeutic confirmation. Phase IV trials are postmarketing studies. Since Phase III pivotal studies are typically large scale with very high cost (the cost in some of Phase III studies could be even more than the total cost spent from the very beginning of a compound to the end of Phase II studies), Phase II studies are one of the most critical components before entering into Phase III confirmatory stage. In addition, regulatory agencies generally recommend that a full dose range of the test therapy has been explored prior to initiating Phase III and that identification of minimum effective dose (MinED) be encouraged (ICH 1994). Given the substantial resources invested in research and development, a well-designed Phase II study could be very beneficial to both patients and sponsors, and accordingly reduce the time to market launch.

In most cases, the Phase II features a learning period when the patient population may be first exposed to the test therapy for evidence of clinical benefit and risk; it is also a period to explore and recommend alternative doses to be tested in the later phase (e.g., Phase III) confirmatory clinical trials. In the development of a first-in-class new chemical entity, the first Phase II study is referred to as a proof-of-concept (PoC) study, and after PoC, the following Phase II studies are called dose-ranging studies. PoC is often faster and cheaper as it only requires a well-tolerated dose of test therapy plus a placebo control group. A dose–ranging study, however, needs multiple doses of the test therapy to characterize the dose-response relationship. Therefore, a classical Phase II development program usually consists of a small-scale PoC trial followed by moderately sized dose-ranging studies.

Sometimes it is desirable to combine the PoC and the dose-ranging studies into one single study with clearly defined objectives, hypotheses and decision criteria. The advantage of such a design is to first make a *go/no go* decision based on PoC. If the decision is *go*, then the same study would sequentially provide dose-ranging information to identify the dose that could achieve the target treatment effect and help design the next study. When well designed and analyzed, the combined study saves development time by providing a range of efficacious doses going forward. The disadvantage is more investment before the concept is proven. In case the test drug does not work, this translates into larger development costs.

Dose ranging studies are too often designed with a very small number of doses (Grieve 2011). A dose ranging trial typically includes a placebo (representing zero dose), and several doses of the test product. One very common design is a four group study with placebo, low dose, medium dose and high dose groups of the drug candidate under development. In practice, usually the first dose ranging study covers a wide dose range, and then the next dose ranging study will be designed with a narrower dose range and tease out the target doses to be assessed in Phase III. Dose range in a clinical trial design is defined as the ratio of the highest dose divided by the lowest dose included in the design. For example, if the doses in trial A are placebo, 20, 40, and 80 mg, then the dose range is 4 (=80/20). If the doses in trial B are placebo, 0.1, 1, and 10 mg, then the dose range is 100 (=10/0.1). Although the doses used in trial A are higher than doses used in trial B, in fact trial B has a much wider dose range—25 times wider than trial A.

3.2 A Motivating Example

In a clinical development program for a test drug to treat osteoarthritis (Ting 2008), the first dose–response study included placebo, 80, 120, and 160 mg of the test drug as seen from Figure 3.1. Results from this study indicate that all three doses are efficacious (Figure 3.1a). These results also show that 80 mg may have already been at the high side of the dose–response curve. A second study was designed to explore the 40 mg dose, and results from this second study are given in Figure 3.1b. With these findings, the project team started Phase III studies using a dose range of 40–20 mg to confirm the long-term efficacy of this test drug.

An End-of-Phase II meeting with the Food and Drug Administration (FDA) was held after the long-term Phase III study started. FDA commented that the minimum effective dose (MinED) was not found yet. Hence the project team designed a third dose–response study to explore the lower dose range. Doses included in the third study are 2.5, 10, and 40 mg. As shown in Figure 3.1c, the third study successfully established the dose–response relationship. Based on these results, it is clear that the Phase III studies are designed with a range of doses that are too high. The impact of this process can cause a major delay in the development program and considerable waste of resources.

The key lesson learned from this particular development program is that a very low dose was not explored in early Phase II. In other words, if a dose at or

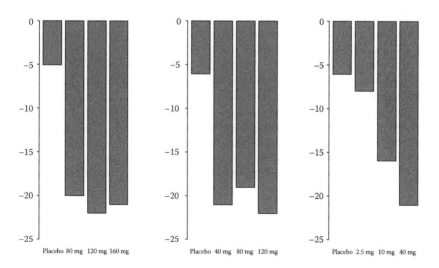

FIGURE 3.1
Dose–response results from the three studies of osteoarthritis drug.

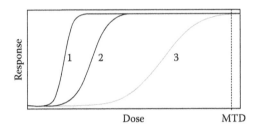

FIGURE 3.2
Several possible dose response curves.

lower than 10 mg was explored in the first study, a narrower dose range could have been established using the second study. Then the Phase III studies can be designed with the appropriate dose range.

In general, the most difficult challenge in designing the first dose ranging trial in clinical development of a new medicinal product is the selection of doses to be included in the design. Typically in the clinical development of a new treatment for chronic diseases, dose-ranging trials are designed during the Phase II stage. At this stage, the best guidance available is data obtained from preclinical experiments or from Phase I biomarkers. Even under the assumptions that the maximum tolerated dose (MTD) was correctly estimated, and there is a monotonic efficacy dose–response relationship, there can still be many possible shapes of dose–response curves.

Figure 3.2 presents an example of three possible dose–response curves. Dose–response curve 1 is based on the rabbit response models, curve 2 is based on mouse models, and curve 3 is from the dog models. The vertical dashed gray line (at the right side of Figure 3.2) represents the MTD. The question then becomes, what range of doses should be considered in the upcoming dose ranging clinical trial? From this perspective, the primary challenge in the dose ranging trial design is to consider which range of doses needs to be studied. In practice, the most difficult question in early Phase II development is "what is the dose range that provides the steepest increase in drug efficacy?" For example, if the MTD is determined as 1000 mg based on Phase I data. At this stage, can the best drug efficacy be found between 0.001 and 0.01 mg? between 0.01 and 0.1 mg? between 0.1 and 1 mg? between 1 and 10 mg? between 10 and 100 mg? or between 100 and 1000 mg?

3.3 Study Design Considerations

It is common that a dose–response model is used in analyzing dose-ranging trials. Sometimes the MCP-Mod (Bretz et al. 2005, Pinheiro et al. 2006) is also applied to help making inferences from clinical trial data. However, from the study design point of view, use of a model could be based on additional assumptions. It is well known that "All models are wrong, some are useful." There are many possible dose–response relationships, so it could be very difficult to guess which one is more appropriate in the design stage given the limited dose–response information. Even a particular model is guessed correctly, the parameters associated with this model could be very difficult to get. Designing dose-ranging trials based on a wrong model, or wrong parameters could potentially lead to very undesirable consequences. In data analysis, after the blind is broken, efficacy dose–response data are observed. At this point, use of a model that *fits the data well* is a reasonable thing to do.

In the design of a first dose ranging clinical trial, the major challenge is that there are too many unknowns. In fact, there is very little information regarding product activities at various doses. In order to design this study, some assumptions are necessary. In this chapter, two fundamental assumptions are needed:

1. The MTD obtained from previous studies is correct.
2. The underlying efficacy dose–response relationship is monotonic, at least between the range of placebo and MTD.

Based on the authors' experiences, any additional assumptions about the shape of the underlying dose–response relationship could be potentially misleading. Hence although modeling could be a useful tool for data analysis, it is not appropriate to use models in study design unless information is available to make reasonable assumptions for a dose–response model. The question raised in this chapter—number of doses vs. dose range—is a design problem in nature, the emphasis is not on data analysis.

In fact, when a model is used in designing a dose-ranging trial, the underlying assumption could be relatively strong—not only does that the selected model have to be right, but the parameters in that model have also to be all correct. In practice, these assumptions could be incorrect, especially when the range of active doses is guessed wrong. The major mistake we learned from the motivating example while designing the first dose ranging trial (placebo, 80, 120, and 160 mg) and the second trial (placebo, 40, 80, and 120 mg) was because the model/parameters used for designing these trials were wrong.

Again, some assumptions have to be made before designing a dose-ranging trial, but most of assumptions are in fact, wrong. Over the years, experiences indicate that the two critical assumptions as stated earlier (correct MTD and monotonicity) are absolutely necessary for study design (even though these two assumptions were not met in some practical situations) because without any one of the two assumptions, such a design is not possible. Furthermore, any additional assumption could potentially lead to very expensive failures. Given this background, the authors would warn readers to check every additional assumption very carefully. For this same reason, the discussion regarding study design in this chapter is basically model free.

On the other hand, models are useful in data analysis. Hence in evaluating the performance of various types of designs discussed in this chapter, E_{max} models are used to perform the simulations. One main reason that models are appropriate in these simulations is that the true underlying model is assumed to be already known, before data are generated.

It is critical to separate thinkings during the study design stage, and that of data analysis. Models can be useful for analyzing data because when the data are available, appropriate models can be selected to fit the observed data. However, at the time of designing a trial, if there are too many unknowns, adding assumptions would add potential risks to the design. In the progression of science, expensive mistakes are often made not because "we didn't know," but because "we thought we knew." The real world motivating example reflected that we thought we knew the model, and designed studies according to the model which eventually led to very expensive failures.

3.4 Finding Minimum Effective Dose

Minimum effective dose (MinED) is a very important concept in product development. Unfortunately, there is not a universally accepted definition of this term (Thomas and Ting 2008). Finding MinED is critical because it is generally believed that toxicity increases as dose increases. Hence a lower dose that is effective implies it could be safer than higher doses.

Clinical results obtained from Phase III studies are used to support a product registration. Based on these data, if a dose is efficacious and safe, then that dose can be approved, and the product label specifies the dose or the range of doses the product is approved for. However, after approval, if excessive adverse events are observed from the low dose of the product, then a question as to whether this low dose is low enough could be raised. At this time, it would be very challenging to ask what the MinED is for this product. Therefore, it is critical to find MinED in Phase II, before it is too late.

Drug price in the United States is not directly associated with dosage. However, in many other countries, the price of a medicinal product could be linked to its dose. In other words, if the dose of the product is higher, then it is more expensive. Now suppose a certain product is approved in such a country, and the price is set based on the lowest approved dose. If later it was detected that the MinED for the product is in fact, lower than the approved dose. Then a dose reduction is needed at a postmarket stage. Accordingly, the price of this product at a lower dose has to be further lowered, also.

When this happens, there is potentially a huge impact on the earnings from such a product. Therefore, from a marketing strategy point of view, it would be critical to identify the MinED before the product enters its Phase III development.

Finding MinED is a Phase II practice. As discussed earlier, if MinED was not identified at Phase II, it would be very difficult and very expensive to find the MinED at a later stage of product development. In fact, Phase II could be the only opportunity to characterize the relationship between doses and product efficacy. Therefore, it is very important to pay more attention to Phase II, and to carefully design dose-ranging clinical trials during this critical phase of clinical development for a new product.

Although there is a definition of MinED from ICH E4—"MinED is the *smallest dose with a discernible useful effect,*" the interpretation and implementation of such a definition varies a lot in practice. The first question is the understanding of "a useful effect." It is difficult to achieve consensus about the smallest magnitude of an effect that is useful. One approach is to follow a well-established treatment effect such as lowering 5 mmHg of blood pressure in developing an antihypertensive product. Another is to consider the concept of minimally clinically important difference (MCID). This is the magnitude of the treatment effect above and beyond placebo effect that is large enough to be perceived by a subject. It is a commonly used concept in outcomes research.

However, the implementation of a discernible effect could be more difficult. Various authors recommend different ways of calling a dose as MinED based on the difference in treatment effect from a given dose as compared with placebo. Some authors suggest the use of a dose–response model to help estimate MinED (Filloon, 1995, Pinheiro et al. 2006). Others (Hochberg and Tamhane 1987, Wang and Ting 2012) suggest the use of pairwise comparisons.

Even among those authors recommending use of pairwise comparisons to find MinED, there could be many differences, also. For example, some may argue that multiple comparison adjustment could be necessary, and some may not. Whether multiple comparison is needed or not, the understanding of "a dose that delivers treatment effect which is different from placebo" could still vary a lot, depending on individual's point of view. For example, let the useful effect defined earlier be denoted as δ, then for a given dose, a point estimate and a confidence interval can be constructed for the true treatment difference δ. Suppose a positive value indicates a benefit effect, then the treatment response of the given dose of test product subtracts the placebo response is expected to be positive. Hence it is hoped that the point estimate of the treatment difference could be as much as δ. This leaves a wide range of interpretations—for MinED, should that dose be such that the point estimate is greater than δ? Should that dose be such that the lower confidence limit to be greater than 0? Should that dose be such that the upper confidence limit to be greater than δ? Or should that dose be such that the lower confidence limit to be greater than δ?

The implementation of this concept of *the smallest dose with a discernible useful effect* could be very difficult, depending on the various interpretation of this definition. However, in practice, one feasible way of finding a MinED in a dose-ranging clinical trial is to find a dose that is lower than MinED (and, of course, above placebo). Regardless which practical definition the project team uses to identify MinED, if there is a dose deemed to be not efficacious, and there is a dose that is higher than such a dose could be considered as efficacious. Then it can be considered that the nonefficacious dose is below MinED, and the efficacious dose is above MinED. From a practical point of view, this information would usually be sufficient to help the team to design Phase III clinical trials.

Of course in the selection of Phase III doses, many other aspects will also have to be included in consideration; for example, safety or toxicity findings, pharmacokinetic properties, formulations, ..., etc. But the information about MinED would be one of the most critical deliverables from Phase II in supporting any Phase III development plans.

3.5 Practical Limitations in the Number of Doses Used in a Design

In order to deal with this challenge, some authors suggest that more doses could help (Krams et al. 2003)—that is, in the first dose-ranging study, adding more doses to the study design. However, the number of doses to be used in a dose-ranging trial is usually limited by practical and logistical considerations. For example, Table 3.1 list the number of patients from each center of each treatment group in a multicenter dose ranging study.

When there is a zero cell (which means there is no patient recruited for that treatment group within the specific center), the treatment-by-center interaction becomes nonestimable using analysis of variance or analysis of covariance. Statistically speaking, it is preferred to perform an analysis on a set of balanced data (i.e., an equal number of patients in each cell). Apparently, as more doses are included in a design, the risk of imbalance increases,

TABLE 3.1

Number of Patients in a Center from a Multicenter Trial

	Placebo	Low	Medium	High
Center 1	6	7	6	8
Center 2	1	1	0	1
Center 3	4	2	3	2

and the likelihood of zero cells also increases. Therefore, it may not be desirable or practical to use too many treatment groups in any given study design. Typical dose-ranging studies already use at least three treatment groups— a placebo group, a high-dose group, and another dose that is in between these two groups. In some situations, an active control is also employed in a dose-ranging trial. Given these many treatment groups already included in a study, it may not be practical to add very many doses into the same study design.

In fact, a major limiting factor for using too many doses is formulation. Most of dose-ranging trials take place during Phase II and, in general, patients recruited in Phase II tend to be outpatients. These patients do not stay in the hospital or in the clinical research unit. Outpatients visit the clinic only at protocol designated time points, and they are not under the investigator's supervision when they are not in the clinic. In Phase I trials, study subjects stays in the clinic, and the health care providers could dose the subject with flexibility in various dose strengths. For example, the test medications can be prepared in powder form or in solutions. This way a wide variety of doses can be delivered to the subjects. However, in an outpatient setting, drugs need to be formulated with fixed doses for patients to take while they are outside of the clinic. At this time, drugs are prepared in tablets or capsules with fixed strengths, and are put in bottles or kits so that it will be convenient for patients to use on their own. Hence in a Phase II dose ranging design, the dose strength is limited by the preformulated tablets or capsules. For example, if the test drug is formulated with 5 and 50 mg tablets, then many different doses can be studied—5, 10, 50 mg, any multiples of 5 or 50 mg, or combinations thereof. However, there is no way to study, say, 18 or 2.7 mg. Therefore, doses selected in a dose ranging design are limited by the available formulations.

Another practical difficulty in running double-blind clinical trials is the blinding procedure. For example, in the aforementioned case study medications are formulated to be 5 and 50 mg tablets, and a dose-ranging trial is designed with placebo, 5, 15, and 50 mg doses. Suppose the 50 mg tablets look different from the 5 mg tablets. Then during the actual trial, each patient will need to take four bottles from the investigator at every visit. A patient randomized to the placebo group will need four bottles of placebos. A patient randomized to the 5 mg group will take one bottle of placebo tablets corresponding to the 50 mg dose, one bottle with true 5 mg tablets, and two bottles of placebo matching the 5 mg tablets. A patient randomized to the 15 mg group will take one bottle of placebo matching 50 mg tablets, and three bottles of true 5 mg tablets. Finally, a patient randomized to 50 mg group will take one bottle of true 50 mg tablets and three bottles of placebo tablets corresponding to the 5 mg tablets. On top of these issues, if an active control is used in the study, and the dosing frequency of the active control is different from the dosing frequency of the study treatment, it will add further complexity to the study conduct.

In clinical practice, drug supply is a major effort. There are usually supply chain experts to help prepare, package, and ship study medications to the investigators. For a trial with too many doses, the supply chain can become overburdened, and the process could be more expensive, and more likely to cause errors. Given all of these practical difficulties, it is very seldom to see a clinical trial with more than six or seven treatment groups in practice.

In our opinion, for the first dose ranging study design, it is more important to cover a wide dose range, than simply adding more doses to a narrow range of doses. This is both practical and scientific. Basically speaking, a trial with 4–5 test doses, plus a placebo control will deliver a good understanding of where the test medication is most active, if the dose range is wide enough, and the dose spacing is reasonable. Hamlett et al. (2002) proposed to use a binary dose spacing (BDS) for dose allocation. Over the years, BDS has been successfully applied in many dose ranging designs (e.g., Ting 2009, Wang and Ting 2012).

In order to determine the doses using BDS approach, it is assumed that the MTD is T, thus the design space is $[0,T]$. Without loss of generality, T can be taken as 1. It is assumed that the dose–response relationship is monotonic. It is further assumed that the number of dose groups is known. Given this setting, doses are chosen from the interval $[0, 1]$. Suppose we want a design with three treatment groups, including the placebo, a low dose, and a high dose. The placebo dose is taken to be zero. The challenge then is to select a low dose and a high dose, keeping in mind that we may not want a high dose too close to 1 (the assumed MTD). An intuitive approach might be to split this interval into half, giving the two intervals $[0, 1/2]$ and $(1/2, 1]$ and select a dose in each of these intervals. A natural choice is to select the midpoint of each interval (split each interval into half), giving the test doses, $x_1 = 1/2^2$ and $x_2 = 3/2^2$. Note that if we choose doses in the upper end (greater than $1/2$ and approaching 1) then these doses tend to be toxic and is generally not of primary interest in the dose-selection process. On the other hand, activities of the lower doses are very important information for drug development. Hence the basic idea for BDS is to search for lower end of the dose range.

In another case, suppose we want to consider a design with four treatment groups including placebo, low, medium, and high doses. In order to avoid selecting too high a dose that can be toxic, we may want to keep $3/2^2$ as the high dose, that is, leave the interval $(1/2, 1]$ unchanged. Since we want to use of some low doses to help identify the MinED, we can divide the lower interval $[0, 1/2]$ into half, giving the new intervals $[0, 1/2^2]$ and $(1/2^2, 1/2]$. We then select the midpoints of these two intervals respectively—split these two intervals into halves—giving the three test doses $x_1 = 1/2^3$, $x_2 = 3/2^3$, and $x_3 = 3/2^2$. We continue in this fashion by splitting the lower interval into half and taking the midpoint of each interval, until all the doses are allocated.

In general, let there be m test doses in addition to placebo, then the test doses are given by $x_1 = 1/2^m$, $x_2 = 3/2^m$, $x_3 = 3/2^{m-1}$, $x_4 = 3/2^{m-2}, \ldots$, $x_m = 3/2^2$. Since this method of dose allocation is based on dividing dose intervals into halves, it is denoted as the binary dose spacing (BDS).

Other ways of subdivision are also possible. Instead of splitting the initial interval into $[0, T/2]$ and $(T/2, T]$, we may use a $1/p$ cut—$[0, T/p]$, $(T/p, T]$—where $p \geq 2$ and repeat the procedure described earlier. The BDS design is nonparametric because it does not depend on any underlying model other than the two assumptions: knowledge of MTD and nondecreasing dose–response relationship. It provides a way to identify doses that are below the MTD and a means to potentially test for the MinED.

In addition to BDS, other algorithm could be useful, for example, log dose spacing, Fibonacci series, and modified Fibonacci series approach can also be considered (Penel and Kramar 2012), and the approach suggested by Quinlan and Krams (2006). However, it is important to note that most of these methods propose dose spacing from lower doses to higher doses. In practice, dose spacing should be considered starting with MTD, and move down to lower doses. Hence when applying any of these algorithms, the dose allocation would be reversed (from high to low, instead of from low to high).

3.6 Simulation Procedure and Results

To investigate the performance of the wider dose range, simulation studies were conducted under various case scenarios as seen from Table 3.2.

Table 3.2 summarizes 16 cases under different dose levels, dose range (100, 50, 25, 10) and number of dose arms (4, 5, 6, or 7 dose arms). The total sample size is fixed to be 240 arbitrarily for each simulation study, which results in 60 subjects/arm for the 4-arm design, 48 subjects/arm for the 5-arm design, and 40 subjects/arm for the 6-arm design. In the 7-arm design, an unbalanced allocation ratio is used to have 240 subjects in total, which include 40 subjects in placebo and highest dose (100 mg) arm and 32 subjects in the rest of the arms. Generally speaking, analysis methods for dose-ranging study could be categorized into two classes—namely, multiple comparisons (MCP) and modeling approaches (Bretz et al. 2005, Pinheiro et al. 2006). In this chapter, all simulation studies were conducted under modeling framework and an R package, 'DoseFinding' (Bornkamp et al. 2012). Although simulation studies were performed based on a three-parameter E_{max} model, in fact dose–response simulation studies could be performed without using any model. For example, Ting (2009) uses a nonparametric approach to perform the simulations. The key purpose of these simulations is to describe the design considerations on a Phase II dose range trial, instead of discussing the model itself. All simulation studies assumed that the true dose–response curve follows a three-parameter E_{max} model as described in the following.

TABLE 3.2

Simulation Schema

No. of Arms	n/arm (Total N)	Cases	Dose (mg)	Dose Range
4	60 (240)	Case 1.1	0, 1, 10, 100	100
		Case 1.2	0, 2, 15, 100	50
		Case 1.3	0, 4, 25, 100	25
		Case 1.4	0, 10, 25, 100	10
5	48 (240)	Case 2.1	0, 1, 4, 25, 100	100
		Case 2.2	0, 2, 10, 50, 100	50
		Case 2.3	0, 4, 10, 25, 100	25
		Case 2.4	0, 10, 25, 50, 100	10
6	40 (240)	Case 3.1	0, 1, 4, 15, 50, 100	100
		Case 3.2	0, 2, 10, 25, 50, 100	50
		Case 3.3	0, 4, 10, 25, 50, 100	25
		Case 3.4	0, 10, 15, 25, 50, 100	10
7	40*2 + 32*5 (240)	Case 4.1	0, 1, 4, 10, 25, 50, 100	100
		Case 4.2	0, 2, 4, 10, 25, 50, 100	50
		Case 4.3	0, 4, 10, 15, 25, 50, 100	25
		Case 4.4	0, 10, 15, 25, 50, 60, 100	10

Notes: (1) 7-Arm design uses unbalance sample size allocation with 40 arm in highest dose and placebo arms and 32/arm in all other does arms, (2) Dose range = highest test dose lowest test dose, (3) 1000 simulation runs for each case.

$$y_i = E_0 + E_{max} \times \frac{\text{Dose}_i}{ED_{50} + \text{Dose}_i} + \epsilon_i \tag{3.1}$$

where

i is the subject ID

y_i is the observed continuous response variable for the ith patient

E_0 is the effect when dose is 0 mg

E_{max} is the maximum effect attributable to the drug

ED_{50} is the dose that produces half of E_{max}

Dose is the dose level

ϵ_i is the random error for the ith patients

Within each of these 16 cases, 4 true ED_{50} values (0.5, 2.5, 10, and 40 mg) are assumed and evaluated. Therefore, there are 64 simulation scenarios in total. The simulation trials were repeated 1000 times for each of the 64 scenarios. The true basal effect E_0 (true response value when dose = 0 mg) is assumed to be 0 for all case scenarios. The true E_{max} is set to be 25. ϵ_i is assumed independently and normally distributed with a mean of 0 and a standard deviation of 15.

3.6.1 Assessment of Statistical Power and False Positive Rate

This section evaluates the statistical power and the false positive rate. The hypotheses underlying the simulation study are

$$H_0 : E_{\max} = 0$$

$$H_1 : E_{\max} \neq 0$$

The statistical power is defined as the percentage of trials that met the statistical criteria within 1000 simulation trials under H_1, which is defined as the estimated 95% confidence interval for $E_{\max}$ parameter exclude 0. On the other hand, the two-sided false positive rate is defined as the percentage of trials that met the same statistical criteria within 1000 simulation trials under H_0, that is, when the underlying true $E_{\max}$ is 0.

Figure 3.3 shows the simulation results for the statistical power (under $H_1, E_{\max} = 25$) for all the 64 cases. The upper-left panel shows the graphical summaries on statistical power for the different dose range (DR = 10, 25, 50, and 100) under each of the four assumed true ED_{50} (0.5, 2.5, 10, and 40 mg)

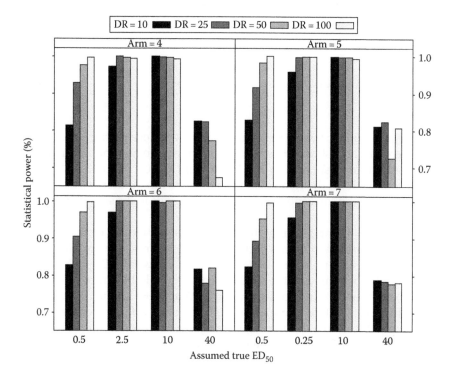

FIGURE 3.3
Graphical summary of the evaluation on statistical power.

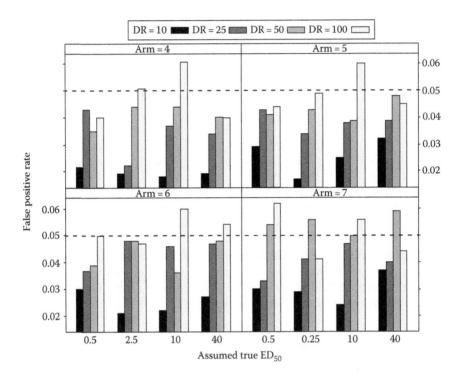

FIGURE 3.4
Graphical summary of the evaluation on false positive rate.

for the 4-arm design; the upper-right panel shows those for the 5-arm design; the lower-left panel shows those for the 6-arm design and the summaries for 7-arm design are displayed in the lower right panel. Figure 3.4 shows the simulation result for the two-sided false positive rate (under H_0, $E_{max} = 0$) of the trials for all the corresponding 64 cases scenarios in the same presentation order.

When $ED_{50} = 0.5$ and 2.5 mg, the statistical power clearly increases in general as the dose range increases under each of the four design options in terms of the different arm numbers. Note that especially when $ED_{50} = 0.5$ mg, the trend is pretty obvious and the statistical power could be improved substantially from about 80% (when dose range is 10) to $\geq 99\%$ (when dose range is 100). When $ED_{50} = 10$ or 40 mg, the statistical power is fairly close and generally comparable on the different dose range cases. When $ED_{50} = 40$ with the number of arm in 5, 6, and 7, the statistical power is comparable in general on the different dose range cases.

However, when $ED_{50} = 40$ with 4-arm design, the statistical power decreases as the dose range increases. In these power analyses, it looks like that for ED_{50} to be 2.5 or 10, the performance of all settings are satisfactory. When $ED_{50} = 0.5$ mg, it is clear that the performance depends on the dose range. This is one of the most critical points in dose ranging designs—project teams should be careful not to miss a very potent product, which delivers meaningful activities at very low doses. When $ED_{50} = 40$ mg, the 4-arm designs do not perform well. This should be expected because none of the proposed doses could offer a good estimate around the 40 mg dose.

Nevertheless, the power estimates under the 5-arm, 6-arm, and 7-arm settings when $ED_{50} = 40$ mg are not satisfactory. Therefore, an additional set of simulations are performed with a total sample size of 420 in order to evaluate the power and false positive rates. Results from these simulations are presented in Figure 3.5. From these results, it appears that the power estimates improved when sample size increased (as expected).

Two-sided false positive rates are generally controlled well. Most of these rates are under 5%. In some of those cases, it is observed that the false positive rate could be slightly inflated (at most 1% inflation). Considering that most

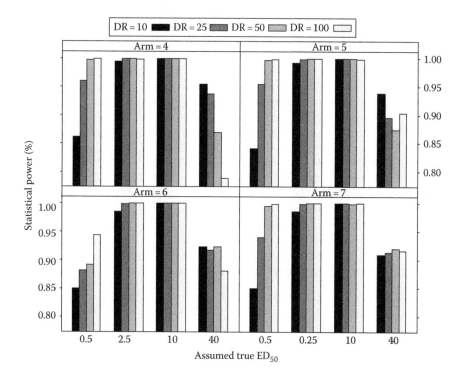

FIGURE 3.5
Graphical summary of the evaluation on statistical power with a total sample.

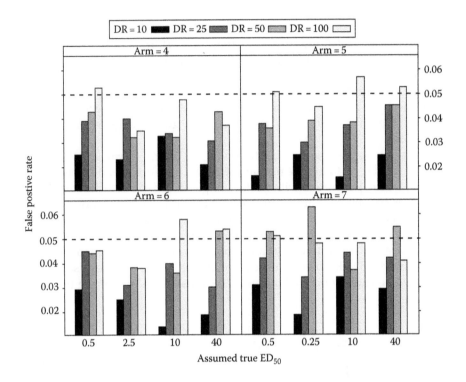

FIGURE 3.6
Graphical summary of the evaluation on two-sided false positive rate with a total sample size of 420.

phase II studies are still in learning phase, this might be acceptable in real world applications. Note that the false positive rate tends to be high in the 7-arm designs. This could be due to the smaller sample size per group. Again, when total sample size increased to 420, the false positive rates also improved slightly (Figure 3.6).

3.6.2 Assessment of E_{max} Model Fit Performance

The performance of E_{max} model is assessed from two perspectives: (1) the percentage of models failing to converge and (2) the percentage of models with unreasonable estimated ED_{50}—this is specified as estimated $ED_{50} < 0$ mg or $ED_{50} > 100$ mg. The performance assessment is summarized in the Table 3.3.

When $ED_{50} = 0.5$ mg and 2.5 mg, the percentage of models failing to converge clearly increases in general as the dose range decreases under each of the four design options in terms of the different arm numbers. Note that especially when $ED_{50} = 0.5$ mg, the trend is pretty obvious. When $ED_{50} = 10$ or 40 mg, the overall percentage of the models failing to converge is pretty

TABLE 3.3

Assessment of E_{max} Model Fit Performance

		Assumed True ED_{50} (mg)							
		% of Models Failing to Converge				% of Models with Unreasonable Estimated ED_{50}			
Dose (mg)	Dose Range	0.5	2.5	10	40	0.5	2.5	10	40
0, 1, 10, 100	100	0.2	0.4	0.3	9.3	0	0	0.1	14.8
0, 2, 15, 100	50	1.8	0.2	0.2	4.0	0	0	0.0	13.4
0, 4, 25, 100	25	5.9	0.0	0.0	1.5	0	0	0.0	12.4
0, 10, 25, 100	10	15.9	2.1	0.0	2.3	0	0	0.0	14.4
0, 1, 4, 25, 100	100	0.0	0.0	0.4	2.4	0	0	0.1	13.0
0, 2, 10, 50, 100	50	1.6	0.0	0.0	3.0	0	0	0.2	18.4
0, 4, 10, 25, 100	25	7.2	0.1	0.0	2.0	0	0	0.2	11.5
0, 10, 25, 50, 100	10	15.1	3.3	0.0	2.0	0	0	0.2	14.9
0, 1, 4, 15, 50, 100	100	0.1	0.1	0.0	3.5	0	0	0.0	15.1
0, 2, 10, 25, 50, 100	50	2.8	0.0	0.1	1.2	0	0	0.0	12.7
0, 4, 10, 25, 50, 100	25	8.5	0.1	0.3	2.0	0	0	0.0	16.2
0, 10, 15, 25, 50, 100	10	15.1	2.6	0.0	2.2	0	0	0.0	12.8
0, 1, 4, 10, 25, 50, 100	100	0.2	0.0	0.1	1.2	0	0	0.0	15.3
0, 2, 4, 10, 25, 50, 100	50	4.6	0.0	0.1	3.1	0	0	0.0	15.9
0, 4, 10, 15, 25, 50, 100	25	9.5	0.1	0.0	1.7	0	0	0.1	15.4
0, 10, 15, 25, 50, 60, 100	10	16.1	3.9	0.0	2.8	0	0	0.0	14.4

Notes: (1) Total sample size in each design scenarios are $N = 240$. Sample size is 60/arm in 4-arm design, 48/arm in 5-arm design, and 40/arm in 6-Arm design. 7-arm design uses unbalance sample size allocation with 40/arm in highest dose and placebo arms, and 32/arm in all other dose arms, (2) Dose range = highest test dose/lowest dose, (3) Unreasonable estimate ED_{50} is defined as estimated $ED_{50} <$ mg or $ED_{50} < 100$ mg.

close and comparable on the different dose range cases. Also note that in the 4-arm design when $ED_{50} = 40$ mg, the nonconvergence rate slightly increases as the dose range increases. This is reasonable because when $ED_{50} = 40$ mg the 4-arm designs provide only four points to help estimate the E_{max} model curve, and hence they perform relatively poorer as compared to designs with more arms. In addition, when dose range increases, the allocated doses are farther away from 40 mg. Therefore as dose range increases, percentages of nonconvergence increases under the 4-arm designs.

When $ED_{50} = 0.5$ and 2.5 mg, the percentage of models with unreasonable estimated ED_{50} are all 0. When $ED_{50} = 10$ or 40 mg, the overall percentage of the models with unreasonable estimated ED_{50} is pretty close and comparable on the different dose range cases for 5 to 7-arm design. In the 4-arm design when $ED_{50} = 40$ mg the percentage of the models with unreasonable

TABLE 3.4

Summary Statistics of Estimated ED_{50}s from E_{max} Model Simulations (Based on Converged Models with Reasonable ED_{50} Estimates)

| | Assumed True ED_{50} (mg) | | | | | | | | | | | |
| | Mean | | | | Standard Deviation | | | | Median | | | |
Cases	0.5	2.5	10	40	0.5	2.5	10	40	0.5	2.5	10	40
Case 1.1	0.5	2.8	11.2	37.2	0.2	1.4	6.5	21.4	0.5	2.5	9.8	0.5
Case 1.2	0.5	2.8	11.6	38.3	0.3	1.3	6.4	21.7	0.5	2.5	10.2	0.5
Case 1.3	0.6	2.7	11.3	40.3	0.5	1.3	5.8	22.4	0.6	2.5	10.1	0.6
Case 1.4	1.0	2.8	10.9	40.3	1.0	1.8	5.0	20.5	0.7	2.6	10.0	1.0
Case 2.1	0.5	2.9	11.8	37.6	0.3	1.9	7.4	21.6	0.5	2.5	10.1	0.5
Case 2.2	0.6	2.8	11.8	38.3	0.4	1.3	7.2	22.2	0.5	2.5	10.1	0.6
Case 2.3	0.7	2.7	11.4	39.6	0.6	1.2	5.9	21.5	0.6	2.5	10.1	0.7
Case 2.4	1.1	2.9	11.1	39.4	1.3	2.0	6.6	21.7	0.7	2.5	9.7	1.1
Case 3.1	0.6	2.8	12.4	38.6	0.3	1.5	8.9	22.2	0.5	2.5	10.2	0.6
Case 3.2	0.5	2.9	11.6	38.6	0.4	1.5	7.2	21.6	0.5	2.5	9.7	0.5
Case 3.3	0.6	2.7	11.8	38.3	0.5	1.3	7.0	21.7	0.5	2.5	10.0	0.6
Case 3.4	1.0	2.8	10.8	40.8	1.2	1.9	5.7	21.4	0.7	2.5	9.7	1.0
Case 4.1	0.5	2.7	11.9	38.3	0.3	1.3	7.7	22.9	0.5	2.5	9.9	0.5
Case 4.2	0.6	2.8	11.6	37.6	0.4	1.8	6.8	21.6	0.5	2.5	10.1	0.6
Case 4.3	0.7	2.6	11.6	39.2	0.7	1.4	5.8	21.9	0.6	2.4	10.3	0.7
Case 4.4	1.1	2.8	11.0	39.9	1.3	2.1	5.7	22.2	0.8	2.5	9.7	1.1

estimated ED_{50} slightly increases as the dose range increases. The reason for this phenomenon is similar to the discussion provided in the bottom of the previous paragraph.

3.6.3 Assessment of Estimation of ED_{50}

Table 3.4 summarizes mean, standard deviation and median of E_{max} model estimated ED_{50}s based on those converged models with reasonable ED_{50} estimates. Means and medians are closer to the true value and the standard deviations get smaller as dose range increases when $ED_{50} = 0.5$ mg. The differences are not very obvious under the other three assumed ED50s.

3.7 Discussions and Conclusions

It is always a huge challenge in designing the first dose-ranging clinical trial. At this stage, there is very limited information about the product activity because not much of clinical data from patients are available. In designing these trials, project team members need to consider the number of treatment groups, the range of doses to be studied, the total sample size, the number of subjects allocated to each dose, and the specific doses to be tested.

In facing with this challenge, investigators tend to allocate more doses into the first dose ranging design in a hope that some of these doses could help locate the product activities. However, there is always a tradeoff between the number of doses and the number of subjects per dose given that the total sample size is fixed. Therefore, when increasing the number of doses, the precision of estimate for each dose decreases (because the sample size for each dose have to be reduced). In practice, we believe that, although more doses may help, it is more important to consider which doses to be studied, and a wider range of doses could be more important than simply adding doses in a narrow dose range. From this perspective, we suggest a wider dose range need to be considered first, and then determine how many doses will be needed to cover this wider dose range. Practically, there are only limited number of doses that can be studied in a single clinical trial. Hence dose allocation is very critical in designing dose ranging trials.

In this chapter, we performed simulations to study the performance of many settings of dose range and number of doses under various assumed ED_{50} values. Based on the simulation results, it appears that a wider dose range will improve study power if the ED_{50} is low. Among the number of treatment arms (4, 5, 6, or 7), it looks like three test doses (the 4-arm design) could be insufficient at some cases. The performance increases when more than three doses are studied. On the other hand, six test doses (the 7-arm design) may not necessarily deliver better results than the four test doses (the 5-arm design) or five test doses (the 6-arm design) comparisons. One concern is that when the total sample size is fixed, the 7-arm design offers a smaller sample size per group, and the precision would be sacrificed. Hence from a practical point of view, a dose ranging design including four to five test doses (in addition to placebo) that cover a wide dose range may be very useful in designing the first dose ranging clinical trial. Note that a nonparametric simulation without using any does–response model could also lead to the same conclusion. The main point of this chapter is that in an early Phase II trial design, a wider dose range is recommended, regardless of the number of doses to be used in the study design.

In a study design where the number of doses is determined, we suggest to allocate doses using the binary dose-spacing method. In practice, the selected doses are restricted by the existing formulations. A motivating example is introduced in this chapter to demonstrate that it took three dose ranging studies where finally the dose range with the steepest increase in efficacy response can be identified. For this example, if the first study uses the treatments in the third study, plus the 160 mg, then one study will help define the dose range, and move to Phase III using 20, 40 mg, and placebo.

References

Bornkamp, B., Pinheiro, J. C., and Bretz, F. (2012). DoseFinding: A R Package for planning and analyzing dose finding experiments. http://cran.r-project.org/web/packages/DoseFinding/DoseFinding.pdf.

Bretz, F., Pinheiro, J. C., and Branson, M. (2005). Combining multiple comparisons and modeling techniques in dose-response studies. *Biometrics*, 61:738–748.

Filloon, T. G. (1995). Estimating the minimum therapeutically effective dose of a compound via regression modelling and percentile estimation. *Statistics in Medicine*, 14:925–932; discussion 933.

Grieve, A. P. (2011). Adaptive trial designs: Looking for answers. *International Clinical Trials*, November 2011, issue 22:16–20.

Hamlett, A., Ting, N., Hanumara, C., and Finman, J. S. (2002). Dose spacing in early dose response clinical trial designs. *Drug Information Journal*, 36(4):855–864.

Hochberg, Y. and Tamhane, A. (1987). *Multiple Comparison Procedures*. Wiley, New York.

ICH. (1994). Dose-response information to support drug registration E4. *International Conference on Harmonisation of Technical Requirements for Registration of Pharmaceuticals for Human Use*. http://www.ich.org/fileadmin/Public_Web_Site/ICH_Products/Guidelines/Efficacy/E4/Step4/E4_Guideline.pdf.

Krams, M., Lees, K. R., Hacke, W., Grieve, A. P., Orgogozo, G. M., and Ford, G. (2003). Acute stroke therapy by inhibition of neutrophils (ASTIN): An adaptive dose-response study of UK-279,276 in acute ischemic stroke. *Stroke*, 34:2543–2548.

Penel, N. and Kramar, A. (2012). What does a modified-fibonacci dose-escalation actually correspond to? *BMC Medical Research Methodology*, 12:103.

Pinheiro, J. C., Bornkamp, B., and Bretz, F. (2006). Design and analysis of dose finding studies combining multiple comparisons and modeling procedures. *Journal of Biopharmaceutical Statistics*, 16:639–656.

Quinlan, J. A. and Krams, M. (2006). Implementing adaptive designs: Logistical and operational consideration. *Drug Information Journal*, 40:437–444.

Thomas, N. and Ting, N. (2008). Minimum effective dose. In *Encyclopedia of Clinical Trials*, D'Agostino, R. B. Sullivan, L. and Massaro, J., Eds., Wiley-Blackwell, Hoboken, NJ.

Ting, N. (2008). Confirm and explore, a stepwise approach to clinical trial designs. *Drug Information Journal*, 42:545–554.

Ting, N. (2009). Practical and statistical considerations in designing an early phase II osteoarthritis clinical trial: A case study. *Communications in Statistics—Theory and Methods*, 38(18):3282–3296.

Ting, N. (2011). Phase 2 clinical development in treating chronic diseases. *Drug Information Journal*, 45:431–442.

Wang, X. and Ting, N. (2012). A proof-of-concept clinical trial design combined with dose-ranging exploration. *Biopharmaceutical Statistics*, 26:403–409.

4

Thorough QT/QTc Clinical Trials

Yi Tsong

CONTENTS

4.1 Introduction

The QT interval is measured from the beginning of Q wave to the end of T wave on an electrocardiagram (ECG) tracing (see Figure 4.1). It reflects the duration of ventricular depolarization and subsequent repolarization. For some drugs, significant prolongation of the absolute QT interval has been associated with the precipitation of a potentially fatal cardiac arrhythmia called torsades de pointes (TdP) and can degenerate into ventricular fibrillation, leading to sudden cardiac death (Moss, 1993). Over the past one and half decades, these cardiac adverse events have resulted in a number of patient deaths and were one of the most frequent adverse events (Wysowski et al., 2001) of a drug that led to either its removal from the market (e.g., terfenadine and cisapride) or the placement of restrictions on its use (e.g., ziprasidone and ranolizine.) With theses experiences, the regulatory agencies now request that each pharmaceutical company conduct at least one thorough QT (TQT) study when submitting an application of any new nonantiarrhythmic drug. The ICH E14 guidance for *The Clinical Evaluation of QT/QTc Interval Prolongation and Proarrhythmic Potential for Nonantiarrhythmic Drugs* was released in May 2005. The abbreviation QTc denotes the QT interval corrected for the heart rate (HR).

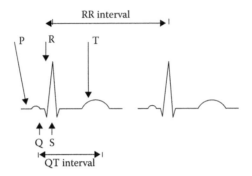

FIGURE 4.1
ECG waves. The P wave represents depolarization occurring within the atria of the heart. The QRS complex represents depolarization occurring within the ventricles of the heart. The T wave represents repolarization occurring within the ventricles of the heart. The QT wave is the period of time it takes to complete both depolarization and repolarization within the ventricles and is measured from the beginning of the Q wave to the end of the T wave.

ICH E14 guidance recommends that the TQT study be conducted generally in early clinical development after some information about the pharmacokinetics of the drug has been obtained. The objective of a TQT study is to determine whether the drug has a threshold pharmacologic effect on cardiac repolarization, as detected by QTc prolongation. The TQT study is a randomized, double blinded, placebo- and active-controlled, crossover or parallel-arm study of healthy subjects, with single or multiple doses of the test and active control drugs. The guidance states that *A negative TQT/thorough QTc study* is one in which the upper bound of the 95% one-sided confidence interval for the largest time-matched mean effect of the drug on the QTc interval excludes 10 ms, a margin of regulatory concern. If the data collected provide enough evidence to show that the drug does not prolong the QTc interval when compared with placebo, the TQT study is claimed as a *negative study*. A negative TQT study supports a conclusion that the drug does not prolong the QTc interval to a clinically significant for proarrhythmic potential. When a TQT study is not negative, that is, if at least one upper bound of the one-sided 95% CI of the time-matched difference exceeds the threshold of 10 ms, the study is claimed to be positive or nonnegative. In this case, a positive TQT study indicates that more intense evaluation of the potential for QTc prolongation is necessary during subsequent drug development.

With the objective of the test to reject the trial to confirm that the response of the test treatment differs no more than a prespecified margin from the placebo, one may have the concern that the trial population may respond unexpectedly to an active treatment with a known prolongation effect of QT interval. For this reason, E14 recommends that "The positive control should have an effect on the mean QT/QTc interval of about 5 ms. Often a positive control treatment with a well known QT interval prolongation and dose-response profile will be used and an assay validation assessment performed often consists of a quantitative assessment test and a qualitative profile identification."

In this chapter, we describe the statistical approaches and issues in the design and analysis of a TQT clinical trial. We organize the rest of the chapter as follows. The approaches for confirmative noninferiority test of drug-induced prolongation of QT interval are given in Section 4.2. The quantitative approaches for validation of study are given in Section 4.3. Regular fixed and flexible sample size designs of the trial are presented in Section 4.4. Measurements of QT interval and the methods of QT interval correction by pulse rate will be given in Section 4.5. Finally, a summary and discussion is given in Section 4.6.

4.2 Assessment of Treatment-Induced Prolongation of QTc Interval

The ECG records the polarization activities of a human heart. Figure 4.1 (Li et al., 2004) displays an ECG of two heartbeats. QRS complex represents depolarization occurring within the ventricles of the heart. T wave represents repolarization occurring within the ventricles of the heart. The beginning of the Q wave to the end of the T wave defines the QT interval and represents the total time required for both ventricular depolarization and repolarization to occur. The duration of this QT interval is used as a biomarker for the risk of TdP. The time between two consecutive R waves, the RR interval, is inversely proportional to the HR. It is known physiologically that QT interval prolonging accompanies the physiological increases in the RR interval. When a drug alters HR, it is important to distinguish an effect of drug on the QT interval versus an artifact due to change in the RR interval (inverse of HR). Therefore, QT interval needs to be corrected before used for the analysis of drug effect on QTc (corrected QT interval). It is desired that the resulting QTc intervals and RR intervals are uncorrelated. Commonly used correction formulas are based on either linear or nonlinear regression models for QT–RR relationship (Ahnve, 1985; Funk-Brentano and Jaillon, 1993; Moss, 1993; Hnatkova and Malik, 1999; Malik, 2001; Li et al., 2004; Wang et al., 2008). We will introduce some of the correction methods in details in Section 4.4. Conventionally, comparisons between treatments are carried out on the mean interval change of QTc adjusted for the HR. Optional single stage approaches to accommodate HR correction in the comparisons will be discussed later in Section 4.4.

The objective of a TQT study is to assess if the test treatment induces medically significant increase in QTc interval on the health subjects. ICH E14 stated that as a reference, 10 ms is chosen as the cutoff size of increase. If the test treatment increases QTc interval, the maximum increase often happened around the time of maximum concentration of dose, Tmax, observed in early bioavailability study. Let $\mu_T(j)$ and $\mu_P(j)$ denote the mean QTc (or baseline adjusted QTc) interval of subjects who received test and placebo treatment at time point j, respectively. Let $\delta_{TP}(j) = \mu_T(j) - \mu_P(j)$ be the test and placebo difference of mean QTc interval at time point j, $j = 1$ *to* J be time points selected around Tmax, the ICH E14 maximum-change approach can be statistically represented as testing the following hypothesis:

$$H_0 : Max_{j=1 \text{ } to \text{ } J} [\delta_{TP}(j)] \geq 10$$

against (4.1)

$$H_1 : Max_{j=1 \text{ } to \text{ } J} [\delta_{TP}(j)] < 10$$

In practice, it is represented by testing the following null hypotheses by time points:

$$H_0(j) : \delta_{TP}(j) \geq 10$$

against (4.2)

$$H_a(j) : \delta_{TP}(j) < 10 \text{ at all } j = 1 \text{ } to \text{ } J.$$

The trial is concluded negative (i.e., *no test treatment induced prolongation of QT interval*) when all J null hypotheses in (4.2) are rejected. TQT trials are mostly carried out in single clinical center with healthy subjects. The null hypotheses are rejected by showing that

$$T_j = \frac{\widehat{\delta}_{TP}(j) - 10}{e\left(\widehat{\delta}_{TP}(j)\right)} < t\left(n^*; 0.05\right)$$ (4.3)

at each time point, where $\widehat{\delta}_{TP}(j)$ is the unbiased estimate of $\delta_{TP}(j)$, $e\left(\widehat{\delta}_{TP}(j)\right)$ is the standard error of $\widehat{\delta}_{TP}(j)$, and $t(n^*; 0.05)$ is the fifth percentile of t distribution of degrees of freedom n^* (determined by the sample sizes of test and placebo group and design model).

Equivalently, one may reject each of the null hypotheses of (4.2) by showing that the upper confidence limit U of the 90% confidence interval of $\delta_{TP}(j)$ is less than 10 ms at each time point:

$$U = \widehat{\delta}_{TP}(j) + t\left(n^*; 0.05\right) e\left(\widehat{\delta}_{TP}(j)\right) < 10$$

It has been shown that this intersection-union test approach controls type I error rate α at 0.05 and it is approximately 5% when QT interval prolongation at single time point (Zhang and Davidian, 2008). Furthermore, it was shown by Berger (1989) to be uniformly most powerful among all monotone, α-level tests for the linear hypotheses (4.2). However, it is conservative with low power to reject the null hypothesis of (4.1) when J is large.

There were many proposals to improve the power for testing hypotheses (4.1). Boos et al. (2007) showed that $Max_{j=1\ to\ J}\widehat{\delta}_{TP}(j)$ is a biased estimate of $Max_{j=1\ to\ J}\delta_{TP}(j)$ and $E\left[Max_{j=1\ to\ J}\widehat{\delta}_{TP}(j)\right] \geq Max_{j=1\ to\ J}\delta_{TP}(j)$. The inequality exists when $\widehat{\delta}_{TP}(j) > 0$. But $\sqrt{n}\left[Max_{j=1\ to\ J}\widehat{\delta}_{TP}(j) - Max_{j=1\ to\ J}\delta_{TP}(j)\right] \longrightarrow$ $N\left(0, 2\left(\sigma_1^2 + \sigma^2\right)\right)$ as $n \longrightarrow \infty$, where n is the number of subjects, σ_1^2 and σ^2 are the intrasubject and intersubject variances at each time point in a crossover clinical trial. Let $\delta_{TP}(j_0) = Max_{j=1\ to\ J}\delta_{TP}(j)$, Boos et al. (2007) showed that the size of bias is

$$
\left[\sum_{j=1,\neq j_0}^{J} \delta_{JP}(j) \int_{a_j}^{\infty} \Phi_J(z, R_j) - Max_{j=1\ to\ J}\delta_{TP}(j)\right]
$$

$$
\times \int_{-\infty}^{a_{j_0}} \Phi_J(z, R_{j_0}) + \sqrt{\frac{2\left(\sigma_1^2 + \sigma^2\right)}{n}\sum_{j=1}^{J}\int_{a_j}^{\infty} z_j \Phi_J(z, R_j)}, \tag{4.4}
$$

where $a_j = \{a_{jj'}\}$ is a vector with $a_{jj'} = \sqrt{n}\left[\delta_{TP}(j') - \delta_{TP}(j)\right]/(2\sigma)$ for $j \neq j'$, $j, j' = 1$ to J and $a_{jj} = \infty$; $R_K = \{r_{j,j',j''}\}$ with $r_{j,j',j''} = \sigma^2/(\sigma_1^2 + \sigma^2)$ for $j' \neq j''$ and $r_{j,j',j''} = 1$ for $j',j',j'' = 1$ to J; $\Phi_J(z, R_j)$ is the density of J-variate normal distribution with R_j as its covariance matrix. They provided also the second moment of $Max_{j=1\ to\ J}\widehat{\delta}_{JP}(j)$ as follow,

$$
E\left\{Max_{j=1\ to\ J}\left[\widehat{\delta}_{TP}(j)\right]\right\}^2 = \sum_{j=1}^{J}\delta_{TP}(j)^2 \int_{a_j}^{\infty}\Phi_J(z, R_j)
$$

$$
+ \sqrt{\frac{2\left(\sigma_1^2 + \sigma^2\right)}{n}\sum_{j=1}^{J}\int_{a_j}^{\infty} z_j \Phi_J(z, R_j)}
$$

$$
+ \frac{2\left(\sigma_1^2 + \sigma^2\right)}{n}\sum_{j=1}^{J}\int_{a_j}^{\infty} z_j^2 \Phi_J(z, R_j) \tag{4.5}
$$

and proposed three approaches of bias-corrected confidence interval of $Max_{j=1\ to\ J}\widehat{\delta}_{TP}(j)$. However, $Var\left[Max_{j=1\ to\ J}\widehat{\delta}_{TP}(j)\right]$ is a function of $\delta_{TP}(j)$. Boos et al. (2007) proposed the *bias correction* methods for the confidence interval of $Max_{j=1\ to\ J}\delta_{TP}(j)$. Tsong et al. (2009) showed that all three proposed approaches lead to inflation of type I error rate. Formula (4.5) leads also to the problem of unequal variances of $Max_{j=1\ to\ J}\widehat{\delta}_{TP}(j)$, $Max_{j=1\ to\ J}\widehat{\delta}_{TT}(j)$ (the maximum estimate of mean difference between two test groups), and $Max_{j=1\ to\ J}\widehat{\delta}_{PP}(j)$ (the maximum estimate of difference between two placebo groups). Hence, nonparametric test procedure including permutation or randomization tests may also lead to inflation of type I error rate.

The second approach was proposed by Garnett et al. (2008). They proposed to assess treatment-induced prolongation of QT interval by comparing the predicted maximum $\Delta\Delta$QTc (treatment QTc difference adjusted by baseline measurement) with 10 ms. It is done by modeling the relationship between concentration (log(concentration)) and $\Delta\Delta$QTc (or log($\Delta\Delta$QTc)) using a linear model. Tsong et al. (2008) raised the issue on model fitness for the purpose of prediction. The linearity fails when the drug-induced QTc response reaches maximum prior to or after the maximum of drug concentration. Meng (2008) proposed criteria to assess the lag time of QTc response. On the other hand, in order to assess the maximum drug-induced prolongation of QTc interval through concentration–response modeling, one needs to have $\Delta\Delta$QTc (or log($\Delta\Delta$QTc)) at individual subject level. TQT clinical trials are designed either with parallel arms or randomized crossover design. Using the design model, statistical comparison of mean treatment difference will be made with adjustment for other design factors. One may not derive treatment difference at individual subject level based on the analysis with the design model. Concentration–response modeler often took the crossover trial data as if collected in a paired trail and ignored other design factors of the clinical trial. For example, with a crossover clinical trial, period and sequence factors of the design are incorrectly eliminated by the modeler when pairing the treatments of the individual subject. Furthermore, no individual subject $\Delta\Delta$QTc (or log($\Delta\Delta$QTc)) may be obtained when TQT is a parallel arm trial.

Liu and Shen (2009) proposed to fit the repeated QTc measurements of test and placebo groups with polynomial regression lines and derive a confidence bend of the mean difference of the two curves. They further proposed to assess the test-induced QTc prolongation by examining the 90% upper bound of the bend against 10 ms. The approach may be applicable to data collected at parallel arm trials, given sufficient measurements around the peak QTc response of the test group. This requires data collected at many time points at the design stage. Such an approach may be difficult to apply when the data are collected in crossover trials.

4.3 Validation Test

In a TQT/thorough QTc trial, positive control treatment is used for the purpose of clinical trial validation (i.e., assay sensitivity) only. They are particularly useful when the outcome of the trial is negative. ICH E14 guidance states that "The positive control should have an effect on the mean QT/QTc interval of about 5 ms (i.e., an effect that is close to the QT/QTc effect that represents the threshold of regulatory concern, around 5 ms)." In current practice, validation is carried out by a statistical hypothesis testing and an

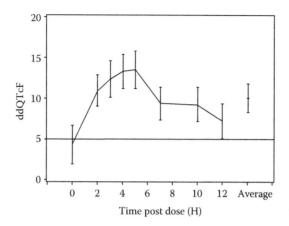

FIGURE 4.2
General time course of Moxifloxacin effect of a double-blind randomized study (mean baseline adjusted Moxifloxacin—placebo difference and its 90% confidence interval)

informal profile validation procedure. For example, Moxifloxacin, a most frequently used positive control treatment, has a well-documented profile (Figure 4.2). Profile validation is carried out by comparing the profile of TQT with the standard profile.

For the nonprofile validation test, let $\mu_{C(j)}$ and $\mu_{P(j)}$ denote the mean QTc interval at jth time point after adjusted for baseline measurement. Let $\delta_{CP}(j) = \mu_{C(j)} - \mu_{P(j)}$. A statistical hypothesis testing is also used for nonprofile validation of the TQT. The hypotheses of interest are

$$H_0 : Max_{j=1 \; to \; J}\delta_{CP}(j) \leq 5\,\text{ms}$$

versus $\hfill (4.6)$

$$H_a : Max_{j=1 \; to \; J}\delta_{CP}(j) > 5\,\text{ms}$$

In practice, in a setup similar to the noninferiority test of treatment-induced QT prolongation, the conventional approach is to test the J sets of hypotheses,

$$H_0(j) : \delta_{CP}(j) \leq 5\,\text{ms}$$

against $\hfill (4.7)$

$$H_a(j) : \delta_{CP}(j) > 5\,\text{ms} \; for \; j = 1 \; to \; J.$$

Validation holds if at least one of the null hypotheses is rejected. Equivalently, one may do that by showing that the lower limit of at least one confidence interval of $\delta_{CP}(j)$ is greater than 5 ms, that is,

$$L = \widehat{\delta}_{CP}(j) + t(n^*; \alpha^*)e(\widehat{\delta}_{CP}(j)) > 5. \tag{4.8}$$

For at least one j, $j = 1$ to J with an properly adjusted type I error rate α^*.

Based on the same discussion given in Section 4.2, $Max_{j=1 \ to \ J}\widehat{\delta}_{CP}(j)$ is a biased estimate of $Max_{j=1 \ to \ J}\delta_{CP}(j)$. An appropriate bias correction method is yet to be developed for testing hypotheses (4.6). On the other hand, testing (4.7) using $\alpha = 0.05$ for each individual test would lead to the inflation of family type I error rate through multiple comparisons. A simple conservative approach would be using Bonferroni adjustment with $\alpha^* = 0.05/J$ for each individual test in the conventional intersection-union test for hypotheses (4.7) or estimate each individual confidence interval in (4.8). Power of the validation test can be improved using many Bonferroni modified (stepwise) procedures. For example, let $p_{(1)} \geq p_{(2)} \geq \cdots \geq p_{(j)}$ be the ordered p-values of the J tests, and $H_{0(1)}, H_{0(2)}, \ldots, H_{0(j)}$ be the corresponding ordered null hypotheses. Using Holm's procedure (1979), the study is validated if $p_{(j)}$, the smallest p-value, is less than $0.05/J$. Validation is failed if $p_{(j)} \geq 0.05$. On the other hand, using Hochberg's procedure (1988), the study is validated if $p_{(j)} < 0.05/j$ for any $j = 1, \ldots, J$. Furthermore, when the covariance structure of measurements across time points of the positive control and placebo is known, power can be further improved in Holm's procedure. For example, Tsong et al. (2010b) explored the issue with Moxifloxacin as the positive control treatment.

In order to compensate the conservativeness of Bonferroni-type adjustment for multiple comparisons, Zhang (2008) proposed a global test by testing a single mean difference across all time points in order to improve power over the multiple comparison approach as described earlier. The problem with the global test is that it changes the hypothesis of interest and averaging large with small differences may lead to further reduction of power for validation when the number of time points is large.

As mentioned earlier, Moxifloxacin was used in most TQT trials and subjects studied in TQT are uniformly healthy young men or women. The information of study results of Moxifloxacin is rich enough to build a stable prior for Bayesian approach for the validation test. Given the fact that hypotheses (4.1) and (4.6) are tested independently, we may carry out validation test using Bayesian approach regardless of what approach was used for noninferiority assessment of the test treatment (Dong et al., 2012).

4.4 Design of Thorough QT Trials

For a typical TQT study, a randomized three or four treatment group design is considered. Subjects studied in the trials are healthy subjects with no prior EKG problems. For demonstration purposes, we focus on a three treatment group case. The three treatment arms are (1) T — test drug, (2) C — positive control, and (3) P — placebo. A TQT study may have either a crossover design or a parallel design. Due to large intrasubject variability, baseline measurements are recommended for OT correction by RR interval. For a parallel study, a whole day of time-matched baseline measurements on the day prior to dosing is customarily collected in order to help adjust for within-subject diurnal variability. For a crossover design, however, it is not essential to have a full day time-matched baseline at the previous day for each period; baseline measurements at each period, just before treatment, might be enough. This is because in a crossover trial, diurnal variability of QTc is accounted for by the design itself, since each subject receives each of all treatment groups in the postdose periods at exactly the same time points. It is recommended (Patterson et al., 2005; Chow and Chang, 2008) that replicate ECG at each time point should be collected, and this certainly applies to the baseline ECGs collection. Within each period, the average of all the measurements before dosing can be used as the baseline value.

A parallel group design may have to be used for drugs with long elimination half-lives, or with metabolites, or a large number of dose groups to be evaluated. With a parallel arm design, subjects will be randomized to each one of the four groups. The choice of time points as well as the number of time points to evaluate the drug's QTc effect should be driven by the pharmacokinetic profile of the drug. The duration of a parallel study might be relatively shorter, and the dropout rate might be smaller compared with a crossover study.

A crossover design is often used in TQT trial, because it needs much fewer subjects than the parallel arm designs for the same study power, since the variability of the estimated treatment differences with the same subject acts as her/his own control is smaller than that for a parallel design with the same number of subjects. One of the major concerns of a crossover trial is the potential effect of one treatment carried over to the periods follow behind. Hence, it is important to have a sufficient washout time between any two periods. For a crossover design, a sequence is a prespecified ordering of the treatments to the periods. The randomization of clinical trial is achieved by randomly assigning the subjects one of the sequences of treatments. For example, the following sequences may be considered for a three-arm trial:

$$\text{Latin Square 1}: \textit{Sequence } \#1 : T/P/C;$$
$$\textit{Sequence } \#2 : C/T/P;$$
$$\textit{Sequence } \#3 : P/C/T;$$
$$\text{Latin Square 2}: \textit{Sequence } \#4 : T/C/P;$$
$$\textit{Sequence } \#5 : C/P/T;$$
$$\textit{Sequence } \#6 : P/T/C;$$

Here, T, C, and P are the test, positive control, and placebo treatment, respectively.

A sequence balanced complete crossover trial design for more than K (>2) treatments is often represented by a group of Latin squares. The first three sequences and the last three sequences given above form two different Latin squares. For a study with 48 subjects, a complete balanced crossover design is to have 6 subjects randomized to each of the above six sequences. The most frequently used crossover design in TQT trial is Williams design (Williams, 1949). The direct-by-carryover interaction is significant if the carryover effect of a treatment depended on the treatment applied in the immediately subsequent period. If the study model includes period, sequence, treatment, first-order carryover, and direct-by-carryover interaction as fixed effects, the design based on one or two Williams squares might not have sufficient degrees of freedom to assess direct-by-carryover interaction (Williams, 1949). If this is the case, one might consider using a design with repeated multiple Latin squares, Williams designs, or complete orthogonal sets of Latin squares. For the three treatments, with 48 subjects, the Williams design is also a completely orthogonal set of Latin squares (Chen et al., 2007).

Let us consider the linear models for the data analysis of the designs. Let us first assume Z_{kij} denote the difference between QTc of the subject i of treatment group k ($k = $ T, C, and P) and baseline QTc value, at time point j. For the parallel arm trial, depending on the covariance structure of Z_{kij}, many linear models may be used to model the relationship of Z_{kij} and the factors. Let k_0 denote the baseline of treatment k and Z_{k_0ij} denote the baseline QTc value. In most situations, one needs to assume that there is interaction between treatment and time and unequal variances across time, and the data need to be modeled at each time point separately as follows:

$$Z_{kij} = R_{kj} + \beta Z_{k_0ij} + \varepsilon_{kij}, \tag{4.9}$$

where $R_{kj} + \beta Z_{k_0ij}$ is the fixed effects part of the model, treatment + baseline QTc; ε_{kij} is the random error with variance σ_{kj}^2 for the kth treatment group. Note that Z_{k_0ij}, the time-matched baseline QTc value, is included as a covariate to prevent conditional bias as described in Section 4.2. When the baseline QT measurement is a single predose value, $Z_{k_0ij} = Z_{k_0i}$. The coventional approach for testing hypotheses (4.2) and (4.7) does not include the covariate Z_{k_0ij} in the model.

Under the assumption of equal variance at each time point, the following linear model may be used:

$$Z_{kij} = R_k + T_j + RT_{ij} + \beta Z_{k_0ij} + \varepsilon_i(k) + \varepsilon_{kij}, \tag{4.10}$$

where $R_k + T_j + RT_{ij} + \beta Z_{k0ij}$ is the fixed effects part of the model, treatment + time + treatment by time + baseline QTc; $\varepsilon_i(k)$ with variance σ_i^2 is a random effect for the ith subject in the kth treatment group, and ε_{kij} is the random error with variance σ^2. Note that the random effect $\varepsilon_i(k) + \varepsilon_{kij}$ in (4.10) is the same as the random error ε_{kij} in model (4.9) for a given time j under equal variance in treatment assumption.

Let us consider the mixed effects model of the Williams design. Let Z_{kijl} be QTc, the corrected QT measurement adjusted for baseline QT of the kth treatment of the ith subject observed at the jth time point of the lth period; $i = 1, \ldots, n, j = 1, \ldots, J$, and $l = 1, \ldots, L$. The model of this crossover trial is

$$Z_{kijl} = \mu + \tau_k + t_j + \eta_l + \tau_k \times t_j + \tau_k \times \eta_l + b_i + \varepsilon_{ij}, \tag{4.11}$$

where μ is the overall mean response, $\tau_k + t_j + \eta_l + \tau_k \times t_j + \tau_k \times \eta_l$ are the fixed effect of treatment + time + period + treatment–time interaction + treatment–period interaction, $b_i \sim N(0, \sigma_s^2)$ are independent and identically distributed normal random subject effect with mean 0 and variance σ_s^2, and $(\varepsilon_{i1}, \ldots, \varepsilon_{iJ})$, $i = 1, \ldots, n$ are independent and identically distributed as

$$\begin{pmatrix} \varepsilon_{i1} \\ \vdots \\ \varepsilon_{iJ} \end{pmatrix} \sim \text{iid } N_J \left(0, \begin{bmatrix} \sigma_{11} & \sigma_{12} & \cdot & \sigma_{1J} \\ \sigma_{21} & \sigma_{22} & \cdot & \cdot \\ \cdot & \cdot & \cdot & \cdot \\ \cdot & \cdot & \cdot & \cdot \\ \cdot & \cdot & \cdot & \cdot \\ \sigma_{J1} & \cdot & \cdot & \sigma_{JJ} \end{bmatrix} \right).$$

Note that $\sigma_{jj}^2 = \sigma^2$ when the variances are equal across all time points and treatment.

Tsong et al. (2013b) proposed the sample size determination to accommodate for the multiple comparisons due to the two set of tests and at multiple time points. The biggest disadvantage of a parallel design is the large number of subjects required. One may consider an improvement with a two-stage adaptive design by testing either hypotheses (4.1) or (4.6) at stage 1. The other hypotheses will be tested only if the first set of hypotheses is rejected. The similar design was proposed for assessment of therapeutic bioequivalence by Tsong et al. (2004). Through the adaptive design, one may drop either test or positive control treatment to achieve the improvement of efficiency. For a crossover trial, it is more complicated to address the consistency issue of data across the stage after dropping an arm. In order to keep the trial blind, Tsong et al. (2013) proposed to maintain the same design but replaced the dropped treatment by placebo.

For a test treatment that requires long exposure before reaching the steady stage, the treatment-induced QT prolongation may only be assessed with a parallel arm design. However, the positive control Moxifloxacin requires only less than 3 h to reach maximum response. In order to maximize the efficiency, a hybrid design was proposed, combining parallel arm and crossover trials into one (Zhang, 2009; Tsong, 2013).

4.5 Measurement and Correction of QT Interval

The population-based correction (PBC) method would assume that subjects under the same treatment (on or off drug) share the same but unknown factor to be estimated. PBC correction method fits a regression model to pooled QT-RR data from all subjects in the current trial. The individual-based or subject-specific correction (IBC) method often fits a regression model to QT-RR data from each individual subject's QT measured at baseline or of placebo against RR data. Under the assumption of constant relationship between QT interval and RR interval, the *slope* of the individual subject estimated using baseline or placebo data will be used on the data collected under treatment to determine the corrected QT interval in the assessment of the treatment effect. These are sometimes called off-drug correction method. On the other hand, correction factors may be determined using data collected at off-drug as well as on-drug periods of the same subject. Though it is a one-step modeling, correction factor for off-drug may be different from the factor for on-drug period. These methods are often called off-drug/on-drug (OFF–ON) correction method (Li et al., 2004). With either PBC or IBC method, there are also variations in assuming that the correction factor may be the same regardless of on or off drug. In contrast, a fixed correction (FC) method uses a *FC* formula that is not derived from the current trial data.

    There are many proposed fixed QT correction (FC) methods in the medical literature. What is considered the best method of correction among them is debatable. Bazett's formula (1920) is probably the most commonly used by clinicians, because it is practical, simple, and convenient. Based on his empirical experience, Bazett assumed the relationship $QT = \alpha \times RR^{1/2}$. Hence, $QTc = QT \times (1000/RR)^{1/2}$. Fridericia (1920) proposed a different empirical model, $QT = \alpha \times RR^{1/3}$, and $QTc = QT \times (1000/RR)^{1/3}$. It is often considered as an improvement of Bazett's correct because it improves the QTc consistency from population to population. In addition, Framingham study (1992) used $QTc = QT \times (1000/RR)^{0.154}$ and Schlamowitz (1946) proposed to use $QTc = QT \times (1000/RR)^{0.205}$ for correction. The FC method did not acknowledge the variation of the QT versus RR relationship from population-to-population or subject-to-subject. These limitations were pointed out by Malik (2001) and Desai et al. (2003). Batchvarov et al. (2002) showed that there were evidences that each individual has slightly different QT profile against RR. It supported the interest to define an improved data-driven correction method to include random effects for each individual.

Let us consider the QT data collected in a crossover design. Let y_{kij} denotes the QT or log(QT) measurement during the kth treatment period, for subject i, at the time point tj, where $k = P$ for placebo, T for test, k_0 for baseline of treatment k, $i = 1, \ldots, n$ and $j = 1, \ldots, J$, the number of measurements per subject, respectively. Let x_{kij} denotes the RR/1000 or log(RR/1000) measurements within the kth treatment periods, respectively. The FC methods stated above can be represented by

$$y_{kij} = \alpha_{kij} - \beta x_{kij}, \tag{4.12}$$

where β is the FC factor derived from dataset other than the current study and α_{kij} is QTc or log(QTc). For example, for the FC introduced earlier, log(QT) and log(QTc) were used in the model (4.12). On the other hand, Hodges et al. (1983) and Rautaharju et al. (1993) pointed out that the relationship between the QT interval and the heart rate (HR) is approximately linear. For this relationship, let y_{kij} represents QT and x_{kij} represents (HR-60) in (4.12).

A PBC method is one that uses the following linear model to determine the correction factor β,

$$y_{kij} = \alpha_{kij} - \beta_k x_{kij} + \varepsilon_{kij}, \tag{4.13}$$

where β_k may be estimated using model (4.13) with data of complete study. In TQT studies, it is often assumed that $\beta_k = \beta$.

The PBC or FC methods can be used even when data are only collected from sparse time points. The PBC method assumes that the underlying true QT-RR (or QT-HR) relationships are the same for all subjects under the same treatment condition (off drug or on drug) or different treatment conditions, while the FC method assumes that a specific and completely known QT-RR relationship exists or all the trial subjects regardless of treatment conditions. Since there may be substantial intersubject variation in QT-RR relationship, the PBC and FC methods may lead to using under- or overestimates of correction of individual QT to a degree that may bias the assessment on QT prolongation in central tendency analysis or categorical analysis.

A TQT study is usually powered to detect a small QT interval prolongation (e.g., 0.01 s or 10 ms). A sensitive QT study requires an accurate QT interval correction method to ensure correct attribution of a small QT interval prolongation to potential sources such as treatment or random noise. It has been demonstrated that the QT-RR relationship exhibits a substantial intersubject variability but a low intrasubject variability in human, leading some authors to suggest the use of IBC method whenever data warrant (Batchvarov and Malik, 2002; Malik et al., 2002, 2004). The IBC method assumes different QT-RR relationships for different individual subjects. The IBC method is feasible in TQT studies, which have QT-RR data from multiple ECGs taken at multiple time points. However, in practice, the RR range in the QT-RR data from an individual subject may often not be sufficiently wide to allow for an accurate estimation of the QT-RR relation.

In order to accommodate the population-to-population and subject-to-subject variation of the relation between QT and RR, Malik (2001) proposed the following IBC method:

$$QT_{ij} = \alpha_i \times RR_{ij}^{\beta_i} \tag{4.14}$$

$$QT_{c_i} = QT_{ij} \times \frac{1000^{\tilde{\beta}_i}}{RR_{ij}^{\tilde{\beta}_i}}, \tag{4.15}$$

where $\tilde{\beta}_i$ is the least square estimate of β_i that utilized the baseline or placebo QT and RR data from subject i only. In other words, Malik (2001) proposed to estimate α_i and β_i in the following linear model:

$$y_{pij} = \alpha_i + \beta_i x_{pij} + \varepsilon_{pij} \tag{4.16}$$

using only the baseline or placebo log(QT) and log(RR/1000) or QT and (HR-60) data. The same $\tilde{\beta}_i$ is used to correct QT observed when the subject is under treatment. Note that Malik's correction factor is estimated with intra-subject averaged over all time points. It is derived under the assumption that the relationship between QT and RR remains the same over all time points and even under treatment.

Without the assumption of invariant QT/RR relationship and intrasubject averaged over all time points, a general linear model for IC method can be represented as follows. Let y_{kij} denotes QT (or log(QT)) measurement during the kth treatment, for subject i at the time point j, $k = P$ for placebo, T for test, k_0 baseline of treatment k, $i = 1, \ldots, n_k$ and $j = 1, \ldots, J$. J is the number of measurements per subject. Let x_{kij} denotes the RR/1000 (or log(RR/1000)) measurements with the kth treatment, respectively. Most correction methods stated above can be represented by the general fixed effect linear model:

$$y_{kij} = \alpha_{kij} + \beta_{kij} x_{kij} + \varepsilon_{kij}. \tag{4.17}$$

α_{kij} is log (QT) or log(QTc)s (intercepts) for the kth treatment of subject i at j time point, respectively, and β_{kij} is the corresponding slopes between QT and RR/100 or between log(QT) and log(RR/1000), and ε_{kij} is i.i.d.N(0, σ_0^2). Note that (4.17) is analyzed assuming QT measurement follows a lognormal distribution when log(QT) and log(RR) are used. Various FC, PBC, and IBC can all be represented by (4.17). For example, for an FC, $\beta_{kij} = \beta$ a constant derived from other study. For an PBC, $\beta_{kij} = \beta$ or β_k, factors estimated from data of current study. For an IBC intrasubject averaged over all time points, $\beta_{kij} = \beta_{ki}$, estimated for each individual using the data of current study.

For a gender-specified PBC, $\beta_{kij} = \beta_k$. Furthermore, when gender difference is of concern, one may analyze the data with gender group in model.

The linear or log-linear correction model was further extended to nonlinear models such as

$$y_{kij} = \alpha_{kij} + \beta_{kij}\left(1 - x_{kij}^{\gamma}\right) + \varepsilon_{kij} \qquad (4.18)$$

for its flexibility (Malik et al., 2004). But a nonlinear model such as (4.18) requires even more data to have good estimate for the increased number of parameters. Ring (2009) proposed some multilevel modeling in order to obtain better individual correction factor. It faces the same problem of requiring more data than practically available.

These correction methods have been compared under different models in numerous studies using real data (Ahnve, 1985; Funk-Brentano and Jaillon, 1993; Moss, 1993; Hnatkova and Malik, 1999; Malik 2001; Malik et al. 2004). The corrections were aimed at removing the dependency of QTc intervals on RR intervals within subjects. The conclusions of the comparisons are dependent on subject composition and the observed range of RR intervals. Most of the comparisons were focused on the model error, which is the intrasubject mean difference between two subsets with each subject. With these constrain, the *error* term would be partially attributed to the data grouping and the outcome of the *error* term may not entirely correspond to the actual biases of the correction methods.

Wang et al. (2008) studied the IBC, PBC, and FC correction methods in order to quantify the biases and variances of the QTc intervals using the IBC, PBC, and FC correction methods conditional on the observed RR intervals. It is easy to see that FC method can be unreliable in comparison to PBC or IBC, because it is derived from source of QT-RR data, other than the current study data although it has smallest variance conditioning on a given RR value. The biased variance of FC may lead to an inflation of type I error rate when used in treatment comparison. Although the Federicia's correction factor has been shown relatively reliable in testing as shown in the numerical example of Wang et al. (2008) as well as in the literature, the use of unvalidated FC factors is discouraged in practice generally.

PBC has smaller variance than IBC conditioning on a given RR, though the difference is usually not large. Wang et al. (2008) showed that the conventional intrasubject averaged PBC is a conditional biased estimate of the true underlying QT interval. But a modified subject-averaged QTc at each time point becomes conditionally unbiased. This leads to the recommendation of using the modified subject-averaged QTc when the data were collected at sparse time points. They also recommended to include RR as covariate in the model for the comparison of treatment effect when using either PBC or IBC may have the effect to further reduce the remaining correlation of QTc interval with RR interval.

Note that with PBC or IBC if the correction factor β_{kgij} is estimated with (4.12) using data of the kth treatment of the current study, $\widehat{\alpha}_{kgij}$, the corrected QT is a random variable instead of an observation. When treatment comparison is carried out after QT is corrected in a two-step setting, the standard error of $\widehat{\alpha}_{kgij}$ needs to be included in the estimation error of the mean treatment difference to avoid standard error bias of comparison.

Li et al. (2004) pointed out that there are three limitations of the approaches. First, the variance of QT interval is usually proportional to its magnitude and the validity of equal variance assumption for the measurement error of QT interval of the models is questionable. Second, subject-specific estimate of $\tilde{\beta}_i$ is often not practical, because the number of ECGs is often small for individual subject. Third, in Malik's IBC method, the underline assumption is that the relationship between QT and RR is consistent for each subject regardless of the potential changes caused by active treatment. Under these concerns, Li et al. (2004) proposed a linear mixed effects regression model for log-transformed QT and RR

$$y_{k,gij} = \alpha_{k,g} + a_{gi} + (\beta_{k,g} + b_{gi})x_{k,gij} + \varepsilon_{k,gij}, \tag{4.19}$$

where $\alpha_{t,g}$ and $\alpha_{p,g}$ are the log(QTc)s (intercepts) for the treatment and placebo periods of gender g, respectively, and $\beta_{t,g}$ and $\beta_{p,g}$ are the corresponding slopes between log(QT) and log(RR). Subject random variables a_{gi} and b_{gi} are the same across placebo and test treatment periods. Therefore, Li et al. (2004) assumed that treatment is assumed to shift only the slope and the intercept in the population average level, but not the random effects for each subject. The random effect $\{a_{gi}, b_{gi}\}$ is i.i.d.$N(0, d)$ and $d = \begin{pmatrix} \sigma_a^2 & \rho\sigma_a\sigma_b \\ \rho\sigma_a\sigma_b & \sigma_b^2 \end{pmatrix}$ is a 2× 2 covariance matrix and $\varepsilon_{k,gij}$ is i.i.d.$N(0, \sigma_0^2)$. All the α's and β's in the model are estimated jointly by maximizing restrictive maximum likelihood (REML). The true QTc based on the mixed effect model proposed by Li et al. (2004) is hence

$$E(QTc_{k,male}) = e^{\alpha_{k,1}}, E(QTc_{k,female}) = e^{\alpha_{k,2}}. \tag{4.20}$$

Dang and Zhang (2013) performed comparisons of the various modeling using data available to FDA. For details of the comparison, please refer to the article.

Model (4.17) may lead to the one-step analysis of TQT studies to incorporate QT correction for RR or HR. But on the other hand, assessment of treatment-induced QT prolongation is a noninferiority test with a margin in original scale. Modeling log-transformed data leads to complication in one-step correction-treatment comparison analysis as discussed in Section 4.5.

Now let us consider applying various correction methods to treatment comparison in (4.1) and (4.6). When an FC method is used, that is, QTc

$\alpha_{kij} = y_{kij} + \beta x_{kij}$, for a given β, $Z_{kij} = exp(\alpha_{kij} - \alpha_{k_0ij})$ for time-matched baseline, or $Z_{kij} = exp(\alpha_{kij} - \alpha_{k_0i})$ for single pre-dose baseline.

When a PBC method is used, $\alpha_{kij} = y_{kij} - \beta x_{kij} - \varepsilon_{kij}$. If the model fits with log-transformed QT values, $Z_{kij} = exp\left(\alpha_{kij} - \alpha_{k_0ij}\right) = exp\left[\left(y_{kij} - \beta_{kij}x_{kij} - \varepsilon_{kij}\right) - \left(y_{k_0ij} - \beta_{kij}x_{k_0ij} - \varepsilon_{k_0ij}\right)\right]$, one-step modeling (4.9) or (4.10) including QT correction directly becomes extremely complicated in the composite form of an additive and a multiplicative model. A conventional two-step approach would use actually $Z_{kij} = exp\left(\widehat{\alpha}_{kij} - \widehat{\alpha}_{k_0ij}\right) = exp\left[\left(y_{kij} - \widehat{\beta}x_{kij}\right) - \left(y_{k_0ij} - \widehat{\beta}x_{k_0ij}\right)\right]$ in model (4.9) or (4.10) and improperly leave out the estimation error of β. On the other hand, if a linear model of PBC method fits well with QT and RR or HR data

$$Z_{kij} = \left(\alpha_{kij} - \alpha_{k_0ij}\right) = \left(y_{kij} - \beta_{kij}x_{kij} - \varepsilon_{kij}\right) - \left(y_{k_0ij} - \beta_{kij}x_{k_0ij} - \varepsilon_{k_0ij}\right)$$
$$= \left(y_{kij} - y_{k_0ij}\right) - \beta_{kij}\left(x_{kij} - x_{k_0ij}\right) - \left(\varepsilon_{kij} - \varepsilon_{k_0ij}\right). \quad (4.21)$$

Substitute Z_{kij} in (4.9) or (4.10) without the covariate term βZ_{k_0ij}, the linear model provides a 1-step approach for treatment comparison of the TQT studies such that

$$\left(y_{kijl} - y_{k_0ijl}\right) = \beta\left(x_{kijl} - x_{k_0ijl}\right) + R_{kj} + \varepsilon_{kij} \quad (4.22)$$
$$\left(y_{kijl} - y_{k_0ijl}\right) = R_k + T_j + (RT)_{ij} + \beta\left(x_{kijl} - x_{k_0ijl}\right) + \varepsilon_i(k) + \varepsilon_{kij}. \quad (4.23)$$

Similarly, when a linear IBC method is applied, $Z_{kij} = \left(\alpha_{kij} - \alpha_{k_0ij}\right) = \left(y_{kij} - \beta_{kij}x_{kij} - \varepsilon_{kij}\right) - \left(y_{k_0ij} - \beta_{kij}x_{k_0ij} - \varepsilon_{k_0ij}\right) = \left(y_{kij} - y_{k_0ij}\right) - \beta_{kij}\left(x_{kij} - x_{k_0ij}\right) - \left(\varepsilon_{kij} - \varepsilon_{k_0ij}\right)$, where α_{kij} and α_{k_0ij} are QTc, x_{kij} and x_{k_0ij} are RR of treatment k and baseline of treatment k. The one-step linear model is represented by

$$\left(y_{kij} - y_{k_0ij}\right) = \beta_{kij}(x_{kij} - x_{k_0ij}) + R_{kj} + \varepsilon_{kij} \quad (4.24)$$

and

$$\left(y_{kij} - y_{k_0ij}\right) = \beta_{kij}\left(x_{kij} - x_{k_0ij}\right) + R_k + T_j + (RT)_{ij} + \varepsilon_i(k) + \varepsilon_{kij}. \quad (4.25)$$

Again, when incorporating the correction method in the analysis, for an FC method, $Z_{kij} = exp(\alpha_{kij} - \alpha_{k_0ij})$ using Bazett's or Fridericia's correct method. When a PBC or IBC method is used with a linear model on log(QT) and log(RR), we have the same difficulty in linear modeling of (4.11). When PBC is used with linear model on QT and RR values, the one-step treatment comparison model including PBC becomes

$$\left(y_{kijl} - y_{k_0ijl}\right) = \mu + \beta\left(x_{kijl} - x_{k_0ijl}\right) + \tau_k + t_j + \eta_l + \tau_k * t_j + \tau_k * \eta_l + b_i + \varepsilon_{ij}. \quad (4.26)$$

With IBC, the model becomes,

$$\left(y_{kijl} - y_{k_0 ijl}\right) = \mu + \beta_{kijl}\left(x_{kijl} - x_{k_0 ijl}\right) + \tau_k + t_j + \eta_l + \tau_k * t_j + \tau_k * \eta_l + b_i + \varepsilon_{ij}.$$

$$(4.27)$$

4.6 Discussion and Conclusions

The history of statistical design and analysis of TQT studies is rather short. TQT is typically carried out at prephase III stage. Before the publication of ICH E14 and FDA drug regulatory requirement, the study is most frequently carried out by clinical pharmacologist without much involvement of statisticians. However, as a part of the NDA requirement for most of new drugs, efficient designs and stringent statistical analysis procedures were proposed in the NDA submission and research publication. This chapter covers some of the basic considerations in design and analysis of the new area of regulatory statistics.

Acknowledgments

This chapter was prepared with many inputs from statistical reviewers of CDER QTc team of the Office of Biostatistics. The authors thank Drs. Stella Machado and Joanne Zhang of CDER, FDA, for the support and discussion on the development of this chapter.

References

Ahnve S. (1985). Correction of the QT interval for heart rate: Review of different formulae and the use of Bazett's formula in myocardial infarction. *Am. Heart J.* 109:568–574.

Batchvarov V, Malik M. (2002). Individual patterns of QT/RR relationship. *Cardiac. Electrophysiol. Rev.* 6:282–288.

Batchvarov VN, Ghuran A, Smetana P, Hnatkova K, Harries M, Dilaveris P, Camm AJ, Malik M. (2002). QT-RR relationship in healthy subjects exhibits substantial intersubject variability and high intrasubject stability. *Am. J. Physiol. Heart Circ. Physiol.* 282:H2356–H2363.

Bazett JC. (1920). An analysis of time relations of electrocardiograms. *Heart* 7:353–367.

Berger RL. (1989). Uniformly most powerful tests for hypotheses concerning linear inequalities and normal means. *J. Am. Statist. Assoc.* 84:192–199.

Boos DD, Hoffman D, Kringle R, Zhang J. (2007). New confidence bounds for QT studies. *Statist. Med.* 26:3801–3817.

Chen L, Tsong Y. (2007). Design and analysis for drug abuse potential studies: Issues and strategies for implementing a crossover design. *Drug Info. J.* 41: 481–489.

Dang Q, Zhang J. (2013). Validation of QT interval correction methods when a drug changes heart rate. *Therap. Innov. Regul. Sci.* 47(2):256–260.

Desai M, Li L, Desta Z, Malik M, Flockhart D. (2003). Variability of heart rate correction methods for the QT interval: Potential for QTc overestimation with a low dose of haloperidol. *Br. J. Clin. Pharmacol.* 55(6):511–517.

Dmitrienko A, Smith B. (2003). Repeated-measures models in the analysis of QT interval. *Pharm. Stat.* 2:175–190.

Dong X, Ding, X, Tsong, Y. (2013). Bayesian approach for assay sensitivity assessment of thorough QT clinical trial. *J. Biopharm. Statist.* 23(1):73–81.

Fridericia LS. (1920). Die Systolendauer im Elekrokardiogramm bei normalen Menschen und bei Herzkranken. *Acta Med. Scand.* 53:469–486.

Funk-Brentano C, Jaillon P. (1993). Rate-corrected QT intervals: Techniques and limitations. *Am. J. Cardiol.* 72:17B–23B.

Hnatkova K, Malik M. (2002). "Optimum" formulae for heart rate correction of the QT interval. *Pacing. Clin. Electrophysiol.* 22:1683–1687.

Hochberg Y. (1988). A sharper Bonferroni procedure for multiple tests of significance. *Biometrika.* 75:800–802.

Hodges M, Salerna D, Erline D. (1983). Bazett's QT correction reviewed: Evidence that a linear correction for heart rate is better. *J. Am. Coll. Cardiol.* 1:694.

Holm S. (1979). A simple sequentially rejective multiple test procedure. *Scan. J. Statist.* 6:65–70.

International Conference on Harmonisation, ICH (E14) Guidance. (2005). The clinical evaluation of QT/QTc interval prolongation and proarrhythmic potential for non-antiarrhythmic drugs. Geneva, Switzerland: International Conference on Harmonisation, May 2005. Available at http://www.ich.org/.

Jones B, Kenward MG. (2003). *Design and Analysis of Cross-Over Trials*, 2nd edn. London, U.K.: Chapman & Hall.

Li L, Desai M, Desta Z, Flockhart D. (2004). QT analysis: A complex answer to a 'simple' problem. *Stat. Med.* 23:2625–2643.

Liu W, Bertz F, Hayter AJ, Jamshidian M, Wynn HP, Zhang Y. (2009). Simultaneous confidence bands for regression analysis. *Biometrics*, DOI: 10.1111/J.1541-0420.2008.01192.x.

Malik M. (2001). Problems of heart rate correction in assessment of drug-induced QT interval prolongation. *J. Cardiovasc. Electrophysiol.* 12:411–420.

Malik M, Camm AJ. (2001). Evaluation of drug-induced QT interval prolongation. *Drug Safety* 24:323–351.

Malik M, Farbom P, Batchvarov V, Hnatkova K, Camm AJ. (2002). Relation between QT and RR intervals is highly individual among healthy subjects: Implications for heart rate correction of the QT interval. *Heart* 87:220–228.

Malik M, Hnatkova K, Batchvarov V. (2004). Differences between study-specific and subject-specific heart rate corrections of the QT interval in investigations of drug induced QTc prolongation. *Pace* 27:791–800.

Meng Z. (2008). Simple direct QTc-exposure modeling in thorough QT studies. *2008 FDA/Industry Workshop*, Rockville, MD.

Moss AJ. (1993). Measurement of the QT interval and the risk associated with QTc interval prolongation: A review. *Am. J. Cardiol.* 72:23B–25B.

Patterson S, Agin M, Anziano R et al. (2005). Investigating drug induced QT and QTc prolongation in the clinic: Statistical design and analysis considerations. *Drug Inform. J.* 39:243–266.

Rautaharju PR, Zhou SH, Wong S, Prineas R, Berenson G. (1993). Function characteristics of QT prediction formulas. The concept of QTmax and QT rate sensitivity. *Comput. Biomed. Res.* 26:188–204.

Ring A. (2009). Impact of delayed effects in the exposure-response analysis of clinical QT trials. *DIA Cardiovascular Safety, QT, and Arrhythmia in Drug Development Conference*, April 30–May 1, 2009, Bethesda, MD.

Schlamowitz I. (1946). An analysis of the time relationship within the cardiac cycle in electrocardiograms of normal man. I. The duration of the QT interval and its relationship to the cycle length (R-R interval). *Am. Heart J.* 31:329.

Tsong Y. (2013). On the designs of thorough QT/QTc clinical trials. *J. Biopharm. Statist.* 23(1):43–56.

Tsong Y, Shen M, Zhong J, Zhang J. (2008b). Statistical issues of QT prolongation assessment based on linear concentration modeling. *J. Biopharm. Statist.* 18(3):564–584.

Tsong Y, Sun A, Kang S-H. (2013). Sample size of thorough QTc clinical trial adjusted for multiple comparisons. *J. Biopharm. Statist.* 23(1):57–72.

Tsong Y, Yan L, Zhong J, Nei L, Zhang J. (2010b). Multiple comparisons of repeatedly measured response: Issues of validation testing of thorough QT trials. *J. Biopharm. Statist.* 20(3):654–664.

Tsong Y, Zhang J, Wang S-J. (2004). Group sequential design and analysis of clinical equivalence assessment for generic nonsystematic drug products. *J. Biopharm. Statist.* 14(2):359–373.

Tsong Y, Zhang J, Zhong J. (2009). Letter to editor: Comment on new confidence bounds for QT studies by Boos DD, Hoffman D, Kringle R and Zhang J. *Statist. in Medicine*, 2007: 26:3801–3817. *Statist. Med.* 28:2936–2940.

Tsong Y, Zhong J, Chen WJ. (2008a). Validation testing in thorough QT/QTc clinical trials. *J. Biopharm. Statist.* 18(3):529–541.

Tsong Y, Zhong J, Zhang J. (2010a). Multiple comparisons of repeated measured response: Assessment of prolongation of QT in thorough QT trials. *J. Biopharm. Statist.* 20(3):613–623.

Wang Y, Pan G, Balch A. (2008). Bias and variance evaluation of QT interval correction methods. *J. Biopharm. Statist.* 18(3):427–450.

Williams EJ. (1949). Experimental designs balanced for the estimation of residual effects of treatments. *Aust. J. Sci. Res.* 2:149–168.

Wysowski DK, Corken A, Gallo-Torres H, Talarico L, Rodriguez EM. (2001). Postmarketing reports of QT prolongation and ventricular arrhythmia in association with cisapride and Food and Drug Administration regulatory actions. *Am. J. Gastroenterol.* 96(6):1698–1703.

Zhang J. (2008). Testing for positive control activity in a thorough QTc study. *J. Biopharm. Statist.* 18(3):517–528.

Zhang J, Machado SG. (2008). Statistical issues including design and sample size calculation in thorough QT/QTc studies. *J. Biopharm. Statist.* 18(3):451–467.

Zhang J. (2009). Moxifloxacin and placebo can be given in a crossover fashion during a parallel-designed thorough QT study. *DIA Cardiovascular Safety, QT, and Arrhythmia in Drug Development Conference*, April 30–May 1, 2009, Bethesda, MD.

5

Controversial (Unresolved) Issues in Noninferiority Trials

Brian L. Wiens

CONTENTS

5.1 Noninferiority Trials

Noninferiority clinical trials have become standard in the development of new drugs, biologics, and devices. When a placebo-controlled trial is impossible due to the ethical or practical limitations of not treating a medical condition with effective available therapy, a novel product can be indirectly demonstrated to be effective by demonstrating the efficacy is similar to, or better than, that of a known effective treatment, the active control.

Although simple in concept, there are numerous potential pitfalls with design, analysis, and interpretation of a noninferiority study. In this chapter, we will discuss a few of these issues, with emphasis on issues that have been recently evolving, or to which new approaches have been recently proposed.

Assumed for many active control, noninferiority trials is that the active control used in the trial has been compared to placebo in multiple historical trials and consistently shown to be superior to placebo in these trials with a benefit of similar magnitude. Deviation from this assumption leads to issues in designing the noninferiority trial, as will be discussed in this chapter. In some noninferiority comparisons, there may not be a historical comparison to placebo. An obvious example is when an active treatment is compared to placebo to demonstrate lack of an important harm. Although much of the information in this chapter can be applied to such comparisons, these are not the focus of the chapter. Instead, the chapter will focus on comparisons of an experimental treatment to an active control to demonstrate efficacy.

Two concepts will be discussed repeatedly in this chapter: assay sensitivity and the constancy assumption. Assay sensitivity is the ability of a trial to distinguish an effective treatment from a less effective or ineffective treatment (International Conference on Harmonisation, 2000). A desired outcome of a noninferiority trial is that the two treatments being compared have identical efficacy. However, if two treatments have identical apparent efficacy, a question arises: was identical efficacy observed because the two treatments provide identical efficacy, or because the clinical trial was unable to distinguish treatments with different efficacy? Thus, assay sensitivity is a property of a clinical trial, not an outcome of a trial or an analysis, that can be reported. The constancy assumption references implied prior comparisons of the active control to placebo. If the constancy assumption holds, the efficacy of the active control to placebo was identical (except for expected variation) in prior comparisons and would again be identical if placebo had been included in the noninferiority trial. A similar concept to the assay sensitivity and the constancy assumption is the historical evidence of sensitivity to drug effect (HESDE), which is a property of prior clinical trials that compared the active control to placebo and should have consistently demonstrated a benefit of the active control (Food and Drug Administration, 2010).

5.2 Issues in Noninferiority Trials

In this section, we will discuss four issues in noninferiority trials: missing data, choice of analysis set, multiple comparisons, and adaptive designs. These issues have several threads in common. Besides the obvious of being issues in noninferiority clinical trials, all have fairly well-established solutions in superiority trials and less well-established solutions in noninferiority trials.

In noninferiority trials, the statistical hypotheses tested are of the following form:

$$H_0 : \mu_E - \mu_C \leq -\delta$$
$$H_1 : \mu_E - \mu_C > \delta$$

where

μ_E and μ_C are the population means of the experimental and active control treatments, respectively

δ is the noninferiority margin, a prespecified quantity stating how much less efficacious the experimental treatment can be and still be considered noninferior

This assumes that larger values of μ_E and μ_C are preferred, without the loss of generality. Of course, the hypotheses could also be written with other parameterizations: a ratio of means, a difference of proportions, or in terms of odds ratios or hazards ratios, as some common examples, depending on the form of the efficacy data. The value δ is developed from and defended by appealing to historical comparison of the active control to placebo, HESDE and the constancy assumption. Tests are one-sided, typically at the level alpha $= 2.5\%$ for efficacy assessments (although alpha $= 5\%$ is occasionally used). A test is typically reported by constructing a two-sided 95% confidence interval on the difference $\mu_E - \mu_C$ and concluding noninferiority if the lower bound exceeds the value $-\delta$. It is uncommon to report p-values when testing noninferiority hypotheses.

5.2.1 Missing Data

Analysis of clinical trials in the presence of missing data is always an issue (Little and Rubin, 2002). Great strides have been made in developing methodology to provide appropriate interpretation of such data, but these methods are developed, implicitly or explicitly, for superiority trials. How well these methods adapt to noninferiority trials is an open question.

Missing data are often classified according to the missing data mechanism, the reasons leading to nonobservance of the information. Data can be classified as missing completely at random (MCAR), missing at random (MAR), or missing not at random (MNAR). Brief definitions of these are as follows:

1. MCAR: Missingness is unrelated to any other information, observed or unobserved.
2. MAR: Missingness is related only to observed information, either baseline or outcome information.
3. MNAR: Missingness is related to unobserved information, including unobserved outcome information.

Simple examples include

1. MCAR: A subject discontinued from a trial due to a job-related relocation by a spouse
2. MAR: A subject discontinued from a trial due to an adverse event, and the adverse event was recorded in the clinical trial database
3. MNAR: A subject discontinued from a trial due to a sudden loss of efficacy, and the efficacy endpoint was not reported to the investigator and thus not recorded in the clinical trial database

Data that are MCAR can be omitted from the analysis without impacting the credibility of the conclusions. Thus, it is tempting to argue that any missing data are MCAR, but the credibility of such arguments may not always be strong enough to support the conclusion.

Data that are MNAR are most problematic for two reasons. First, such data are difficult to account for in an analysis. Second, it is impossible to know whether data are MNAR or not because by definition the data are unobserved.

Data that are MAR can be accounted for in an appropriate analysis, but proper care must be taken and simplistic strategies like ignoring the missing data are not appropriate. Because it is both difficult to do an analysis with data that are MNAR, and difficult to prove whether data are MNAR, a common recommendation for the analysis of clinical trials with missing data is to assume that missing data are MAR and report some sensitivity analyses that make allowance for some level of missingness that is MNAR (National Research Council, 2010).

5.2.1.1 Analysis Methods for Data with Missingness

MAR analyses can be accomplished throughly appropriate imputation of missing data. This methodology has been thoroughly developed for superiority testing but has not been considered in detail for noninferiority testing. First, a simple single imputation strategy such as last observation carried forward (LOCF), in which the last observed value among longitudinally collected data is imputed for any missing data at subsequent timepoints, is not an appropriate imputation strategy, for superiority or for noninferiority analyses (Wiens et al., 2013). More complicated analyses including multiple imputation have not been studied in detail for noninferiority analyses to our knowledge.

Analysis of longitudinal data, with an MAR mechanism, may be accomplished with a repeated measures model. Such an analysis partially accounts for the missing data by modeling the covariance between observations within a subject and using that to inform about the missing data points. (This is not an imputation strategy, but a strategy to account for missing data through a model.) When the correlation between outcomes at different timepoints is high, this strategy can provide a powerful analysis. Additionally, this has been explicitly studied for noninferiority trials and demonstrated to provide both adequate protection of the type I error rate and improved power compared to analysis of observed data and, at times, compared to LOCF (Yoo, 2010; Wiens and Rosenkranz, 2013).

To the extent that baseline covariates predict missingness, inclusion of such covariates in the model will support an analysis when some data are MAR. Inclusion of baseline covariates in a model also provides improved power if the covariates predict outcomes, both by accounting for any baseline differences in the covariate after randomization and by reducing the variance of the estimator.

A caveat must be added to these recommendations on analyses in the presence of missing data. To maintain assay sensitivity and support the constancy assumption, an analysis that matches or mimics the analysis of the historical comparisons of the active control to placebo is preferred. If analyses of such studies do not include useful covariates in the model, or do not employ a repeated measures model, a quandary exists. Should the analysis appeal to the constancy assumption and match the historical comparisons, or should the analysis be changed to more appropriately protect against data that are MAR? There is no correct answer to this question, and most analysts will probably choose to report both analyses, hoping for consistency. A related issue is that details of the comparison of the active control to placebo may not be publicly available, if the active control was developed by a different sponsor than the one developing the experimental treatment. A partial solution to this is for editors of medical journals to require specifics in the publication of any comparison of a treatment versus a placebo, enough specifics to allow a new sponsor to design an experimental to the active treatment in a subsequent noninferiority trial. (This is recommended even though it is recognized that the word count may be exceeded. Publishing results of many trials with insufficient detail for any is not an admirable goal for a medical journal.)

Finally, any analysis of clinical trial data will have, by design, numerous sensitivity analyses. These analyses will each change the assumptions from the primary analysis. The goal is to demonstrate that the conclusions of the study are not dependent on specific assumptions of the primary analysis. Analysis of data with different sets of covariates, different covariance structures for repeated measurements, and different imputation strategies for data imputation methods are examples of sensitivity analyses.

A tipping point analysis is a simple sensitivity analysis that, with binary data, involves imputing all sets of potential outcomes (Yan et al., 2009). If there are m_E and m_C subjects with missing primary outcome data in the experimental and control groups, respectively, there will be $(m_E + 1) \times (m_C + 1)$ potential ways to impute successes and failures. Although this might be a large number of combinations, modern computers can make quick work of the calculations, and results can be presented graphically in a grid of size $(m_E + 1) \times (m_C + 1)$, with shading or coloring indicating which combinations result in various conclusions (noninferiority, superiority, or neither). With continuous data, the tipping point analysis is not as easy, both because there are infinitely many potential combinations of outcomes and because the mean and variance are affected separately by the imputation. A simplified tipping point analysis can be constructed for continuous data by, first, imputing the same value for each subject in a treatment group and, second, using the imputed data to calculate a difference in means but not the standard error of that difference. While not completely satisfying, such a sensitivity analysis can inform about how different imputation must be in the two groups before there is a material change in the conclusions. The hope, of course, is that the difference in imputation must be dramatic to affect the interpretation and, thus, that there must be a lot of data that are MNAR to change the conclusions about the relative treatment effects.

5.2.1.2 Example

As an example, consider a noninferiority study in which the primary comparison is of the proportion of subjects who achieve an efficacy goal, or a success rate. With 200 subjects in the study, 100 received the investigational treatment and 100 received the active control. A noninferiority margin of 0.1 was chosen. At the conclusion of the study, the outcome data are as shown in Table 5.1.

A naive analysis, ignoring all subjects with missing data, compares rates of 86.8% and 90.4% (79/91 and 85/94). A test statistic proposed by Farrington and Manning indicates that noninferiority was demonstrated. However, this analysis ignores missing data, and such missing data could affect the conclusions of the study. A worst-case analysis imputes failure for all subjects with missing data in the investigational group and success for all subjects with missing data in the active control group. Such an analysis does not conclude noninferiority, using the Farrington and Manning test statistic. However, a best-case analysis, imputing success for all subjects with missing data in the investigational group and failure for all subjects with missing data in the active control group, concludes superiority of the investigational treatment over the active control, using the Fisher exact test. Thus, an accounting for

TABLE 5.1

Example Data from a Clinical Trial with Missing Data

	Active Control	Investigational
Success	79	85
Failure	12	9
Missing	9	6
Total	100	100

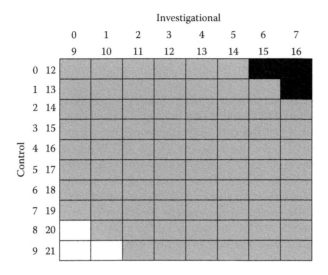

FIGURE 5.1
Tipping point plot for the example data in Table 5.1. The top row on the horizontal axis and the left column on the vertical axis indicate the number of subjects with missing data for whom failure was imputed. The second row/second column indicate the total number of failures used for analysis, after the observed failures are added to the imputed failures. Each box is shaded to represent a conclusion of superiority (white), noninferiority (gray), or neither (black).

the subjects with missing data is important, as the conclusions of the study can be affected.

The tipping point graphic is presented in Figure 5.1. The figure has 70 squares representing all possible imputations of missing data. The horizontal axis shows the results if $0, 1, \ldots, 6$ subjects with missing data in the investigational arm are imputed as failures, and the vertical axis shows the results if $0, 1, \ldots, 9$ subjects with missing data in the control arm are imputed as failures. Each of the 70 combinations is shaded, with black indicating that noninferiority was not concluded, gray indicating that noninferiority was concluded, and white indicating that superiority of the investigational treatment was concluded. Over most of the graphic, the shading is gray, so unless missingness status is highly related outcome, and the relationship is opposite in the two treatment arms, the conclusion of noninferiority will hold. The worst-case and best-case imputations described earlier are corner points in the graphic, and at those corners, the conclusion changes. An examination of the reasons for missingness, baseline data, outcome data including safety concerns, and any other available information can be used to determine whether the extremes of the figure are a plausible representation of the data.

If the tipping point graphic presents with much of the space shaded black (or much shaded white), the conclusions of the study are more difficult to defend, and additional information about the nature of the missing data, and the subjects with missing data, is required to defend an interpretation.

5.2.2 Analysis Sets

The choice of primary analysis set in superiority studies is well established. An intent to treat (ITT) analysis set is typically primary, with few exceptions. However, this is not the case with noninferiority studies, which have often used different primary analysis sets (Hung and Wang, 2014).

The definition of an ITT analysis set is not universal. In general, all subjects who are randomized to treatment (in a randomized study) will be included in the ITT analysis set (Wiens and Zhao, 2007). If some randomized subjects are excluded for reasons based on postrandomization events, the resulting analysis moves away from the goal of including all subjects and the resulting analyses may be less convincing (International Conference on Harmonisation, 1998).

The ITT analysis set provides several benefits to a superiority study. Most notably, excluding data from subjects based on postrandomization events allows for the introduction of bias, since the exclusion can be based on treatment effect. Such data may be MAR or MNAR and may not be appropriately addressed by an analysis that essentially assumes they are MCAR. Knowing that all subjects will be included in the ITT analysis encourages full follow-up of all subjects, reducing the amount of missing data. Of importance to statistical theorists, randomization is the basis of inference, and maintaining all subjects in the study supports resulting inference. From a practical standpoint, including subjects in the analysis who are not fully compliant is often thought to lessen the apparent difference in treatments, resulting in a reduced chance to demonstrate a statistically significant difference and therefore being a conservative analysis.

A per protocol (PP) analysis set has historically been preferred as a primary or co-primary analysis set for noninferiority trials. The purpose is essentially to preserve assay sensitivity. An ITT analysis, by including subjects who are noncompliant with the protocol-required actions or who have missing data, may make the treatment groups appear to be more similar. By excluding such subjects from the analysis, the hope is to preserve assay sensitivity and thereby improve confidence in the conclusions of the study.

Other analysis sets have also been proposed. An as-treated (AT) analysis is similar to an ITT analysis, with two general exceptions (Food and Drug Administration, 2010). First, an AT analysis will generally analyze subjects according to the treatment actually received rather than to the randomized treatment, if there is a difference, whereas an ITT analysis will follow the randomization schedule. Second, an AT analysis may omit some data obtained after the premature discontinuation of study treatment in subjects who discontinue before the scheduled conclusion of the protocol-mandated regimen. Whether such data are omitted, and how much, may depend on the pharmacokinetics of the treatment, the availability of alternative therapies, and perhaps other factors identified prospectively.

Our concern with the PP analysis is that it may not preserve assay sensitivity. Additionally, it may introduce bias, and the magnitude and direction of such bias is difficult to predict.

Consider the following simplistic (but hypothetical) example. A clinical trial of a novel oral antibiotic is compared to an approved oral antibiotic with known efficacy in an uncomplicated respiratory tract infection (sinusitis or streptococcal pharyngitis, as examples). Some subjects will have a nearly immediate cessation of symptoms, due to the antibiotic administered, the natural resolution of a self-limiting disease, or a combination of the two. Such subjects will not have motivation to continue the protocol-mandated study activities, as they receive no benefit from them and it is onerous to continue with follow-up visits. Other subjects will perceive no immediate benefit, due to lack of efficacy, a resistant microorganism, or some other factor. Such subjects will similarly have little incentive to continue the protocol-mandated study activities, as they perceive no benefit and symptoms remain. Such subjects may seek alternative therapies outside of the protocol, including receiving other approved antibiotic therapy from a different physician. In either situation, subjects may be excluded from PP analyses due to noncompliance with the protocol. If the experimental antibiotic has less efficacy than the approved active control, exclusion rates may be similar for the two treatment groups, but reasons may be different. If accurate information on reasons for noncompliance is not obtained, a PP analysis may, on the surface, appear to support a conclusion of noninferiority, as will an ITT analysis that includes all subjects. In this case, the PP analysis does not preserve assay sensitivity. The reason is that assay sensitivity is compromised by missing data, and a PP analysis does nothing to reduce missing data or appropriately account for it in the analysis.

In the preceding example, there are numerous actions that a sponsor could take to prevent or account for the problem. The two general themes are preventing missing data and obtaining accurate information on subjects who are noncompliant. Enrolling subjects likely to be compliant is one piece of advice, advice that is easier to give than to implement. Although it is never possible to perfectly predict which subjects will be compliant with protocol-mandated activities, site personnel can be trained to preferentially enrol subjects thought to be more likely to be compliant. These may include subjects without transportation issues, subjects who are excited about scientific research, and subjects who have been educated about the importance of complete and accurate information. Additionally, subjects can be requested more strongly to return for prespecified primary efficacy assessments, to minimize the burden on a subject and still collect data of most importance. The protocol can be written so that an early treatment failure is defined as an outcome, with rescue therapy available for subjects who request or require it. Subjects who are noncompliant can be quizzed for reasons, but in a manner that elicits accurate information. Again, this will require training of site personnel and, probably, education of study subjects to avoid a judgemental attitude while drawing out important information. In the preceding example trial, a phone call to a study subject who did not appear for an efficacy assessment may obtain useful information on reasons for noncompliance.

Appropriate analysis methods for missing data also should be implemented, as discussed in a previous section. In the preceding example, a worst-case imputation could be used by assigning all subjects who received the experimental treatment and had missing data to be failures and all subjects who received the active control and had missing data to be successes. (Note that this is an extreme point in a tipping point analysis, as described in the prior section.) Although such an analysis is possibly appropriate in this example, in general, it presents a very high hurdle and is not very specific. Preventing missing data is the best option, but obtaining at least partial information on reasons for noncompliance (again, by nonjudgmental quizzing of noncompliant subjects) can obtain some information that may make the worst-case imputation unnecessary for at least some subjects with missing data (and may support worst-case imputation for other such subjects).

Appropriate methods for analyses in the presence of missing data have been developed in the past couple of decades. These methods should be used, instead of resorting to a PP analysis (Rosenkranz, 2013). Again, the reader is directed to the previous section for more discussion of analyses of missing data in noninferiority trials.

5.2.3 Multiple Comparisons

Multiple comparisons are a concern in any clinical trial in which multiple endpoints are assessed, which is to say, multiple comparisons are a concern in any clinical trial (Dmitrienko et al., 2010). When multiple endpoints are assessed as potentially offering evidence of efficacy, the probability that at least one shows evidence goes up unless the required level of evidence is increased for any single endpoint. Additionally, multiple comparisons arise when more than two treatment groups are compared. Unique to noninferiority trials, multiple comparisons arise when an investigational treatment is assessed for superiority over an active control after noninferiority has first been demonstrated.

5.2.3.1 Testing Multiple Hypotheses in a Noninferiority Trial

When multiple hypotheses are considered for noninferiority, a noninferiority margin must be developed for each comparison. When comparing multiple investigational treatments to a single control for a single endpoint, the same noninferiority margin will often be used for each comparison. An exception is when the various investigational treatments have different characteristics—different routes or frequency of administration, say, or different safety profiles—resulting in different clinically important differences (see Wiens, 2002, for a discussion).

To the extent that study success is defined as rejecting at least one null hypothesis among a family of two or more primary hypotheses, the multiple comparisons methods developed for superiority trials can be applied to noninferiority trials with only minor modifications. Because *p*-values are not typically reported for noninferiority trials, testing procedures that rely on adjusted alpha levels can be accomplished by adjusting the confidence bounds of the confidence interval—for example, reporting a two-sided 97.5% confidence interval when a one-sided alpha of 1.25% is otherwise required for a given multiple testing procedure is an example of the simple Bonferroni adjustment for multiple comparisons.

The Bonferroni adjustment can be used for most situations in which multiple hypotheses are tested, but when the number and the complexity of the hypotheses increase, a more sophisticated multiple comparison procedure will often be preferred. Logical restrictions can often be found in dose–response studies, in which a lower dose is tested for noninferiority only if the higher dose is first shown noninferior. Logical restrictions can also be found when one or more secondary endpoints is tested for noninferiority only if noninferiority is first demonstrated for the primary (or at least one coprimary) endpoint. Such analyses are natural applications for family-based gatekeeping and tree-structured testing procedures (Dmitrienko et al., 2007), or for graphical procedures (Bretz et al., 2009).

Another complexity unique to noninferiority trials is in the analysis of a three-arm clinical trial, which compares an experimental treatment, an active control, and a placebo (or negative control) in a single trial. Such a design cannot be implemented if a negative control is unethical, a common situation when an active control is available, but when possible, this design can provide useful information such as internal validation of assay sensitivity. A three-arm trial will generally require that two hypotheses be rejected for a successful study: the experimental treatment must be superior to the negative control and must be noninferior to the active control. Thus, two null hypotheses must be rejected for a successful study, and both can be tested at the full alpha level. Whether a third null hypothesis must be rejected, showing the superiority of the active control over the negative control, is not universally agreed. Obviously, it is preferred to have this outcome, but a trial may be successful if this hypothesis is not rejected, depending on the prespecified criteria in the protocol and any regulatory guidelines.

5.2.3.2 Superiority after Noninferiority

One other source of multiple comparisons that is unique to noninferiority trials will be discussed, namely, demonstration of superiority of the experimental treatment over the active control conditional on first demonstrating noninferiority (Hung and Wang, 2014).

When only one endpoint comparing two treatment groups is considered, testing superiority after first demonstrating noninferiority is straightforward. A fixed sequence testing procedure can be used to justify testing both null hypotheses (the null of inferiority and the null of equality) at the full alpha level. Although it has been argued that testing can proceed in either order—testing superiority only after first showing noninferiority, or testing noninferiority only after first failing to show superiority—statistical theory better justifies only one direction of consideration (Morikawa and Yoshida, 1995; Dunnett and Gent, 1996; Wiens, 2001). This has little practical consequence, however.

When multiple endpoints or multiple treatment groups are considered, testing of superiority and noninferiority becomes more complicated. A question will be the order in which testing proceeds—testing noninferiority on all endpoints first, or testing superiority immediately after concluding noninferiority on an endpoint, or some mixture of these two strategies. Two issues arise: the loss of power to demonstrate noninferiority after a failed test of superiority on a prior comparison and the predicted power to demonstrate superiority on a given endpoint. If testing proceeds among hypotheses using a fixed sequence procedure, all hypotheses are tested at the full alpha level. Testing continues until a null hypothesis is not rejected, at which point all testing stops and subsequent untested null hypotheses cannot be rejected. Thus, if there is a priori belief that the experimental treatment and active control will be essentially identical on a given efficacy endpoint, it makes little sense to plan a test of superiority on that endpoint before noninferiority has been considered for other endpoints. A tree-based method that allows for splitting of alpha to simultaneously test superiority on one endpoint and noninferiority on other endpoints, such as proposed by Dmitrienko and colleagues, will have an impact on power, even though a test will be allowed (Dmitrienko et al., 2007).

A conclusion of superiority can be helpful, even if, as is often the case, regulators are reluctant to give overt claims of superiority. The question of assay sensitivity, for example, is answered if superiority is demonstrated, and controversies about the choice of noninferiority margin also are irrelevant. However, such help can be obtained with almost as much benefit without using an alpha-preserving strategy, but instead by presenting the bound of the resulting confidence interval in the analysis.

In most situations, it will be preferred to test noninferiority on all comparisons of interest before testing superiority on any comparison. This is especially true when there is little prior belief in the superiority of an experimental treatment over the active control or when the regulatory benefit of a superiority conclusion is limited.

5.2.3.3 Example

Consider a study with four treatment arms. One arm, AC, is an approved product administered according to the approved posology that has been

TABLE 5.2

Treatment Groups in Example

Group	Treatment	Posology
AC	Active control	Standard
AC*	Active control	Novel
I	Investigational	Standard
I*	Investigational	Novel

demonstrated superior to placebo. A second arm, AC*, is the approved product administered according to a novel, less onerous posology (e.g., lower dose, less invasive administration, or less frequent administration). The third arm, I, is an investigational product administered according to the approved posology of the active control. The fourth arm, I*, is the investigational product administered according to the novel posology of arm AC*. Table 5.2 lists the four treatment groups.

A simple method to control multiple comparisons is to use the Bonferroni procedure. With six pairwise comparisons, a type I error rate of $0.05/6 = 0.0083$, two-sided, is required. Thus, noninferiority in any pairwise comparison can be assessed by constructing two-sided 99.27% confidence intervals on the difference, and concluding noninferiority of one group to the other if the margin, δ, is outside of the confidence interval. However, this method is inefficient and ill-defined. It is inefficient because other, more powerful procedures can be used, and it is ill-defined because only one of the treatments, AC, has been previously compared to placebo, and any noninferiority comparisons involving only the other three groups are difficult to justify.

A more complicated, but probably better, method to control multiple comparisons is to use a gatekeeping approach (Dmitrienko et al., 2003). The hypotheses that make up each family will change with individual circumstances, but in this case, perhaps the first family will consist only of the hypothesis to demonstrate noninferiority of I to AC, because this is the comparison that demonstrates efficacy of a new treatment. The second family, tested only if the primary comparison demonstrates noninferiority of I to AC, might consist of the hypotheses to demonstrate noninferiority of AC* to AC and of I* to AC; these are secondary because they demonstrate efficacy of a treatment used in a new manner. Hypotheses that compare among the three investigational groups, AC*, I, and I*, can be considered, but in such comparisons, the noninferiority hypotheses must be carefully considered since none of these have been compared to placebo, and, thus, a noninferiority margin developed for AC might not be applicable. For this example, we will assume such comparisons are not part of an alpha-preserving multiple comparisons strategy. A third family might consider superiority of one of the three experimental groups (AC*, I or I*) to AC. An advantage of this procedure compared to the Bonferroni-based procedure is the ability to use larger values of alpha

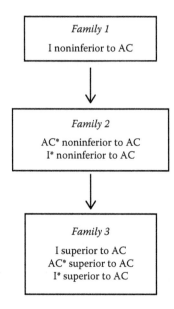

FIGURE 5.2
Gatekeeping procedure for example in Table 5.2.

for pairwise comparisons of most interest: I to AC and AC* to AC. A disadvantage might be the inability to make any conclusions about AC* if I is not demonstrated noninferior to AC in the first family of hypotheses. The gatekeeping procedure is illustrated in Figure 5.2; see Dmitrienko et al. (2003) for further details.

Consider another complexity, adding a secondary endpoint for which a noninferiority margin is prospectively identified. To utilize a family-based gatekeeping approach, the designer must decide whether the comparison of the secondary endpoint, I to AC, belongs between Family 1 and Family 2, or between Family 2 and Family 3, or is added to the two hypotheses in Family 2.

More complex multiple comparisons procedures for this example can be considered with tree-structured gatekeeping procedures (Dmitrienko et al., 2007) or graphical procedures (Bretz et al., 2009). Such procedures can take advantage of logical relationships in the hypotheses. A logical relationship in this example is that AC* cannot be concluded superior to AC unless it has first been concluded noninferior to AC. The simpler gatekeeping procedure in Figure 5.2 actually allows such a comparison, possibly reducing power for logical comparisons if alpha is assigned to a hypothesis that cannot be rejected. Such procedures can also supply an alternative to family-based gatekeeping by allowing flexibility in the order in which hypotheses are tested.

5.2.4 Adaptive Designs

Adaptive designs are designs that can change during the course of the study, based on accumulating data from the study, using rules developed before the study began (Hung and Wang, 2014). With proper planning and conduct, an adaptive design can improve the efficiency of a clinical trial, get effective treatments to patients in need, and reduce the number of clinical trial subjects who receive inferior treatment during clinical trials. Many adaptations have been proposed, including adaptations to the treatments being studied, the primary endpoint, or (probably most common) the planned sample size of the clinical trial. While these have been proposed primarily for superiority trials, some have been studied for noninferiority trials as well (Food and Drug Administration, 2010).

Sample sizes are commonly adjusted through group sequential designs. In a group sequential design, the accumulating data from the study are summarized at preplanned points during the course of the study. If, at any preplanned point, the data support a conclusion, the study is discontinued (Jennison and Turnbull, 2000). Alternatively, if the data suggest that the planned sample size is either insufficient or excessive to demonstrate a difference, the total sample size can be increased or decreased (Gallo et al., 2006). The strength of evidence necessary to stop a study early is very high early in the study, and the boundaries are created so that the probability of concluding a benefit at any point (one or more of the preplanned points) is controlled at the alpha level. Benefits of a group sequential design include a potentially smaller total sample size, a potentially faster time to market, and exposure of fewer subjects to inferior treatment (or the same number of subject, for a shorter time period). Risks of a study with a group sequential design include the potential to introduce bias after an unblinded summary of interim data and a potential to enrol more subjects (leading to higher cost and longer time to market) if the early summaries do not strongly support a conclusion of efficacy, due to adjustments necessary to control the type I error rate.

Group sequential designs are less common in noninferiority trials, even though the mathematics works out essentially identically to a superiority trial. There are two reasons for being less common. First, if two treatments have identical efficacy, there is no ethical imperative to discontinue the study, because no subjects are receiving inferior treatment. Second, if it is possible to demonstrate noninferiority with a sample size smaller than anticipated for the final analysis, it may be possible to demonstrate superiority with the full planned sample size. Thus, while the efficiency argument remains for a sponsor, other aspects differ, and a sponsor is advised to approach a group sequential design in a noninferiority trial with caution.

A form of group sequential design that may be underutilized in noninferiority trials is a futility analysis. With a futility analysis, a decision is made based on accumulating evidence to stop a trial because there is very little chance of having a successful outcome. The primary motivator for a futility analysis in a superiority study is efficiency—avoiding expenditures on a product that has little chance of demonstrating benefit. In a noninferiority trial, there is an additional motivator, namely, that it is unethical to continue treating subjects with an investigational treatment that will most likely not be concluded similar to a treatment with known benefit.

Other adaptations are possible. Sample size reestimation, for example, has been studied in noninferiority trials. Other adaptations have not been studied in detail and should be approached with caution. Examples of adaptations that have not been studied in detail in noninferiority trials (to our knowledge) are seamless phase II/III trials, adding or dropping an arm, changing a primary endpoint [see, among other references, Gallo et al. (2006), Liu et al. (2002), and Chow and Chang (2008)]. In many situations, such adaptations are used to speed the development process, and opportunities to speed the development process for a product that must be confirmed in an active control, noninferiority trial, may be more limited than for products that can be confirmed in placebo-controlled superiority trials.

5.3 Summary

We have discussed several issues in noninferiority trials that differ from superiority trials and/or on which there is not currently a consensus on the best practice for adoption. In the near future, some of these issues may be resolved, so the reader is advised to keep current with statistical literature and regulatory guidance.

References

Bretz, F., Maurer, W., Brannath, W., and Posch, M. 2009. A graphical approach to sequentially rejective multiple test procedures. *Statistics in Medicine*, 28:586–604.

Chow, S.-C. and Chang, M. 2008. Adaptive design methods in clinical trials—A review. *Orphanet Journal of Rare Diseases*, 3:11.

Dmitrienko, A., Offen, W., and Westfall, P. 2003. Gatekeeping strategies for clinical trials that do not require all primary endpoints to be significant. *Statistics in Medicine*, 22:2387–2400.

Dmitrienko, A., Tamhane, A.C., and Bretz, F. 2010. *Multiple Testing Problems in Pharmaceutical Statistics*. Chapman & Hall/CRC, Boca Raton, FL.

Dmitrienko, A., Wiens, B.L., Tamhane, A.C., and Wang, X. 2007. Tree-structured gatekeeping tests in clinical trials with hierarchically ordered multiple objectives. *Statistics in Medicine*, 26:2465–2478.

Dunnett, C.W. and Gent, M. 1996. An alternative to the use of two-sided tests in clinical trials. *Statistics in Medicine*, 15:1729–1738.

Food and Drug Administration. 2010. Guidance for Industry: Non-inferiority clinical trials. US Health and Human Services, Washington, DC.

Gallo, P., Chuang-Stein, C., Dragalin, V., Gaydos, B., Krams, M., and Pinheiro, J. 2006. Adaptive designs in clinical drug development—An executive summary of the PhRMA working group. *Journal of Biopharmaceutical Statistics*, 16:275–283.

Hung, H.M.J. and Wang, S.-J. 2014. Emerging challenges of clinical trial methodologies in regulatory applications, in Young, W and Chen, D. (eds.), *Clinical Trial Biostatistics and Biopharmaceutical Applications*, CRC Press, Boca Raton, FL, pp. 3–40.

International Conference on Harmonisation. 1998. Statistical principles for clinical trials E-9. Downloaded from http://www.ich.org/fileadmin/Public_Web_Site/ICH_Products/Guidelines/Efficacy/E9/Step4/E9_Guideline.pdf on August 11, 2013.

International Conference on Harmonisation. 2000. Choice of control group and related issues in clinical trials E-10. Downloaded from http://www.ich.org/fileadmin/Public_Web_Site/ICH_Products/Guidelines/Efficacy/E10/Step4/E10_Guideline.pdf on August 11, 2013.

Jennison, C. and Turnbull, B. 2000. *Group Sequential Methods with Applications to Clinical trials*. Chapman & Hall/CRC, Boca Raton, FL.

Little, R.J.A. and Rubin, D.B. 2002. *Statistical Analysis with Missing Data*. Wiley, New York, New York.

Liu, Q., Proschan, M.A., and Pledger, G.W. 2002. A unified theory of two-stage adaptive designs. *Journal of the American Statistical Association*, 97:1034–1041.

Morikawa, T. and Yoshida, M. 1995. A useful testing strategy in phase III trials: Combined test of superiority and test of equivalence. *Journal of Biopharmaceutical Statistics*, 5:297–306.

National Research Council. 2010. *The Prevention and Treatment of Missing Data in Clinical Trials*. The National Academies Press, Washington, DC.

Rosenkranz, G.K. 2013. Analysis sets and inference in clinical trials. *Therapeutic Innovation and Regulatory Science*, 47:455–459.

Wiens, B.L. 2001. Something for nothing in noninferiority/superiority testing: A caution. *Drug Information Journal*, 35:241–245.

Wiens, B.L. 2002. Choosing an equivalence limit for noninferiority or equivalence studies. *Controlled Clinical Trials*, 23:2–14.

Wiens, B.L. and Rosenkranz, G.K. 2013. Missing data in non-inferiority trials. *Statistics in Biopharmaceutical Research*, 5:383–393.

Wiens, B.L. and Zhao, W. 2007. The role of intention to treat in analysis of noninferiority trials. *Clinical Trials*, 4:286–291.

Yan, X., Lee, S., and Li, N. 2009. Missing data handling methods in medical device clinical trials. *Journal of Biopharmaceutical Statistics*, 19:1085–1098.

Yoo, B. 2010. Impact of missing data on type 1 error rates in non-inferiority trials. *Pharmaceutical Statistics*, 9:87–99.

Section II

Adaptive Clinical Trials

6

Adaptive Designs in Drug Development

Sue-Jane Wang and H.M. James Hung

CONTENTS

There have been experiences indicating potential insufficiency of conventional nonadaptive fixed designs for clinical trials; in particular, some argue that the trial design should be expected to answer many study questions and not just those defined by the primary study objective. Subsequently, the level of difficulty in conducting a trial to address several study questions increases significantly. As a result, adaptive design is viewed as one alternative trial design that not only can address several study questions but also save the time and costs necessary for drug development. This is the subject of this chapter.

6.1 Background

An earlier attempt of using adaptive methods in controlled clinical trials seemed to begin with the play-the-winner rules that were proposed in 1960s to 1970s, for example, a deterministic formulation of Zelen (1969), a probabilistic formation of Wei and Durham (1978). The play-the-winner rule is recognized as an example of a response-adaptive procedure under which the randomization ratio changes for consecutive patient cohorts on the basis of clinical response or clinical outcome.

Then the sample size reassessment approaches surfaced in 1990s, although the related work might have appeared in 1940s, for example, Stein (1945). In 1990s, Shih and Gould (1990) proposed sample size reestimation without unblinding the treatment group identification. Wittes and Brittain (1990) considered an approach based on unblinded internal pilot data.

Around the same time, the newer adaptive designs with multistage testing evolved, for example, Bauer (1989), Bauer and Kohne (1994). The attractive features of the multistage testing have drawn many trial practitioners' attention and have been incorporated into newer trial designs (Zeymer et al., 2001).

These adaptive designs and methodologies have been expanded from aiming at a traditional single controlled trial to considering the entire drug development program. The terms, Phase I, Phase II, Phase III, and Phase IV, commonly used to distinguish what are expected from the accumulating knowledge in a traditional drug development program have been reconsidered to make the development program more efficient or speedy. Thus, there is a strong desire to combine phases to form a single controlled trial. For instance, the terms Phase I/II/III, Phase Ia/Ib, Phase Ib/IIa, Phase IIa/IIb, Phase IIb/III, and Phase IIIb/IV have been seen in regulatory submissions (Wang, 2010a), where each term described is in reference to a single clinical trial.

The main feature from these *newer* terms is to pursue a drug development program in a two-stage or multistage manner to anticipate that the maneuver can be much flexible, for example, Muller and Schafer (2001). Some people find the feature appealing and promising, noting that it increases the efficiency of drug development by reducing the total sample size and the total program duration. Others view it troubling or as an attempt to rush through the drug development without sufficient and pertinent information for properly assessing the benefit–risk ratio of an investigational medical product.

It is these views people hold as their arguments for considering or not considering adaptive designs in planning their clinical trials.

6.2 Design Elements for Adaptation

In the context of an adaptive design, the design elements eligible for adaptation in an ongoing trial need to be prespecified. A list of commonly considered design elements is given as follows:

- Study sample size or statistical information that includes number of clinical events in an event-driven trial
- Dose regimen in a first-in-human trial, a dose-ranging trial, a dose-response trial, a dose-finding or a dose-selection trial
- Study inclusion/exclusion criteria
- Study (sub)population or subgroup
- Primary study endpoint or key secondary endpoint(s)
- Primary study objective in an active controlled trial
- Final analysis method
- Final multiplicity adjustment method
- Appropriate baseline covariate form
- Combination of statistical information and other design elements listed

Although sample size is a commonly used term, statistical information can better describe when adaptation occurs in an interim. That is, statistical information refers to not only the number of subjects (the sample size) but also the outcome data that are readily available for these subjects eligible for adaptation consideration at an interim analysis. For the design element of dose regimen, we distinguish the dose regimen among dose-ranging, dose-response, dose-finding, and dose-selection trials. Dose-ranging trials are pure exploratory trials. Depending on the efficacy endpoint used, a dose-response trial may be an exploratory trial or a confirmatory trial. The term dose finding is frequently used in exploratory adaptive design trials for future planning of confirmatory trials. Dose-selection trials are about adaptation to, say, one or two dose regimens in confirmatory adaptive design trials.

The adaptation of study design elements described can be performed in a blinded fashion or an unblinded fashion where treatment assignment is incorporated in the adaptation rules, algorithms, or criteria, such as the interim observed treatment effect estimates. That is, to know the interim observed treatment effect estimates requires the knowledge of response in the treated arm and the response in the control arm in an interim analysis. These adaptations may be pursued in an exploratory trial or a confirmatory trial. Confirmatory trials described in ICH E9 (1998) are sometimes known as adequate and well-controlled (A&WC) trials (FDA, 2002).

In a prospectively planned adaptive design clinical trial, the underlying principles that distinguish an adaptive design exploratory trial from an adaptive design A&WC trial rely upon the study objectives, the amount of available prior data, and the degree of uncertainty on key design aspects. To bridge the gap between pure exploration and A&WC investigation, the principles of a more focused exploration should not be overlooked if an experimental treatment cannot build on prior knowledge, internal or external to the specific drug development program.

6.3 Adaptive Design versus Reactive Revision

Adaptive design described in US Food and Drug Administration (FDA) draft guidance on adaptive design (FDA, 2010) requires potential design modifications be prospectively planned. The commonly seen scenarios, however, may not be forthcoming in terms of prospective planning.

In traditional fixed design trial planning, protocol amendments are common means to continue design modifications, which often rely on emerging data external and potentially internal to the trial data, suggesting further *planning* actions may be needed. However, it should be clear that use of internal trial data for adaptation without prospective planning can make the trial not adequate and not well controlled. The concerns can escalate if interim unblinded data information is used for adaptation and disseminated intentionally or unintentionally.

We will refer *the inappropriate use of protocol amendments for major design modifications and changes without proper planning in design* to *reactive revision*. This section contrasts a few adaptive design frameworks, briefly describes modeling and simulation tools for consideration with an adaptive protocol design, and gives two case examples to illustrate the risks in drug development with reactive revisions.

6.3.1 Adaptive Design

Among the adaptive design controlled trials that are prospectively planned, the utilities of a two-stage design for final statistical inference is the distinction between the before and after the adaptation (Wang and Bretz, 2010). The first stage may include more than one interim analysis. Under the null hypothesis of no treatment effect, the second stage data are independent of the first stage data. To maximize the value of the data collected, it is desired to combine the data from both stages.

By combining the data, the role of a single two-stage adaptive design clinical trial may be better suited, depending on its context in the entire drug development program. For instance, the role may be exploratory in nature or confirmatory by design. In this context, one can classify the role of a two-stage adaptive design trial into a learning trial or a confirmatory trial (FDA, 2010). Besides the exploratory framework and the confirmatory framework, we will introduce the term *learn-and-confirm* that appears to be of great interest (Wang, 2010b), which often blurs the line between exploration and confirmation. The *learn versus confirm* approach and the modeling and simulation (M&S) suitable for adaptive protocol design planning will also be briefly introduced.

6.3.1.1 Adaptive Design Learning Trial

An adaptive design learning trial provides wide flexibility to generate a hypothesis or hypotheses. The main purpose of an adaptive design learning trial is to increase the probability of correct selection on design elements, for example, a plausible dose regimen, a plausible primary efficacy endpoint, or a plausible patient subpopulation, for planning an A&WC confirmatory trial.

When multiple doses are investigated in a two-stage adaptive design exploratory trial, the study design is usually not statistically powered to detect a nonzero treatment effect at a specific dose regimen. Stage 1 data are generally used to learn and at the same time cross-validate the selected dose regimen(s) in Stage 2 in an early development program. This is mostly seen in exploratory dose-ranging adaptive design trials, where often several more doses used in conventional exploratory fixed design trials are studied, for example, Bornkamp et al. (2007). These adaptive design dose-ranging trials aim to select the minimum number of doses for confirmatory trial planning, via either design-based or model-based adaptation (Bornkamp et al., 2007). The sample sizes used in such trials are generally based on feasibility consideration.

An adaptive design learning trial may be to explore tolerability or early safety on dose regimens in a first-in-human trial, to explore activity or efficacy of a dose regimen in a dose-ranging trial, or to explore patient (sub)population or primary efficacy endpoint in an early proof of concept trial. Using dose-ranging exploratory adaptive design trial as an example, the ultimate intents to adaptively explore dose regimen(s) in Stage 1 include (1) to minimize false-negative selections, that is, by inappropriately terminating early further investigation of useful dose regimens, and (2) to minimize false-positive selections, that is, by cherry-picking dose regimen(s) for Stage 2 testing and/or estimation.

Some people argue the inefficiency of including several more doses in such an adaptive design trial. However, trial efficiency may be more appropriately evaluated via the entire drug development program rather than in a single adaptive design trial. When the false selections are carefully investigated, the design aims to increase the probability of correct dose selection, not necessarily focusing on making a Type I error or a Type II error in a hypothesis testing framework (Wang et al., 2011a). This probability is predicated on an earlier correct internal go decision for further development of an experimental compound.

6.3.1.2 Adaptive Design Confirmatory Trial

Unlike adaptive design learning trials, a two-stage adaptive design clinical trial that is inference based is often referred to as an adaptive design confirmatory trial. Essentially, studywise Type I error probability needs to be controlled strongly (Zelen, 1969; Reflection paper, 2007). It is sometimes known to be an inferential-seamless adaptive design. A confirmatory adaptive trial should also fulfill A&WC investigation described in 21CFR314.216 (FDA, 2010).

6.3.1.3 Learn and Confirm Adaptive Design Trial

Far more often than not, the two-stage adaptive designs that are not well understood are of a third framework. In the following we explain when an adaptive design is considered not well understood. While maximizing the information by combining data from both stages, it may not be of interest in committing to a strong control of the studywise Type I error rate across the two stages but only to the hypotheses selected from Stage 1 and it is desired to be considered as a bona fide confirmatory or A&WC adaptive design trial. This framework considers multiplicity adjustment only by accounting for the final hypothesis or hypotheses selected from Stage 1. We refer to this framework a learn-and-confirm approach (Wang, 2010b; Wang et al., 2010). A key question under the learn-and-confirm framework is whether strong control of Type I error probability is at issue if such a trial aims for confirmatory evidence. We discuss the inappropriateness of the learn-and-confirm framework in Section 6.4.

6.3.1.4 Learn versus Confirm

One may take advantage of saving the white space between Stage 1 and Stage 2 in a two-stage adaptive design trial; that is, only apply for the institutional review board (IRB) once without the intent to combine data from both stages or more explicitly the phase 2 trial captured in Stage 1 portion and phase 3 trial captured in Stage 2 portion. As one adaptive design trial, it is in fact two trials: Stage 1 is one trial and Stage 2 is another trial. These trials are sequentially conducted. Thus, the white space between *stages* in an adaptive design trial can be eliminated to make it operationally seamlessly.

Statistical inference will be based solely on data obtained from Stage 2 whose design builds upon learning from Stage 1 data. Stage 2 design and analysis deal with its own multiplicity of hypotheses satisfying an A&WC

investigation. Unlike the learn-and-confirm framework, this framework separates learning phase from confirming phase and is referred to as *learn versus confirm* adaptive design. It is sometimes referred to as an operationally seamless two stage adaptive design.

6.3.2 Adaptive Protocol Design

Drug development often consists of a sequence of clinical trials. Prior to implementing a clinical trial, it may be useful intuitively to conduct simulation studies that incorporate the postulated mechanistic model of the compound and the biological model of the disease or empirical data obtained from early phase drug development. These models and prior knowledge provide the opportunity to create clinical scenarios of a varying degree of uncertainty and imprecision serving as a gamete of potential disease-treatment-associated clinical outcomes for scenario comparison.

The scenarios are then experimented via modeling and/or simulation studies. Based on the operating characteristics shown from the modeling or simulation studies, one can form a series of critical go/no-go decisions. Consequently, the comparison enlists an educated choice among clinical scenarios simulated for designing potentially more efficient and informative trials (Wang, 2009; Brenda et al., 2010; Wang and Bretz, 2010).

6.3.2.1 Statistical Information Planning for Adaptation in Confirmatory Trials

For planning the statistical information of a confirmatory trial, a common wisdom is careful planning without compromising the control of the overall Type I error rate, frequentist, and Bayesian alike, for example, Lee and Zelen (2000); Simon (1982, 1994); Staguet et al. (1979); Hung et al. (2007). Careful planning on sample size or statistical information in a confirmatory trial relies upon proper exploration of what should be a reasonably postulated treatment effect and the standard deviation of the response variable from learning trials, which are mostly Phase II trials. These exploratory trials, are crucial for developing a new molecular entity or a new chemical entity, as no information can be borrowed from other drug development programs for their sample size planning.

When reasonable exploration is in place, there are generally two types of adaptation on statistical information in confirmatory trials. An unblinded statistical information (i.e., knowledge of the estimate of the interim treatment effect size) reassessment could be of interest if there is no obvious threshold for defining a clinically important treatment effect (Muller and Schafer, 2001). For instance, the original sample size may be planned as 145 patients per arm to detect an effect size of 0.33 at 80% power with a two-sided 0.05 level test. It is also possible to preplan for the opportunity when 50% of the patients have completed the study to reassess and increase the sample size to a maximum of no more than twofold the initially planned sample size, say, to 290 patients per arm. The planned maximum sample size of 290 patients would not only maintain the desired power when there are 216 subjects per arm based on an effect size of 0.27, it also accounts for more uncertainty if the resulting effect size of 0.23 is the true effect size and it is still clinically acceptable albeit smaller than 0.27, the lower limit of the approximate range.

On the other hand, suppose there is a reasonable consensus that at least nine points improvement would be considered clinically relevant. In such a case, a blinded statistical information reassessment that is intended to estimate the pooled variance of the primary efficacy outcome, irrespective of which treatment a patient receives, is an alternative adaptive approach. This consensus assumption of 9 points improvement implies that the standard deviation would be between 27 points and 33 points under the standardized effect size of 0.33–0.27. If the sample size range is 145–290 patients per arm based on at most twofold sample size increase in a reassessment strategy, the blinded approach can account for a larger standard deviation, approximately 38–39 points, than 33 points.

6.3.2.2 Modeling and Simulation

With an adaptive design, suppose the choice of patient population and primary efficacy endpoint are sensible for the proposed indication in a drug development program. To make Phase II exploration more relevant to the drug development program, the adaptive paradigm incorporating modeling and scenario simulation, such as modeling and simulation approaches discussed in adaptive design dose-ranging research (Bornkamp et al., 2007), with some attempt to make statistical inference seems to present a plausible paradigm for an informed exploration. One important feature of the adaptive design dose-ranging research is the demand for dose(s) being selected to meet the prespecified clinically relevant effect (ICH E4, 1994). While rigor in the selection criteria is encouraged, no demand of strong familywise error control would be imposed on exploratory adaptive design trials (Wang, 2007, 2010a; FDA, 2010).

Useful information obtained via exploratory adaptive design clinical trials can effectively lead to a Phase III confirmatory drug development program that may also incorporate adaptive design tools. The role of simulation and sometimes empirical modeling for a Phase III confirmatory program presents abundant opportunities for modern protocol design via clinical scenario planning. Clinical scenarios for a Phase III program considered in Bretz and Wang (2010) consist of a combination of effect sizes, variability, placebo response, and Type I and Type II errors for at least two adequate and well-controlled trials. The models if used may help to improve the precision of a treatment effect estimate at the analysis stage for a Phase III program (Wang, 2009).

6.3.3 Reactive Revision

In many cases, protocol amendments are changes of major study design elements as a result of reacting to internal/external information of an ongoing trial often without prospective specification. That is, this is adaptation on the fly. While it may be agreed that only external information is used for modification, it would be challenging to justify that no information from an ongoing trial is used for design and/or analysis modification. We refer to such nonprospectively planned changes reactive revision. In what follows, we will give two examples of reactive revisions on sample size in an ongoing trial to illustrate the impacts on Type I error probability.

6.3.3.1 A Case of a Two-Arm Active Controlled Trial

In a two-arm randomized open-label multicenter active controlled trial, the primary objective is to assess if a new anti-infective would be noninferior to its active comparator on clinical response by 1 week following an experimental anti-infective for drug development. The study assumes 60% clinical response rate at 1 week with the active control agent targeting a noninferiority margin of 20% for the new treatment at 90% power. Per-arm sample size of approximately 200 patients was planned assuming 65% patients are clinically evaluable. After trial initiation and approximately 50% of the outcome data were collected, the pharmaceutical drug sponsor requested an increase in sample size based on the observed treatment difference in a protocol amendment. It was unclear what was the motivation of the protocol amendment at approximately 50% information time in an open-label study. But, this proposal to modifying the sample size in an amendment of an ongoing trial is essentially a change to the planned noninferiority margin.

For simplicity, consider that the noninferiority margin can be defined using the control event rate (π_C) or the control effect ($\pi_C - \pi_P$) from the historical placebo-controlled data based on $\delta = \lambda \pi_C$ or $\delta = \lambda(\pi_C - \pi_P)$ for some $0 < \lambda < 1$. Hung and Wang (2004) expose a potentially serious problem with noninferiority testing when the noninferiority margin δ may have been directly or indirectly influenced by examination of the concurrent noninferiority trial data. One may have used $^*\delta = \lambda p_C$ as if it were the equivalence of the noninferiority margin, where $p_C = (1 - w)\pi_C^* + w\hat{\pi}_C$, a weighted average of the estimate π_C^* from the external data and the estimate $\hat{\pi}_C$ from the current noninferiority trial data. The expectation of $\hat{\pi}_C$ may differ from π_C. This is possible, for instance, when the patient population in the noninferiority trial may have a better response rate, because the standard medical care has been improved over time. On the other hand, if the effect of the control deteriorates over time or has not been vigorously studied because measurements are imprecise or drugs are only modestly effective (Temple, 1983), then the expectation of $\hat{\pi}_C$ may be smaller than π_C. Let the expectation of $\hat{\pi}_C$ be $\pi_C + \delta_C$, for some unknown δ_C.

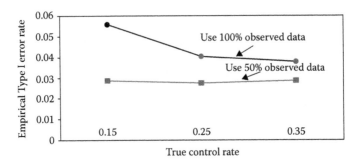

FIGURE 6.1
Impact on Type I error rate one cannot rule out when the (interim) observed data are used to update the noninferiority margin.

Hung and Wang (2004) illustrate via simulation studies to shed some light into the Type I error rate as a result of using λ such that $\lambda p_C = \delta$ based on examination of the concurrent noninferiority trial data. Again, for simplicity, assume that $\delta_C = 0$ and the constancy condition holds (Wang et al., 2002). By setting $\delta = 0.05$, λ is not a fixed constant. In the simulation, we set $\pi_C = 0.15$, 0.25, 0.35 and $w = 0.5, 1.0$. It is clear from Figure 6.1 that the Type I error rate of the noninferiority test is inflated when λ changes as a function of p_C and can inflate more than double the nominal significance level of one-sided 0.025, also see Wang et al. (1997, 2002).

6.3.3.2 A Case of a Two-Arm Superiority Trial

WIZARD was a large outcome placebo-controlled trial to assess the efficacy of azithromycin in preventing the progression of clinical coronary artery disease, measured by the first occurrence of death due to all causes, nonfatal myocardial infarction (MI), coronary revascularization, or hospitalization for angina (the primary efficacy composite endpoint), in subjects who had an MI at least 6 weeks prior to randomization (Dunne, 2000). The study was originally designed as a group sequential trial with two interim analyses and a final analysis to detect approximately 25% hazard reduction, assuming 8% event rate per year in subjects at least 6 weeks after an MI. Although the data monitoring committee (DMC) cautioned prior to trial initiation, based on their previous experience, that the 25% hazard reduction might be overly optimistic, the study plan did not change and a total of 522 events was planned to give a 90% study power for a 0.05 level test (Cook et al., 2006).

There were two major modifications to the statistical analysis plan (Cook et al., 2006). For illustration purposes, we focus on the second design modification. Specifically, the study was resized at 83% (=432/522) information time and added a third interim analysis. Re-sizing the trial to target an 18.5% hazard reduction with 90% unconditional power updated the total study size

TABLE 6.1

Data Information of the Two Interim Analyses of WIZARD

Interim Analysis (IA)	# of Events D (1^0 Endpoint)	1^0 Endpoint Information Time (t)	Observed Log-Rank z-Score z_1	Conditional Power
First	270	52%	1.99	88%
Second	432	83%	1.44	18%

Note: z_1 is the interim observed z-value for the second primary endpoint.

from 552 events to approximately 1038 events to satisfy at least 75% conditional power based on the new sample size. Sample size reassessment after interim analysis was not a preplanned goal for WIZARD; thus, the change on sample size in a group sequential trial after two interim analyses have been performed is a reactive revision.

Table 6.1 (Table 2 in Wang et al., 2012) presents the summary data of the two interim analyses of WIZARD (Cook et al., 2006). The summary data are used to highlight how difficult and tricky the sample size reassessment can be if it relies on the results of the interim analyses when they are not prospectively planned maneuvers. For exploration, let us pretend that the second primary endpoint (all-cause mortality or recurrent MI) that was used for the interim analyses is the most important endpoint to evaluate and that the interim analyses were performed only to calculate conditional power to determine whether to increase the sample size. We can then discuss possible implications if the sample size reassessment is performed at 52% information time versus at 83% information time.

The statistical literature contains several valid statistical methods for sample size reassessment based on an observed treatment effect of a confirmatory trial; see the references cited in Chen et al. (2004) including an earlier work (Shun et al., 2001). The method by Chen et al. (2004) may recommend sample size reassessment if the conditional power based on the observed effect size with the originally planned sample size is at least 50%. When the adaptation occurs, both the weighted Z-statistic and the unweighted Z-statistic must achieve statistical significance to conclude that an experimental treatment is effective.

Based on the current trend of z_1, the conditional power is about 88% and 18% at the two interim analyses, respectively, see Table 6.1. To achieve the conditional power at 90% level, the number of events needed is 570 based on the reassessment at 52% information time and 1995 at 83% (Wang et al., 2012). Chen et al. (2004) would recommend either not to increase or to increase just enough number of events at 52% information time, since the conditional power is more than 50%. If the second interim analysis results were to be used for sample size reassessment, the conditional power evaluated under the current trend at 83% information time would have been too low (18% < 50%) to be promising for upsizing.

In retrospect, the results from WIZARD help to see important insights concerning the risks that may incur when there is a lack of understanding about the treatment effect. Sample size reassessment may not be the solution to rescuing a confirmatory trial that is not properly planned.

6.4 Statistical Inference Issues with an Adaptive Design

In principle, statistical inference is expected to adhere to strong control of the studywise Type I error rate for a conventional nonadaptive design confirmatory or A&WC trial. An adaptive design confirmatory trial should be no exception, namely, *One Trial One Studywise Type I Error Rate* (Wang et al., 2010). The seminal article of a two-stage adaptive design confers this principle (Bauer and Kohne, 1994).

Later advancement in statistical methodologies also follows this very principle *One Trial One Studywise Type I Error Rate* in designing and analyzing confirmatory adaptive design trials. The list of publications has been growing since 1994. A less discussed multiplicity adjustment needed is in the setting where the key secondary endpoint may rise to a labeling claim in a group sequential or adaptive design clinical trial motivated and illustrated via simulation studies by Hung et al. (2007). The overviews of multiplicity in adaptive design clinical trials throughout a drug development program including exploratory and confirmatory trials can be found in Wang et al. (2011a).

The potential advantage of a two-stage or multistage adaptive design confirmatory clinical trial that combines data from all stages for ultimate statistical inference is its prespecified flexibility. When the adaptation attempts to facilitate better learning and formalize the ultimate hypothesis through learning, if the learning data is to be used to make statistical inference for confirmatory evidence, it is often argued that the relevant statistical hypotheses are only those hypotheses selected in Stage 1 for final statistical inference.

While the learning data from Stage 1 is to be combined with the data from Stage 2 to test the selected hypothesis for final statistical inference, the learning data is used once to select the hypothesis and may possibly be used for efficacy testing in Stage 1 depending on the prespecified adaptation rules. In addition, the learning data is also used to test the selected final hypothesis; therefore, it should be subjected to studywise or familywise Type I error control. We will use *learning-free* Type I error rate to refer to the Type I error rate associated with only the selected final hypotheses in Stage 2 without considering necessary adjustment of familywise Type I error rate that needs to be controlled.

Wang et al. (2010) explored the impacts on Type I error rate with inappropriate use of *learn and confirm* in confirmatory adaptive design trials.

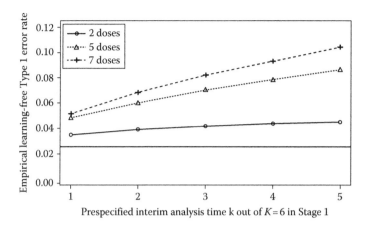

FIGURE 6.2
Select among treatment dose groups based on clinical endpoint, cutoff $D = 0$.

The adaptive selection scenarios explored in Wang et al. (2010) to illustrate the impacts on Type I error rate range from treatment selection to endpoint selection when sample size is unchanged. In the scenarios considered, the empirical learning-free Type I error rate due to the selection process will depend on the number of treatment arms investigated initially, the prespecified interim analysis time, the cutoff threshold used in the adaptive adhoc rules, the selection criterion, and whether biomarker endpoints or clinical endpoints are used for selection at an interim analysis.

As an example, Figure 6.2 (Figure 1 of Wang et al. 2010) depicts the range of empirical learning-free Type I error rate as a function of the interim analysis time and the number of treatment arms investigated initially when a liberal cutoff threshold is employed in an interim treatment selection. When the cutoff threshold accounts for the uncertainty in the numerical difference between treatment estimates, the optimism in adaptation would gradually be downplayed, resulting in a smaller magnitude of the inflated Type I error rate, see for example, Figure 2 of Wang et al. (2010). The statistical inference issue described here can be similarly addressed for endpoint selection (see Section 3.3 of Wang et al., 2010), (sub)population selection, or study objective selection (Wang et al., 1997).

6.5 Adaptive Monitoring-Logistics Models and Issues

In implementing an adaptive design clinical trial, it is of paramount importance to assure the integrity of the trial and to protect trial data for an unambiguous and objective interpretation of the study finding (Brenda, 2010;

Guidance, 2006; Wang et al., 2005). Henceforth, any interim adaptation requires establishment of written charters and procedures for adaptive monitoring, interim analysis, adaptive recommendation, and adaptive decision to assure validity and integrity of trial results. The different trial monitoring-logistics models will depend on the principles as to whether complete independence of the interactions among the involved parties is feasible via established firewalls (Wang et al., 2005, 2011b).

6.5.1 Types of Models Seen in Regulatory Submissions

In regulatory submissions, we have seen four general types of monitoring-logistics models proposed to implement an adaptive design clinical trial. The first two types are models not involving DMC: the sponsor-only model (SOI) that is internal and completely contained within the sponsor, and, the independent statistics analysis center (ISAC) model that is either academic affiliated or through contract research organizations (CROs). The remaining two types are models involving DMC: the DMC model and the combination model combining the DMC and ISAC functions.

Figure 6.3 (Figure 1 of Wang et al., 2011b) presents the general communication flow at the time of interim analysis where adaptation may occur. The interaction flow among the involved parties is based on their roles for implementing the adaptive design clinical trial, including the sponsor functions; the functions of the ISAC and CROs hired by the sponsor; the roles of the DMC, the chair, and the members of the steering committee. For a detailed description on the interaction and communication flow, and for regulatory experiences and issues encountered using the four types of adaptive monitoring-logistics models, the readers are referred to Wang et al. (2011b).

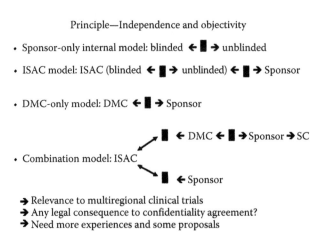

FIGURE 6.3
Adaptive monitoring-logistics models proposed in regulatory submissions.

6.5.2 Firewalls and Confidentiality

Operational bias is a key concern in implementing an adaptive design clinical trial. The operational bias refers to the problems with the study conducts, for example, due to dissemination of interim unblinded data or their analysis results, adaptive decision, or subjective decision due to discretionary authority given. Operational bias can critically affect the validity of study conclusions. Therefore, it is critical to limit access to unblinded data in adaptive design clinical trials. In principle, the firewalls established for the interaction flow among the parties should define the objectivity of the adaptive monitoring process and implementation of the adaptive trial for protecting the trial integrity. The established firewalls should allow the role of managing trial conducts be separated from the role of DMC, and, the trial processes and adaptive monitoring procedures should be clearly described.

Often in regulatory submissions, other than stating the existence of the confidentiality agreement of the sponsor's internal personnel who will be unblinded to the interim data and analysis results, little has been discussed about the ramifications if the unblinded internal personnel do not follow the confidentiality agreement. It is well known that the consequences of unblinded interim information leakage not only introduce the operational bias due to trial misconduct but also make the final study results uninterpretable.

It is important to point out that regardless of which trial logistics model may be more appropriate in a particular scenario, the potential for compromising the trial integrity is always of great concern, such as uncontrolled or unavoidable verbal communication undocumented. Thus far, no single advice as to how to best assure trial integrity in use of adaptive design for confirmatory evidence can be given. Often, it rests upon the appropriateness of the firewalls and whether the firewalls established can be strictly followed.

In the context of group sequential design, many authors have articulated the roles of a DMC, issues on firewalls and confidentiality, and some recommendations, for example, Guidance (2006), Ellenberg et al. (1993, 2002), DeMetz et al. (2006), and Herson (2009). From US FDA draft guidance on adaptive design (FDA, 2010), group sequential design is a type of adaptive design, whose null hypothesis or hypotheses do not change. Giving or requesting the DMC to act on a greater role with increased responsibilities in an adaptive design setting will undoubtedly raise greater challenges on the independence and objectivity of an DMC (Guidance, 2006).

Chow et al. (2012) propose some solutions to address the concerns of the additional responsibilities of the DMC for adaptive design clinical trials. Spencer et al. (2012) propose some solutions with implementation of an adaptive dose-selection design acknowledging operational challenges. As the operational bias is a key issue to success of an adaptive design trial implementation, we believe the emerging consensus from every involved party in implementing an adaptive design clinical trial is yet to be formed. It should be obvious that the more complex adaptive design confirmatory trial is proposed, the larger challenges there will be. There is no doubt that much more experiences are needed via submitted adaptive design proposals, exploratory or confirmatory, for regulatory reviews and experience building.

6.6 Summary

This chapter gives a brief literature background motivation of pursuing adaptive design clinical trials throughout a drug development program by pharmaceutical industry. In Section 6.2, design elements eligible for adaptation are introduced. The types of adaptive design framework are explained in Section 6.3. Furthermore, modeling and/or simulation are briefly mentioned as tools for adaptive clinical trial protocol design planning. Not prospectively planned adaptation falling under the category of *reactive revision* is used as a metaphor to illustrate the risk and challenges with such an approach using two practical case examples seen in a confirmatory clinical trial. In Section 6.4, we address statistical inference issues with an adaptive design. In particular, we give the extent of inflation that can occur for adaptive dose selection when a learning-free Type I error is intended as an adaptive design confirmatory clinical trial. In Section 6.5, we present types of adaptive monitoring-logistics models proposed or seen in regulatory submissions and shared some regulatory experiences. We also raise challenges and discuss the firewalls and confidentiality issues faced with in implementing adaptive design clinical trials.

Acknowledgments

The contents are based on lectures given at Deming conference and the regulatory science research conducted by the authors. Thanks are due to Dr. Lisa LaVange for the editorial comments that helped improve the presentation of this chapter.

Disclaimer

This chapter reflects the views of the authors and should not be construed to represent the views or policies of the US Food and Drug Administration.

References

Bauer P (1989) Multistage testing with adaptive designs (with Discussion). *Biometrie und Informatik in Medizin and Biologie* 20:130–148.

Bauer P, Kohne K (1994) Evaluation of experiments with adaptive interim analysis. *Biometrics* 50:1029–1041.

Benda N, Brannath W, Bretz F, Burger HU, Friede T, Mauer W, Wang SJ (2010) Perspectives on the use of adaptive designs in clinical trials. Part II. Panel discussion. *Journal of Biopharmaceutical Statistics* 20:1098–1112.

Bornkamp B, Bretz F, Dmitrienko A, Enas G, Gaydos B, Hsu CH, Konig F et al. (2007) Innovative approaches for designing and analyzing adaptive dose-ranging trials. *Journal of Biopharmaceutical Statistics* 17:965–995.

Bretz F, Wang SJ (2010) From adaptive design to modern protocol design for drug development: Part II. Success probabilities and effect estimates for phase III development programs. *Drug Information Journal* 44(3):333–342.

Chen YH, DeMets DL, Lan KK (2004) Increasing the sample size when the unblinded interim result is promising. *Statistics in Medicine* 23:1023–1038.

Chow SC, Corey R, Lin M (2012) On the independence of data monitoring committee in adaptive design clinical trials. *Journal of Biopharmaceutical Statistics* 22:853–867.

Cook TD, Benner RJ, Fisher MR (2006) The WIZARD trial as a case study of flexible clinical trial design. *Drug Information Journal* 40:345–353.

DeMetz DL, Furberg CD, Friedman LM (2006) *Data Monitoring in Clinical Trials: A Case Studies Approach.* New York: Springer.

Dunne MW (2000) Rationale and design of a secondary prevention trial of antibiotic use in patients after myocardial infarction: The WIZARD (Weekly Intervention With Zithromax [azithromycin] for Atherosclerosis and Its Related Disorders) trial. *Journal of Infectious Disease* 181(Suppl 3):S572–S578.

Ellenberg SS, Fleming TR, DeMets DL (2002) *Data Monitoring Committees in Clinical Trials: A Practical Perspective.* New York: John Wiley & Sons.

Ellenberg SS, Geller N, Simon R, Yusuf S (eds) (1993) Proceedings of 'Practical Issues in data monitoring of clinical trials', Bethesda, Maryland, January 27–28, 1992. *Statistics in Medicine* 12:415–646.

FDA (2002) Food and Drug Administration, Health Human Services, Code of Federal Regulation, 21CFR312.23(a). Available at http://www.accessdata.fda.gov/scripts/cdrh/cfdocs/cfcfr/cfrsearch.cfm?fr=312.23.

FDA (2010) U.S. Food and Drug Administration, Draft guidance for industry on adaptive design clinical trials for drugs and biologics. Released on February 25, 2010 for public comments. Available at www.fda.gov/downloads/Drugs/GuidanceComplianceRegulatoryInformation/Guidances/UCM201790.pdf.

FDA (2006) Guidance for clinical trial sponsors: On the establishment and operation of clinical trial data monitoring committees. March 2006. Available at http://www.fda.gov/downloads/RegulatoryInformation/Guidances/UCM126578.pdf.

Herson J (2009) *Data and Safety Monitoring Committees in Clinical Trials.* New York: Chapman & Hall/CRC Press, Taylor & Francis, Boca Raton, FL.

Hung HMJ, Wang SJ (2004) Multiple testing of non-inferiority hypotheses in active controlled trials. *Journal of Biopharmaceutical Statistics* 14(2):327–335.

Hung HMJ, Wang SJ, O'Neill RT (2007) Statistical considerations for testing multiple endpoints in group sequential or adaptive clinical trials. *Journal of Biopharmaceutical Statistics* 17:1201–1210.

ICH E4 (1994) International Conference on Harmonization (ICH) Guidance (ICH) Topic E4: Dose-response information to support drug registration. Available at http://www.ich.org/fileadmin/Public_Web_Site/ICH_Products/Guidelines/Efficacy/E4/Step4/E4_Guideline.pdf.

ICH E9 (1998) International Conference on Harmonization (ICH) Guidance (ICH) Topic E9: Statistical principles for clinical trials. Available at http://www.ich.org/fileadmin/Public_Web_Site/ICH_Products/Guidelines/Efficacy/E9/Step4/E9_Guideline.pdf.

Lee SL, Zelen M (2000) Clinical trials and sample size considerations: Another perspective. *Statistical Science* 15(2):95–110.

Muller HH, Schafer H (2001) Adaptive group sequential designs for clinical trials: Combining the advantages of adaptive and of classical group sequential approaches. *Biometrics* 57:886–891.

Reflection paper (2007) Reflection paper on methodological issues in confirmatory clinical trials planned with an adaptive design. European Medicines Agency EMA CHMP/EWP/2459/02 October 2007. Available at http://www.ema.europa.eu/docs/en_GB/document_library/Scientific_guideline/2009/09/WC500003616.pdf.

Shih WJ, Gould AL (1990) Interim analysis for sample size reestimation without unblinding treatment group identification in double-blind clinical trials. Abstract no. 27, Society for Clinical Trials Eleventh Annual Meeting, Toronto, Ontario, Canada, May 6–9, 1990.

Shun Z, Yuan W, Brady WE, Hsu H (2001) Type I error in sample size re-estimations based on observed treatment difference. *Statistics in Medicine* 20:497–513.

Simon R (1982) Randomized clinical trials and research strategy. *Cancer Treatment Reports* 66:1083–1087.

Simon R (1994) Some practical aspects of the interim monitoring of clinical trails. *Statistics in Medicine* 13:1401–1409.

Spencer K, Colvin K, Braunecker B, Brackman M, Ripley J, Hines P, Skrivanek Z, Gaydos B, Geiger MJ (2012) Operational challenges and solutions with implementation of an adaptive seamless phase 2/3 study. *Journal of Diabetes Science and Technology* 6:1296–1304.

Staquet MJ, Rozencweig M, Von Hoff DD, Mugia FM (1979) The delta and epsilon errors in the assessment of clinical trials. *Cancer Treatment Reports* 63:1917–1921.

Stein C (1945) A two-sample test for a linear hypothesis whose power is independent of the variance. *Annals of Mathematical Statistics* 16:243–258.

Temple R (1983) Difficulties in evaluating positive control trials. In: *Proceedings of the Biopharmaceutical Section of the American Statistical Association*, American Statistical Association, Alexandria, VA, pp. 48–54.

Wang SJ (2007) Discussion of the White Paper of the Pharmaceutical Research and Manufacturers of America (PhRMA) working group on adaptive dose-ranging designs. *Journal of Biopharmaceutical Statistics* 17:1015–1020.

Wang SJ (2009) Commentary on experiences in model/simulation for early phase or late phase study planning aimed to learn key design elements. *Statistical in Biopharmaceutical Research* 1(4):462–467.

Wang SJ (2010a) Editorial: Adaptive designs: Appealing in development of therapeutics, and where do controversies lie? *Journal of Biopharmaceutical Statistics* 20(6):1083–1087.

Wang SJ (2010b) Perspectives on the use of adaptive designs in clinical trials: Part I. Statistical considerations and issues. *Journal of Biopharmaceutical Statistics* 20(6):1090–1097.

Wang SJ, Bretz F (2010) From adaptive design to modern protocol design for drug development: Part I. Editorial and Summary of Adaptive Designs Session at the Third FDA/DIA (Drug Information Association) Statistics Forum. *Drug Information Journal* 44(3):325–331.

Wang SJ, Hung HMJ, O'Neill RT (2005) Uncertainty of effect size and clinical (genomic) trial design flexibility/adaptivity. *The Proceedings of American Statistical Association, Biopharmaceutical Section,* [CD-ROM]. Alexandria, VA: American Statistical Association.

Wang SJ, Hung HMJ, O'Neill RT (2010) Impacts of type I error rate with inappropriate use of learn for confirm in adaptive designs. *Biometrical Journal Special Issue* 52(6):798–810. DOI:10.1002/bimj.200900207.

Wang SJ, Hung HMJ, O'Neill RT (2011a) Regulatory perspectives on multiplicity in adaptive design clinical trials throughout a drug development program. *Journal of Biopharmaceutical Statistics* 21:846–849.

Wang SJ, Hung HMJ, O'Neill RT (2011b) Adaptive design clinical trials and logistics models in CNS drug development. *European Neuropsychopharmacology Journal* (International Society for Central Nervous System (CNS) Clinical Trials and Methodology (ISCTM) journal) 21:159–166.

Wang SJ, Hung HMJ, O'Neill RT (2012) Paradigms for adaptive statistical information designs: Practical experiences and strategies. *Statistics in Medicine* 31: 3011–3023.

Wang SJ, Hung HMJ, Tsong Y (2002) Utility and pitfall of some statistical methods in non-inferiority active controlled trials. *Controlled Clinical Trials* 23(1):15–28.

Wang SJ, Hung HMJ, Tsong Y, Cui L, Nuri W (1997) Changing the study objective in clinical trials. In: *Proceedings of the American Statistical Association, Biopharmaceutical Section.* Alexandria, VA: American Statistical Association.

Weil LJ, Durham S (1978) The randomized play-the-winner rule in medical trials. *Journal of American Statistical Association* 73:840–843.

Wittes J, Brittan E (1990) The role of internal pilot studies in increasing the efficiency of clinical trials. *Statistics in Medicine* 9:65–72.

Zelen M (1969) Play the winner rule and the controlled clinical trials. *Journal of American Statistical Association* 64:131–146.

Zeymer U, Suryapranata H, Monassier JP, Opolski G, Davies J, Rasmanis G, Linssen G et al., for the ESCAMI Investigators (2001) The Na^+/H^+ exchange inhibitor eniporide as an adjunct to early reperfusion therapy for acute myocardial infarction. *Journal of the American College of Cardiology* 38:1664–1650i.

7

Optimizing Group-Sequential Designs with
Focus on Adaptability: Implications of
Nonproportional Hazards in Clinical Trials

Edward Lakatos

CONTENTS

7.1 Introduction

This chapter focuses on the design and postdesign implications of survival trials with interim analyses for which the assumptions of constant hazards may be violated. It is frequently stated that a simple linear regression line can often provide a reasonable model for data that are not linear. A similar interpretation of the constant hazards assumption can be very problematic. In spite of the fact that failure to recognize nonproportional hazards can have serious consequences, its presence often remains undetected. Thus, it is important to recognize possible nonproportional hazards situations and understand their consequences. Additionally, the consequences of nonproportional hazards can provide insight into issues when the hazards are well behaved.

In 1986, Lakatos proposed the use of Markov models for calculating sample size for survival trials for which the constant hazard assumption might not apply. The Markov model not only allowed both the control and experimental hazards to vary over time, but also allowed for nonproportional hazards (including treatment lags), time-dependent rates of enrollment, loss to follow-up, and noncompliance. In 1988, Lakatos extended that Markov model for calculating sample size for the log-rank statistic; the 1986 paper was for analyses that would be based on a comparison of proportions.

Lakatos and Lan (1992) used simulations to compare the Markov approach with two other popular approaches to sample size determination for the log-rank statistic: Rubinstein et al. (1980) was developed for exponentially distributed survival curves, while Schoenfeld (1981) only required proportional hazards. It was not unexpected that the Markov approach could perform far better when the proportional hazards assumption was violated. But perhaps surprisingly, the Markov approach also outperformed the other methods under proportional hazards as well as exponential settings. The reason for this will be explained later. Neither the Rubinstein nor Schoenfeld methods addressed noncompliance or time-dependent rates of enrollment.

Lakatos (2002) further developed the Markov approach for group-sequential trials and showed how the Markov model could be used for predicting the accumulation of events over real (calendar) time. Those event-prediction methods appear to be the only ones currently available that allow for time-dependent rates of failure, loss-to-follow-up, enrollment, and noncompliance. Lakatos further developed the Markov approach to include adaptive and enrichment designs, and designs that change eligibility requirements midtrial (these methods are currently unpublished, but have been used in a number of trials). In recent years, regulatory authorities often require primary endpoint follow-up of patients who have *dropped out* of the trial. This classification, off drug in study (ODIS), from a mathematical modeling point of view, is identical to noncompliance.

The remainder of this chapter is divided into three sections. Section 7.2, through a series of examples, presents a heuristic discussion of some implications of nonconstant and nonproportional hazards. Section 7.3 presents a group-sequential design for sample size re-estimation, particularly relevant for nonproportional hazards. Section 7.3 also discusses what constitutes a good group-sequential design. Section 7.4 provides a rudimentary explanation of the Markov model for survival trials and discusses some advantages of the Markov approach.

7.2 Examples of the Impact of Nonconstant and Nonproportional Hazards

Five examples of the impact of assuming constant or proportional hazards, when either of these assumptions is violated, are presented in this section. The first example shows how sample size is effected when constant hazards is violated, as well as when proportional hazards is violated. The second example examines the impact of departure from nonproportionality on the design of an event-driven trial. Sample size evaluation for group-sequential trials is discussed in Example 3, and sample size re-estimation is the focus of the fourth and fifth examples.

7.2.1 Example 1: Power with a Declining Control Group Rate and Treatment Lag

7.2.1.1 Impact of Nonconstant Hazards on Sample Size

Reference Trial: A Cardiovascular trial with Major Adverse Cardiac Events (MACE) as the primary endpoint (Cannon et al. 2004).

When statisticians calculate sample sizes for survival trials, they usually reference prior trials for the placebo rate and sometimes for the treatment effect. For each of the examples in this section, information is extracted from prior trials.

A sketch of the survival curve for MACE events for a statin arm of a large trial, adapted from Cannon et al. (2004), is shown in Figure 7.1.

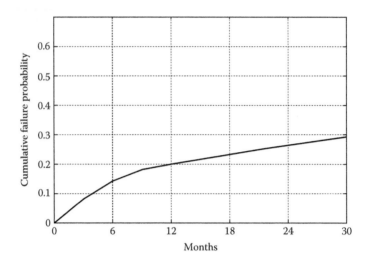

FIGURE 7.1
Sketch of the survival experience of a statin arm of the PROVE-IT trial.

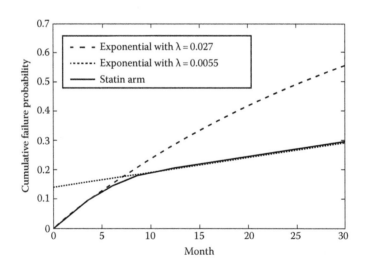

FIGURE 7.2
Sketch of the survival curve of a statin arm of the PROVE-IT trial compared with two exponential curves fitted to the earlier and later portions of the statin curve.

Since it is difficult to visually assess whether the survival of this curve follows an exponential model, exponential curves were fitted to two separate portions of the statin curve (Figure 7.2).

The hazard λ for the curve fitted to months 0–5 of the statin curve has $\lambda \approx 0.027$; for months 9–30, $\lambda \approx 0.0055$. The hazard for the earlier portion

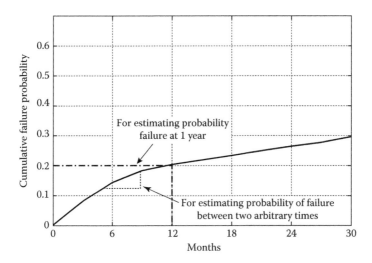

FIGURE 7.3
Estimating failure probabilities from graph.

of the trial is approximately five times as large as for the latter portion. The implications of assuming the hazard is constant in designing a new trial are now explored.

Suppose that every patient in the new trial being designed will be taking a statin as background therapy, and that patients will, in addition, be randomized to placebo or a new experimental cholesterol-lowering drug. Statisticians often take the 1-year failure rate, indicated in Figure 7.3 by the dot-dashed lines, as a basis for sample size calculations. In doing so, the working assumption is that the failure rate is constant throughout the trial. From Figure 7.3, the 12-month failure rate is about 20%.

Rather than assuming that the failure rate is constant, the time-dependent failure rate function can be estimated from these survival curves. Just as one can use the survival curves to calculate the failure rate for the 12-month period [0, 12] under the assumption of constant hazards during that period, one can also calculate the conditional probability of failing in any given interval $[t_1, t_2]$ on the x-axis using the corresponding probabilities of failure $[F(t_1), F(t_2)]$ on the y-axis. Here, $F(t)$ is the cumulative probability of failing at time t.

The result is the piecewise constant approximation to the hazard curve shown in Figure 7.4.

A procedure for smoothing the hazard function is applied. Since annual failure rates are more familiar, for each of the *pieces*, the hazard has been replaced by the annual failure probability. The resultant piecewise constant estimated annual probability of failing curve is displayed in Figure 7.5.

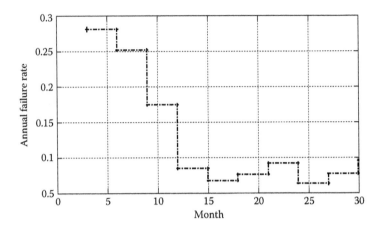

FIGURE 7.4
Piecewise constant estimate of the failure rate function.

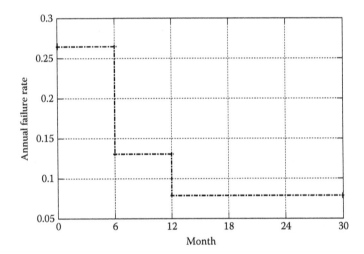

FIGURE 7.5
Piecewise smoothed version of failure rate function.

The smoothing process has resulted in three constant levels. The conditional failure rate (on an annual basis) during the first 6 months is about 26.3%, is about 13.4% during the next 6 months, and is about 8% from month 12 to month 30. Far from constant, the failure rate is over three times as large during the early months postrandomization compared with the latter portion of the Kaplan–Meier (KM) curves (Kaplan and Meier 1958).

Sample sizes under the following three sets of assumptions (1–3) are now calculated. Assumptions common to all three sets of sample size calculations:

- Recruitment length: 18 months
- Trial length: 36 months
- Power: 90%
- Significance level: 0.025 one-sided

Further assumptions

1. Constant risk, constant treatment effect as in Table 7.1.
2. Declining failure rate, constant treatment effect as in Table 7.2.
3. Declining failure rate, treatment lag as in Table 7.3.

TABLE 7.1

Assumed Failure Rates Assuming
Constant Hazards

Failure Rate	Treatment Effect
20% entire trial	15% entire trial

TABLE 7.2

Assumed Failure Rates Captured from
PROVE-IT Kaplan–Meier Plots

Failure Rate	Treatment Effect
26% months 1–6	15% entire trial
13% months 7–12	
8% months 13-end	

TABLE 7.3

Assumed Failure Rates Captured
from PROVE-IT Plots and Assumed
Treatment Lag

Failure Rate	Treatment Effect
26% months 1–6	5% months 1–3
13% months 7–12	10% months 4–6
8% months 13-end	15% months 7-end

TABLE 7.4

Summary of Sample Size Calculations

Control Failure Rate	Treatment Effect	N	Events
Constant	Constant	3508	1294
Declining	Constant	5095	1315
Declining	Treatment lag	8240	2158

Summary of results presented in Table 7.4.

The fact that the required number of events in each of the first two rows of Table 7.4 are reasonably close to one another is expected. The small difference is computational error; the required number of events in those two cases should be identical since the treatment effect is assumed to be constant. In the discussion that follows, 1300 will be taken as the nominal number of events for rows 1 and 2. The calculations that take into account the declining event rate in the control group reveal that 45% more patients (5095) will be needed to achieve the desired 1300 events in 36 months. If only 3508 patients are enrolled, then it will take about 62 months for the 1300 events to occur. The reason it takes so much longer than the 45% would seem to indicate is because the failure rate during the remainder of the trial is more likely to be the 8% derived by taking the time-dependent nature of the survival curves into account as compared with the 20% failure rate based on the unrealistic assumption of constant risk.

The number of events accumulated in 36 months assuming $N = 3508$, but assuming a declining control failure rate with constant treatment effect is 906 events. In other words, if we design assuming a constant placebo failure rate when in fact the failure rate is declining, then the calculated sample size is 3508, but the number of events in 36 months will only be 906. It will take about 62 months before the desired 1300 events occur.

When the proportional hazards assumption is violated, the implications are more profound. Examination of the original KM plots from PROVE-IT shows that the curves do not begin to separate for about 2 years, a threshold lag. The treatment lag assumed in Table 7.3 is far more modest. First, it assumes the lag period is only 6 months long, after which the drug is fully effective. In addition, it assumes a gradual increase in effectiveness during that 6-month period. Even with this modest lag, the required number of events increases by 64%, from 1315 to 2158. The number of patients required is 8240, which is well over twice the sample size obtained under the constant hazards assumption.

7.2.1.2 Nonproportional Hazards

The remaining examples in this section focus on nonproportional hazards. Under the proportional hazards assumption, the hazard curves for each of

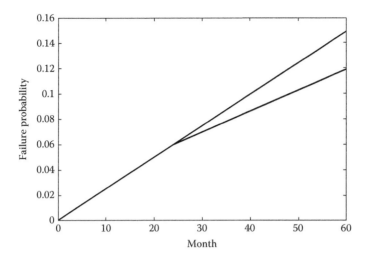

FIGURE 7.6
Constructed sketch showing the general shape of survival curves when there is a threshold treatment lag.

the treatment arms need not be constant, but the hazard ratio and, in turn, the treatment effect must be. The implications of assuming proportional hazards when that assumption is not justified are now explored. Two types of nonproportional hazards alternatives are examined. A treatment lag refers to the situation in which the full benefit of treatment does not appear until some period of time from randomization. The effect of treatment can gradually begin to emerge, or emerge suddenly at the end of the lag period. The latter is called a *threshold lag*. There are numerous examples in the literature (e.g., Bernier et al. 2004, Cooper et al. 2004, Roach et al. 2008, Rubins et al. 1999, Sacks et al. 1996, Scandinavian Simvastatin Survival Study Group 1994, The Lipid Research Clinics Study Group 1984), one of which will be examined in Example 4. The general pattern of the survival curves exhibiting a threshold lag is displayed in the stylized (actual survival curves are rarely straight) lines in Figure 7.6.

The opposite phenomenon, often referred to as an antilag, occurs when initially stronger treatment effects diminish substantially as time from randomization increases. Figure 7.7 shows stylized survival curves in which the treatment effect is completely gone after 24 months (there is no further separation of the curves after that time).

In Example 2, a trial in which the treatment effect diminishes over time is used to show how problems can arise with the common *events-driven trial* if the nonproportionality is ignored. In Example 3, two trials with identical overall proportions failing in both the control groups and the experimental groups are constructed: one has a declining failure rate while the other has an

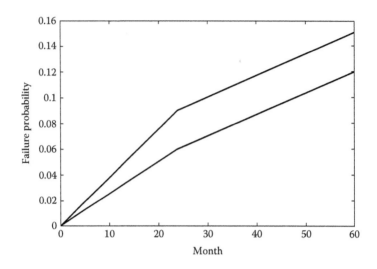

FIGURE 7.7
Constructed sketch showing the general shape of survival curves when the treatment effect diminishes over time.

increasing rate. This example shows how the form of nonproportional hazards can result in a decrease in the group sequential sample size as compared with a trial with no interims. Examples 4 and 5 show how nonproportional hazards can cause undesirable effects in adaptive trials.

7.2.2 Example 2: Events-Driven Trial—Power Decreases with Increased Follow-Up

7.2.2.1 Impact of Nonproportional Hazards on Sample Size

Nonproportional hazards for which the treatment effect diminishes over time. Reference Trial: The Beta Blocker Heart Attack Trial (BHAT) (The Beta-Blocker Heart Attack Trial Research Group 1982).

The BHAT trial has provided insight into a variety of statistical issues since the early 1980s. The smoothed hazard ratio curve, derived by applying the STOPP® graph reader module to each of the survival curves from BHAT, is presented in Figure 7.8.

One of the interesting aspects of this hazard ratio curve is that there is very little, if any, treatment effect after 18 months. With the hindsight afforded by this STOPP analysis, it is apparent that the original survival curves (see, e.g., DeMets et al. 2006) do not continue to separate after 18 months (it should be noted that, as in almost all survival curves, the far right side of the original graph is based on a rather small proportion of the original cohort [406 out of 3706 patients]). A common assumption is that the treatment effect is constant, that is, proportional hazards.

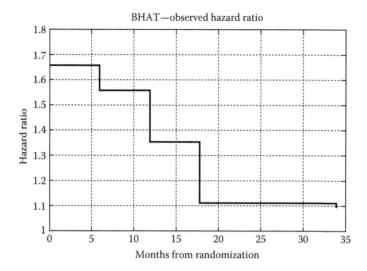

FIGURE 7.8
Smoothed piecewise constant hazard ratio function derived from the BHAT survival curves.

One of the most common survival trial designs is the *event-driven trial*. Suppose a new trial is being designed for which the treatment effect is likely to resemble that shown in Figure 7.8. The basic concept behind the event-driven trial is that power is determined by the number of events rather than the number of patients. While this is certainly true, when nonproportional hazards is possible, it is not the full story. In the discussion that follows, careful attention as to which time frame applies is essential. In survival trials, there are two time frames: the study is carried out in calendar time (CT), with an official start and end date, and data are recorded according to calendar time. Survival data are analyzed in terms of time from randomization. Hazard functions are also defined in terms of time from randomization (RT) not calendar time. If there is a period of time (RT scale) during which there is no treatment effect, that period will reduce the power of the log-rank statistic. More specifically, assume that from some time forward (RT) there is no treatment effect. The situation is similar to that of Figure 7.8, but that the hazard ratio would be 1 rather than being slightly larger. If that period of time is included in the calculation of the log-rank statistic, then the statistic will be diminished. Heuristically, statistics compare the size of the signal (the estimate of treatment effect) to the noise (variance). The period of no treatment effect contributes nothing to the signal, but the additional events increase the noise, resulting in a smaller statistic. In other words, the period of no treatment effect *dilutes* or *waters down* the statistic. The same is true if there is a period of strong treatment effect followed by a much weaker treatment effect.

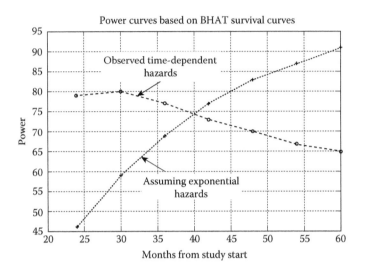

FIGURE 7.9
Comparing power based on exponential assumption with power based on the actual shape of the observed survival curves.

With a hazard ratio curve of the underlying form of Figure 7.8, once all of the patients have been exposed for 18 months (CT), continued follow-up, in general, serves to diminish the power. The original survival curves show variability after month 18 so some minor deviation from this principle should be expected. Figure 7.9 displays power for trials of various lengths whose underlying hazard ratio curve is as depicted in Figure 7.8. The calculations are performed under two different sets of assumptions: (1) the treatment effect is constant (labeled *exponential*), and (2) the treatment effect is as derived from the hazard ratio graph of Figure 7.8. For a trial of length 24 months, the power using the observed nonconstant hazard ratio is about 79%, while it is about 46% under the assumption of a constant hazard ratio. Under the unrealistic assumption of a constant hazard ratio, the longer trial continues, the greater the power. Under the empirically observed hazard ratio of Figure 7.8, increasing the length of the trial to 30 months marginally increases the power, but then the power begins to decline. At around 4.5 years, the power under the unrealistic exponential assumption appears to be well over 80%, while the actual power has diminished to about 68%.

It is often tempting, for cost considerations, to design a trial with a small sample size. Using the exponential model provides comfort because it will indicate that if the trial continues long enough, and if the drug works, then the power will increase and the trial will eventually succeed. If the nonproportionality resembles that of Figure 7.8, that comfort is false because the longer the trial continues, the more the power will diminish, even if the drug works very well. The best strategy for this situation is to design the trial with a sufficient sample size to end within 30 months when the power is still relatively high.

It should be noted that the very small treatment effect observed after 18 months in Figure 7.8 does not indicate that the drug is no longer beneficial. Because of the strong treatment effect observed during the first 18 months after randomization, the treatment groups may no longer be balanced, with the sickest patients still at risk in the active group, but their counterparts are no longer at risk in the control group. In such a situation, in order to maintain the slightly beneficial observed treatment effect (or even no treatment effect at all), the drug may still have a very strong actual treatment effect. In order to determine whether there is actual benefit to continued treatment, a re-randomization design is needed.

Finally, it is important to realize that noncompliance or ODIS has an effect on the hazard ratio that is similar to what was observed in Figure 7.8. As trials progress, it is often the case that more and more patients stop complying with their randomized treatments. Generally, patients who do not comply do not receive the benefit of the treatment they are no longer taking. The treatment effect, averaging over the trial population, diminishes over time, and the power decreases relative to a trial without noncompliance. The effect is similar to what was observed in Figure 7.8, although the extent may be less pronounced. Thus, even if the treatment itself maintains its strength (i.e., the proportional hazards setting), noncompliance can cause relative diminution of the power as the trial progresses. In this case, the event-driven trial may fail, even though the drug is effective.

Implications of nonproportional hazards in the general shape of BHAT will be revisited in Example 5.

7.2.3 Example 3: Negative Inflation of Sample Size for Group-Sequential Trial

A group-sequential trial for which the sample size is smaller than the same trial with no interim analyses, see Lakatos (2002).

An example similar to that presented in Table 7.5 was presented by Lakatos (2002) and verified by simulations. Here, two trials have been constructed, both with 12 months of uniform enrollment and total trial length of 60 months each. The treatment effect is decreasing in trial 1 (i.e., hazard ratio getting closer to 1) and increasing in trial 2. The example was constructed so that the cumulative failure rates for the entire trial in each arm are the same in both trials. Consequently, sample size methods that do not take the nonproportional hazards into account must produce the same sample size for both trials. The *Fixed* sample size row shows that sample size methods, such as STOPP, that do take the nonproportionality into account produce different sample sizes, even though the overall failure rates are identical in both trials.

TABLE 7.5

Two Trials with the Same Overall Proportions Failing in Each Group, but the Required Sample Size Adjustment for Group-Sequential Boundaries Go in Opposite Directions

		Trial 1		Trial 2	
		Annual Failure Rates		Annual Failure Rates	
		Control	Exp.	Control	Exp.
Assumptions	Months 0–24	0.1	0.025	0.1	0.075
	Months 25–36	0.1	0.08	0.1	0.056
	Months 37+	0.1	0.08	0.1	0.03
	Entire trial	0.409	0.229	0.409	0.229
Results		Sample sizes			
	Fixed	324		422	
	O'Brien–Fleming	282		447	
	Pocock	240		550	

In this example, however, the focus is on group-sequential sample size and power issues. In trial 2, the sample size increases as the design goes from fixed sample to O'Brien–Fleming to Pocock. In trial 1, the opposite occurs–the sample size decreases as the design goes from fixed sample to O'Brien–Fleming to Pocock. This unusual finding is actually easy to understand. In trial 1, the treatment effect is strongest (largest hazard ratio) during the first 24 months of the trial. The Pocock boundary, which is easy to exceed at the earliest interim analyses, is able to take advantage of the strong early treatment effect. The O'Brien–Fleming boundary (O'Brien and Fleming 1979), which is much more difficult to exceed at early looks, is less able to take advantage of the strong early treatment effect. And the fixed sample design can only reject at the end of the trial and thus cannot take advantage of period during which the treatment effect is strongest.

Some methods for calculating sample size for group-sequential trials calculate an inflation factor associated with a given boundary and then multiply the fixed sample design sample size by that inflation factor to obtain the group sequential sample size. The example shows that for a given boundary the *inflation factor* can vary dramatically, from the strong deflation in trial 1 to the strong inflation in trial 2, based on the nonproportionality of the hazard functions. And the examples presented in this section show that there are trials with early treatment effects that diminish over time, as well as trials with treatment effects that may not begin to emerge until the trial is well underway. Any trial with noncompliance, or dropouts whose events must be included in the final analysis (ODIS), or treatment lag will experience nonproportional hazards, and appropriate sample size methods, such as those in STOPP, are needed for accurate sample size evaluation.

7.2.4 Example 4: Adaptively Increased Sample Size Decreases Power-Treatment Lag

Impact of Nonproportional Hazards on Adaptive Designs I. Treatment Lag. Reference Trial: CARE, Sacks et al. (1996).

Suppose we wanted to design a trial to test a new cholesterol-lowering agent to be administered in addition to background statin therapy similar to the pravastatin arm of the CARE trial. The KM plots for the CARE trial that exhibit the threshold lag pattern (the general shape of Figure 7.6) could be used to inform the design of the new trial. As with many trials of treatments shown to lower cholesterol, there is a treatment effect lag before the treatment begins to impact cardiovascular events. In this trial, there is no apparent separation of the KM curves for about 2 years (see Sachs et al. 1996).

Suppose we want the trial to be completed in 5 years. The hazard ratio curve derived using STOPP is shown in Figure 7.9.

As is common in many cardiovascular trials, the *change in risk* is given as "−24%" on the original KM plots. Typically, this is based on the Cox proportional hazards model, and the published curves are for the 5-year period of the trial. The 24% reduction reflects an *averaging* of 2 years of no treatment effect with 3 years of a strong treatment effect. The implicit assumption of proportional hazards requires that the underlying hazard ratio curve be constant or nearly constant rather than the STOPP derived curve of Figure 7.10. If one performed an interim at 3 years using the Cox proportional hazards model with the same proportional hazards assumptions that led to the 24%

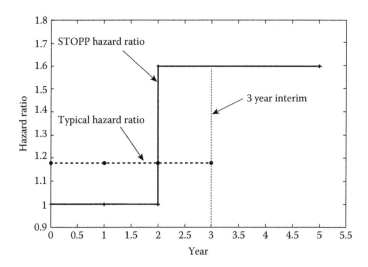

FIGURE 7.10
Two derived hazard ratio curves from the CARE trial, one assuming proportional hazards and the other using STOPP, assuming piecewise exponential hazards.

reduction for the full trial, the calculated hazard ratio would be about 1.18. The estimate at 3 years, assuming proportional hazards, would then represent an *averaging* of the same 2 years of no treatment effect with only 1 year of a strong treatment effect. This is displayed in Figure 7.10 as *typical hazard ratio*. Figure 7.10 also shows the time-dependent hazard ratio using STOPP: there are two distinct levels of the STOPP derived hazard ratio corresponding to the obvious two periods of the KM plots. While the untenable proportional hazards assumption of a constant hazard ratio would lead to an estimated treatment effect of at about 1.18 for the first 3 years, in actuality it is likely to be 1.0 for the first 2 years because of the treatment lag in cardiovascular events typical of cholesterol-lowering drugs. After 2 years, the survival curves separate with a hazard ratio of about 1.6. Thus, using the 1.18 hazard ratio is likely to substantially underestimate the treatment effect for the remainder of the trial.

For sample size re-estimation, a number of issues arise.

1. If the interim hazard ratio (about 1.18) based on the proportional hazards assumption is used to calculate conditional power or to recalculate pretrial estimates of power, the calculated number of events needed for the desired power is likely to be much greater than is actually needed. This could result in the sponsor abandoning the trial. Suppose, however, the decision is to increase the number of events and, in turn, the sample size in order to complete the trial within the originally planned 5 years. In order to increase the events without increasing the planned length of the trial, the number of patients must be increased. Such a strategy would likely dramatically reduce the power and possibly result in a failed trial. In turn, the sponsor might abandon the new drug. The reason is that all of the patients enrolled after the 3-year interim (calendar time) will remain in the no-treatment effect portion of the KM plots (randomization time) of Figure 7.9 for the remainder of the trial. Consequently, they will only serve to reduce the size and power of the final log-rank statistic.

2. The effect of the adjustment to account for the sample size re-estimation is less transparent. The data collected after the interim will be downweighted as a result of the adaptive sample size recalculation (assuming that the recalculation decision to increase the sample size is taken). In the original CARE trial, virtually all of the data that contributed to the positive (nonzero) estimate of the hazard ratio occurred after 2 years. So most data that led to the positive result in the original CARE trial will be downweighted in the adaptive trial. Only the portion of the data collected prior to the 3-year interim (calendar time) but after the 2-year lag (randomization time) will both escape the downweighting and contribute to the positive

treatment effect. This will be a small portion of the patients random-ized before the interim—this happens in part because the recruitment process limits the exposure time (randomization time) at the time of the interim (calendar time).

One of the difficulties of sample size re-estimation when there is nonproportional hazards of the form seen in CARE is that at the time of the 3-year interim only a small proportion of the enrolled patients will have entered the period of positive treatment effect (the period between 2 and 3 years [randomization time]). This phenomenon can be seen in virtually any KM plot by comparing the proportion at risk at the left side of the plot (4159 = 2078 + 2081) (in the original KM plots) to the proportion at the right (1754 = 854 + 900). So the vast majority of pre-interim data will represent the period of *no treatment effect*. At the time of the interim, the data representing the period when the treatment effect begins to emerge will be very sparsely represented, making estimation of the treatment effect treacherous. Yet trials designed for adaptive sample size re-estimation rely heavily on this compro-mised estimate. The group-sequential approach to sample size re-estimation presented in Section 7.3 eliminates this problem.

7.2.5 Example 5: Adaptively Increased Sample Size Decreases Power-Anti-Lag

Impact of Nonproportional Hazards on Adaptive Designs II. Reference Trial: BHAT (The Beta-Blocker Heart Attack Trial Research Group 1982).

The derived hazard ratio curve for BHAT is presented in Figure 7.8. The derived time-dependent hazard ratio curve for CARE is presented in Figure 7.10. In Example 4, we noted that the CARE publication presented the 5-year treatment effect as −24%, which appears to have been derived from a Cox proportional hazards model and that a similar calculation at the 3-year interim would result in an estimated hazard ratio of about 1.18. One can sim-ilarly derive an estimated hazard ratio of about 1.33 for BHAT based on the survival curves estimated also at a 3-year interim. Both of these have been plotted in Figure 7.11. On the basis of this plot, BHAT would appear far more promising than CARE at the interim, all other factors being the same. If the original design numbers of events for both of these trials were inade-quate, preference would likely be given to increasing the number of events for BHAT. The observed time-dependent hazard curves derived using STOPP (Figure 7.12) tell a starkly different story. Increasing the number of events in the trial with similar hazard ratio to BHAT is likely to result in failure since the additional data would represent mainly the period during which the treat-ment effect had largely vanished (months 18+). Here, a better strategy is to start the trial with a larger sample size since those additional patients would contribute to the early postrandomization period during which the treatment effect was most pronounced. However, increasing the sample size is also

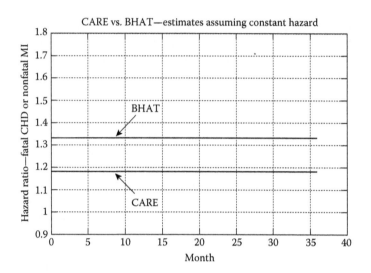

FIGURE 7.11
Hazard ratio curves from CARE and BHAT assuming constant hazards.

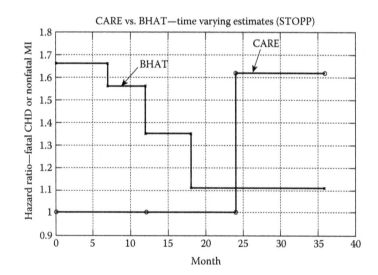

FIGURE 7.12
Time-dependent hazard ratio curves from BHAT and CARE derived from Kaplan–Meier plots.

likely to increase the length of follow-up, and the earliest patients would then also reduce the power as they are followed longer. The balance between these two competing forces must be carefully evaluated. In this situation, it would have been much better to begin the trial with a far larger sample size, thus avoiding this problem.

Note that even though the CARE and BHAT nonproportionality of the hazards work in opposite directions (CARE has a treatment lag, BHAT a diminishing treatment effect), in both cases, starting the trials with large sample sizes is a much better approach than increasing the sample sizes at an interim analysis.

But consider the time-dependent hazard ratios derived using STOPP (Figure 7.12). A quite different picture emerges from these curves: one should be more optimistic about the success of CARE than BHAT. And the strategies for optimal design are quite different based on the unrealistic Figure 7.11 as compared to the time-dependent hazard ratios based on actual data (Figure 7.12). As in the discussion of CARE with plans to conclude the trial in about 5 years of study start, sample size re-estimation at the 3-year interim is contraindicated. A much more reasonable strategy would be to start the trial with a generous sample size and use a group-sequential design to potentially stop early for favorable efficacy (see Example 6).

For BHAT-like trials, extending the trial to accommodate a larger sample size is also problematic (cf. Example 4). Here, one should try to enroll as many patients as possible in a short period of time so that the balance of patients in the early period of strong efficacy is much greater than in the late period of very weak efficacy. The trial should be as compact as possible. One could also consider a re-randomization design to test whether continued treatment after 18 months is warranted.

7.3 Group-Sequential Design for Sample Size Re-Estimation

In this section, a method for sample size re-estimation called *designing for a range of alternatives* is presented. The presentation is through an example. Some reasons why one might prefer this approach to some of the usual approaches are discussed. Two of the main reasons are (1) it avoids projections based on interim estimates of the treatment effect which invariably are quite unreliable, and (2) it avoids problems relating to nonproportional hazards discussed in the previous section in which adding patients actually decreases the power.

7.3.1 Example 6

Designing for a range of alternatives: a group-sequential approach to sample size re-estimation.

This example describes methodology used for an actual trial, although parameter assumptions have been changed.

Suppose that an investigational new drug (call it X11) was developed to address the fact that patients on statin therapy still have high risk of MACE. The mechanism of action is to further modify the patient's lipid profile.

TABLE 7.6

Declining Hazards Derived
from PROVE-IT Plots

Failure Rates
26% months 1–6
13% months 7–12
8% months 13-end

The new trial will randomize patients, all of whom will be taking statin therapy, to additionally take X11 or placebo. The PROVE-IT trial, discussed in Example 1, provides a good reference for background failure rates of patients taking statin therapy. The time-dependent rates derived using STOPP are given in Table 7.6.

Although lipid modification therapy is generally accompanied by a treatment lag, for simplicity here, a constant treatment effect will be assumed. In the actual trial, enrollment rates were developed that included a gradual increase from study start, as well as slower periods during the summer and holidays. For this example, it was simply assumed that the maximum rate of enrollment would be reached 10 months after study start, with a linear increase to that rate (10% of maximum for month 1, 20% for month 2,. . .,90% for month 9, and 100% for the remainder of the enrollment period). The enrollment pattern is described in relative rather than absolute numbers since the sample size has not yet been determined, and the recruitment pattern will be used in that determination.

A single trial with two-sided significance level of 0.01 was agreed to. Also expected were 5% per year loss-to-follow-up and 5% per year noncompliance rates. The lengths of the enrollment period and full trial are assumed to be 18 and 36 months, respectively.

The smallest clinically meaningful difference (SCMD) is assumed to be 20%, while the more optimistic anticipated treatment effect (ATE) is assumed to be 35%.

The group-sequential boundary was a custom boundary initially developed to allow termination at 18 months at the two-sided 0.001 significance level and at 36 months at the two-sided 0.01 level. (A continuous spending function was later fitted to meet these criteria.) The 18-month termination was not considered by the sponsor as early termination, but rather the original target end of trial. If convincing evidence of benefit was not achieved at 18 months, then the trial would automatically expand up to 36 months. The 24-month interim was added later.

Some operating characteristics of the trial based on these assumptions are displayed in Tables 7.7 and 7.8, for target sample sizes 3300 and 3475, respectively.

TABLE 7.7

Operating Characteristic for Trial with $N = 3300$

Look Time (Month)	Timing			Cumulative Alpha Spent (One-Sided)	Cumulative Power Treatment Effect			
	Information Fraction	Event for 25% Treatment Effect	Boundary		20% (SCMD)	25%	30%	35% (ATE)
18	0.5	378	3.29	0.0005	20	41	69	88
24	0.75	567	2.77	0.003	59	83	97	100
36	1	756	2.71	0.005	77	94	100	100

TABLE 7.8

Operating Characteristic for Trial with $N = 3475$

Look Time (Month)	Timing			Cumulative Alpha Spent (One-Sided)	Cumulative Power Treatment Effect			
	Information Fraction	Events for 25% Treatment Effect	Boundary		20% (SCMD)	25%	30%	35% (ATE)
18	0.5	398	3.29	0.0005	22	46	73	90.6
24	0.75	597	2.77	0.003	62	85	98	100
36	1	796	2.71	0.005	80	95	100	100

These tables document a rather different approach to sample size estimation and re-estimation than is typical. Here, the sponsor is given two options to choose from (Table 7.7 or 7.8), although presumably these options would be developed with the sponsor, and there is no need to limit the number of options to two. If a maximum sample size of 3300 is chosen (Table 7.7) and if the optimistic 35% treatment effect is true, then there is a good probability (88%) that the trial will demonstrate significance at the 18-month interim. The significance, here at the two-sided 0.001 level, reflects regulatory concerns and is considered strong evidence of benefit. These tables were constructed, again, to address regulatory concerns, to exclude early termination that would not provide sufficient data for that understanding: follow-up must be sufficiently long to provide an understanding of both the benefit and the safety profiles (cf. Pocock 2006). If, as is more usually the case, the most optimistic treatment effect is not borne out, the trial can continue as long as 36 months, with 77% power to detect the 20% SCMD. If the actual treatment effect is modestly larger, power increases dramatically, up to 94% for a 25% treatment effect, and nearly 100% for treatment effects in excess of 30% at 36 months. If an intermediate treatment effect such as 25% is true, the chances of the trial terminating at 24 months (85%) are good, while the chances of statistical significance at 36 months are very likely.

The 77% and 88% powers are key points of the design presented in Table 7.7 because the former provides power at the SCMD should the trial go to completion, while the latter provides power for terminating at the earliest possible time, should the most optimistic treatment effect assumption hold. If one desires a more robust design, in the sense that the 77% and 88% powers are replaced by the typical 80% and 90% powers, the maximum sample size of 3475 can be used; see Table 7.8.

Using a group-sequential design in this way provides the same or better protection against a too small sample size as what is usually referred to as adaptive sample size re-estimation. But it requires no projections based on highly variable interim estimates of treatment effect. As was noted earlier, sample size increases using typical methods can have unintended highly deleterious effects in the presence of some nonproportional hazard alternatives, which are often not recognized. This does not appear to be a problem with this group-sequential approach. And if an interim estimate of the treatment effect is not the same as the true underlying effect (it never is), this group-sequential approach does not suffer from what is a major problem with re-estimating the sample size based on an interim estimate of treatment effect. In particular, a bad interim estimate can lead to the wrong target sample size to adapt to. But with this group-sequential approach, if the interim estimate is high, then the trial is likely to terminate for benefit. We will never know if it is too high, but we will know that the chance of a type I error is no more than 0.001 two-sided at the first interim and 0.01 two-sided at the final, and somewhere in between at the second interim. This is certainly sufficient evidence for benefit. Further, the group-sequential approach does not lose efficiency as a result of using a weighted method for sample size re-estimation. As discussed earlier, in some nonproportional hazards situations, the portion of the curve that is downweighted may be the portion that contributes most to reaching statistical significance.

What the group-sequential design is not able to do is allow unlimited expansion of the sample size. Although theoretically the usual adaptive sample size re-estimation does allow unlimited expansion, real-world considerations render this impossible. There is never a blank check, and the same considerations for the group-sequential approach that led to the tension between the SCMD (adequate power) and the ATE (cost) are at play in adaptive sample size re-estimation. One of the really nice features of the group-sequential approach is that if the DSMB does not believe that there is sufficient evidence of benefit for early termination, the sample size is automatically increased without ever involving the sponsor. In contrast, adaptively increasing the sample size at an interim requires that the DSMB provide its recommendation to the sponsor. If the sponsor agrees, then the investigators are notified. Often, this is interpreted as evidence that the treatment effect may be substantially less than expected. In turn, investigators may be reluctant to enroll some patients, and stock analysts may downgrade the stock.

Tables 7.7 and 7.8 present sample size/power, not as a single number, but as 12 cross-tabulated numbers each. The latter is far more informative. Analogously, when presenting the results of a survival trial, reporting the proportions surviving in each treatment arm is simply not adequate. Rather, the standard is to present the far more informative Kaplan–Meier curves.

7.3.2 What Constitutes an Optimal Design?

When Pocock introduced the group-sequential concept (Pocock 1977), he stated that "the main objective of a group sequential procedure is to reduce the number of patients on an inferior treatment by early termination of the trial." Consistent with this objective, Pocock, in that 1977 paper, suggested minimizing the average sample number (ASN) as a means of selecting the best boundary. More recent thinking, as illustrated in recent papers by Pocock (2005, 2006), contrasts markedly from that original objective. In turn, one should reconsider whether minimization of the ASN should still be considered an appropriate approach. Although Pocock's recent papers appeared after the trial discussed at the beginning of this section was designed, many of Pocock's recommendations are consistent with the basis underlying the *designing for a range of alternatives* approach.

The recommendations in Pocock's 2006 paper can be useful for choosing one boundary over another. The focus will be on some underlying themes of that paper:

- "What in practice constitutes a sensible boundary?"
- "How can we deter some DMC's from stopping trials too soon for benefit?"
- To terminate a trial early for benefit "one needs proof beyond a reasonable doubt that a treatment benefit is sufficiently marked to alter future clinical practice."

Emanating from these themes are some Pocock's recommendations:

- "The [Pocock boundary] is clearly unsuitable for benefit since its stopping P-values around 0.02 or greater are liable to provide insufficient strength of evidence."
- "The O'Brien and Fleming boundary was introduced to overcome this problem and indeed in early interim analyses it does . . ." Pocock continues by stating that the O'Brien–Fleming boundary is "too lenient at its last interim look. . ." In the case of a five-look O'Brien–Fleming boundary, to terminate early for benefit at the last interim "one only needs to achieve $P < 0.023$." Pocock concludes "In reality, this would be a crazy thing to do since one is denying the opportunity to achieve a more convincing result with the final data, just when the trial is entering the home stretch."

First, to clarify the issues, a visual means of comparing group-sequential boundaries will be presented. Group-sequential boundaries deal with a multiple comparisons problem. Under the null, looking at the data multiple times increases the chances of a type I error or false positive. One of the most familiar ways of addressing multiple comparisons is the Bonferroni method, in which portions of the alpha are allocated to each of several tests. Lan and DeMets (1983) introduced the alpha-spending function. An alpha-spending function uniquely defines a group-sequential boundary by dictating how much alpha will be spent, or allocated, at each interim analysis. Examining the amount of alpha allocated to each of the target look times is a revealing way to understand the characteristics of a boundary. Figure 7.13 displays how the original O'Brien–Fleming (left bar of each pair) and Pocock boundaries, each with five equally spaced looks, allocate the alpha. Based on how they allocate alpha, the two boundaries are starkly different, with the Pocock boundary allocating the largest portion of alpha to the earliest look, with decreasing amounts to each successive look. The O'Brien–Fleming does just the opposite, allocating almost no alpha to the first look and increasing amounts to each successive look. This heavy bias of the Pocock boundary toward stopping at the earliest looks and the opposite for the O'Brien–Fleming make it clear why the ASN of the Pocock is so much smaller than that of the O'Brien–Fleming.

Consider the amount of alpha allocated to the first look (information fraction 0.2). It is dramatically different for these two boundaries. Pocock (2006)

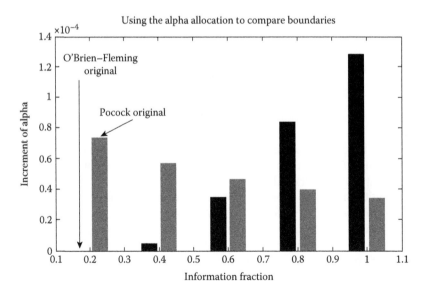

FIGURE 7.13
Comparing the original O'Brien–Fleming and Pocock boundaries.

stated that the Pocock boundary (1977) is far too lenient, permitting early termination when the strength of evidence is insufficient. Pocock further states that, in contrast, the O'Brien–Fleming boundary overcomes this problem—"in early interim analyses it does provide an appropriately stringent boundary." It does this by allocating very little alpha to the first look (cf. Figure 7.13). Suppose in an actual trial the Pocock boundary was chosen as the monitoring boundary and that a DMC observed a boundary crossing at the first look. Suppose in addition, understanding that the strength of evidence was insufficient, the DMC ignored the boundary crossing and let the trial continue. All of the alpha allocated to the first look would be lost. And the Pocock boundary allocates a large portion to that first look. The fact that there seems to be almost universal agreement that the Pocock boundary at the first interim analysis is *far too lenient*, leads to the conclusion that, based on the first interim analysis alone, the Pocock boundary is suboptimal.

One could improve the Pocock boundary by taking all of the excess alpha, say beyond what O'Brien–Fleming allocates to the first look, and redistribute it to the right. There are many ways to redistribute that alpha, but for simplicity we redistribute all of it to the second look. Call this Pocock Variant 1, or Pv1 (see Figure 7.14).

How does this modification affect the ASN? For simplicity, we use Pocock's original definition of the ASN. For comparability, the same alternative will be used for the remainder of this example. Assume a placebo failure rate of 20% and a 15% treatment effect. The sample size for a fixed sample

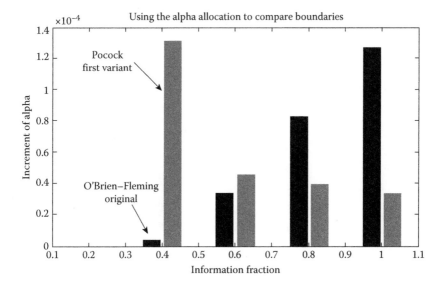

FIGURE 7.14
Comparing the original O'Brien–Fleming boundary with the first variant of Pocock boundary.

design with these parameters is 4866. The ASNs for the original Pocock PvO (Pocock version original) is 3333, and for the original O'Brien–Fleming (OBFvO) is 3692. The ASN for the improved version of Pocock's boundary Pv1 is 3445. It should not be surprising that the ASN for Pv1 increases, moving away from PvO and toward OBFvO. As the alpha is redistributed to the right, the probability of terminating at look 1 is diminished, while the probability of terminating at look 2 is enhanced. On the basis of minimizing the ASN, PvO is superior to Pv1, while on the basis of Pocock's 2006 recommendation, the opposite is true. From observing how PvO and Pv1 allocate alpha, it is clear why Pv1 is preferred to PvO, while in contrast, why PvO appears better than Pv1 with respect to minimizing the ASN. Minimizing the ASN for the boundaries PvO and Pv1 rewards going in the wrong direction.

Although we could continue with similar arguments regarding this new boundary, Pv1, and how it allows early termination at the second look with *insufficient strength of evidence*, we move on to Pocock's point (2) above which focuses on weaknesses in the O'Brien–Fleming Boundary. In particular, Pocock (2006) states that "one only needs to achieve $P < 0.023$ at the fourth analysis to stop the trial for benefit. In reality, this would be a crazy thing to do since one is denying the opportunity to achieve a more convincing result with the final data, just when the trial is entering the home stretch." Figure 7.15 compares the original five-look O'Brien–Fleming, OBFvO, with a variant that allocates very little alpha to the fourth look.

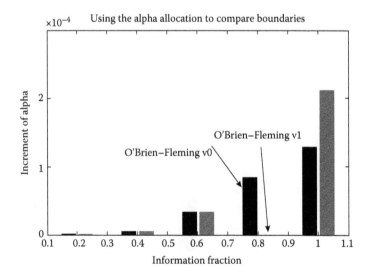

FIGURE 7.15
Comparing original O'Brien–Fleming boundary with the first variant of O'Brien–Fleming boundary.

The original five-look O'Brien–Fleming, OBFvO, allocates a considerable amount of alpha to the fourth look. Pocock states rather emphatically that it would be misguided to terminate early at the fourth look. In the modified version, OBFv1, very little alpha is allocated to the fourth look. If we were to redistribute that alpha to the left, we would risk making the boundary too lenient at some earlier look, facilitating early termination when there is likely to be insufficient strength of evidence. Redistribution to the left would result in an inferior boundary. Redistribution to the right results in OBFv1, a boundary that virtually eliminates the chances of making what Pocock calls the *crazy* decision unless there is overwhelming evidence of benefit at the fourth look. If the evidence at the fourth look is overwhelming, the advantage of waiting until the end to provide more convincing evidence is of limited value.

The modified O'Brien–Fleming boundary, OBFv1, is superior, as the single modification addresses Pocock's concern, while at the same time increases the chances of success should the trial not terminate early, by allocating the additional alpha to the final analysis. But, as is easy to see, the redistribution of alpha to the right increases the ASN, in this case from 3692 to 3923. In other words, minimizing the ASN would choose the boundary that facilitates the *crazy* option over Pocock's preferred option that makes it much more difficult to terminate at the fourth interim, while providing more power for the final analysis.

As an additional note, Pocock (2006) emphasizes, as do many other experts in the area, that group-sequential boundaries should be viewed as guidelines rather than rules. DMCs should feel free to overrule a boundary crossing, if "proof of benefit beyond a reasonable doubt" has not been established, or if any other concerns, such as a questionable risk-to-benefit ratio, exist. It is important to realize that viewing boundaries as *guidelines* is one-directional. No one is recommending that a failure to cross an efficacy boundary can be over-ruled and the trial terminated early for efficacy. The one-directional nature always results in terminating later than an initial crossing. If one could incorporate the probability of treating a boundary as a guideline by ignoring potential boundary crossings, that action can only result in an increased ASN. The *guidelines* interpretation of boundary crossings is also in conflict with minimizing the ASN.

That statistical boundaries should be viewed as guidelines infuses the decision process with a degree of flexibility that is likely to be very difficult to capture in a formula such as the ASN. That is an important reason for having DMCs. But more importantly, as shown in this section, minimizing the ASN is not consistent with current thinking regarding when to stop a trial for benefit.

Variants of Pocock's original ASN have been proposed, but the intent of those variations is to minimize the number of patients. As was shown earlier, considerations such as Pocock (2006) presented argue for terminating later in preference to earlier.

There have been some suggestions that a balance should be sought between the competing goals of obtaining evidence *beyond a reasonable doubt* and *minimizing patient exposure*. Balance entails a compromise of each position. Compromising *beyond a reasonable doubt* means early termination when there still remains a reasonable doubt of benefit. Many subsequent decisions have to be made. The sponsor will have to decide whether there are adequate data to submit to regulatory authorities, or whether another trial is needed, or whether the drug should be abandoned. In turn, regulatory authorities will have to decide whether to approve the therapy when there still exists reasonable doubt as to its efficacy. The practicing physician will then have to decide whether to prescribe, perhaps for decades, a treatment for which there is reasonable doubt that it provides any benefit to the patient. And, if aware, the patient will have to decide whether to use a therapy for which doubt exists regarding its efficacy. Perhaps the worst situation is for a dying patient to learn after months, years, or decades of taking some drug, that that drug actually provided no benefit, and his life might have been spared had he taken a different drug that actually provided the sought after benefit. Montori et al. (2005) performed a systematic review of trials stopped early for benefit and concluded that "RCTs stopped early for benefit ... show implausibly large treatment effects, particularly when the number of events is small. These findings suggest clinicians should view the results of such trials with skepticism." Trials should not be terminated before proof of benefit *beyond a reasonable doubt* has been achieved. Once that has been achieved, and all of the other elements critical to a decision to terminate early have been considered, the decision should be made at the earliest possible time.

In Example 6, our original design had only one interim and one final analysis. This appears to address the problem Pocock found with the fourth interim of the five look O'Brien–Fleming boundary. But our original rationale was different from Pocock's. In designing the trial of Example 6, alpha was viewed as a scarce resource. Whether significance will be obtained at the end of the trial is never known in advance. This led us to initially design the trial with a single interim analysis, conserving any alpha that would have been allocated to an intermediate interim and instead allocating it to the final analysis. At the pretrial DMC meeting, the DMC raised the concern that the expected time period between the single interim and final analyses could be too long and not provide sufficient protection for the enrolled patients should strong efficacy emerge. Based on this concern, an additional interim was added. Whether or not to include such an interim depends on the specifics of the trial, including the length of time for which no interim is planned.

A review of Pocock's (2006) criteria for when to stop a trial for benefit, reflecting 30 years of experience in group-sequential methods since he introduced the concept, provides a much different view as to what makes a good boundary compared with his proposal 30 years earlier of minimizing the ASN. What appeared at the outset (1977) to be an obvious objective of minimizing the number of patients exposed to an inferior treatment now seems at odds with current thinking. If the ASN is at odds with those current recommendations, what formula should replace it? The fact that boundaries should be viewed as guidelines, as only one element of the much larger set of considerations on which the termination decision is made, provides some insight. In other words, the guidelines perspective suggests that there is at work a far more complex set of issues for evaluating and comparing boundaries than is probably possible in a formula. It would be far better to examine how competing boundaries can achieve the objectives as set forth by Pocock rather than try to find a formula that can replace that thoughtful evaluation.

7.4 Understanding the Markov Model

A comprehensive discussion of the Markov model for designing clinical trials (Lakatos 1986, 1988) is beyond the scope of this section. Rather, a few basic concepts will be explained to provide a glimpse of its power and flexibility: (1) the relationship between standard survival analyses and the Markov model, (2) the ability of the Markov model approach to handle complex survival trials with many simultaneous survival-type processes, (3) its ability to handle simultaneous (and time-dependent) processes, and (4) advantages of its ability to model the entire survival experience rather than a one- or two-parameter summary.

7.4.1 Relationship between Standard Survival Analyses and the Markov Model

If the failure rate for some population is 75% per year, the 2-year failure rate cannot be 150%. At the end of the first year, only 25% will have survived, and only that 25% will be at risk of failure at the start of the second year. So the proportion failing during the second year is $0.25 \times 0.75 = 0.1875$. Adding this to the proportion failing the first year gives the 2-year failure rate: $0.1875 + 0.75 = 0.9375$. The same result can be obtained using only survival probabilities: $0.25 \times 0.25 = 0.0625$ is the 2-year survival probability. Subtracting this from 1.0 gives the 2-year failure probability just calculated.

This is a standard approach to calculating the probability of surviving, or equivalently failing, in some interval of time for which we have estimates of the probabilities in each of possibly many subintervals that comprise the entire interval. This is a *product rule*. In Figure 7.16, there are two 1-year subintervals comprising the 2 years. Going from left to right, at time 0 all patients are still at risk. The probability of surviving to the end of year 1 is the probability of being at risk at time 0 times the probability of surviving to the end of year 1 conditional on being at risk at the beginning of year 1. This is $1 \times 0.25 = 0.25$. And the probability of surviving to the end of year 2

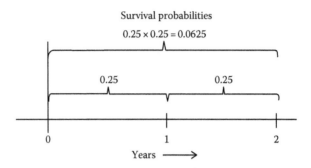

FIGURE 7.16
Calculating the cumulative survival probabilities using conditional survival probabilities for each of the subintervals.

is the probability of surviving to the end of the year 1 times the probability of surviving the next year (year 2) conditional on having survived to the end of the year 1. This is $0.25 \times 0.25 = 0.0625$. Both life tables and the Kaplan–Meier product limit estimator use the same basic approach. Suppose the probability of surviving 50 years is desired. For life table analyses, we commonly divide a period of interest, here 50 years, into periods of equal lengths, say 1 year. Individual estimates of the conditional probabilities of survival for each of the 50 years, conditional on having been at risk at the beginning of that specific year, are obtained. Then, as given earlier, we successively obtain cumulative estimates of surviving to the end of each of the 50 years, given that the person was alive (at risk) at the beginning of that year. The Kaplan–Meier estimate is calculated similarly, but rather than using intervals of equal length, intervals demarcated by the unique event times of the data set are used; estimates of conditional probabilities of survival for each such interval are also obtained from the data.

Returning to the example depicted in Figure 7.16 and assuming the failure rate is constant over subintervals of equal length (the conditional probabilities), then the probability of surviving in each half year is $\sqrt{0.25} = 0.5$, or for a month is $0.25^{1/12} = 0.890899$. This is easily verified using the standard formula $S(t) = e^{-\lambda t}$ relating the cumulative survival probability to the hazard. The use of the notation λ rather than $\lambda(t)$ implicitly assumes that the hazard is constant. Although this chapter focuses on time-dependent hazards, the simpler case of constant hazards will be discussed first.

The right two columns of Table 7.9 display the results of successively applying the conditional monthly survival probability 0.890899, row by row. Initially, at the time of randomization, all patients are at risk; none have failed, so the probability of being at risk is 1. The conditional monthly survival probability, as calculated earlier, is 0.890899, which when multiplied by the probability of being at risk gives the cumulative probability 0.890899 of

TABLE 7.9

Cumulative Failure and At-Risk Probabilities for Control and Experimental Arms of Trial

	Experimental Arm		Control Arm	
Month	Failed	At Risk	Failed	At Risk
Rand	0	1	0	1
1	0.081084	0.918916	0.109101	0.890899
2	0.155594	0.844406	0.206299	0.793701
3	0.224062	0.775938	0.292893	0.707107
4	0.286978	0.713022	0.370039	0.629961
5	0.344793	0.655207	0.438769	0.561231
6	0.39792	0.60208	0.5	0.5
7	0.446739	0.553261	0.554551	0.445449
8	0.4916	0.5084	0.60315	0.39685
9	0.532823	0.467177	0.646447	0.353553
10	0.570704	0.429296	0.68502	0.31498
11	0.605513	0.394487	0.719384	0.280616
12	0.6375	0.3625	0.75	0.25
13	0.666893	0.333107	0.777275	0.222725
14	0.693903	0.306097	0.801575	0.198425
15	0.718722	0.281278	0.823223	0.176777
16	0.74153	0.25847	0.84251	0.15749
17	0.762488	0.237512	0.859692	0.140308
18	0.781746	0.218254	0.875	0.125
19	0.799443	0.200557	0.888638	0.111362
20	0.815705	0.184295	0.900787	0.099213
21	0.830649	0.169351	0.911612	0.088388
22	0.84438	0.15562	0.921255	0.078745
23	0.856999	0.143001	0.929846	0.070154
24	0.868594	0.131406	0.9375	0.0625

being at risk at the end of the first month. Recorded in the last column of the month 1 row is the cumulative probability 0.890899 of being at risk, and in the middle row $1 - 0.890899 = 0.109101$ of having failed. Now, starting from the end of month 1, the probability of being at risk is 0.890899 (this is also the cumulative survival probability recorded in the month 1 row), and multiplying by the probability of surviving to the end of month 2, conditional on surviving to the end of month 1 (this conditional probability is assumed here to be the same 0.8909899 for all months), the unconditional (or cumulative) probability of being at risk at the end month 2 is 0.793791. The failure probability, just to its left in Table 7.9, is $1 - 0.793791 = 0.206299$. In a similar fashion, we continue to multiply the previous result (the probability of still being at risk) by the conditional probability of surviving in the next month, obtaining the cumulative probability of being at risk the end of that month. At the end of 1 year, the cumulative probability of still being at risk from the rightmost column is 0.25, consistent with our original assumption of a 75% failure rate per year. And at 24 months, the last entry in the same column is 0.0625, consistent with our calculation performed in the very first paragraph of this section ($0.25 \times 0.25 = 0.0625$). In this case, the at-risk column gives the cumulative probability of survival based on the assumed monthly conditional survival probabilities.

The last two columns of Table 7.9 were calculated for the control group. Assuming a 15% yearly treatment effect, the 1-year failure rate for the experimental arm is $(1 - 0.15) \times 0.75 = 0.6375$. The two columns under the header *experimental arm* of Table 7.9 were calculated assuming this failure rate just as was done for the two columns under the header *control arm*. In this very simple setting, the Markov model, discussed later, will reproduce Table 7.9 exactly.

To calculate sample size for a 2-year trial with simultaneous patient entry (simultaneous entry is assumed here for simplicity; for staggered entry, see Lakatos (1986, 2002)), we could use a sample size formula for a comparison of proportions or use other formulas intended for the log-rank statistic (e.g., Rubinstein et al. 1981, Schoenfeld 1981).

Both the Rubinstein and Schoenfeld formulas use a single parameter, the hazard, to characterize the survival for each arm. The log-rank statistic compares the entire survival curve, and similarly, the Markov approach to sample size for the log-rank (Lakatos 1988) uses all of the intermediate information in Table 7.9 in a way that is very similar to the way the log-rank statistic is calculated. Why this is important will be discussed later.

7.4.2 Markov Model to Generate Table 7.9

The Markov model approach of Lakatos (1986) is applied separately to each treatment arm, just as was done earlier in calculating the survival and failure probabilities in Table 7.9.

7.4.3 Constructing Table 7.9 Using a Two-State Markov Model

Similar to the usual survival analyses, such as Kaplan–Meier, logrank, Cox model, etc., the Markov model is a time from randomization model. Consider again the last two columns of Table 7.9, the failed and at-risk columns of the control arm. The two columns were calculated using straightforward intuitive calculations. Beginning at the time of randomization, all patients are initially at risk. A constant failure (respectively, survival) rate of 0.75 (0.25) per year was assumed, giving rise to a constant conditional monthly survival rate of 0.890899. Each month we multiplied the probability of still being at risk by the conditional probability of surviving for that month to find the cumulative probability of being at risk at the beginning of the next month. The *Failed* column gives the cumulative failure probability, and is $1 -$ the survival probability. Both the failure and at risk are cumulative from time of randomization. The conditional probabilities enter the Markov model through the transition matrices, which will be discussed next.

A Markov model with two states (failed state, at-risk state) to generate Table 7.9 is now presented. Only the control group is discussed; modeling for the experimental arm is similar. Let t denote time from randomization in months, so the sequence $\{t_0, t_1, \ldots, t_{24}\} = \{0, 1, \ldots, 24\}$ are the months in the left column of Table 7.9. The t_i subdivide the period of the trial into equal subintervals $[t_{i-1}, t_i], i = 1, \ldots, 24$. First note that for any given row of the table, the sum of the last two columns is 1. At any given time, or row, either a patient has failed or is still at risk. So these two categories comprise the entire control group at that time. For example, for the row labeled month 3, the distribution of the patients into the two categories *failed* and *at-risk* is $[0.292893, 0.707107]$. This is referred to as the distribution at t_3. At the top of those two columns is the distribution at randomization $[0, 1]$. For the Markov model, these categories are referred to as *states*. In this simple model, at any time during the trial, a patient must be in one and only one of the two states (mutually exclusive and exhaustive): the failed state or the at-risk state. The distribution at any time t is denoted by

$$D_t = \left[p_f(t), p_a(t)\right]' \tag{7.1}$$

where M' denotes transpose of the matrix M, $p_f(t)$ and $p_a(t)$ are the cumulative probabilities of having failed and being at risk, respectively, at time t.

As time progresses, patients will transition from the at-risk state to the failed state, and they do so with conditional probability p_f^i, where p_f^i is the probability of failing during the interval $[t_{i-1}, t_i]$. Form the transition matrices

$$T_{i-1,i} = \begin{array}{c} \\ F \\ A \end{array}\begin{bmatrix} F & A \\ 1 & p_f^i \\ 0 & 1 - p_f^i \end{bmatrix} \tag{7.2}$$

Then the distribution at time t_i is given by

$$D_{t_i} = T_{i-1,i} D_{t_{i-1}} \tag{7.3}$$

In the example giving rise to Table 7.9, the conditional failure rate was constant over time. Thus, p_f^i is also constant over time and for the example,

$T_{i-1,i} \equiv T_i$ is constant as well. Starting with the initial distribution $D_{t_0=0} = [0,1]'$, and successively applying (7.3),

$$D_1 = T_{0,1} * D_0$$
$$D_2 = T_{1,2} * D_1$$
$$\vdots = \vdots$$
$$D_{24} = T_{23,24} * D_{23} \tag{7.4}$$

This sequence of D_i's so obtained are the successive rows of the last two columns in Table 7.9. In particular, for $i = 3$,

$$D_{t_i} = T_{i-1,i} \times D_{t_{i-1}}$$
$$D_3 = T_{2,3} \times D_2$$
$$D_3 = \begin{bmatrix} 1 & 1 - 0.890899 \\ 0 & 0.890899 \end{bmatrix} \times \begin{bmatrix} 0.206299 \\ 0.793701 \end{bmatrix} = \begin{bmatrix} 0.292893 \\ 0.707101 \end{bmatrix} \tag{7.5}$$

Each entry in $T_{t_{i-1,i}}$ defined in (7.2) has a specific interpretation.

1. $T_{t_{i-1,i}}(1,1)$ is the conditional probability of transitioning from the failed state to the failed state, which is 1. Once a patient has failed, that patient remains in the failed state for the remainder of the trial.
2. $T_{t_{i-1,i}}(1,2)$ is the conditional probability of transitioning from the at-risk state to the failed state, which is p_f^i.
3. $T_{t_{i-1,i}}(2,1)$ is the conditional probability of transitioning from the failed state to the at-risk state, which is 0.
4. $T_{t_{i-1,i}}(2,2)$ is the conditional probability of transitioning from the at-risk state to the at-risk state (or simply remaining in the at-risk state) which is $1 - p_f^i$.

So standard matrix multiplication, successively applied as in Equations 7.4 and 7.5, can be used to generate Table 7.9.

It was noted earlier how the final row of Table 7.9, which gave the 2-year probabilities of failure in each of two treatment arms, could then be used in standard sample size formulas. How to expand the Markov model (7.1 through 7.3) in two ways is now discussed: simultaneous processes and time-dependent failure rates.

7.4.3.1 Simultaneous Processes

7.4.3.1.1 Censoring Process

In most survival trials, censoring plays an important role. There are numerous processes that can be considered censoring processes. Loss to follow-up is a censoring process, as is dying from a disease not included in the primary endpoint. Suppose the primary endpoint for the clinical trial that gave rise to Table 7.9 is MACE. Suppose also that patients are at high risk of dying from cancer, with a cancer failure rate of 75% per year. These rates are unrealistically high but are used intentionally for purposes of illustration. If a person dies of cancer before experiencing the primary cardiovascular endpoint, then the time of the cardiovascular endpoint will be censored. It is important to understand that both the cardiovascular and cancer processes here are co-equal processes from the point of view of natural processes. If a person has a MACE event prior to dying of cancer, then the patient will be removed from the trial, and the time of the cancer death will be censored. Just as cancer censors MACE, MACE censors cancer. Both processes have equal footing in nature as well as in the Markov model. Most sample size calculations treat them quite differently. Typically, the sample size is first calculated without regard to the censoring due to cancer. Then, the 2-year probability of loss due to censoring (0.9375) is calculated and the sample size is inflated to compensate for this loss: $N_{adj} = N/(1 - 0.9375)$. The Markov model is designed specifically to deal with multiple simultaneous processes. Consider a three-state model, where, in addition to the at-risk and failed states, a censoring (or loss) state is included. Without formally writing down the details of this model, Table 7.10 presents the successive distributions resulting from this three-state Markov model of the control group. Since the yearly failure rates for the two processes are both 0.75, the monthly conditional probabilities for censoring due to cancer are the same as those for MACE: 0.0890899. The cumulative probability of a MACE event at the end of 1 month is 0.1060, the same as in Table 7.9. In addition, the cumulative probability of being censored by the end of 1 month is also 0.1060. However, the probability of still being at risk at the end of 1 month is not the same as in Table 7.9 because both the censoring and MACE processes remove patients from the at-risk process. In the month 2 row, the cumulative probability of having a MACE event is no longer the same as the corresponding cell of Table 7.9. This is due to the fact that the at-risk population is diminished much more quickly in Table 7.10 because, in addition to the MACE process of Table 7.9, the cancer process is reducing the at-risk population of Table 7.10.

At month 24, the cumulative MACE rate is 0.4984, considerably lower than the corresponding MACE rate of 0.9375 of Table 7.9. In addition, the 2-year cumulative cancer rate is also 0.4984, much less than the calculated 2-year rate of 0.9375.

If a 15% treatment effect is assumed in the three-state model with censoring (Table 7.10) as was assumed in the two-state uncensored model giving rise

TABLE 7.10

How Censoring Effects Cumulative At-Risk Probabilities
of Trial in Table 7.9

Month	Censored (Cancer)	Failed (MACE)	At-Risk
Rand	0	0	1
1	0.1060	0.1060	0.7881
2	0.1895	0.1895	0.6211
3	0.2553	0.2553	0.4895
4	0.3071	0.3071	0.3858
5	0.3480	0.3480	0.3040
6	0.3802	0.3802	0.2396
7	0.4056	0.4056	0.1888
8	0.4256	0.4256	0.1488
9	0.4414	0.4414	0.1173
10	0.4538	0.4538	0.0924
11	0.4636	0.4636	0.0728
12	0.4713	0.4713	0.0574
13	0.4774	0.4774	0.0452
14	0.4822	0.4822	0.0357
15	0.4860	0.4860	0.0281
16	0.4889	0.4889	0.0221
17	0.4913	0.4913	0.0175
18	0.4931	0.4931	0.0138
19	0.4946	0.4946	0.0108
20	0.4957	0.4957	0.0085
21	0.4966	0.4966	0.0067
22	0.4973	0.4973	0.0053
23	0.4979	0.4979	0.0042
24	0.4984	0.4984	0.0033

to Table 7.9, the sample size can be calculated. In the presence of censoring, the number of patients needed for the logrank, which is based on the successive distributions of the failed (MACE) and at-risk columns, is 942 patients, with 433 events. The censored state (second column) is considered a nuisance variable and does not directly enter the calculations. This is similar to the way the log-rank statistic is calculated based on the number of patients at risk and failed at each distinct event time. The fact that the required number of patients is different comes about because of the changes in the last two columns of Table 7.10 due to censoring. In the no censoring case of Table 7.9, the number of patients is 481 and the number of events 433. The number of events is the same in the two models because the treatment effect is constant over time (i.e., proportional hazards). Although the number of events in the three-state model is still 433, the censoring results in an increase in the number of patients from 481 to 942, an increase of 85%. The typical approach to calculating sample size when there is censoring (such as loss to follow-up) is to calculate the probability of censoring over the entire trial and inflate the sample size to adjust for the loss (Rubinstein et al., 1981, do not have this problem). Here, calculated in isolation, the 75% loss rate over 2 years is 93.75%. So the typical adjustment would inflate the number of patients to $481/(1 - 0.9375) = 481 * 16 = 7696$, a 1500% increase compared with the 85% increase using the Markov model. The logic for the typical adjustment is that with all but 1/16th of the patients being censored, the sample size must be increased by a factor of 16 to have enough patients to meet the unadjusted sample size requirements. This logic is flawed. The two processes, cancer and MACE, do not function in isolation. This is certainly true in the natural processes occurring in the clinical trial. While the Markov model calculates how the simultaneous processes work in conjunction with one another, the typical approach to adjusting for censoring ignores this important interaction and assumes the processes occur in isolation of one another.

7.4.4 ODIS and Noncompliance Process

Without loss of generality, only the experimental arm is discussed. In most large trials, some portion of the randomized patients will drop out of the trial. Increasingly, regulatory authorities, consistent with the intent-to-treat principle, are requiring follow-up of such patients for the primary endpoint. Such patients are no longer taking their randomly assigned medications and thus no longer receive the full benefit of those medications. Exactly how much benefit is lost is difficult to know, but it is often assumed that once patients stop taking their medications they no longer receive any benefit from the medications they are not taking. Similar to censoring, those patients who will have stopped taking medications by the end of the study do not all terminate those medications at the end of the study, nor at the beginning. Rather, patients, one-by-one drop out during the course of the trial. When a patient in the active group drops out and consequently terminates study medication, their failure rate is assumed to revert to the unmedicated state. (Other assumptions for the change in their failure rates are possible.) The experimental arm of the trial will gradually have more and more patients no longer taking their medications so that the average event rate for the entire experimental arm will gradually increase. This violates the basic Markov property that the probability of failure for a patient in a given state at a given time depends only on that state and time, and not on how that patient arrived in that state at that time. Here, patients who have discontinued therapy have a higher failure rate than those who have not. The solution is to create another state for these patients who are no longer taking their medications. That is how the Markov model proposed by Lakatos (1986) handled ODIS (when that paper was written, the term *noncompliers* was used, rather than ODIS, but the underlying principles are the same). The Markov model, which easily handles ODIS, uses separate states for experimental arm patients who are still taking randomized therapy versus those who are not. Gradually, as the trial progresses, the drop-out process transitions patients from the *at-risk-on-therapy* state to the *at-risk-off-therapy* state. While the failure rates from these two states are different, it is the same for all patients within either of the two states. The Markov model does not restrict what failure rates are assigned to these two states.

Other Approaches to ODIS. Current commercially available software often offer simulation methods for more complex clinical trial modeling. However, none appears to have offered a means of simulating ODIS. The problem is that simulation methods require prespecification of the trial parameters. As described in the preceding paragraph, the failure parameter for the experimental arm of a trial with ODIS is constantly changing as more and more patients stop taking their medications. Lachin and Foulkes (1986) offer a model-based approach to adjusting sample size for noncompliance/ODIS. Their basic failure model is an exponential model, and they adjust the failure rate parameter to account for ODIS. However, the exponential model assumes the failure rate is constant, which is not consistent with the gradually changing failure rate inherent in the ODIS process. Their approach is to assume all dropouts occur at the time of randomization. This is a conservative approach, but other than comparing with the Markov model, there appears to be no way of assessing how conservative the approach actually is. To simulate the situation, one needs to prespecify a failure rate for, say, the experimental arm. But that failure rate is constantly changing due to ODIS.

7.4.5 Time-Dependent Processes

In Section 7.2, a number of examples of trials whose results demonstrated nonconstant and/or nonproportional hazards were given. In that section, it was shown how the Kaplan–Meier curves from prior trials can be used to quantify the nonconstant hazard functions. Earlier in this section, a formal Markov model was presented. The example used there assumed that the conditional failure probability was constant over time. To model nonconstant failure rate, simply replace the constant conditional failure probabilities with the time-varying probabilities.

7.4.6 Provide Better Sample Size and Power Estimates by Using the Entire Projected Survival Experience

The Markov approach allows calculation of sample size for the logrank that is (1) similar to the way the logrank is actually computed and (2) is appropriate for any nonproportional hazards alternative (i.e., treatment effects that change over time). While the logrank is optimal under proportional hazards assumptions, it is valid without that assumption. Lakatos and Lan (1992) used simulations to compare the Markov approach to sample size for the logrank (1988) with the exponential-based approach (using Rubinstein et al. 1981) and the proportional hazards-based approach (Schoenfeld 1981). Simulations were performed under three sets of assumptions: exponential assumptions, proportional hazards assumptions, and nonproportional hazards assumptions. It was expected that the Markov approach would perform better under nonproportional hazards. However, the Markov approach also performed better than the other approaches when the underlying data were generated under exponential and/or proportional hazards assumptions. Here is the reason.

While the following concepts can be explained using the partial likelihood for the Cox model (1972), using the log-rank statistic as introduced by Mantel (1966) is more straightforward. When Mantel introduced what is now called the log-rank statistic for censored survival data, it was in terms of 2×2 tables. For simplicity, assume no ties. Merge the event data from the two treatment groups, rank the event times, and at each such distinct event time form a conditional 2×2 table. For each such table, calculate the observed minus expected.

Here, for simplicity, it is assumed that there is exactly one failure at the time of the ith ranked failure (d_i, the number failed in group 1, is 0 or 1). The number of patients still at risk at the time of the ith ranked failure are given in the last column of Table 7.11. The expected, under the null hypothesis, is that the 1 total failure for this 2×2 table occurs in proportion to the patients allocated to group 1, that is, $1 \times n_{i1}/n_i$, where the 1 is added for emphasis. Note that under equal allocation, the expected, at the beginning of the trial is $n_{i1}/n_i = 1/2$. However, if the treatment is effective, the balance of patients still at risk in the two groups changes as the trial progresses. The term n_{i1}/n_i can be derived from the *at-risk ratio* n_{i1}/n_{i2} and vice versa. In Table 7.2, at randomization, the at-risk ratio is 1:1, whereas at the bottom of Table 7.2 it is approximately 2:1.

Table 7.9, generated using the Markov approach, can be used to approximate, one row at a time, the observed minus expected that will be calculated

TABLE 7.11

Conditional 2×2 Table for Log-Rank Statistic

	Failed	At-Risk	Total
Group 1	d_i	$n_{i1} - d_i$	n_{i1}
Group 2	$1 - d_i$	$n_{i2} - 1 - d_i$	n_{i2}
Total	1	$n_i - 1$	n_i

once the trial data has been collected. And it is crucial that that calculation of the observed minus expected reflect the changing value of the expected as the trial progresses. In Schoenfeld's derivation of his sample size formula, he makes the assumption that the at-risk ratio, and hence the expected, is constant throughout the trial. In contrast, Schoenfeld's derivation of the distribution of the logrank in the same paper makes no such assumption. Both the Schoenfeld and Lakatos approaches to sample size for the log rank are based on Schoenfeld's derivation of the distribution of the log-rank statistic. However, the Markov model allows Lakatos to estimate how the expected will change over the course of the trial, while Schoenfeld assumes that the expected will be constant.

7.4.7 Conclusions

A number of examples from actual clinical trials were given that exhibited various types of departures from constant and proportional hazards. These departures, if ignored, were shown to have a variety of unfortunate consequences for the design of trials. These include inappropriate sample sizes, the misuse of event-driven trials, and adaptive designs for which an increase in sample size could lead to decreased power, etc.

The *designing for a range of alternatives* approach to sample size re-estimation was introduced. This approach has the advantages of not depending on interim estimates that often are highly unreliable, potentially leading to incorrect and costly decisions. In addition, if the smaller sample size proves inadequate, the sample size is automatically increased without any interaction with the sponsor. This can be important not only because it facilitates maintaining the confidentiality between the DMC and the sponsor, but because if the sponsor accepts a DMC recommendation to increase the sample size, the investigators and stock analysts most often interpret this as an indication that the treatment effect is smaller than anticipated. This may lead investigators to hesitate to enroll sicker patients some of whom may be the best patients to demonstrate the efficacy of the drug. Stock analysts may downgrade the sponsor's stock, causing financial problems for the company and, in turn, possibly for the trial.

Armed with an understanding of the possible shape of the hazard and/or hazard ratio curves, many of these pitfalls can be minimized or avoided entirely. In order to do that, appropriate software, such as STOPP, is needed.

Optimization of group-sequential designs was also addressed. The typical approach to optimization is minimization of the ASN. Pocock originally proposed this approach in 1977 when he introduced the group-sequential approach. It was shown how current thinking as represented by Pocock (2006) 30 years later argues against using the ASN. It was also argued that no formulaic approach to optimization is likely to work because the *guidelines* nature of boundaries.

References

Bernier J, Domenge C, Ozsahin M. 2004. Postoperative irradiation with or without concomitant chemotherapy for locally advance head and neck cancer. *New England Journal of Medicine* 350:1945–1952.

Cannon CP, Braunwald E, McCabe CH et al. 2004. Intensive versus moderate lipid lowering with statins after acute coronary syndromes. *New England Journal of Medicine* 350:1495–1504.

Cooper JS, Pajak TF, Forastiere AA et al. 2004. Postoperative concurrent radiotherapy and chemotherapy for high-risk squamous-cell carcinoma of the head and neck. *New England Journal of Medicine* 350:1937–1944.

DeMets DL, Furberg CD, Friedman LM, eds. 2006. *Data Monitoring in Clinical Trial: A Case Studies Approach.* Springer, pp. 66.

Kaplan EL, Meier P. 1958. Non-parametric estimation from incomplete observations. *Journal of the American Statistical Association* 53:457–481.

Lachin JM, Foulkes MA. 1986. Evaluation of sample size and power for analyses of survival with allowance for nonuniform patient entry, losses to follow-up, noncompliance, and stratification. *Biometrics* 42:507–519.

Lakatos E. 1986. Sample sizes for clinical trials with time-dependent rates of losses and non-compliance. *Controlled Clinical Trials* 7:189–199.

Lakatos E. 1988. Sample sizes based on the logrank statistic in complex clinical trials. *Biometrics* 44:229–241.

Lakatos E. 2002. Designing complex group sequential trials. *Statistics in Medicine* 21(14):1969–1989.

Lakatos E, Lan KKG. 1992. A comparison of sample size methods for the logrank statistic. *Statistics in Medicine* 11:179–191.

Lan KKG, DeMets DL. 1983. Discrete sequential boundaries for clinical trials. *Biometrika* 70:659–663.

Montori VM, Devereaux PJ, Adhikari NKJ et al. 2005. Randomized trials stopped early for benefit: A systematic review. *JAMA* 294(17):2203–2209.

O'Brien PC, Fleming TR. 1979. A multiple testing procedure for clinical trials. *Biometrics* 35:549–556.

Pocock SJ. 1977. Group sequential methods in the design of clinical trials. *Biometrika* 64:191–199.

Pocock SJ. 2005. When (not) to stop a clinical trial for benefit. *JAMA* 294(17): 2228–2230.

Pocock SJ. 2006. Current controversies in data monitoring for clinical trials. *Clinical Trials* 3(6):513–521.

Roach M III, Bae K, Speight J et al. 2008. Short-term neoadjuvant androgen deprivation therapy and external-beam radiotherapy for locally advanced prostate cancer: Long-term results of RTOG 8610. *Journal of Clinical Oncology* 26:585–591.

Rubins HB, Robins SJ, Collins D et al. 1999. Gemfibrozil for the secondary prevention of coronary heart disease in men with low levels of high-density lipoprotein cholesterol. *New England Journal of Medicine* 341:410–418.

Rubinstein LV, Gail MH, Santner TJ. 1981. Planning the duration of a comparative clinical trial with loss to follow-up and a period of continued observation. *Journal of Chronic Diseases* 34:469–479.

Sacks FM, Pfeffer MA, Moye LA et al. 1996. The effect of pravastatin on coronary events after myocardial infarction in patients with average cholesterol levels. *New England Journal of Medicine* 335:1001–1009.

Scandinavian Simvastatin Survival Study Group. 1994. Randomised trial of cholesterol lowering in 4444 patients with coronary heart disease: The Scandinavian Simvastatin Survival Study (4S). *Lancet* 19; 344(8934):1383–1389.

Schoenfeld D. 1981. The asymptotic properties of non-parametric test for comparing survival distributions. *Biometrika* 68:316–318.

The Beta-Blocker Heart Attack Trial Research Group. 1982. A randomized trial of propranolol in patients with acute myocardial infarction. I. Mortality results. *JAMA* 247:1707–1714.

The Lipid Research Clinics Group. 1984. The Lipid Research Clinics Coronary Primary Prevention Trials (LRC-CPPT). I. Reduction in incidence of coronary heart disease. *JAMA* 251:351–364.

8

Group Sequential Design in R

Keaven M. Anderson

CONTENTS

8.1 Introduction

This chapter demonstrates the gsDesign R package using two examples of trials planned with a single futility analysis. If you wish to replicate the calculations, you will need to install gsDesign 2.8-3 or later, which is available from the CRAN website http://www.cran.r-project.org. While considering a single interim analysis may seem overly restrictive, this can be the most important analysis in a group sequential design in terms of limiting the patient volunteers required as well as research expenses for trials which have little chance of helping patients. In addition, designs with more analyses are simple variations of those shown here. The chapter is intended to both demonstrate the essential R coding and how to evaluate design bounds. The chapter was written using a combination of LaTeX and the `knitr` R package (Xie 2014); this allowed combining R commands and text markup in the source document for the chapter. By applying the `knitr` package to the source, R commands were run and a pure LaTeX file was produced for typesetting. Note that files with the extension .rnw in the package distribution are source files that can be used with LaTeX and `knitr` to produce a typeset document with results from embedded R code.

The same group sequential/futility boundaries will be applied to two trial designs:

- A trial to demonstrate noninferiority of a new treatment compared to an active control for a trial with a binary endpoint.
- A trial to demonstrate superiority of a new treatment compared to control with a time-to-event primary endpoint.

These two designs are used to show how much in common there is in designing trials with different endpoint types as well as noninferiority versus superiority trials.

The chapter begins with sample size derivation for fixed designs. Next, we provide a quick review of the theory behind group sequential designs followed by the R commands required to derive such designs. Most of the rest of the chapter is dedicated to displays of design characteristics and factors to consider in selecting group sequential designs. However, we also include a section demonstrating a simulation for the group sequential design with a binary endpoint in order to demonstrate how easy and fast it is to do a large simulation to confirm the asymptotic theory approximations. Finally, in the discussion, we summarize some suggestions for how to choose futility criteria and the timing of a futility analysis.

8.2 Fixed Designs

8.2.1 Noninferiority Design with a Binary Endpoint

The first trial is a two-arm noninferiority trial with a binary outcome. Assuming both arms have a success rate of 0.677, we wish to have 90% power to rule out noninferiority when the experimental arm event rate is 0.07 lower than control $(0.677 - 0.07 = 0.607)$ with one-sided Type I error of $\alpha = 0.025$. The sample size for a fixed design with no interim analysis to satisfy these requirements is derived as follows:

```
require(gsDesign)
n.fix <- nBinomial(p1 = 0.677, p2 = 0.677, delta = 0.07)
n.fix
```

```
## [1] 1873
```

The sample size method for this computation was proposed by Farrington and Manning (1990), which is an extension to noninferiority designs of the method by Fleiss et al. (1980). This did not require arguments `alpha=.025` (1-sided Type I error) or `beta=.1` (Type II error or 1−power), since they are default values. The value `n.fix` can be used to generate various group sequential designs. `nBinomial` does not round up to an even number, 1874. The `ceiling` function can be used to round up for a fixed design.

```
ceiling(n.fix/2) * 2
```

```
## [1] 1874
```

Note that this `nBinomial` call returns the total sample size, not the sample size per group demonstrated as follows when a 2:1 (experimental:control) is specified:

```
ceiling(nBinomial(p1=.677, p2=.677, delta=.07,
                  ratio=2, outtype=2))
```

```
##      n1    n2
## 1  686  1372
```

8.2.2 Superiority Design with a Time-To-Event Endpoint

We use the Lachin and Foulkes (1986) method to derive a fixed design sample size for a trial with the following characteristics:

- A Type I error of .025 (one-sided) and Type II error of .1 (90% power).
- An exponentially distributed primary endpoint with a median time-to-event of 6 months in the control group and 10 months in the experimental group (hazard ratio of .6). While the sample size routine

allows a piecewise exponential failure distribution and a stratified population, we demonstrate here with a simple, and commonly used, exponential time-to-event. In this case, the parameter `lambdaC` represents the failure rate λ for the control group. A median m is converted to a failure rate by $\lambda = \log(2)/m$.

- An exponential dropout rate with a 5% dropout rate after 12 months. The exponential rate here is computed by $\eta = -\log(.95)/12$.
- Thirty-six months of total study duration (`T` = 36) with a minimum follow-up of 12 months (`minfup` = 12). Implicitly, this means 24 months of enrollment (`T` - `minfup`). Note that piecewise constant enrollment rates are allowed in order to account for enrollment ramp-up. This enrollment ramp-up option is not demonstrated here; see the help file for `nSurv`.

```
n.fixs <- nSurv(lambdaC=log(2)/10, eta= -log(.95)/12,
                hr=.6, T=36, minfup=12)
nevents <- n.fixs$d
n <- n.fixs$n
nevents
```

```
## [1] 160.2
```

```
n
```

```
## [1] 237.6
```

Thus, a fixed design with these characteristics requires a total sample size of 238 and the analysis will be well powered when 161 events have been observed. The following shows all sample size information for this design:

```
n.fixs
```

```
## Fixed design, two-arm trial with time-to-event
## outcome (Lachin and Foulkes, 1986).
## Solving for:  Accrual rate
## Hazard ratio                    H1/H0=0.6/1
## Study duration:                    T=36
## Accrual duration:                   24
## Min. end-of-study follow-up: minfup=12
## Expected events (total, H1):       160.2
## Expected sample size (total):      237.6
## Accrual rates:
```

```
##          Stratum 1
## 0-24       9.899
## Control event rates (H1):
##          Stratum 1
## 0-Inf     0.0693
## Censoring rates:
##          Stratum 1
## 0-Inf     0.0043
## Power:                    100*(1-beta)=90%
## Type I error (1-sided):   100*alpha=2.5%
## Equal randomization:         ratio=1
```

Note that accrual rates are computed to make the trial fit into the desired timelines. If it is more realistic to fix enrollment and vary the duration of enrollment as developed by Kim and Tsiatis (1990), this can be done as follows:

- The parameter gamma provides enrollment rates and R provides enrollment duration for each rate; in this case (when T is not specified as in the previous example), the final enrollment rate period is extended to obtain adequate power. We assume enrollment ramps up with 2 patients per month for 6 months and then 5 per month thereafter (gamma = c(2,5),R=c(6,6)).
- We still assume 12 months minimum follow-up at the end (minfup = 12) and the same exponential distributions for time-to-event and dropout as before.

```
n.fixs2 <- nSurv(lambdaC=log(2)/10, eta= -log(.95)/12,
                 hr=.6, gamma=c(2,5), R=c(6,6), minfup=12)
sum(n.fixs2$R) # enrollment duration
```

```
## [1] 46.21
```

```
n.fixs2$T      # trial duration
```

```
## [1] 58.21
```

Another option is to fix enrollment rates and duration, allowing the follow-up duration to vary in order for the trial to be adequately powered. This variation is left to the reader, if interested; see help(nSurv).

8.3 Group Sequential Design Theory

Group sequential designs allow multiple, preplanned analyses of an outcome at different points in time during the course of a study. See, for example, books by Jennison and Turnbull (2000) or Proschan et al. (2006). In this section, we briefly review the asymptotic distribution theory applied for many group sequential designs.

We begin with what we will call a *natural parameter* of interest, which we will denote by δ. For our two examples, δ is the difference in response rates (experimental minus control) for the binomial trial design or minus the logarithm of the hazard ratio for the time-to-event study (experimental/control). In both cases, a positive δ indicates a benefit with experimental treatment. We could also use, for example, the log-relative risk or log-odds-ratio for the natural parameter for a trial comparing two treatment arms for a binomial outcome. For a continuous outcome, we could let δ be the difference in mean change from baseline for two treatment arms. For k distinct, sequential analyses of the data from the trial, we assume we have an asymptotically efficient estimator $\hat{\delta}_i$ with statistical information $\mathcal{I}_i$ (variance $1/\mathcal{I}_i$) and an asymptotically efficient test statistic Z_i, $1 \leq i \leq k$. The group sequential design presented here is based on the assumption that the test statistics $Z_1, \ldots, Z_k$ are approximately distributed according to the multivariate normal distribution with, for $1 \leq i \leq j \leq k$,

$$\mathrm{E}\{Z_i\} = \sqrt{\mathcal{I}_i}\delta \tag{8.1}$$

$$\mathrm{Cov}\{Z_i, Z_j\} = \sqrt{\mathcal{I}_i/\mathcal{I}_j} \tag{8.2}$$

where $0 < \mathcal{I}_1 < \cdots < \mathcal{I}_k$. See, for example, Jennison and Turnbull (2000) for many cases where the asymptotic normal assumption outlined is used. Since Z_i and $\hat{\delta}_i$ are both asymptotically efficient, we have, for $1 \leq i \leq k$,

$$Z_i \approx \sqrt{\mathcal{I}_i}\hat{\delta}_i. \tag{8.3}$$

For the binomial event rate comparisons we consider, n_i will represent the sample size at analysis i, $1 \leq i \leq k$. For the time-to-event case, n_i represents the number of events included in analysis i, $1 \leq i \leq k$. In either case, we rely on an approximate relationship

$$\mathrm{Var}\{\hat{\delta}_i\} = \frac{\psi^2}{n_i} = \frac{1}{\mathcal{I}_i} \tag{8.4}$$

or, equivalently,

$$\mathcal{I}_i = \frac{n_i}{\psi^2}. \tag{8.5}$$

Note that this all assumes the variance of $\hat{\delta}_i$ is approximately the same for all underlying δ values.

We will arbitrarily assume that we wish to test a null hypothesis H_0: $\delta = \delta_0$ versus H_1: $\delta = \delta_1$ for some real values $\delta_0 < \delta_1$. We generally assume that $\delta_0 = 0$ will correspond to superiority testing. For our examples, this would be the response rate difference (experimental minus control) or minus log hazard ratio (experimental/control) greater than 0. Extending this, $\delta_0 < 0$ would correspond to a noninferiority test. Now we consider a *standardized effect size,* which we denote as

$$\theta = \frac{(\delta - \delta_0)}{\psi}. \tag{8.6}$$

The hypotheses mentioned are equivalent to H_0: $\theta = 0$ versus H_1: $\theta = \theta_1$ where $\theta_1 = (\delta_1 - \delta_0)/\psi > 0$. Letting $\hat{\theta}_i = \left(\hat{\delta}_i - \delta_0\right)\big/\psi$ or $\hat{\theta}_i = \left(\hat{\delta}_i - \delta_0\right)\big/\hat{\psi}_i$, we have for $1 \leq i \leq k$,

$$\mathrm{Var}\{\hat{\theta}_i\} \approx \frac{1}{n_i} \tag{8.7}$$

and

$$E\{Z_i\} \approx \sqrt{n_i}\theta. \tag{8.8}$$

A simple, general form for a sample size calculation for a trial with a single analysis after n_{fix} observations (or events) is

$$n_{fix} = \left(\frac{Z_{1-\alpha} + Z_{1-\beta}}{\theta_1}\right)^2. \tag{8.9}$$

This is an exact sample size for a one-sided test of the mean equal to 0 for a set of independent and identically distributed normal random variables with variance ψ^2 under the assumption that the true mean is $\theta_1 \neq 0$. This assumes one-sided Type I error α and Type II error β where $1 - \beta$ is the power for the test. $Z_{1-\alpha}$ represents the inverse of the standard normal distribution function evaluated at $1 - \alpha$. Analogously, the amount of statistical information required to power a trial is

$$\mathcal{I}_{fix} = \left(\frac{Z_{1-\alpha} + Z_{1-\beta}}{\delta_1 - \delta_0}\right)^2. \tag{8.10}$$

We will sometimes refer to $t_i = \mathcal{I}_i/\mathcal{I}_k = n_i/n_k$ as the proportion of (statistical) information at analysis i, $i = 1, \ldots, k$.

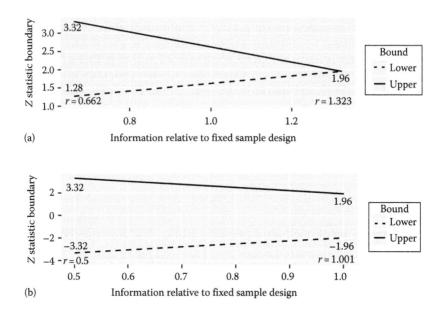

FIGURE 8.1
Examples of (a) asymmetric and (b) symmetric group sequential design bounds for trials with a single interim analysis.

For a group sequential design, we perform repeated testing with the statistics $Z_1, \ldots, Z_k$ assumed to be distributed as given. We set testing cutoffs $l_i < u_i$, $i = 1, \ldots, k - 1$. For asymmetric designs, we would generally have $l_k = u_k$ and reject the null hypothesis that $\theta = 0$, in favor of $\theta = \theta_1 > 0$ (Figure 8.1a). For symmetric, 2-sided testing, we would have $u_i > 0$ and $l_i = -u_i$, $1 \leq i \leq k$ (Figure 8.1b). For $1 \leq i \leq k$, the trial is stopped at analysis i to reject H_0 if $l_j < Z_j < u_j, j = 1, 2, \ldots, i - 1$ and $Z_i \geq u_i$. If the trial continues until stage i, H_0 is not rejected at stage i, and $Z_i \leq l_i$, then H_1 is rejected in favor of H_0, $i = 1, 2, \ldots, k$. Note that if $l_k < u_k$, there is the possibility of completing the trial without rejecting H_0 or H_1.

Next we examine the probability of crossing bounds. For $i = 1, \ldots, k$, the probability that the first boundary crossed is the upper bound at analysis i is

$$\alpha_i(\theta) = P_\theta \left\{ \{Z_i \geq u_i\} \bigcap_{j=1}^{i-1} \{l_j \leq Z_j < u_j\} \right\}$$

and the probability of first crossing a lower bound at analysis i is

$$\beta_i(\theta) = P_\theta \left\{ \{Z_i < l_i\} \bigcap_{j=1}^{i-1} \{l_j \leq Z_j < u_j\} \right\}.$$

The value $\alpha_i(0)$ is commonly referred to as the amount of Type I error spent at analysis i, $1 \leq i \leq k$. The total upper boundary crossing probability for a trial is denoted in this one-sided scenario by

$$\alpha(0) = \sum_{i=1}^{k} \alpha_i(0).$$

Studies using $\alpha(0)$ as the Type I error probability calculation are said to have a binding futility bound, since the Type I error is inflated above the nominal level if the trial is declared positive when an efficacy bound is crossed after a futility bound was previously crossed.

The probability of first crossing an upper bound at analysis i while ignoring the lower bound, $i = 1, 2, \ldots, k$, is

$$\alpha_i^+(\theta) = P_\theta \left\{ \{Z_i \geq u_i\} \bigcap_{j=1}^{i-1} \{Z_j < u_j\} \right\}. \tag{8.11}$$

In this case, the futility bound defined by l_i, $1 \leq i \leq k$ is termed nonbinding since Type I error is not inflated if a trial is declared positive when an efficacy bound is crossed after previously crossing a futility bound. Again, $\alpha_i^+(0)$ is commonly referred to as the amount of Type I error spent at analysis i, $1 \leq i \leq k$. Thus, when describing an asymmetric design, we must distinguish between designs with binding and nonbinding futility bounds. The total Type I error for a trial with a nonbinding futility bound or no futility bound is

$$\alpha^+(0) \equiv \sum_{i=1}^{k} \alpha_i^+(\theta). \tag{8.12}$$

For power calculations, we still use $\alpha(\theta)$ to compute power accounting for the futility bound.

Letting $t_i = n_i/n_k$, $1 \leq i \leq k$, a parameterization fully specifying a group sequential design is n_k combined with:

$$\mathbf{l} = (l_1, \ldots, l_k), \tag{8.13}$$

$$\mathbf{u} = (u_1, \ldots, u_k), \tag{8.14}$$

$$\mathbf{t} = (t_1, \ldots, t_{k-1}). \tag{8.15}$$

From (8.1) and (8.2), we see that given $\mathbf{l}$, $\mathbf{u}$, and $\mathbf{t}$, the distribution of $Z_1, \ldots, Z_k$ depends only on $\sqrt{n_k}\theta$. We can, thus, deduce that the sample size required to power a group sequential trial is linear in $1/\theta_1^2$, as is the case for the fixed design sample size n_{fix} in (8.9). This implies that the sample size ratio of a group sequential design divided by the corresponding fixed design sample size can be written as

$$\frac{n_k}{n_{fix}} = R(\mathbf{l}, \mathbf{u}, \mathbf{t}).$$

Based on this relation, the common application of the group sequential design package described here is to start with a fixed design sample size n and back-calculate the standardized effect size based on (8.9) as

$$\theta = \left(\frac{Z_{1-\alpha} + Z_{1-\beta}}{\sqrt{n_{fix}}} \right). \tag{8.16}$$

In the binomial and time-to-event cases we have presented, the fact that the value of ψ differs (usually slightly) between the null and alternate hypothesis is taken into account in the fixed sample size equation and (8.9) only holds approximately. We have, nonetheless, chosen to use (8.16) to apply standard group sequential calculations within the gsDesign R package. Other software, for example, EAST™, uses fixed sample size calculations based on a single variance under the null or alternative hypothesis as in (8.9). The assumption in gsDesign is that this gives a less accurate "starting point" for group sequential design sample size calculation. The interested reader is encouraged to examine this using simulation as demonstrated in Section 8.6.

The final topic of this theory overview is spending functions. An α-spending function is a nondecreasing function f on $[0, \infty]$ such that $f(0) = 0, f(t) = \alpha$ for $t \geq 1$. We define a β-spending function g on $[0, \infty)$ analogously as nondecreasing with $g(0) = 0$ and $g(t) = \beta$ for $t \geq 1$. We let θ_1 represent the standardized effect size we wish to power a group sequential trial for with power $1 - \beta$ and one-sided Type I error α where $0 < \alpha < 1 - \beta < 1$. We then set Type I error by implicitly defining for $1 \leq i \leq k$

$$f(t_i) = \sum_{j=1}^{i} \alpha_j(0)$$

for a study with a binding futility bound or

$$f(t_i) = \sum_{j=1}^{i} \alpha_j^+(0)$$

for a study with a nonbinding futility bound. We set Type II error analogously with, for $1 \leq i \leq k$,

$$g(t_i) = \sum_{j=1}^{i} \beta_j(\theta_1).$$

Given values for $\beta_i(\theta_1)$ and either $\alpha_i(0)$ or $\alpha_i^+(0)$, $1 \leq i \leq k$, bounds $l_1, \ldots, l_k$ and $u_1, \ldots, u_k$ are computed in the following using numerical methods presented in Chapter 19 of Jennison and Turnbull (2000).

8.4 Group Sequential Design Derivation

The function gsDesign transforms an input fixed design sample size to a group sequential design with the desired error properties assuming the distributional properties for outcomes in the fixed design. For time-to-event designs, the function gsSurv is a combination of nSurv and gsDesign functions. This section will focus on basic design derivation. The next section will focus on various bound characteristics and how to generate them in text or plot output. Those interested in design without the programming or vice versa can skim material of less interest.

While other options are available, the defaults for gsDesign and gsSurv functions are to produce asymmetric group sequential designs with futility bounds that are nonbinding. The asymmetry allows futility bounds to be very different than efficacy bounds. Since most trials are performed with the *burden of proof* on a new treatment compared to a control, designs with asymmetric bounds are quite common. We will focus not only on trials with a single interim analysis with a very high interim bound to establish efficacy but also an interim futility bound that at least suggests a positive efficacy finding. The potential implications of futility analyses include different possibilities:

- A sample size increase that may be substantial
- A decrease in power
- A futility analysis that is not meaningful, because either the interim futility bar is low or because the analysis is done too late to substantially impact the patient and financial resources required

The basic choice we demonstrate here is based on the combined consideration of these factors. We choose to lower the power from the fixed design examples shown with 90% power to 85% power to allow a meaningful futility bar at a meaningful time point.

We use the default Hwang-Shih-DeCani spending function for both lower and upper spending functions (Hwang et al. 1990). For the lower bound, we use the spending function parameter $\gamma = 3$ to produce an aggressive early

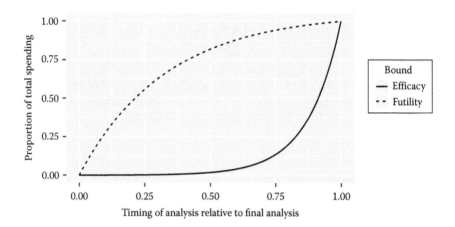

FIGURE 8.2
Hwang–Shih–DeCani spending functions with $\gamma = -8$ (efficacy) and $\gamma = 3$ (futility).

stopping bound for futility. For the upper bound, we use $\gamma = -8$ to set a very stringent bound to consider stopping for a positive efficacy finding at the interim analysis. *Timing* is computed as the interim analysis information divided by the planned final analysis information. For the binomial information, timing is determined by the proportion of the final planned number of observations to be analyzed at the interim analysis. For time-to-event outcomes, timing refers to the planned number of events included in an analysis divided by the planned number of events for the final analysis. We produce a quick plot showing these spending functions using the ggplot2 package; there is a simpler way to produce this plot we will note later, so those not interested in learning a basic ggplot command should, just review the plot and skip the code (Figure 8.2).

```
# load the ggplot2 R package
require(ggplot2)
# x-axis will run from 0-1
timing <- (0:100)/100
# relative spending for efficacy bound with gamma = -8
spend1 <- sfHSD(alpha=.025, t=timing, param=-8)$spend / .025
# relative spending for futility bound with gamma = 3
spend2 <- sfHSD(alpha=.15, t=timing, param=3)$spend / .15
# put these together in a data frame for ggplot
dat <- rbind(data.frame(Timing=timing, Spend=spend1,
                        Bound="Efficacy"),
             data.frame(Timing=timing, Spend=spend2,
                        Bound="Futility"))
# aes specifies x and y variables, lty the line type;
```

```
# geom_line is used for line plots and ylab, xlab customize
# axis labels
ggplot(data=dat,aes(x=Timing, y=Spend, lty=Bound)) +
  geom_line() + ylab("Proportion of total spending") +
  xlab("Timing of analysis relative to final analysis")
```

This design uses a standard, nonbinding futility bound approach where Type I error is computed assuming the trial is not stopped if a futility bound is crossed. A nonbinding futility rule is generally recommended by regulatory authorities. This allows a trial to continue and still control Type I error if the futility bound is not obeyed.

8.4.1 Binomial Design

For the binomial design, we simply plug in the fixed design total sample size and the following group sequential design parameters. Type I error and Type II error are computed using spending functions. The default for gsDesign is to use Hwang-Shih-DeCani spending functions (Hwang et al. 1990). The spending function parameter γ is specified for the futility bound by sflpar=3 and for the efficacy bound by sfupar=-8. Timing of the interim analysis is set for 40% of the final planned sample size using timing=0.4, and the number of analyses is set with k=2. As previously, alpha=0.025 is a default value that need not be specified. However, we now beta=0.15 corresponding to 15% Type II error and 85% power. Note that this has to be done both in the calls to nBinomial and to gsDesign. If you choose somethings other than alpha=0.025, that also needs to be specified in both places. We follow this by printing a textual summary of the designs as well as a plot of the approximate event rate difference that would be required to cross each bound. There is a specifically designed plot function for examining differences in binomial outcomes (experimental rate minus control rate), which will be automatically invoked by setting endpoint="Binomial" along with the null hypothesis rate difference delta0= -0.07 and alternate hypothesis difference delta1=0. Note the expression call for ylab in the plot call and the resulting y-axis label. Also note that the cex argument shrinks text size for the plot so that it fits in a typeset document.

```
n.fix <- nBinomial(p1=.677, p2=.677, delta0=-.07, beta=.15)
y <- gsDesign(k=2, beta=.15, timing=.4, sflpar=3, sfupar=-8,
              n.fix=n.fix, endpoint="Binomial",
              delta0= -.07, delta1=0)
y

## Asymmetric two-sided group sequential design with
```

```
## 85 % power and 2.5 % Type I Error.
## Upper bound spending computations assume
## trial continues if lower bound is crossed.
##
##                    ----Lower bounds----   ----Upper bounds-----
## Analysis   N    Z    Nominal p Spend+   Z    Nominal p Spend++
##        1  820 0.92    0.8212 0.1103 3.54    0.0002 0.0002
##        2 2050 1.96    0.9751 0.0397 1.96    0.0249 0.0248
##    Total                    0.1500                  0.0250
## + lower bound beta spending (under H1):
##  Hwang-Shih-DeCani spending function with gamma = 3
## ++ alpha spending:
##  Hwang-Shih-DeCani spending function with gamma = -8
##
## Boundary crossing probabilities and expected sample size
## assume any cross stops the trial
##
## Upper boundary (power or Type I Error)
##           Analysis
##    Theta       1       2   Total E{N}
##    0.0000 0.0002 0.0187 0.0189 1039
##    0.0749 0.0809 0.7691 0.8500 1814
##
## Lower boundary (futility or Type II Error)
##           Analysis
##    Theta       1       2   Total
##    0.0000 0.8212 0.1599 0.9811
##    0.0749 0.1103 0.0397 0.1500
```

```
plot(y, cex=.6, plottype="xbar",
    ylab=expression(paste(hat(p)[E]-hat(p)[C])))
```

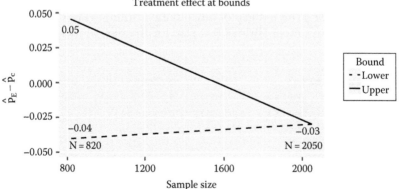

We note the impact of a stringent futility bar and the change in power on the sample size versus a design with no interim. For a fixed design, the sample size is 1874 for 90% power or 1602 for 85% power. By applying the stringent futility bar above and 85% power, we get a final analysis sample size of 2050, a 28% increase over a fixed design with 85% power, and a 9% increase over a fixed design with 90% power.

For event rate differences, either a failure rate or a response rate may be of interest. The default *y*-axis labeling assumes event rates are stated in terms of failure rates and we would want to demonstrate that the experimental rate is lower. For a case where you wish to show a lower failure rate in the experimental group, coding is a little simpler than given earlier. We encourage the reader to try the following code demonstrating this:

```
x <- gsDesign(n.fix=nBinomial(p1=.15,p2=.1, beta=.15),
           delta1=.05, endpoint="Binomial", beta=.15)
plot(x, plottype="xbar")
```

Also, consider the following code to produce a plot of the spending functions more simply than shown previously:

```
plot(y, plottype="sf")
```

See help(plot.gsDesign) for more details on the plot specifications.

8.4.2 Time-To-Event Design

For a time-to-event design, we will consider both the number of endpoints and the sample size for the design with no interim analysis we derived in Section 8.2.2. The basic print command is commented out in the following code. The output is fairly lengthy, and a more pleasantly formatted option more suitable for document preparation will be presented shortly. Here we plot the approximate hazard ratio if a test statistic is at a study boundary versus the number of events planned for each analysis.

```
ys <- gsSurv(lambdaC=log(2)/10, eta= -log(.95)/12, hr=.6,
           T=36, minfup=12, k=2, timing=.4, sflpar=3,
           sfupar=-8, beta=.15)
# use the following for printing the design
# we will summarize using other output options below
# ys
# plot approximate hazard ratio at bounds
plot(ys, plottype="HR", cex=.6, xlab="Number of events")
```

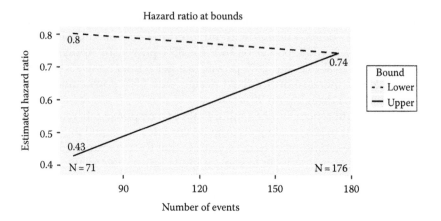

8.5 Showing Design Properties

More compact output suitable for inclusion in a document can be gotten by using the `summary` and `gsBoundSummary` functions. For a high-level summary of a design, use `summary`. In order to break the output into shorter lines for this output, we use `strwrap` here.

```
strwrap(summary(y), width = 55)
```

```
## [1] "Asymmetric two-sided group sequential design with"
## [2] "non-binding futility bound, 2 analyses, sample size"
## [3] "2050, 85 percent power, 2.5 percent (1-sided) Type I"
## [4] "error. Efficacy bounds derived using a"
## [5] "Hwang-Shih-DeCani spending function with gamma=-8."
## [6] "Futility bounds derived using a Hwang-Shih-DeCani"
## [7] "spending function with gamma=3."
```

To output directly to text in the document, do the following using the `knitr` option `results='asis'`: We put the resulting text in italics to distinguish them from surrounding text.

```
cat(summary(ys))
```

Asymmetric two-sided group sequential design with non-binding futility bound, 2 analyses, time-to-event outcome with sample size 260 and 176 events required,

85 percent power, 2.5 percent (1-sided) Type I error to detect a hazard ratio of 0.6. Enrollment and total study durations are assumed to be 24 and 36 months, respectively. Efficacy bounds derived using a Hwang-Shih-DeCani spending function with gamma = −8. Futility bounds derived using a Hwang-Shih-DeCani spending function with gamma = 3.

Next we show boundary characteristics using `gsBoundSummary`.

`gsBoundSummary`(y)

```
##   Analysis                   Value Efficacy Futility
##   IA 1: 40%                     Z   3.5435   0.9198
##     N: 820            p (1-sided)   0.0002   0.1788
##                    delta at bound   0.0457  -0.0400
##        P(Cross) if delta=-0.07     0.0002   0.8212
##          P(Cross) if delta=0       0.0809   0.1103
##     Final                      Z   1.9609   1.9609
##     N: 2050           p (1-sided)   0.0249   0.0249
##                    delta at bound  -0.0295  -0.0295
##        P(Cross) if delta=-0.07     0.0189   0.9811
##          P(Cross) if delta=0       0.8500   0.1500
```

We demonstrate `xprint` to summarize the set of bound characteristics in Table 8.1:

`xprint`(`xtable`(`gsBoundSummary`(y, deltaname="pE-pC"),
   `label="gsBSy", digits=4,`
   `caption="Basic summary of bounds for binomial trial."))`

TABLE 8.1

Basic Summary of Bounds for Binomial Trial

Analysis	Value	Efficacy	Futility
IA 1: 40%	Z	3.5435	0.9198
N: 820	p (one-sided)	0.0002	0.1788
	pE−pC at bound	0.0457	−0.0400
	P(Cross) if pE − pC = −0.07	0.0002	0.8212
	P(Cross) if pE − pC = 0	0.0809	0.1103
Final	Z	1.9609	1.9609
N: 2050	p (one-sided)	0.0249	0.0249
	pE−pC at bound	−0.0295	−0.0295
	P(Cross) if pE − pC = −0.07	0.0189	0.9811
	P(Cross) if pE − pC = 0	0.8500	0.1500

TABLE 8.2

Asymmetric Two-Sided Group Sequential Design with Nonbinding Futility Bound, 2 Analyses, Time-to-Event Outcome with Sample Size 260 and 176 Events Required, 85% power, 2.5% (1-Sided) Type I Error to Detect a Hazard Ratio of 0.6

Analysis	Value	Efficacy	Futility
IA 1: 40%	Z	3.5435	0.9198
N: 198	p (1-sided)	0.0002	0.1788
Events: 71	HR at bound	0.4287	0.8026
Month: 18	Spending	0.0002	0.1103
Trial POS: 48.1%	B-value	2.2411	0.5817
Post IA POS: 80.7%	CP	1.0000	0.2565
	CP H1	0.9986	0.8013
	PP	0.9984	0.3395
	P(Cross) if HR = 1	0.0002	0.8212
	P(Cross) if HR = 0.6	0.0809	0.1103
Final	Z	1.9609	1.9609
N: 260	p (1-sided)	0.0249	0.0249
Events: 176	HR at bound	0.7435	0.7435
Month: 36	Spending	0.0248	0.0397
	B-value	1.9609	1.9609
	P(Cross) if HR = 1	0.0189	0.9811
	P(Cross) if HR = 0.6	0.8500	0.1500

Note: Enrollment and total study durations are assumed to be 24 and 36 months, respectively. Efficacy bounds derived using a Hwang-Shih-DeCani spending function with gamma = −8. Futility bounds derived using a Hwang-Shih-DeCani spending function with gamma = 3.

The "complete" summary of boundary characteristics available is produced as follows for the time-to-event design and is shown in Table 8.2. For the time-to-event design, the effect size approximated at the bound is the hazard ratio of the experimental group relative to control at the boundary. The important difference from that given earlier is the `exclude=NULL` argument, which excludes none of the available boundary summaries from the output.

```
xprint(xtable(gsBoundSummary(ys, POS=TRUE, exclude=NULL),
  label="gsBSys", digits=4,
  caption=summary(ys)))
```

We provide a brief explanation of the available bound characteristics here and follow with more detail later in the chapter. The characteristics in the left-hand column of Table 8.2 are

- The percent of information at interim analyses (sample size in the binomial design; events for the time-to-event design).
- The sample size at each analysis. Note the increase relative from the previously computed fixed design sample size that produces 90% power, 238. For the time-to-event design, we note that even though only 40% of the total planned events are evaluated at the interim analysis, 197 patients out of the total of 260 are expected to be enrolled when the required events have been observed; note that this calculation does not account for further delay between the occurrence of the final event required for the interim and the completion of the interim analysis.
- For a time-to-event endpoint, this will also include the number of events at each analysis and the timing of the analysis.
- In the section for the first interim, the baseline probability of success (POS) is provided. Here, this is based on a default, relatively noninformative prior distribution for the treatment effect (discussed in the following).
- The posterior probability of success (Post IA POS) given the trial continues after the interim analysis. Because the aggressive futility bound requires at least a moderate interim observed treatment effect to continue, the POS has increased substantially from the baseline POS just by knowing the interim futility has been passed and the interim efficacy bar has not. This quantity also depends on the specified prior distribution.

The available summaries in the efficacy and futility bound columns are:

- The Z-statistic at the boundaries.
- The nominal 1-sided p-value at the boundaries. For asymmetric designs, this value corresponds to the probability of the boundary Z-value or smaller for both the futility and efficacy bounds.
- The approximate treatment effect size at the bound. This is labeled *HR* here to represent the approximate hazard ratio required to cross a bound; for the binomial design in Table 8.1, this was labeled pE − pC.
- Incremental error spending at each analysis as defined by the design spending function. These will sum to the total spending of α or β across all analyses.
- The conditional power of a positive efficacy finding if the interim test statistic is at the bound. This is computed both under the alternative hypothesis treatment effect (CP H1) and under the approximated treatment effect assuming a Z-value at the bound (CP).
- The predictive probability or power (PP) of a positive efficacy finding if the interim Z-statistic is at the bound. This is based on updating the prior distribution for the treatment effect based on the interim result.

- The *B*-value at the bound.
- The cumulative probability of crossing the boundary under the null and alternative hypotheses. These calculations assume the trial stops the first time a boundary is crossed. Since the designs here have nonbinding futility bounds, at the end of the trial, the cumulative probability of crossing the upper bound under the null hypothesis is less than the nominal Type I error for the trial.

8.6 Simulating the Binomial Design

We show a simple simulation of the interim analysis under the null hypothesis for the binomial trial so that we may later compare some of the asymptotic approximations in our output to those from a simulation. A simulation of one million trials takes only a few seconds using the built-in R function `rbinom`.

```
# number of simulations
nsim <- 1000000
# control response rate under H0
pC <- .677
# experimental response rate under H0
pE <- .607
# H0 delta value (used later)
delta0 <- pE-pC
# sample size per group at interim
n1 <- ceiling(.4*2358/2)
# simulation of control event count at interim
xC1 <- rbinom(n=nsim,size=n1,prob=pC)
# simulation of experimental event count at interim
xE1 <- rbinom(n=nsim,size=n1,prob=pE)
```

8.7 Individual Characteristics of Bounds

If individual characteristics are desired, one simple way to get them is from an individual row returned by the `gsBoundSummary` function. In the following sections, we show direct calculations. We also show the expected sample size. Concepts are defined in further detail than that is given earlier.

8.7.1 Z-Values

The efficacy and futility bounds at the interim analysis are

```
round(y$upper$bound[1], 3)
```

```
## [1] 3.544
```

```
round(y$lower$bound[1], 3)
```

```
## [1] 0.92
```

Note that since the Z-statistic at the futility bound is greater than 0, a positive interim treatment effect is required to pass the futility bound.

8.7.2 Boundary Crossing Probabilities

Rows of boundary crossing probabilities are labeled with values of theta equal to 0 corresponding to the null hypothesis and a greater value corresponding to the alternate hypothesis. The alternate hypothesis theta-value is determined by n.fix, the sample size required for a fixed design; this is done using Equation 8.16. One objective of the design is to stop early if, in truth, the new treatment is not as good as control. We can see from the following y$lower$prob table that there is a 82% probability of crossing the futility bound at the interim analysis under the null hypothesis. From the y$upper$prob table, we see there is only a 0.02% chance of crossing the efficacy bound at the interim analysis under the null hypothesis.

```
y$lower$prob
```

```
##          [,1]     [,2]
## [1,] 0.8212 0.11031
## [2,] 0.1599 0.03969
```

```
y$upper$prob
```

```
##             [,1]     [,2]
## [1,] 0.0001974 0.08093
## [2,] 0.0187208 0.76907
```

Here is code and the corresponding formatted table of upper boundary crossing probabilities under the null and alternate hypotheses (Table 8.3).

TABLE 8.3

Incremental Upper Boundary Crossing
Probabilities at Each Analysis under the
Null and Alternate Hypotheses

	H_0	H_1
Analysis 1	0.0002	0.081
Analysis 2	0.0187	0.769

```
require(xtable)
tab1 <- y$upper$prob
rownames(tab1) <- c("Analysis 1", "Analysis 2")
colnames(tab1) <- c("H$_0$", "H$_1$")
print(xtable(tab1, caption="Incremental upper boundary crossing
 probabilities at each analysis under the null and alternate
 hypotheses.",
   label="tab:one", digits = c(0, 4, 3)),
   sanitize.text.function=function(x){x})
```

We examine the above normal approximation values used to derive the
design versus the simulation results.

```
# interim Z-values for simulation
z1 <- testBinomial(x1=xE1, x2=xC1, n1=n1, n2=n1, delta0=delta0)
# compare simulation probability of futility under H0
# with normal approximation
sum(z1<y$lower$bound[1])/nsim
```

```
## [1] 0.8215
```

```
pnorm(y$lower$bound[1])
```

```
## [1] 0.8212
```

To get total power for each of the design, we sum upper boundary crossing
probabilities under the alternate hypothesis as follows:

```
sum(y$upper$prob[, 2])
```

```
## [1] 0.85
```

8.7.3 Natural and Standardized Parameters

Here we demonstrate a general approach to approximating the observed natural parameter and standardized parameter given an interim normal test statistic value, Z_i. We assume the natural parameter δ has an estimate $\hat{\delta}_i$ with $E\{\hat{\delta}_i\} \approx \delta$ and $\mathrm{Var}\{\hat{\delta}_i\} \approx \psi^2/n_i$ for some value ψ, $1 \leq i \leq k$. Recall the standardized parameter is $\theta = \delta/\psi$ and its estimate at analysis i as $\hat{\theta}_i = \hat{\delta}_i/\psi$ or $\hat{\theta}_i = \hat{\delta}_i/\hat{\psi}_i$, $1 \leq i \leq k$.

If we let

$$Z_i = \frac{\sqrt{n_i}\hat{\delta}_i}{\psi} \approx \sqrt{n_i}\hat{\theta}_i, \tag{8.17}$$

then Z_i is approximately distributed as $N(\sqrt{n_i}\delta/\psi, 1)$, $1 \leq i \leq k$. In practice, ψ will be replaced with an estimate (see binomial example) or the test statistic will be computed in a completely different fashion, such as a logrank test for the time-to-event example given later. However, we will assume (8.17) can be used as an approximation. This immediately allows us to translate a presumed interim Z_i-value into a corresponding estimate of the standardized treatment effect:

$$\hat{\theta}_i \approx Z_i\sqrt{n_i} \tag{8.18}$$

Based on an input sample size for a fixed design with no interim analyses, gsDesign computes a value θ_1 that corresponds to δ_1 and stores it in the variable delta. Assuming $\psi = \theta/\delta$ is equal to or well approximated by θ_1/δ_1, for any presumed θ-value, we can compute a corresponding

$$\delta = \delta_0 + \frac{\theta(\delta_1 - \delta_0)}{\theta_1}. \tag{8.19}$$

Thus, assuming an interim test statistic Z_i, we would approximate $\hat{\delta}_i$ by

$$\hat{\delta}_i \approx \delta_0 + \frac{Z_i\sqrt{n_i}(\delta_1 - \delta_0)}{\theta_1}. \tag{8.20}$$

Replacing Z_i with an interim boundary value for a group sequential design, we can approximate the observed treatment effect if a test statistic were at that bound using (8.18) for the standardized parameter and (8.20) for the natural parameter, $1 \leq i \leq k$.

8.7.4 Difference in Binomial Success Rates

For a difference in binomial parameters $p_2 - p_1$ (1 = control, 2 = experimental), we assume ξn_i observations in the experimental group and $(1 - \xi)n_i$ in the control group, and observed event rates of $\bar{p}_{1i}$ in the control group and $\bar{p}_{2i}$ at analysis i, $i = 1, 2, \ldots, k$. Then

$$\psi^2 = \frac{p_2(1 - p_2)}{\xi} + \frac{p_1(1 - p_1)}{1 - \xi}. \tag{8.21}$$

In this case, we test with a noninferiority margin $\delta_0 = 0.07$ and Z_i is a Miettinen and Nurminen (1985) test statistic where ψ is replaced by a maximum likelihood estimator $\hat{\psi}_i$, $i = 1, 2, \ldots, k$:

$$Z_i = \frac{\bar{p}_2 - \bar{p}_1 + \delta_0}{\hat{\psi}/\sqrt{n_i}}. \tag{8.22}$$

For the binomial noninferiority trial, we have $p_1 = p_2 = 0.677$, which yields a ψ^2-value of

```
psi2 <- 4 * 0.667 * (1 - 0.677)
sqrt(psi2)
```

```
## [1] 0.9283
```

Without applying the specific binomial formula, we estimate $\psi \approx \theta_1/\delta_1$ as follows. We have $\delta = 0.07$, the noninferiority margin, and θ_1 is stored in y\$delta. It is unfortunate that y\$delta is on what we are referring to as the θ-scale, while 0.07 is on the δ-scale; this is a conflict between gsDesign and standard notation used in textbooks Jennison and Turnbull (2000); Proschan et al. (2006).

```
psi <- 0.07/y$delta
psi
```

```
## [1] 0.9345
```

We use this latter estimate of ψ, since it generalizes to other endpoint types and the two approximations are similar. For the noninferiority design with a binomial outcome, we have the approximation

$$\bar{p}_{2i} - \bar{p}_{1i} \approx \frac{Z_i \psi}{\sqrt{n_i} - \delta_0}. \tag{8.23}$$

At the interim lower bound for the binding and nonbinding designs, we thus approximate the treatment difference by

```
y$lower$bound[1]/sqrt(y$n.I[1]) * 0.07/y$delta - 0.07
```

```
## [1] -0.03998
```

Again, since the timing and bound are the same for the truncated and fully powered designs, these treatment estimates are also the same. We compare this value with the event rate differences from the simulation, which result in Z-statistics near the bound.

```
pdiff <- (xE1-xC1)/n1
range(pdiff[abs(z1-y$lower$bound[1])<.001])
```

```
## [1] -0.04237 -0.04237
```

8.7.5 Approximating the Hazard Ratio at a Bound

Estimating the hazard ratio at a bound for the time-to-event designs is a simple extension of the above. We approximate the treatment effect on the δ-scale as before for the nontruncated and truncated designs, respectively. Since this scale is for the logarithm of the hazard ratio, we exponentiate to get the hazard ratio at the futility bound. The hypothesized hazard ratio under H_1 is 6/10, the ratio of hypothesized median time-to-event in the two groups.

```
# alternative hypothesis hazard ratio
hr1 <- 6/10
hr1
```

```
## [1] 0.6
```

```
# Approximate hazard ratio if interim test statistic is at the
# lower bound. First, use general approximation above
hr <- exp(ys$lower$bound[1]/sqrt(ys$n.I[1]) * log(hr1)/ys$delta)
hr
```

```
## [1] 0.8033
```

Note that the approximation is slightly different than that produced by gsBoundSummary previously. That approximation was based on the Schoenfeld (1981) asymptotic approximation to the distribution of $\hat{\delta}_i$. This can be computed as follows:

```
# Use approximation based on Schoenfeld approximation. This is
# the value used in gsBoundSummary
```

```
gsHR(z = ys$lower$bound[1], i = 1, x = ys)
```

```
## [1] 0.8026
```

The difference between these is that the first bases the ψ-value on the Lachin and Foulkes (1986) sample size compared to the slightly simpler Schoenfeld (1981) method.

8.7.6 B-Values

The B-value at the interim analysis is calculated by $\sqrt{t_1}Z_1$ where Z_1 is the Z-statistic and $t_1 = .4$ is the proportion of statistical information at the interim analysis. The B-values at the interim lower bound for each design are computed as follows:

```
y$timing
```

```
## [1] 0.4 1.0
```

```
y$lower$bound[1] * sqrt(y$timing[1])
```

```
## [1] 0.5817
```

Advantages of B-values are that the expected value is linear in t_i so that deviations in the trend over time can be evaluated by a deviation from linearity. B-values also provide a particularly simple exposition of conditional power; see, for example, Proschan et al. (2006).

8.7.7 Expected Sample Size

Expected sample size under the null and alternative hypotheses is computed in `gsDesign` and returned as follows from the binomial design under consideration:

```
y$en
```

```
## [1] 1039 1814
```

We can use the simulation to approximate the expected sample size under the null hypothesis and check versus the given null approximation (1039):

```
# probability of stopping (bound crossed) in simulation
pstop <- sum(z1<y$lower$bound[1]|z1>=y$upper$bound[1])/nsim
```

```
# interim and final sample size (round up to even number)
n1 <- ceiling(y$n.I[1]/2)*2
n2 <- ceiling(y$n.I[2]/2)*2
# expected sample size
n1 * pstop + n2 * (1-pstop)
```

```
## [1] 1039
```

Neither of these calculations accounts for enrollment not included in an interim analysis if that interim stops the trial; this overrun probably will be accounted for in future releases of the gsDesign package.

The survival endpoint accounts for enrollment that occurs while events accrue for an interim analysis, but does not currently account for the additional enrollment that is expected to accrue between the time enough events have occurred to do the interim and the time the actual interim analysis is performed. Again, this overrun will probably be accounted for in a future release of gsDesign.

```
# need to add experimental and control sample sizes
ys$eNE + ys$eNC
```

```
##         [,1]
## [1,] 196.3
## [2,] 259.5
```

8.7.8 Conditional Power

Conditional power can be computed for an estimated treatment effect or for a given, fixed treatment effect. We will compute conditional power assuming a test statistic at the lower bound at the interim analysis. The default is to compute for the observed treatment effect, the null hypothesis treatment effect, and the alternate hypothesis treatment effect. The basic call to compute these conditional power values is

```
ycp <- gsCP(x = y, i = 1, zi = y$lower$bound[1])
ycp$upper$prob
```

```
##         [,1]    [,2]    [,3]
## [1,] 0.2565 0.03749 0.8013
```

The `theta` values corresponding to the estimated treatment effect, null hypothesis treatment effect, and alternate hypothesis treatment effect are obtained by

```
ycp$theta
```

```
## [1] 0.03213 0.00000 0.07491
```

Selecting the treatment effect used to compute conditional power is problematic. We see here that the conditional power assuming $\theta = \theta_1$ yields a conditional power of 80.1% while assuming the $\theta = \hat{\theta}_1$, the interim trend, yields a conditional power of only 25.7%. An interim estimate of treatment effect is quite unreliable, making the latter choice risky. Choosing $\theta = \theta_1$ assumes the interim trend was random. Perhaps, this is also risky. This leads to the concept of predictive power discussed in the next section.

8.7.9 Predictive Power

Predictive power is computed by averaging conditional power over a posterior distribution for the standardized effect size parameter θ. The posterior distribution can be computed based on a prior distribution for θ and the interim test statistic. Example calculations for this will assume the default normal prior distribution for θ. We consider the design we have stored in y and denote the standardized effect size for the alternate hypothesis by θ_1 =y$delta. The default prior distribution for θ is assumed to be $N(\mu = \theta_1/2, \sigma = 10/\sqrt{n_{fix}}$ where n_{fix} is the fixed design sample size required to power the trial as desired. This is generated as follows:

```
prior <- normalGrid(mu = y$delta/2, sigma = 10/sqrt(y$n.fix))
summary(prior)
```

```
##             Length Class  Mode
## z           185    -none- numeric
## density     185    -none- numeric
## gridwgts    185    -none- numeric
## wgts        185    -none- numeric
```

The prior variance is equivalent to that of an estimate of θ based on a sample size that is 1% of the fixed design sample size for the treatment comparison of interest. The values produced in prior include a grid z on which the prior distribution is computed, the prior density values at those points in density, gridwgts which can be used for integrating any function over the given grid, and wgts which can be used to integrate the density over the chosen set of points. Of these, we really only need z (the grid for integration) and wgts (density adjusted by numerical integration constants in gridwgts).

The next step is to compute predictive power for interim `i=1` with an interim test statistic at the lower bound `zi=y$lower$bound[1]`:

```
gsPP(x=y, theta=prior$z, wgts=prior$wgts, i=1,
     zi=y$lower$bound[1])
```

```
## [1] 0.3395
```

Note that predictive power can vary considerably based on the prior distribution assumption. However, if there is good prior knowledge, this provides a good way to predict success by combining prior information with an interim result. As noted, the default in `gsBoundSummary` is a relatively noninformative prior which should be useful for prediction when there is little prior information.

8.7.10 Probability of Success and Predicted Probability of Success

At the beginning of a trial, the probability of a positive trial might be predicted by averaging the probability of a positive trial over a prior distribution for the treatment effect. This might be termed the predictive power at the beginning of the trial or the POS for the trial. Using the same prior distribution we used for PP, this is computed in the following.

```
gsPOS(x = y, theta = prior$z, wgts = prior$wgts)
```

```
## [1] 0.481
```

At the time of an interim analysis, the sponsor of a trial is generally only informed of whether or not to continue the trial. Normally, this would correspond to not crossing a stopping boundary, although there are cases where a trial continues in spite of crossing a boundary due to, say, secondary endpoint or safety considerations. Here we compute the conditional or updated POS, given that a boundary has not been crossed at an interim analysis.

```
gsCPOS(x = y, theta = prior$z, wgts = prior$wgts, i = 1)
```

```
## [1] 0.8074
```

8.8 Discussion

We have shown many characteristics of a group sequential design boundary that may be useful in helping to choose interim analysis bounds. When selecting a bound, study designers are encouraged to consider the trade-offs in different characteristics such as maximum sample size, probability of early stopping under the null hypothesis, β-spending, and interim timing.

For trials with more than one interim analyses, the reader is encouraged to evaluate piecewise linear spending using the function `sfLinear` or one of the two parameter spending functions, which are summarized in `help(sfLogistic)`. The flexibility provided enables you to customize boundary characteristics more than traditional one-parameter spending functions, such as the Hwang-Shih-DeCani spending function considered here. See Anderson and Clark (2010) for further details on this approach.

In practice, the reader will want to plan an appropriate interim analysis plan for his or her trial. Some variations that the reader may wish to consider when designing a trial or when trying to understand the recommendations here are

- If the median control time-to-event is shorter or longer, what is the implication on when to plan an interim analysis?
- If the power is maintained at 90% for a group sequential design, what are the sample size implications of a less stringent futility bar? Consider both the probability of early stopping under the hypothesis of no treatment difference and also the total sample size.
- What are the implications of an earlier futility analysis in terms of enrollment at the interim, how stringent the interim bar can be in terms of an observed difference, conditional power, Type II error at the interim, and the increase in sample size required to maintain the desired power? Try things like `timing=.3` or `sflpar=-1` to see the impact.

With these considerations, you should be able to decide on designs that provide appropriate interim analysis guidance.

References

Anderson, K. M. and Clark, J. B. (2010). Fitting spending functions. *Statistics in Medicine*, 29:321–327.

Farrington, C. P. and Manning, G. (1990). Test statistics and sample size formulae for comparative binomial trials with null hypothesis of non-zero risk difference or non-unity relative risk. *Statistics in Medicine*, 9:1447–1454.

Fleiss, J. L., Tytun, A., and Ury, H. K. (1980). A simple approximation for calculating sample sizes for comparing independent proportions. *Biometrics*, 36:343–346.

Hwang, I. K., Shih, W. J., and DeCani, J. S. (1990). Group sequential designs using a family of type 1 error probability spending functions. *Statistics in Medicine*, 9:1439–1445.

Jennison, C. and Turnbull, B. W. (2000). *Group Sequential Methods with Applications to Clinical Trials*. Chapman & Hall/CRC, Boca Raton, FL.

Kim, K. and Tsiatis, A. A. (1990). Study duration for clinical trials with survival response and early stopping rule. *Biometrics*, 46:81–92.

Lachin, J. M. and Foulkes, M. A. (1986). Evaluation of sample size and power for analyses of survival with allowance for nonuniform patient entry, losses to follow-up, noncompliance, and stratification. *Biometrics*, 42:507–519.

Miettinen, O. and Nurminen, M. (1985). Comparative analysis of two rates. *Statistics in Medicine*, 4:213–226.

Proschan, M. A., Lan, K. K. G., and Wittes, J. T. (2006). *Statistical Monitoring of Clinical Trials. A Unified Approach*. Springer, New York.

Schoenfeld, D. (1981). The asymptotic properties of nonparametric tests for comparing survival distributions. *Biometrika*, 68:316–319.

Xie, Y. (2014). *Dynamic Documents with R and knitr*. CRC Press, Boca Raton, FL.

Section III

Clinical Trials in Oncology

9

Issues in the Design and Analysis of Oncology Clinical Trials

Stephanie Green

CONTENTS

9.1 Introduction

The history of clinical trials perhaps goes back as far as biblical times, as described in Daniel 1:11–16: "Then Daniel said to the steward whom the chief of the eunuchs had appointed over Daniel, 'Test your servants for ten days; let us be given vegetables to eat and water to drink. Then let our appearance and the appearance of the youths who eat the king's rich food be observed by you, and according to what you see deal with your servants.' So he hearkened in this matter, and tested them for ten days. At the end of ten days it was seen that they were better in appearance and fatter in flesh than all the youths who ate the king's rich food. So the steward took away their rich food and the wine they were to drink, and gave them vegetables" (Daniel 1:11–16; The Holy Bible, 1962) Rudimentary, but some principles of clinical trials are recognizable.

The next identifiable trial did not occur until 1747, when James Lind performed the first planned, controlled experiment in human subjects when he tested six approaches to the treatment of scurvy at sea in two patients apiece, from which he identified citrus as a cure (Lind, 1753). Sadly, his observations were not followed up with as much rigor as the initial experiment, plus the cost of lemons was high, so were not effectively implemented until decades later (Thomas, 1997; Cook, 2004).

Numerical methods and principles of clinical studies developed during the 1800s, but there was little to test in the way of treatment options until the twentieth century. Early investigations in cancer treatment involved surgery and radiation therapy (radical mastectomy for primary breast cancer, oophorectomy for advanced breast cancer, radiation for head and neck cancer), and systemic treatment was introduced in the 1940s (nitrogen mustard for lymphoma, folic acid antagonists in leukemia) (DeVita and Rosenberg, 2012). The first modern randomized clinical trial tested streptomycin as treatment for tuberculosis in 1946, which successfully demonstrated efficacy of the agent. Cancer trials started soon after, with the first National Cancer Institute (NCI) sponsored trial in the United States being a comparison of chemotherapy regimens in acute leukemia (Zubrod, 1982). There is now a substantial history of important clinical trials in cancer, such as those demonstrating cures for childhood cancers, the NSABP trial indicating that lesser surgery than radical mastectomy should be used, and the trial of imatinib in chronic myelogenous leukemia (CML) which ushered in the era of targeted cancer therapies.

Randomized clinical trials are blunt tools but are the best method available for proving the usefulness (or not) of new treatments. These trials and the trials leading up to them must be designed and conducted with care. Numerous design and analysis issues need to be considered to assure that results will be interpretable and will reliably inform treatment decisions.

9.2 Design

9.2.1 Phase 1 and Phase 1/2 Design

The primary aim of a phase 1 design is to identify a dose for further testing of a new agent. Multiple dose levels are administered and assessed for toxicity, tolerability, and feasibility. Most designs for cancer trials are predicated on the assumption that antitumor benefit increases with increasing dose. Thus, the primary aim is to estimate the maximum tolerated dose (MTD), with efficacy assessment an informal secondary objective. Issues in the design of phase 1 trials arise from limited information available on new agents. Little may be known about safety at this stage of development, so caution concerning starting dose and aggressiveness of escalation is very important. However, excessive caution may result in long trials and excessive numbers of patients being treated at nonefficacious doses. Optimal trade-offs are difficult to identify.

9.2.1.1 Classic 3+3 Design

The most common design still used for cancer agents is the $3 + 3$ design, in which patients are enrolled in cohorts of three. For each new dose level tested, if there are no dose limiting toxicities (DLTs), the next three patients are treated at the next higher dose level. If there is one DLT, the next three are treated at the same dose level and escalation continues if 0 or 1 experiences a DLT. If there are two or more DLTs at a dose level (either $\geq 2/3$ or $\geq 2/6$), this level is concluded too toxic. In this case, if six patients have been treated at the previous dose level, this is specified as the MTD; if only three have been treated, three additional patients are accrued and if none of these experiences a DLT, this is specified as the MTD. Properties of 3+3 design are not ideal. MTD estimates are biased and sensitive to both starting dose and shape of the dose–toxicity curve (Storer, 1989). The number of patients treated at low doses tends to be high, which means too many may be receiving ineffective treatment with respect to efficacy. In addition, the design is inflexible, in that there is no standard modification for the design if 16%–33% is not the desired target toxicity level.

9.2.1.2 Algorithm-Based Designs

Numerous designs have been proposed to improve upon the old standard. Accelerated titration (Simon et al., 1997) is one approach to making phase

1s more efficient. The concept is to escalate with relatively large increases in dose in one patient per cohort until a DLT is observed and then switch to a more conservative scheme. Potential benefits are more dose levels examined using fewer patients and higher probability of receiving an active dose. Other approaches involve use of up and down algorithms that better target the correct MTD (Ivanova et al., 2003; Storer, 2011). An example is the k-in-a-row design, for which de-escalation occurs if the most recent patient experiences a DLT, escalation occurs if the most recent K patients all received the same dose and none had a DLT, and otherwise dose remains the same. Advantages of these designs are ease of implementation and improved properties due to use of more of the cumulative trial information to decide whether to increase, decrease, or stay the same.

9.2.1.3 Model-Based Designs

Another class of designs are model based, with continual reassessment methods (CRMs) (O'Quigley et al., 1990) and modifications the most commonly used. For these designs, a target probability π of DLT is specified and the dose–toxicity relationship is assumed to follow a nondecreasing function $f(d) =$ population probability of a DLT at dose d. The true MTD is then precisely defined as the dose d^* for which $f(d^*) = \pi$. A typical CRM design accrues patients in cohorts of 1–3. After each cohort is assessed, the MTD is reestimated using Bayesian methods and the next cohort is treated according to the dose closest to but not higher than the new estimate (subject to not skipping doses). Accrual is stopped after either N patients in a row are treated at the same dose or until a maximum sample size is reached. Simulations indicate design properties generally are better than for the 3+3 design. One variation on CRM is escalation with overdose control (EWOC, no relation to Star Wars), which considers the probability of DLT for all doses, not just the MTD, with the aim of restricting the probability of overdosing. Simulations suggest this is achievable, but at the expense of longer trial time (Tighiouart et al., 2005). See Chapter 11 for a detailed discussion of EWOC designs.

9.2.1.4 Designs When Monotonicity Is Not Assumed

Simple assumptions of increasing effect with increasing dose do not always apply. For instance, if two agents in a combination are being escalated, then the dose relationship would follow a partial ordering (increasing toxicity with escalation of one agent when the other is fixed, but no assumptions when dose is increased for one and decreased for the other). Schemes to identify combinations with probability of DLT close to the target probability have been proposed, including CRM designs for two agents (Yuang and Yin, 2008) and up and down designs followed by isotonic regression for estimation of DLT probabilities subject to the partial ordering across the two-dimensional dose grid (Ivanova and Wang, 2004).

New biological agents provide other examples. Efficacy may be expected to decrease or plateau instead of continuing to increase with dose; in this case, escalating to the MTD and assuming that this will be the best choice would not be optimal. In this setting, joint assessment of both response (tumor shrinkage or other early endpoint reflecting antitumor activity of the agent) and toxicity may be useful in recommending a dose for further study, as discussed in Section 9.2.1.5.

9.2.1.5 Phase 1/2 Designs

Designs jointly assessing toxicity and response to determine a dose for further testing are phase 1/2 designs. The aim is to identify doses with acceptable toxicity and response. Approaches are available for both a three-outcome case (toxicity, no response, response without toxicity) and for two two-outcome cases. One approach (Thall and Cook, 2004) for the latter is to define toxicity–response contours, which are sets of probabilities of equal acceptability. The contours are defined by eliciting clinical opinion on minimum acceptable response probability given 0 probability of DLT (point $(p_r, 0)$), maximum acceptable probability of DLT given probability 1 of response (point $(1, p_t)$), and one additional point of the same acceptability (point (p_1, p_2)). Contours closer to the ideal $(1, 0)$ are more desirable than those further away (Figure 9.1). The general approach is, after each assessment, to identify acceptable doses given the current data, treat the next cohort of patients at either the next untested dose or the dose corresponding to the most desirable contour, whichever is lower. The study is stopped early if there

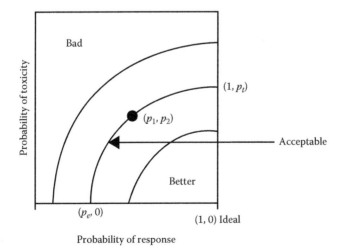

FIGURE 9.1
Efficacy–toxicity contours defining sets of equally desirable outcomes.

are no acceptable doses. If not stopped early, the most desirable acceptable dose at the end of the trial is chosen for further study. Acceptability is defined using Bayesian methods; a dose is acceptable if it is reasonably likely that the response probability is $>p_e$ and toxicity probability is $<p_t$. Simulations were promising, including modest sample size when there is no acceptable dose and rapid identification of dose in the setting of increasing and then decreasing response. Similar designs have been proposed for continuous biomarkers (Bekele and Shen, 2005).

The typically small sample sizes and limited possible dosing levels of phase 1 cancer studies preclude highly accurate estimation of the MTD. Improvements over the standard 3+3 design are needed; yet, the standard 3+3 is still the most widely used, with newer designs making only slow inroads into practice (Penel et al., 2009). Although a practical drawback of many of the newer designs is the need for simulations to determine design properties and to calibrate errors, these have potential to be useful design options in various settings and are worth considering.

9.2.2 Phase 2 Design

The primary aim of phase 2 designs is to preliminarily assess efficacy of new agents or combinations in order to determine if further, more definitive study is warranted. Trials of modest size are designed to limit time and resources needed to stop development of regimens not likely to be useful or to begin definitive assessment in a reasonable time frame. Phase 2 issues are related to these modest sizes and time pressures. Short-term outcomes such as RECIST response (Eisenhauer et al., 2009), the most common definition related to tumor shrinkage for solid tumors, are often used. These may or may not provide adequate information on which to base a decision to launch a trial with a long-term endpoint, which typically is progression- or disease-free survival (PFS, DFS) or overall survival (OS). Another issue is that insufficient information on outcome for standard of care may be available to design a convincing single arm trial using hypotheses derived from reports in the literature. On the other hand, small randomized trials typically have both high false positive and false negative rates, making these unreliable as well.

9.2.2.1 Single Arm Designs

If outcome for a particular population is well characterized, for example, if outcome for patients on standard of care has been uniformly poor for a long time, then a single arm trial is useful. A high level of activity in such patients could be sufficient to inform a decision to study the regimen further. With increasing availability of new treatments, this is less common than in the past, but if appropriate, the usual setup is to choose null and alternative hypotheses, $H_0: p = p_0$ vs $H_A: p = p_A$, where p_0 and p_A are response probabilities not of interest and of interest respectively based on historical activity in the disease.

An interim analysis to allow for an early stop due to lack of efficacy is typically included. N_1 patients are accrued. If at least r_1 responses are observed, then an additional N_2 patients are accrued. At the end of the trial, if at least r_2 responses in the $N_1 + N_2$ patient are observed, and if toxicity is acceptable, then continued development of the regimen is recommended. Various ways to choose sample size and stopping rules have been proposed; for example, a commonly used approach is to minimize the expected sample size for given level and power (Simon, 1989). Note this and other minimization approaches require that the precise number of patients be enrolled at each stage; otherwise, optimality is lost. If precise numbers cannot be enrolled, a practical approach is to test the alternative ($p = p_A$ vs $p < p_A$) at the interim analysis at a conservative level, and the null at the level α at the final analysis (Green and Dahlberg, 1992).

In some settings, it is important to consider both probability of response and probability of early treatment failure. A number of designs for three level outcome (such as response, early fail, no response/no early fail) have been developed, which require both that the response be sufficiently high and that early failure be sufficiently low, as a potentially better measure of effectiveness of the treatment. Another setting where this approach may be used is in *window* designs, for which newly diagnosed patients are given new treatments for a short period of time prior to starting standard treatment under the assumption that activity of new treatments might better be assessed in patients without prior treatment. Due to the potential for delay in standard treatment putting patients at risk, early stopping rules for early failure are incorporated (Chang et al., 2007).

If populations are less well characterized but still only a moderate range of outcomes is considered likely, a three-level decision rule is a possibility (Storer, 1992). For this design, if the control probability is assumed to be in the range of p_L to p_H, the regimen is considered promising if the hypothesis $H_1: p < p_H$ is rejected, unpromising if the hypothesis $H_0: p < p_L$ is not rejected, and of uncertain promise (further testing needed) otherwise. If uncertainly instead is related to distribution of two subpopulations, each of which are reasonably well characterized (e.g., two disease subtypes for which similar treatments are used but one is more aggressive disease) but percent of each that will be enrolled is uncertain, a design allowing for different proportions of poor risk patients can be used. For these designs, nulls are specified for each group and the test statistic is conditional on the number of patients accrued to each group (London and Chang, 2005; Jung et al., 2012). Subset results are not tested with these designs, but for a heterogeneous population, this strategy is potentially more useful than assuming a single null probability.

If response within marker subgroups is of interest for a targeted agent and little is known about expected response in marker subsets (usually the case), then single arm phase 2s are inadequate to examine the question of whether treatment works better in the subset. A single arm trial cannot distinguish between prognostic factors (marker associated with outcome in general) and

TABLE 9.1

Prognostic vs. Predictive: Not Possible to Distinguish in a Single Arm Study

	Prognostic		Predictive		Single Arm Phase 2 Information for Each	
	Marker +	Marker −	Marker +	Marker −	Marker +	Marker −
Control	45%	15%	30%	30%	30%	
Experimental	60%	30%	60%	30%	60%	30%

predictive factors (differential effect of treatment based on the marker). If the probability of response is higher in marker positive than marker negative, and if there is no information available on response by marker in general to help interpretation, then it is not possible to tell whether marker + is a good prognostic factor for all patients or whether treatment is more effective in marker positive. For instance, consider Table 9.1. If 60% of patients respond in marker positive patients and 30% in marker negative for the targeted agent, and historically 30% of patients respond overall on standard treatment, then this is consistent with treatment being effective only in marker positive patients—if 30% respond to standard treatment regardless. But this is also consistent with the marker being prognostic and treatment being effective in all patients—if half are in each group and 45% and 15%, respectively, respond in marker positive and negative patients on standard treatment, then the treatment difference is the same for each group.

Occasionally, a specific set of patients treated in the past is chosen for comparison with new patients treated on the new regimen, for example, patients from a trial with similar eligibility or a recent sequence of patients with the same diagnosis at an institution. Since results of the control group are already known, sample size considerations are a function of these results, and standard methods for use with prospective control groups do not apply (Dixon and Simon, 1988; O'Malley et al., 2002). In rare populations, this may be the only feasible option for comparison, but results must be interpreted cautiously. Despite best efforts, the patient characteristics in the two groups will differ systematically in both known (such as time frame of the study, amount of follow-up, differences in prognostic factors) and unknown ways.

Differences in outcome for cancer treatments are generally modest, so effect of selection bias when assessing against historical controls is likely to be of similar or larger magnitude than the effect of treatment. An example is high-dose (HD) therapy with peripheral blood stem cell support in breast cancer. Pilot studies appeared highly promising in contrast to results from previous studies with lower dose therapy. However, patients eligible for such aggressive therapy are substantially better risk than patients eligible for standard treatment and so would be expected have better outcome regardless. Thus, historical results were not an appropriate benchmark and, unfortunately, randomized trials did not demonstrate benefit of the HD approach. HD therapy with autologous bone marrow transplant in multiple myeloma provides another example. Pilot results looked very promising compared to results from trials of standard treatment (Barlogie et al., 1995). However, when results from previous trials were restricted to patients who would have been eligible for the HD study, standard treatment results looked substantially better (Green et al., 2012). As for the breast cancer example, the randomized trial did not demonstrate benefit of HD therapy (Barlogie et al., 2006).

9.2.2.2 Randomized Phase 2 Designs

The difficulties inherent in identifying promising regimens based on single arm trials is one of the reasons randomized phase 2 studies have become more common in recent years. Options include selection designs, discontinuation designs, and designs with a control arm.

For a selection design, the aim is to decide which of multiple candidate treatments should be taken forward for further study. The intent is not a definitive comparison, rather to choose a treatment that is not likely much worse than the other candidate treatments. The arm observed best by any amount is chosen. Sample size is chosen such that if one treatment is superior by Δ, then the probability of choosing that treatment is at least π. Versions of this design have been developed for response (Simon, 1987) and PFS (Liu et al., 1993). For example, if $\Delta = 0.15$ and $\pi = 0.9$, then a response trial with 3 candidate agents would require 55 patients per arm. A PFS trial with hazard ratio $\Delta = 1.5$ has a similar sample size requirement, 54 per arm. A limitation of these designs is that a single regimen is always chosen, even if none or more than one seem promising; the designs can be modified to require the chosen agent exceeds a threshold level of activity.

For a randomized discontinuation design, all patients begin treatment with the experimental regimen. Patients who remain stable without serious toxicity at a specified time point are randomized between experimental treatment and either placebo or standard treatment. The run-in phase is used to make the randomized group more homogeneous, allowing for fewer randomized patients to be needed. Limitations of the design include the possibility of a very large number of patients accrued in order to get the required number of randomized patients and lack of interpretability of results. A positive result might suggest the agent has activity, but basically, the design answers the question of whether patients on a new regimen who are doing well should continue on the regimen. It does not necessarily provide a good answer to the question of whether patients should be treated with the regimen in the first place.

For two arm phase 2s with a control arm, these trials can be viewed as underpowered phase 3s with high level, poor power and often with a primary endpoint that is not the one planned for the phase 3. As such, it should not be a surprise that these are also not necessarily reliable. Part of the issue is that many new treatments are ineffective. For instance, with all else being the same in phase 2 and 3, if only 10% of treatments are effective and if randomized phase 2s with level 0.15 and power 0.8 are used, only 37% of positive phase 2s will be true positives. If single arm hypotheses are correct (big if), testing can be done with a modest sample size at more conservative level and power. For level 0.05 and power 0.9, 67% of positives would be true positives. Another reason for lack of reliability is use of short-term endpoints. In particular, while response typically does have an association with longer survival, a difference in the estimates of probability of response typically is not a good predictor of difference in survival distributions. Longer-term outcomes more likely to inform the phase 3 primary outcome measure should be used if feasible.

9.2.2.3 Phase 2 Strategy

The track record for phase 2 prediction of phase 3 success is not good. In a report (Zia et al., 2005) of 43 phase 3 cancer trials done after a positive phase 2 in a setting of the same treatment in the same patient population for the phase 3, experimental arm response was lower in 35 of the 43 phase 3s than in the phase 2s and only 12 of the 43 phase 3s were positive.

Phase 2 strategy will depend on various factors, including availability and reliability of historical information, size and heterogeneity of the patient population, type of error most important to control, and percent of new treatments expected to be effective. If the population is limited and there are multiple candidate treatments, a selection design might be required. If the population is large and historical information is poorly characterized, for instance, unknown standard treatment outcome in subsets of patients with recently identified biologic markers who will receive novel targeted therapy in the phase 2, then the knowledge gain from a larger randomized phase 2 might be in order. If well characterized with low variability, a single arm trial in the same population as the potential phase 3 would be the most efficient approach. If variability or historical characterization is moderate, then design compromises can be considered in the context of failure rate of new treatments, ability to launch phase 3 trials, and extent of unmet need. If the population is large and unmet need is high, a randomized phase 2 with high level and high power will result in lower phase 3 success (more false positives) but will also identify the most useful treatments (fewer false negatives), while if the population is small or if the budget is small, the reverse will improve phase 3 success rate at the expense of missing some effective treatment options. A staged approach, such as using the three-level decision rule and following with an additional phase 2 study if results are in the inconclusive zone, has potential to improve phase 3 yield without always doing randomized phase 2s, but at the expense of delayed decision making.

The limited information available in phase 2 studies, plus the modest fraction of new treatments that are useful in cancer, precludes high accuracy in identifying treatments likely to succeed in definitive trials. Phase 2 approach should be tailored accordingly; in particular, the less information already available, the larger the phase 2 should be.

9.2.3 Phase 3 and 2/3 Design

Randomization is the key feature of phase 3 trials. Randomization assures comparability of patient groups at the time of randomization and, if the trial is properly conducted, allows the conclusion that outcome differences between treatment arms can be ascribed to treatment differences rather than systematic differences between arms unrelated to treatment. Other characteristics that must be addressed are level and power, and tests and testing strategy (including interim testing). In addition, it should be kept in mind that a sound design alone is not sufficient for a successful trial. Flaws in trial conduct and patient noncompliance can introduce biases that compromise the integrity of the trial and render results uninterpretable.

9.2.3.1 Randomization

The primary factors to consider when choosing a randomization scheme are balance and predictability. There is a trade-off with respect to these, in that the most unpredictable scheme is simple randomization—random assignment of each patient with no balancing for prognostic factors—while enforced balance necessarily introduces some level of predictability. Although simplicity and unpredictability are desirable, the risk of chance imbalance compromising interpretation of results is an important consideration, so some method of balancing is typically employed. The most common types are blocked randomization (number of patients on each arm is equal at the end of accrual to each block of assignments), blocked randomization within cells defined by stratification factors, and dynamic allocation (treatment assignment is weighted toward minimizing imbalance on the margins according to a measure of imbalance). Among balancing schemes there are trade-offs. Randomized blocks with small block sizes have the advantage of good treatment balance but can be highly predictable at a single institution if the block size is known (if this method is used, larger or variable sized blocks is recommended). For blocked randomization within cells, cell balance will generally be good if the number of cells is relatively small, but marginal and arm balance may not be good. However, if there are too many factors and there are few patients per cell, blocked randomization within cells will work poorly; for the typically moderate sample sizes used in randomized oncology trials, the number of stratification factors should be restricted (usually 3 or less). For dynamic randomization, marginal balance will be good, but cell and arm balance may not be. A general advantage to balanced randomization schemes is that balance is expected to be better on unmeasured factors than with simple randomization (Aickin, 2001). Another potential advantage is improved efficiency due to reduced variability, although benefits may be modest in practice (e.g., Begg and Kalish, 1984).

A less common approach is adaptive randomization, where the probability of assignment changes according to current efficacy results. Adaptive randomization has advocates, but there are multiple difficulties with this approach including the possibility of too few patients on one arm for a convincing result, loss of power due to unequal randomization, and potential bias if allocation percents change at the same time the type of patient enrolled changes. Another difficulty is that adaptive randomization requires efficacy results to be current, which is difficult if not prohibitive to implement in cancer trials.

In addition to type of randomization, timing of randomization is important. The first point to make is that treatment should start as soon as possible after randomization. Particularly with highly compromised patients, any delay risks the patient becoming unsuitable for treatment. Since such patients still must be included in the analysis, efficiency and interpretability of the trial are reduced. The second is related to the time treatment arms diverge. For instance, if the study design includes randomization for both induction and maintenance treatments, then a choice must be made whether to randomize the sequence up front or to randomize induction at the time of induction and maintenance at the time of maintenance. Different questions are answered depending on the choice. The first answers a question about planned treatment strategy. Patients who do not receive the planned maintenance treatment are still analyzed according to the assigned sequence; if compliance is poor, the comparison may be impossible to interpret. The second asks the induction question with respect to a short-term outcome and the maintenance question with respect to a long-term outcome conditional on having completed induction and willingness of patients to be randomized to maintenance. Two questions are answered, but the question of which sequence is most effective will not be. See Chapter 2 for detailed discussion of randomization methods.

9.2.3.2 Blinding

Use of placebos to blind patients and investigators to treatment assignment can be important in controlling sources of bias in the conduct of the trial, such as investigator's assessments of disease and toxicity plus decisions when to discontinue treatment. For patients, compliance with taking control treatment may be improved and patient-reported outcome (PRO) measures will not be influenced by knowledge of assignment if the blinding works. Furthermore, blinding improves confidence in the study results. However, it often is not feasible to blind a trial, either for practical reasons (e.g., surgery vs radiation therapy or IV vs oral therapy) or due to distinct toxicities associated with each treatment arm, which is quite common in cancer trials.

If a trial is placebo controlled, it can be difficult to tell whether the blinding worked. Testing effectiveness of blinding is tricky. For instance, testing whether or not the probability of a correct answer is >0.5 may not be a useful indication. If all patients who respond guess they are on the experimental arm regardless and all who do not respond guess they are on control, then if the experimental arm has more responders, guessing correctly will be >0.5 but not because patients knew which treatment arm they were on. In any case, testing is not usually reported (2% of blinded trials) and, when it is, results are mixed (14 of 31 reporting successful blinding) (Hrobjartsson et al., 2007).

9.2.3.3 Two Arm Trials

9.2.3.3.1 Basic Two Arm Trial

The most basic phase 3 is a two arm trial either comparing a control arm with a new treatment or comparing two standard treatments. Control arms may consist of no active treatment (rare in cancer due to severity of disease) or standard treatment with or without a placebo. The aim of a controlled trial is to demonstrate superiority of a new treatment, while the aim of a trial with two standards is to determine which, if either, is better. If the new treatment is targeted therapy, primary aims of testing within a marker defined subset may also be incorporated. Since phase 3 primary outcomes are usually PFS, DFS, or OS, statistical considerations are driven by a number of events. Sample size is a function of level, power, treatment difference (typically expressed as a hazard ratio), accrual rate, expected median or event rate on the control arm, and duration of follow-up. For diseases with potential for cure, it is also a function of cure rate. Required sample sizes increase with lower event rates and decrease with longer duration of follow-up. For instance, sample size for a 0.05 level trial with power 0.9 for a hazard ratio of 1.5 can range from 170 to 110 per arm when median on the control arm is 1 year by increasing postaccrual follow-up from 1 year to 5. If median on the control arm is 5 years (1/5th the death rate compared to a median of one if the distribution is exponential), the range of sample sizes changes to 390–220.

Since many cancer types are not common, feasibility decisions must be made concerning eligibility (e.g., expand eligibility for higher accrual rate or restrict eligibility to high risk patients for a higher event rate) and the time/sample size trade-off (long follow-up and wait for events or more patients for earlier events). Expected effect of treatment over time must also be considered; follow-up must be long enough to assess benefit over a suitable time frame.

9.2.3.3.2 Noninferiority

At times the aim of a trial is not to demonstrate superiority but to show non-inferiority (efficacy not substantially worse than on active control). Given the considerable toxicity of many cancer agents, noninferiority is an important concept since less toxic agents with similar efficacy can be of great benefit to patients. Noninferiority trials typically are designed to rule out a modest decrease in efficacy due to the less toxic treatment, with adequate power to conclude the treatment is noninferior when, in fact, the treatment difference is 0. Trials can be set up to test the hypotheses H_0: $\Delta \geq \Delta_0$ vs H_A: $\Delta < \Delta_0$, where Δ_0 is largest acceptable decrease in efficacy, or equivalently to generate the confidence interval for the difference with the intent to conclude noninferiority if the upper limit of the confidence bound is below Δ_0. Appropriate choice of Δ_0, the noninferiority margin, is key. Ideally, a good estimate of improvement of standard treatment over no treatment is available and the margin is chosen to show that at least a part of the benefit due to control is maintained as well as to show that the new treatment is likely similar to control. The ideal is rarely achievable in cancer trials (since trials with a no treatment control arm are rare), so the margin is usually chosen based on clinical grounds. Consequently, it may unclear at the end of the trial if non-inferiority has been demonstrated. Sample sizes for noninferiority trials with margin Δ_0 and power for 0 are similar to sample sizes for superiority trials with power for an alternative of Δ_0, so noninferiority trials are typically large. Chapter 5 addresses issues in noninferiority trials.

9.2.3.3.3 Designs for Targeted Agents

There are three main types of designs for two arm trials of targeted agents: all comers, marker positive, and strategy designs. For the all comer design, patients have markers evaluated but are randomized to the same treatment arms regardless of marker status. A secondary aim of the trial may be to investigate treatment effect within marker subsets. An example is a National Cancer Institute of Canada (NCIC) study of placebo vs erlotinib in nonsmall cell lung cancer (NSCLC) (Shepherd et al., 2005), for which overall benefit led to approval for unselected patients. However, a retrospective analysis indicated benefit was restricted to EGFR+ disease, an observation later verified in other studies. For the marker positive design, only patients with the marker for which the treatment is hypothesized to work are enrolled. An example is the Breast Intergroup Study of trastuzumab in Her2+ breast cancer (Slamon et al., 2001). Accruing to a subset meant utility in patients with a lower level of positivity could not be examined, but with only 20% of breast cancer patients being Her2+ and a strong rationale for restriction, an all-comer approach was not indicated. The result was an exciting success story. For strategy designs, the marker is evaluated for all patients. The patients are randomized to standard treatment vs marker strategy approach (standard treatment for marker negative and standard plus targeted agent for marker positive). An unsuccessful example for this approach is a trial of docetaxel+cisplatin (T+P) vs T+P in ERCC-1 low and T+gemcitabine in ERCC-1 high NSCLC (Cobo et al., 2007) designed under the hypothesis that ERCC-1 patients would preferentially benefit from gemcitabine. The main flaw in the trial was not determining ERCC-1 values in the control group. There was a difference in response between the arms, but without the marker information for control, it was not possible to tell if the difference was due to the strategy or to superiority of gemcitabine in general.

Marker positive trials are the most efficient if the biology is well understood, there is a clear reason to restrict to a subset, and marker prevalence is low (Simon, 2004). Potential drawbacks are that a large number of patients might need to be screened, and that the trial will fail if assumptions concerning the appropriate subset do not hold. The all-comer approach is preferred if the new treatment might also be useful in marker negative patients or if a cut point is not well established, while strategy designs are inefficient and generally are not preferred (Hoering et al., 2008). Hybrid designs allowing for both an overall and marker positive subset comparison can be useful, provided an appropriate testing strategy is employed.

9.2.3.4 Multiarm Trials

Multiarm trials generally have multiple primary goals, which lead to various design challenges. One basic issue is whether the focus is to ask the main clinical question (which treatment regimen should be used to treat patients) or to answer more general questions (e.g., should radiation therapy (RT) be given in addition to chemotherapy in principle). The first requires decision rules for identifying the best arm; the second would allow for pooling across arms (compare arms with RT vs no RT) for treatment conclusions. Other issues involve error rates, since level and power may be addressed in multiple ways. In addition, early stopping rules become complex, with potential for uninterpretable trial results if possible scenarios are not addressed in the design.

9.2.3.4.1 Testing Strategy and Level

If multiple primary comparisons are planned, then a testing strategy needs to be identified. For instance, consider a trial of five treatment arms, with no hypothesized relationships among the arms and all possible comparisons being of potential interest. A major multiple testing problem ensues: there are 10 possible pairwise comparisons, 80 other ways to compare 2 groups of arms and 120 ways of ordering the arms. If all are tested at the 0.05 level, then >30% may be expected to appear significant instead of the usual 5% when there are no differences among the arms. There are various choices for how to control the type one error for the experiment once the specific hypotheses to be tested have been chosen. One might be to choose a single test reflecting the hypothesized relationship among the arms, for instance, if there is a natural ordering to the arms (e.g., by addition of agents). If there are multiple hypotheses, one option sometimes taken is to not control, such as comparing each of three experimental arms to a control arm, with each test done at level α. It could be argued in some cases this is reasonable, but a positive result for only one comparison would likely still be questioned (at least by statisticians). Alternatively, a global test of all arms can be done first, with pairwise comparisons done only if the global test is rejected, or a single hypothesis can serve as a gatekeeper, with additional testing done only if the primary is rejected. These will restrict the probability to less than α that at least one null hypothesis will be rejected when all nulls are true (weak control).

Weak control might not be sufficient, however, such as for regulatory purposes. Strong control is achieved if the probability of incorrectly rejecting any null is at most α, regardless of how many nulls are true. A common conservative approach is to test each of K specified hypotheses at level α/K (Bonferoni testing). Strategies based on the closed testing principle can also be used and are more efficient (Dmitrienko et al., 2009). For these approaches, the sets of possible true nulls are considered; α level tests are associated with each; and rejection of a specific null requires that the test for each set containing that null be rejected. For example, for two hypotheses, the sets of possible true nulls are H_1, H_2, and [H_1 and H_2]. H_1 is rejected if the tests for both H_1 and [H_1 and H_2] are rejected. If, for instance, the test for [H_1 and H_2] consists of a Bonferroni approach, testing at level $\alpha/2$ for each of H_1 and H_2, then a test of H_2 is allowed to be done at level α if H_1 is rejected at level $\alpha/2$, or H_1 is allowed to be tested at level α if H_2 is rejected at level $\alpha/2$. Level is maintained if either both or only one null is true and the procedure is less conservative than the Bonferroni approach. See Section IV (Chapters 13 through 16) of this book for discussions of testing strategies.

9.2.3.4.2 Power

In addition to level, there are also issues with power in multiarm trials, as there is more than one aspect that needs to be considered, such as power for each comparison, power if all alternatives are true, or power to identify the correct treatment arm under various alternative scenarios. Factorial designs provide an example. Consider the simplest factorial design that randomizes treatment A vs no A and treatment B vs no B, so patients are treated with neither A nor B (O), A only (A), B only (B), or both (AB). Testing of A is done by using O and B to compare to A and AB with a stratified test (similarly for testing B). Studies are typically designed assuming there are no interactions, that is, the effect of A is the same with B or without B, and with no level adjustment for multiple tests. A is concluded useful if the test of A vs not A is significant, and similarly for B. The study is designed to have sufficient power, P, for each of these tests. This approach may be sufficient for testing utility of A and B in principle but may not be sufficient for clinical decisions on what treatment to use. Assuming no interaction, O would be recommended if neither test is significant, A if only A vs not A is significant, B if only B vs not

B is significant, and AB if both are significant. Considered from this perspective, level defined as rejecting at least one null hypothesis when both are true is close to 2α and power to reject both nulls when both alternatives are true is close to $P \times P$, neither of which may be considered sufficient for identifying O or AB as the preferred arm. If the assumption of no interaction is incorrect, the properties of a factorial design may become poor. In a simulated example (Green, 2011), if A and B are both effective alone but the combination adds nothing, this approach rejects neither A nor B in about 30% of trials, one of A or B in about half of trials and both in about 20%, none of which reflect the truth. If there is an interaction, then testing may help under scenarios such as the above. However, unless sample size is increased, power for interaction testing is poor, plus the additional testing will both further inflate level and reduce power if there is no interaction. Factorial designs require large numbers of patients if interactions are thought possible; four times the sample size of a two arm trial is needed for power to detect an interaction of the same magnitude as main effects. It should also be kept in mind that the effect of A being *the same* with or without B is a function of the measure of difference. For example, if the effect is the same when summarized as a hazard ratio, it will generally not be the same if summarized as a difference in medians.

A trial in patients with osteosarcoma (Meyer et al., 2005) provides an example of how factorial trials may fail to provide clear answers. Patients were randomized to receive standard chemotherapy (C), C+MTP, C+ifosfamide or C+MTP+ifosfamide, with the aim of assessing whether the addition of MTP, ifosfamide or both to standard treatment would improve event free survival. The trail was designed to use the standard testing strategy (compare MTP vs no MTP and ifosfamide vs no ifosfamide). Using this approach, results suggested that only MTP should be recommended for use in addition to C. However, the observed ordering of the treatment arms with respect to event-free survival was C+MTP+ifosfamide first, then C+MTP, then C and finally C+ifosfamide, suggesting a possible treatment interaction and generating uncertainty as to the appropriate conclusion. Further follow-up reduced the suggestion of interaction but not to the extent it could be ruled out, leaving the role of ifosfamide still unclear (Hunsberger et al., 2008).

9.2.3.4.3 Monitoring and Interpretation

Monitoring and interpretation issues also arise in multiarm trials. An example is a study of standard dose cisplatin (SD) vs HD cisplatin vs HD + mitomycin C (MMC), designed to test whether increasing dose intensity would improve outcome in non–small cell lung cancer patients (Gandara et al., 1993). At the time of the first interim analysis, pairwise tests of SD vs HD and SD vs HD plus MMC were done. The first test rejected the alternative that HD was superior to SD at an appropriately conservative level (in fact, the HD arm appeared worse than SD), while the SD vs HD plus MMC comparison rejected neither the null nor the alternative. These would suggest dropping the HD arm since not superior to SD, and to continue with SD and HD plus

MMC. However, the HD result called into question the whole rationale for the trial. The monitoring committee judged that continuing trial with low dose vs HD + MMC was not appropriate and the study closed (Green et al., 2012). This scenario was not anticipated; so was not built into the interim testing plan, resulting in unknown impact on design properties.

Another example is a colon adjuvant study of observation vs levamisole vs levamisole + 5FU (Moertel et al., 1990). Test results at the end of the trial did not indicate benefit due to levamisole alone but indicated benefit to the combination. The difficulty in interpretation of this trial relates to the role of levamisole. While 5FU was shown to be worthwhile, it was unknown if the addition of levamisole was necessary.

Thus, there are numerous challenges in designing multiarm studies. There are many possible choices for hypotheses to be tested, multiple level and power considerations, and potential interim testing complications. Design properties may be difficult to characterize, particularly since it is likely not all possibilities will be addressed during the design stage, with potential for compromise of interpretation of the trial due to unforeseen results and actions. More assumptions are required than for two arm trials, some of which are likely to be incorrect (e.g., interaction at some level often would be a more reasonable assumption than no interaction) and more patients per arm than for two arm trials are required for adequate power to pick the best arm. If worth the risk, care should be taken to have sufficient power for the right questions and to allow for the possibility that assumptions are wrong.

9.2.3.5 Phase 2/3

One approach to phase 2/3 design is to formally embed a randomized phase 2 into the phase 3 (Hunsberger et al., 2009). Incorporation of a phase 2 basically amounts to including a highly aggressive early stopping rule for unpromising results in the phase 3. Consequently, although there may be sample size savings, the total sample size will be larger than that for a phase 3 with more traditionally conservative stopping rules. Consider power for this approach, which is equal to $P(\text{phase 2 positive} \mid \text{alternative}) \times P(\text{phase 3 positive} \mid \text{alternative and phase 2 positive})$. If typical powers of 0.8 and 0.9 are used for the phase 2 and 3 respectively and if phase 2 positivity does not have much impact on the second term (e.g., if phase 2 sample size is relatively small or phase 2 and 3 primary outcomes are different and poorly related), then power is likely close to 0.72. Sample size must be increased to restore adequate power.

9.2.3.5.1 Screening Designs

Other phase 2/3 designs incorporate two main themes, one being screening designs. The aim of these designs is to start with several candidate treatments, eliminate either some or all but one at an early analysis, and continue with the remaining for a definitive phase 3 comparison. Characteristics of selection designs may include meeting a threshold difference compared to control, stopping rules chosen to minimize exposure to ineffective treatments or to have a reasonable chance of selecting a promising agent if at least one exists, and final tests adjusted for the selection tests done previously. Such designs are particularly attractive if there are multiple promising regimens and it is not feasible to launch multiple phase 3 trials. Potential for efficiency includes improved chance that an agent chosen for further study is active. Screening designs have been proposed specifically for studies with both binomial (Thall et al., 1988) outcomes and time to event outcomes (Schaid et al., 1990), as well as in a more generalized framework using efficient score tests (Stallard and Todd, 2003). Issues with these designs include missing active agents or not choosing the best if only one is taken forward, need for numerical methods and optimality searches, the necessity for the phase 3 part to be larger than a standard phase 3 due to initial testing, and complexity of statistical considerations.

9.2.3.5.2 Model Based Phase 2/3

The second theme involves joint assessment of early and late outcomes. The relationships of early outcomes to final outcome in patients are modeled to inform interim trial decisions (phase 2 aspect) with the final analysis done on the final outcome. One approach is a seamless phase 2/3 design, in which K possible pathways to the final outcome are modeled (Figure 9.2), with treatment effects potentially different for each (Inoue et al., 2002).

For this approach, Bayesian methods for $4K$ parameters are proposed to calculate $\Phi_1 = P(\Delta$ will be >0 if accrual is stopped and the final analysis is done in X months) and $\Phi_2 = P(\Delta$ will be >0 if the trial is completed), where Δ is the treatment difference. Decision rules at interim analysis times are to stop and conclude experimental is better than standard if both $P(\Phi_1 > 0.98)$ and $P(\Phi_2 > 0.98)$ are high; to stop and conclude experimental is not better than standard if either $P(\Phi_1 > 0.98)$ or $P(\Phi_2 > 0.98)$ is small; to start a phase 3 (i.e., add a lot of institutions) if $P(\Phi_2 > 0.98) \geq 0.8$; otherwise continue.

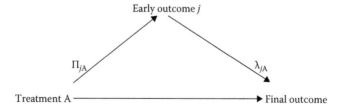

FIGURE 9.2
Association of early outcome j with final outcome on treatment arm A: change in event rate for final outcome after occurrence of early event.

The trial is stopped at the final phase 2 analysis if $P(\Phi_2 > 0.98) < 0.8$. If the phase 3 is initiated, Bayesian monitoring continues using Φ_1; experimental is concluded superior at end of the trial if $P(\Delta > 0) > 0.98$. The sample size is calibrated for appropriate level and power. Other proposals combine aspects of themes, using early outcomes to eliminate unpromising arms and then combining the phase 2 and 3 by modeling the association between short- and long-term outcomes (e.g., Liu and Pledger, 2005).

While phase 2/3 methods may have promise, many are resource intense (lengthy design stage to calibrate properties; computationally intense methods needed for Bayesian updates). In addition, usefulness for model-based designs depends on extent of association between early and late endpoints, consistency of association among treatment arms, correctness of the model of association, and amount of late endpoint information available early. Incorrect assumptions concerning these aspects will limit utility of the approaches. There are also practical issues such as timeliness of availability of early outcomes, ease of opening and closing treatment arms, feasibility of initiating a lot of new clinical sites at time of switching to phase 3, acceptability to Institutional Review Boards (IRBs) of starting a large randomized trial without phase 2 data. And finally, various papers point that the main hope for these designs—efficiency—is not always met. Trials are not necessarily shorter or require fewer patients than the standard phase 2 → phase 3 approach.

9.3 Analysis

9.3.1 Analysis of Primary and Secondary Outcomes

9.3.1.1 Choice of Primary Test

The first analysis issue is choice of primary test, which should reflect the trial design. For single arm trials, this is generally straightforward, less so for randomized trials. A well-balanced randomized trial with a significant log-rank test is considered a convincing standard when the primary outcome is time to event, so log-ranks are often chosen as primary regardless of stratification. However, this negates some of the advantage of stratifying. Since balancing on covariates decreases variance of estimates, not adjusting when stratified randomization was done will result in a conservative test and some loss in efficiency (Weir and Lees, 2003). Furthermore, if proportional hazards (PH) assumptions hold and there are large covariate effects, the unstratified log-rank test does not perform well (Anderson et al., 2006).

Testing balance of covariates before doing the primary test is not recommended. If not prespecified, then doing an additional test when imbalance is detected may inflate level. Furthermore, effect on power of prespecified testing is generally small, so it is an unnecessary complication (Permutt, 1990). As well, lack of significance does not mean adjustment is unnecessary for highly prognostic factors; if stratified randomization is not done, consider prespecifying a test adjusting for known important factors as the primary analysis instead of employing a balance testing strategy.

9.3.1.2 Analysis Population

Use of all randomized patients (or all eligible randomized patients per pre-randomization criteria) according to the assigned arm for the primary analysis of a phase 3 superiority trial is a well-established principle, since elimination of noncompliant patients from the analysis tends to inflate estimates of treatment difference. (*Note*: for the same reason, intent-to-treat analysis may be unsuitable for noninferiority trials, as this will tend to be anticonservative in this setting.)

For single arm trials, restricting to evaluable patients is common. Evaluability may include requirements for a minimum amount of protocol treatment, no major treatment deviations, and at least one postbaseline assessment, under the assumption that this will provide a fairer assessment of treatment benefit. However, typical reasons for cancer patients having a minimal amount of treatment or for not having postbaseline assessments are clinical progression, side effects of treatment outweighing potential benefit of treatment, or death due to disease. Omitting such patients from the analysis is likely to result in optimistic estimates of treatment benefit. On the other hand, including these in the analysis requires assumptions that may or may not be correct. For estimation of the probability of response, patients with unknown response generally should be included in the analysis as assumed nonresponders for the reasons described. This will not necessarily be correct for all patients, so the estimate is biased somewhat low, but is preferable to the larger bias in the other direction.

Restriction to evaluable patients may be done for randomized trials as a sensitivity analysis, as consistency with the primary is helpful when there are compliance issues on the trial but would rarely be done as primary. A particularly striking example of uninterpretable analysis in evaluable patients is a randomized trial of invitation for annual screening for prostate cancer vs no invitation (2-1 randomization), with 46,486 people invited or not (Labrie et al., 2004). The results reported were 74 prostate cancer deaths in 14,231 patients in the control group and 10 of 7,384 in the screening group, which at first glance seems convincing. But this means 24,907 randomized subjects were considered not evaluable: 23,785 of the 31,133 men randomized to screening did not get screened and 1,122 of the 15,353 men randomized to no screening did get screened. Of course, there are numerous potential biases as a function of who got screened and who did not. The as-randomized comparison did not show benefit of screening, resulting in differences of opinion as to the implications of the results (Elwood, 2004).

9.3.1.3 Disease-Free and Progression-Free Survival

Most cancer phase 3s have time to event outcomes as primary, either DFS or PFS, or OS, with OS preferred unless prohibitively long or another outcome has been demonstrated to reflect OS benefit reasonably well. Distributions of these outcomes are estimated with the most frequently cited method in the statistical literature (Ryan and Woodall, 2005), the Kaplan and Meier estimate. This method uses the partial information from patients who have not yet had the event by censoring at the last follow-up time, T, and assuming subsequent outcome will be distributed the same as for other patients without an event by T who are followed after time T. In particular, this means an assumption is made that the censoring mechanism is independent of subsequent outcome.

A potential problem with PFS is inadequate disease assessments. Censorship is not a solution to this problem, since reasons for inadequate assessments are often related to the outcome, so censorship is informative, not independent as required for unbiased estimates. For instance, patients who go off treatment before progression often do not have further disease assessments. If the reason is due to symptomatic deterioration, then censorship at time of treatment will result in an optimistic bias in the PFS estimate. Diligent follow-up and patient retention are critical to avoid such biases. Another issue is assessment schedule, which should be the same for each arm. If disease assessment intervals are not the same, then the treatment arm with the longer interval will appear to have longer PFS despite no difference, since events occurring at the same time will more often be detected later on the arm with the longer assessment interval. A related consideration is the fact that PFS is only known to be within the interval between the date progression is documented and the last prior date with no progression documented. Use of date of documentation produces an overestimate. If the assessment interval is relatively short with respect to expected time to failure, the overestimate is modest, but long intervals are problematic. For instance, if patients discontinue treatment without progression, miss many tumor evaluations, and then die, using the death date is inappropriate. There is potential for better estimation by using information on failure times occurring in other patients during the censoring interval or by using model assumptions. See Chapter 12 on the topic of interval censorship.

9.3.1.4 Overall Survival

Survival does not have disease assessment issues, but even this most basic of outcomes can be compromised. Loss to follow-up, if due to reasons related to expected survival, will result in biased estimates. Data collection problems can also introduce bias, such as delayed event reporting or long intervals between follow-up requests for survival status. Sensitivity analyses may need to be done, such as best and worst cases or modeling of the association between censorship and survival.

Other issues arise due to additional treatment after study treatment discontinuation. Generally, cancer treatments become progressively less effective after each treatment failure, so it is expected that the trial treatment will have greater impact on OS than subsequent treatments. Without this expectation, OS could not be used as a primary outcome, since it is not an option to deny cancer patients additional treatment after treatment failure on a clinical trial. However, in some cases, subsequent treatment can complicate interpretation of results, such as trials in diseases for which potentially curative procedures may be done in patients responding to treatment. Differences in treatment arms will be a function both of direct effects of the treatments, plus indirect effects related to effectiveness in rendering patients eligible for the curative procedure (e.g., transplant after induction therapy in leukemia). Reliable assessment of direct effect only is not achievable. The usual attempt consisting of censorship at time of procedure is not adequate due to nonindependent censoring mechanism (patients suitable for transplant at time T have systematically different characteristics from those who are not). It is best to compare OS without censorship, including both the direct and indirect effects of the treatments.

Other examples are trials that allow crossover treatment. If a trial of combination treatment C+E vs C alone, allows treatment with E after failure on C alone, the question of whether the combination is superior with respect to OS may not be answerable. If E is ineffective, then crossover to E after C could compromise the survival comparison if effective salvage treatment exists—then C+E may incorrectly appear superior to C. On the other hand, if E is effective, crossover to E after C may result in reduced difference between arms and C+E may incorrectly appear not useful. Supportive analyses may be required to explore possible true treatment effect. Various model-based methods have been proposed (Morden et al., 2011) or outside information on expected postfailure outcome on C could be used if available.

9.3.1.5 Patient-Reported Outcomes

Patient reported outcomes (PROs) of disease-related symptoms are often considered important to incorporate into trials, as these are considered to be direct measures of treatment benefit along with overall survival. These and other aspects of quality of life are possibly the most challenging outcomes to address in cancer trials. PROs are subjective and easily influenced by knowledge of which treatment is being received, so ideally, the trial should be blinded. Measurement instruments should have content and construct validity, reliability, and measure aspects different from other outcomes. As well, scales with items that clearly allow assessment of improvement in disease symptoms should be included, not to be confounded with differences related to toxicity; unfortunately, this can be challenging to achieve given the wide variety of both symptoms and side effects experienced by cancer patients. PROs may include emotional functioning, physical functioning, general symptoms, specific symptoms, subscale scores, composite scores and global scores, and for each of these, scores at each time point, difference from baseline of scores at each time point, models of change over time, time to deterioration and duration of improvement. Thus, it is necessary to choose a limited number of key comparisons along with a testing strategy to control type one error for multiplicity, with the remaining comparisons to be considered exploratory.

Missingness is a common problem, with questionnaires typically missed due to feeling poorly, toxicity, progression or death, so, cannot be considered missing at random. Ideally assessments should continue after discontinuation of treatment, but this is usually not achievable. Because missingness is not at random, standard analytic methods will be biased and sensitivity analyses must be done. Regardless of method used, if there is substantial missing information, then interpretation is necessarily nondefinitive. For instance, if patients have stable scores while on treatment, good (high score) risk patients stay on treatment longer than poor (low score) risk patients, and PRO measures are not done after treatment discontinuation, then over time, average PRO scores appear to improve (proportion of good risk patients increases over time) despite no change in scores. Similarly, if dropout patterns are different between two arms, then it may appear the arms are different when they are not. Restricting to complete information is not a solution; in this example, it would appear that patients had high scores that stayed high. Of course the situation is more complicated than this, since scores do not remain stable. In particular, scores tend to worsen prior to discontinuation due to progression or intolerable side effects. Methods that model patterns related to dropout (e.g., change from baseline assessed according to number of assessments completed) can be useful for supplementary analyses if informative missingness is suspected (Pauler et al., 2003). A single method needs to be identified as primary, but if results are shown to be consistent using more than one method, this will add confidence to interpretation of results.

9.3.2 Exploratory Analyses

9.3.2.1 Subsets

Subset testing in studies is often done with the informal aim of examining which type of patient may benefit most from treatment. This is a natural question to ask, but most trials are not adequately designed to be able to answer the question. Typically, many subsets are tested at the conventional 0.05 level, so type one errors are common due to multiple testing. For instance, if a stratification factor has four levels, the probability that at least one of the tests in the four subsets will have $p < 0.05$ is $1 - (0.95)^4 = 0.19$ when there are no differences—much larger than 0.05. On the other hand, subset sample sizes are small, so type two errors are also large due to poor power. A simulation based on a successful 968 patient phase 3 trial with 3 stratification factors, 2 with 3 levels and 1 with 2, provides an example (Fleming, 1995). Despite an overall positive result, in 67% of the simulated trials, estimates for at least one of the eight subsets indicated the experimental arm was of no benefit or worse than control. A classic example of how subset results can go wrong is the 17,000+ patient ISIS-2 trial analysis of astrologic signs, in which Libras and Gemini apparently did not benefit from aspirin, while the other 10 signs did (Collins and MacMahon, 2001). Considering exploratory analyses even in very large trials does not necessarily produce reliable subset results, reliability in much smaller trials must be assumed poor.

An examination of subset reporting in clinical trials publications (Wang et al., 2007) indicates reporting is often not being done appropriately. Following are some points to consider.

- Since number and results of subset analyses done influences interpretation of subset results, complete descriptions should be included in publications.
- If overall results are positive, negative subset results do not prove lack of efficacy. If overall results are negative, positive subset results do not prove efficacy. Discrepant results must be interpreted cautiously and reproduced in subsequent trials.
- Avoid providing p-values for subset analyses. A conservative approach is to not report subset results as interesting unless there is evidence that the treatment effect is different in the subsets (i.e., test for interaction is significant). This includes stratification factors. Stratification does not justify subset tests.
- If subset questions are of interest, design the study to be able to answer them. Limit the number of subsets and use an appropriate scheme for multiple testing.
- If it is considered likely that the treatment should work better in one subset than another, then consider including the subset as one of the primary hypotheses and design the trial accordingly.

9.3.2.2 Competing Risks

Competing risks analyses are useful for outcomes consisting of more than one failure type, of which at least one precludes adequate assessment of another failure type. For more in-depth introduction of competing risks, please refer to Chapter 10 and references therein. An example is DFS, which consists of two competing risks, relapse of disease, and death before relapse. For this and other examples, there may be a desire to analyze the effect of treatment on one failure type while eliminating the risk of other types, in particular for DFS, the effect of treatment on relapse eliminating the competing risk of death. A common but inappropriate analysis approach to this question is to censor for relapse at time of death without relapse. Since events cannot happen after death, it is difficult to interpret what is being estimated.

Censorship in this case means assuming that if the patient had not died at time T after enrollment, then the time to subsequent relapse would have been similar to that of patients who actually survived without relapse up to time T. But the characteristics of a patient who dies at time D cannot be assumed similar to those who did not. A death due to treatment toxicity might imply sensitivity to treatment that would have resulted in better than average time to relapse. More likely, factors leading to death might be also be associated with poor disease outcome (e.g., comorbidities or compromised immune system). Thus, the independence of the censoring mechanism (death) with the outcome (relapse) cannot be assumed and estimates are biased in unknown ways. Another example is time to complete response in hematologic cancers where treatment may be discontinued due to insufficient response. At this point, the treatment regimen is typically changed and disease assessments for the study treatment are discontinued. Here the competing outcomes are response on study treatment and treatment discontinuation without response. In this example, it is not possible to respond on treatment after going off, so censoring is not an option.

There are two main ways of analyzing competing risks. One is to consider cause-specific hazards (failure rates). At each cause-specific failure time, the number of patients still at risk (patients with no failure of any type who are still being followed) on each treatment arm is calculated. If k of the N patients still at risk are from arm A, then the chance that failure is from arm A should be k/N. If there are more failures than expected on arm A, the relative failure rate is concluded higher on A than on B. The second is a subdistribution (also known as cumulative incidence) approach. The distribution of time to each subtype as a first event is estimated for each arm and can be compared to test whether probabilities of specific failure types as a first event differ between the arms. (The sum of these sub-distributions will be equal to the distribution of the composite endpoint.) Gray's test (Gray, 1988) is used when data are censored. The two methods have different interpretations and will not necessarily provide the same result. Difference in failure type F at time T for the subdistribution approach considers the percent of patients who had a first failure due to F by time T on each arm with respect to the total number of patients on each arm. Difference in failure type F at time T for the relative failure rate approach considers number of failures due to F relative to the number of patients still at risk on each arm. If number of patients with F as a first failure by time T is the same on each arm but more patients fail due to other failure types on one of the arms prior to time T, the comparisons will give different results, equal by arm by the subdistribution method and unequal by the relative rate method.

A study of 785 women comparing HD chemotherapy plus peripheral blood stem cell support (transplant) vs intermediate dose chemotherapy as adjuvant therapy in high risk breast cancer provides a useful example (Peters et al., 2005). On this study, 137 relapses were observed on the transplant arm compared to 183 on the intermediate dose arm. This might initially seem to indicate efficacy of the high dose treatment but does not take into account that early deaths without relapse are more common in transplant regimens than in standard chemotherapy regimens. Reduction of relapse at the expense of additional early deaths may not be an acceptable tradeoff. There were 33 toxic deaths after transplant (30 within the first year) on this study and none on control. Total number of events (relapse or death) on the two arms was similar (179 and 193) at the time of analysis, and the DFS comparison was not significant. The initial death rate was higher on transplant than on control, but at 5 years, the survival estimates were the same (71%). Despite rate of relapse being lower on transplant, overall there was no benefit to this treatment, emphasizing the importance of reporting results in the context of all of the competing risks. Clearly, comparisons out of context can be misleading. If DFS is the same on two arms, then if B is better with respect to relapse, A must be better with respect to death.

While analysis of an event component would not be sufficient to demonstrate superiority of a treatment, analysis of all components provides insight into how patients fail and may suggest avenues of investigation. For transplant, enhanced method to reduce the toxic death rate is an obvious an example. For a more in-depth introduction of competing risks, please refer to Chapter 10.

9.3.2.3 Outcome by Outcome Analysis and Surrogate Endpoints

Trials are designed to answer a limited number of questions. It is natural to want to explore other questions, even if imperfectly, to glean as much insight from the trials as possible. One way this has been done is to analyze one outcome by another in hopes of showing causal relationships, such as DFS by dose received to show higher doses improve DFS, or in hopes of identifying useful short-term outcomes, such as survival by side effect to show that a benefit in the short-term outcome will be sufficient to assume a benefit in the long-term outcome of survival. Unfortunately, this type of approach sheds little light on the questions of interest.

One issue is that association of outcomes does not imply a causal relationship. Common factors may be associated with both outcomes. For instance, age is related to both dose and survival; the elderly both die sooner and are less tolerant of treatment. Any association of received dose with survival could potentially be explained by underlying factors, most of which, unlike age, will be unmeasured or unknown. In addition, analysis results may depend heavily on outcome definitions. For the dose example, if dose is defined as total dose received, then patients who relapse and discontinue treatment early will have low total dose; in this case, DFS determines dose, not the other way around. On the other hand, percent planned dose received while on treatment will be biased in the other direction, since patients on treatment for a long time have more opportunity for dose reductions. Different definitions of dose will result in different answers to the question of possible relationship (Redmond et al., 1983). In the end, randomized trials testing different dose levels must be done for a definitive answer.

Time bias is a factor in most outcome by outcome analyses. For instance, the longer on treatment, the longer the survival time and the more likely a rash will occur, so there is an automatic association of the outcomes. The induced correlation of survival and rash does not necessarily imply the same mechanism is producing both or that occurrence of rash is a good predictor of subsequent outcome. A method that reduces time bias is landmark analysis, for which status of the short-term outcome is determined at a particular time (e.g., rash or not by time T among patients still alive at time T) and the longer-term outcome subsequent to this time (e.g., survival after time T) is assessed according to status. Although this mitigates the time bias, it does not address the issue of multiple factors being associated with both outcomes. Cox time-dependent modeling is a better approach to analysis, as the model incorporates the time of occurrence of rash and can incorporate patient characteristics. A time dependent approach was used in a head and neck cancer trial of cisplatin plus placebo vs cisplatin plus cetuximab and demonstrated a strong relationship between skin toxicity and survival on the cetuximab arm (Burtness et al., 2005). Yet even though most patients on the cetuximab arm had skin toxicity, cetuximab was not associated with improved overall survival, so previous speculation from single arm results that rash should be a good efficacy marker (e.g., Herbst et al., 2005) was not supported.

The most challenging issue to address is whether a short-term outcome is a *surrogate endpoint*. On a patient level, this means there is a strong association between the surrogate and the primary outcome of interest (usually survival) after adjustment for treatment, while on a population level, it means the effect of treatment on the surrogate predicts the effect of treatment on the primary. It is relatively easy to demonstrate whether outcomes on a patient level are associated. An example is investigation of prognostic significance of major molecular response (MMR) to TKI agents with respect to longer-term outcome in CML. A landmark analysis done in a trial of imatinib 800 vs imatinib 400 vs imatinib 400 plus interferon (Hehlmann et al., 2011) reported 99% three-year survival in patients with an MMR at month 12 vs 93% three-year survival in patients without an MMR at month 12, suggesting reasonable prediction on the patient level. Demonstration of population level prediction is much more difficult: "The validity of a surrogate endpoint should be judged by the probability that the trial results based on the surrogate endpoint alone are 'concordant' with the trial results that would be obtained if the true endpoint were observed and used for the analysis" (Begg and Leung, 2000). For this a sufficient number of randomized trials must be available for assessment of the extent to which treatment differences with respect to the proposed surrogate are correlated with treatment differences with respect to the primary. An example of demonstration of population level surrogacy is an assessment of 18 randomized adjuvant colon cancer trials with over 20,000 patients (Sargent et al., 2005), in which correlation of DFS and survival hazard ratios was 0.92 in 25 comparisons, and in 23 of the 25, the results were the same for both DFS and survival (either both significant or both nonsignificant). DFS met standards expressed for surrogacy and was accepted by the FDA for use as the primary outcome for colon adjuvant trials.

It is uncommon to have sufficient information for this type of demonstration. Thus, the conclusion of the authors of the CML trial is rather surprising: "...the data demonstrate that MMR...can be used in the future as surrogate markers of survival in patient care and in clinical trials." Although the patient level association was observed, there was no evidence presented suggesting population level association; in fact, their own study did not support this. MMR at month 12 occurred in 54%, 31%, and 34% of patients on the three arms respectively. If MMR were a surrogate, then imatinib 800 would be expected to be superior with respect to survival compared to the other two arms, but this was not observed to be the case (three year survival 96%, 95%, 96% respectively).

Early outcomes may provide convincing evidence of activity of a new regimen but may not be as useful for regulatory approval and establishing a new standard of care (exception: rare disease with limited treatment options). These typically require evidence of direct benefit to patients, where direct benefit is considered to be either longer life or better quality of life. Disease outcomes such as response in themselves are not direct measure of benefit although may result in benefit if disease symptoms and limitations improve. Since patient-reported quality of life in cancer trials is so difficult to collect and analyze, and since it is so difficult to prove surrogacy of other outcomes for survival, survival remains the main measure of benefit. See Chapter 19 for additional discussion of surrogate endpoints.

9.4 Conclusion

Most of the issues in design and analysis of cancer clinical trials are common across all clinical trials. Issues related to missing data, multiple testing strategies, interpretation of exploratory analyses, selection biases, evaluability, and so on are not disease specific. Yet there are aspects that do make cancer trials particularly challenging. Study of life-threatening disease changes perspective on risk–benefit balance and on what types of trial are acceptable to conduct. The assumption that more toxic is more efficacious and that the severity of disease justifies risk of serious toxicity leads to routine acceptance of side effects that are unacceptable in other settings and an aggressive approach to phase 1 testing based on toxicity only. High unmet need for effective treatment in many uncommon cancers leads to consideration of design and strategy compromises to reduce sample size and shorten development time. Various outcomes are used in phase 2, but in the end, definitive benefit to patients is primarily measured by survival time. Appropriate surrogates for survival are difficult to validate, so phase 3 trials are often lengthy, with answers not available for many years. Cancer symptoms are often severe, so quality of life is also important, but this is particularly difficult to assess in cancer studies due to nonrandomly missing data as patients drop off treatment for toxicity, progression, or death. Many markers related to tumor growth and spread have been identified and many agents targeting these have been developed, introducing new design challenges related to the identification of appropriate subsets and to the ever-decreasing patient population sizes. The list goes on, but despite the many challenges, it is gratifying to see new avenues of research unfold and to have a role in the progress being made.

References

Aickin M. (2001) Randomization, balance, and the validity and efficiency of design-adaptive allocation methods. *Journal of Statistical Planning and Inference*. 94: 97–119.

Anderson G. et al. (2006) On use of covariates in randomization and analysis of clinical trials. In J. Crowley and D. P. Ankerst (eds.). *Handbook of Statistics in Clinical Oncology*, 2nd edn. Chapman & Hall/CRC Press, Boca Raton, FL, pp. 167–180.

Barlogie B. et al. (1995) Autologous marrow and blood transplantation. In K. Dicke and A. Keeting (eds.). *Proceedings of the Seventh International Symposium*, Arlington, TX, pp. 399–410.

Barlogie B. et al. (2006) Standard chemotherapy compared with high-dose chemoradiotherapy for multiple myeloma: Final results of phase III US intergroup trial S9321. *Journal of Clinical Oncology* 24:929–935.

Begg C. and Kalish L. (1984) Treatment allocation for nonlinear models in clinical trials: The logistic model. *Biometrics* 40:409–420.

Begg C. and Leung D. (2000) On the use of surrogate end points in randomized trials. *Journal of the Royal Statistical Society: Series A* 163:15–28.

Bekele B. and Shen Y. (2005) A Bayesian approach to jointly modeling toxicity and biomarker expression in a phase I/II dose-finding trial. *Biometrics* 61: 344–354.

Burtness B. et al. (2005) Phase III randomized trial of cisplatin plus placebo compared with cisplatin plus cetuximab in metastatic/recurrent head and neck cancer: A Eastern Cooperative Oncology Group Study. *Journal of Clinical Oncology* 23: 8646–8654.

Chang M., Devidas M., and Anderson J. (2007) One- and two-stage designs for phase II window studies. *Statistics in Medicine* 26:2604–2614.

Cobo M. et al. (2007) Customizing cisplatin based on quantitative excision repair cross-complementing 1 mRNA expression: A phase III trial in non-small-cell lung cancer. *Journal of Clinical Oncology* 25:2747–2754.

Collins R. and MacMahon S. (2001) Reliable assessment of the effects of treatment on mortality and major morbidity, I: Clinical trials. *Lancet* 357:373–380.

Cook G. (2004) Scurvy in the British Mercantile Marine in the 19th century, and the contribution of the seamen's Hospital Society. *Postgraduate Medical Journal* 80:224–229.

DeVita Jr. V.T. and Rosenberg S.A. (2012) Two hundred years of cancer research. *New England Journal of Medicine* 366:2207–2214.

Dixon D. and Simon R. (1988) Sample size considerations for studies comparing survival curves using historical controls. *Journal of Clinical Epidemiology* 41: 1209–1213.

Dmitrienko A., Tamhane A., and Bretz F. (eds.). (2009) *Multiple Testing Problems in Pharmaceutical Statistics*. Chapman and Hall/CRC Press, Boca Raton, FL.

Eisenhauer E. et al. (2009) New response evaluation criteria in solid tumours: Revised RECIST guideline (version 1.1). *European Journal of Cancer* 45:228–247.

Elwood M. (2004) A misleading paper on prostate cancer screening. *Prostate* 61: 372.

Fleming T. (1995) Interpretation of subgroup analyses in clinical trials. *Drug Information Journal* 29:1681S–1687S.

Gandara D.R. et al. (1993) Evaluation of cisplatin in metastatic non-small cell lung cancer: A phase III study of the Southwest Oncology Group. *Journal of Clinical Oncology* 11:873–878.

Gray R. (1988) A class of K-sample tests for comparing the cumulative incidence of a competing risk. *The Annals of Statistics* 16:1141–1154.

Green S. and Dahlberg S. (1992) Planned versus attained design in phase II clinical trials. *Statistics in Medicine* 11:853–862.

Green S. et al. (2012) *Clinical Trials in Oncology*, 3rd edn. Chapman & Hall/CRC Press, Boca Raton, FL.

Green S.J. (2011) Factorial designs for time to event endpoints. In J. Crowley (ed.). *Handbook of Statistics in Clinical Oncology*, 3rd edn. Marcel Dekker, New York.

Hehlmann R. et al. (2011) Tolerability-adapted Imatinib 800 mg/d versus 400 mg/d versus 400 mg/d plus interferon in newly diagnosed chronic myeloid leukemia. *Journal of Clinical Oncology* 29:1634–1642.

Herbst R. et al. (2005) Phase II multicenter study of the epidermal growth factor receptor antibody cetuximab and cisplatin for recurrent and refractory squamous cell carcinoma of the head and neck. *Journal of Clinical Oncology* 24:5578–5587.

Hoering A., LeBlanc M., and Crowley J. (2008) Randomized phase III clinical trial designs for targeted agents. *Clinical Cancer Research* 14:4358–4367.

The Holy Bible, *Revised Standard Version*, World Publishing Company, Cleveland OH, 1962.

Hrobjartsson A. et al. (2007) Blinded trials taken to the test: An analysis of randomized clinical trials that report tests for the success of blinding. *International Journal of Epidemiology* 36:654–653.

Hunsberger S., Freidlin B., and Smith M. (2008) Correspondence. Complexities in interpretation of osteosarcoma clinical trial results. *Journal of Clinical Oncology* 26:3103–3104.

Hunsberger S., Zhao Y., and Simon R. (2009) A comparison of phase II study strategies. *Clinical Cancer Research* 15:5950–5955.

Inoue L.Y., Thall P.F., and Berry D.A. (2002) Seamlessly expanding a phase II trial to phase III. *Biometrics* 58:823–831.

Ivanova A. et al. (2003) Improved up-and-down designs for phase I trials. *Statistics in Medicine* 22:69–82.

Ivanova A. and Wang K. (2004) A non-parametric approach to the design and analysis of two-dimensional dose finding trials. *Statistics in Medicine* 23:1861–1870.

Jung S. H., Chang M., and Kang S. (2012) Phase II cancer clinical trials with heterogeneous patient populations. *Journal of Biopharmaceutical Statistics* 22:312–328.

Labrie F. et al. (2004) Screening decreases prostate cancer mortality: 11 year follow-up of the 1988 Quebec prospective randomized controlled trial. *Prostate* 59:311–318.

Lind J. (1753). *A Treatise of the Scurvy.* Sands, Murray, and Cochran, Edinburgh, Scotland.

Liu P.-Y., Dahlberg S., and Crowley J. (1993) Selection designs for pilot studies based on survival endpoints. *Biometrics* 49:391–398.

Liu Q. and Pledger G. (2005) Phase 2 and combination designs to accelerate drug development. *Journal of the American Statistical Association* 100:493–502.

London W. and Chang M. (2005) One- and two-stage designs for stratified phase II clinical trials. *Statistics in Medicine* 24:2597–2611.

Meyer P. et al. (2005) Osteosarcoma: A randomized prospective trial of the addition of ifosfamide and/or muramyl tripeptide to cisplatin, doxorubicin, and high-dose methotrexate. *Journal of Clinical Oncology* 23:2004–2011.

Moertel C. et al. (1990) Levamisole and fluorouracil for adjuvant therapy of resected colon carcinoma. *New England Journal of Medicine* 322:352–358.

Morden J. et al. (2011) Assessing methods for dealing with treatment switching in randomized controlled trials: Simulation study. *BMC Medical Research Methodology* 11:4.

O'Malley J. et al. (2002) Sample size calculation for a historically controlled clinical trial with adjustment for covariates. *Journal of Biopharmaceutical Statistics* 12:227–247.

O'Quigley J., Pepe M., and Fisher L. (1990) Continual reassessment method: A practical design for phase I clinical trials. *Biometrics* 46:33–48.

Pauler D., McCoy S., and Moinpour C. (2003) Pattern mixture models for longitudinal quality of life studies in advanced-stage disease. *Statistics in Medicine* 22:795–809.

Penel N. et al. (2009) "Classical 3 + 3 design" versus "accelerated titration designs": Analysis of 270 phase 1 trials investigating anti-cancer agents. *Investigational New Drugs* 27:552–556.

Permutt T. (1990) Testing for imbalance of covariates in controlled experiments. *Statistics in Medicine* 9:1455–1462.

Peters W. et al. (2005) Prospective, randomized comparison of high-dose chemotherapy with stem-cell support versus intermediate-dose chemotherapy after surgery and adjuvant chemotherapy in women with high-risk primary breast cancer: A Report of CALGB 9082, SWOG 9114, and NCIC MA-13. *Journal of Clinical Oncology* 23:2191–2200.

Redmond C., Fisher B., and Wieand H.S. (1983) The methodologic dilemma in retrospectively correlating the amount of chemotherapy received in adjuvant therapy protocols with disease-free survival. *Cancer Treatment Reports* 67:519–526.

Ryan T. and Woodall W. (2005) The most-cited statistical papers. *Journal of Applied Statistics* 32:461–474.

Sargent D. et al. (2005) Disease-free survival versus overall survival as a primary end point for adjuvant colon cancer studies: Individual patient data from 20,898 patients on 18 randomized trials. *Journal of Clinical Oncology* 23:8664–8670.

Schaid D., Wieand S., and Therneau T. (1990) Optimal two-stage screening designs for survival comparisons. *Biometrika* 77:507–513.

Shepherd F. et al. (2005) for the National Cancer Institute of Canada Clinical Trials Group. Erlotinib in previously treated non-small-cell lung cancer. *New England Journal of Medicine* 353:123–132.

Simon R. (1987) How large should a phase II trial of a new drug be? *Cancer Treatment Reports* 71:1079–1085.

Simon R. (1989) Optimal two-stage designs for phase II clinical trials. *Controlled Clinical Trials* 10:1–10.

Simon R. et al. (1997) Accelerated titration designs for phase I clinical trials in oncology. *Journal of the National Cancer Institute* 89:1138–1147.

Simon R. and Maitournam A. (2004) Evaluating the efficiency of targeted designs for randomized clinical trials. *Clinical Cancer Research* 10:6759–6763.

Slamon D. et al. (2001) Use of chemotherapy plus a monoclonal antibody against HER2 for metastatic breast cancer that overexpresses HER2. *New England Journal of Medicine* 344:783–792.

Stallard N. and Todd S. (2003) Sequential designs for phase III clinical trials incorporating treatment selection. *Statistics in Medicine* 22:689–703.

Storer B. (1989) Design and analysis of phase I clinical trials. *Biometrics* 45:925–938.

Storer B. (1992) A class of phase II designs with three possible outcomes. *Biometrics* 48:55–60.

Storer B. (2011) Choosing a phase I design. In J. Crowley and A. Hoering (eds.). *Handbook of Statistics in Clinical Oncology*, 3rd edn. Marcel Dekker, New York.

Thall P. and Cook J. (2004) Dose-finding based on efficacy-toxicity trade-offs. *Biometrics* 60:684–693.

Thall P., Simon R., and Ellenberg S. (1988) Two-stage selection and testing designs for comparative clinical trials. *Biometrika* 75:303–310.

Thomas D. (1997) Sailors, scurvy and science. *Journal of the Royal Society of Medicine* 90:50–54.

Tighiouart M., Rogatko A., and Babb J. (2005) Flexible Bayesian methods for cancer phase I clinical trials. Dose escalation with overdose control. *Statistics in Medicine* 24:2183–2196.

Wang R., Lagakos S., and Ware J. (2007) Statistics in medicine-reporting of subgroup analyses in clinical trials. *New England Journal of Medicine* 357:2189–2194.

Weir C.J. and Lees K.R. (2003) Comparison of stratification and adaptive methods for treatment allocation in an acute stroke clinical trial. *Statistics in Medicine* 22:705–726.

Yuang Y. and Yin G. (2008) Sequential continual reassessment method for two-dimensional dose finding. *Statistics in Medicine* 27:5664–5678.

Zia M. et al. (2005) Comparison of outcomes of phase II studies and subsequent randomized control studies using identical chemotherapy regimens. *Journal of Clinical Oncology* 23:6982–6991.

Zubrod C.G. (1982) Clinical trials in cancer patients: An introduction. *Controlled Clinical Trials* 3:185–187.

10

Competing Risks and Their Applications in Cancer Clinical Trials

Chen Hu, James J. Dignam, and Ding-Geng (Din) Chen

CONTENTS

10.1 Introduction

Competing risks are frequently encountered in clinical research, especially in cancer, where individuals under study may experience one of two or more different types (causes) of failure after initial diagnosis and treatment. For each subject, typically we observe the time to failure and its failure type, or time to censoring. Unless one is to simply consider the time to the first failure of any type as the endpoint of interest, special attention and methodology are required for competing risks analysis. This is because the occurrence of first failure either prevents the occurrence of other types of failures (the typical competing risks scenario), or the disease natural history is altered due to the subsequent second-line treatment (in cases where events are not strictly mutually exclusive but are so in the sense that one will be the first failure). For example, localized prostate cancer patients may either die due to prostate cancer or from other causes, due to comorbid conditions or natural causes. Another example arises in locally advanced lung cancer, where the first failure after definite localized treatment (such as radiotherapy) could be in-field failure (local recurrence), out-of-field failure (distant metastases or second primary cancer), or death without disease. While in such setting, individuals may have multiple events (e.g., distant followed by local failure, or vice versa), since the disease natural history is largely dependent on the pattern and sequence of the multiple events as well as the subsequent treatment options, competing risks analysis (e.g., of time to first event and its type) may still offer the most clinically relevant information regarding failure patterns and prognosis. Two principal types of competing risks analysis are often used in a complementary fashion as each addresses important aspects of competing risks observations. In a nutshell, one is based on the cause-specific hazards (CSHs), while the other is based on cumulative incidence (Kalbfleisch and Prentice, 2002). We review their distinctions with an emphasis of theoretical and practical implications and argue the importance of using the right methods to answer different questions of interest. The associated estimable quantities and corresponding statistical inference methods are discussed accordingly. We illustrate the analytic methods through a reanalysis of a Radiation Therapy Oncology Group (RTOG) clinical trial.

10.2 Statistical Formulations and Considerations

10.2.1 Issues of Using Composite Endpoints

One simple and commonly used way to bypass the competing risks problem is to transform the problem into a univariate survival problem of the time to first event regardless of its type, that is, event-free survival (EFS). Examples in cancer clinical trials include disease-free survival (DFS) or progression-free survival, defined as time to cancer recurrence/progression *or* death, whichever occurs first. Overall survival, defined as time to death regardless of cause, can also be viewed as another example of EFS that represents the ultimate clinical benefits. Standard survival analysis techniques, including Kaplan–Meier (KM) estimator (Kaplan and Meier, 1958), log-rank test (Mantel, 1966), and Cox proportional hazard (PH) regression model can be used accordingly.

However, as discussed by Mell and Jeong (2010), a number of issues may arise when using composite endpoints in the presence of competing risks. To ease the understanding, let us consider a randomized clinical trial to evaluate the efficacy of experimental therapy on cancer recurrence, with noncancer mortality being the competing event. If the treatment effectively reduces the cancer recurrence and has no (or even moderately adverse) impacts on noncancer mortality, using EFS endpoint may represent a diluted (or null) effect, which ultimately translates into a reduced precision in estimating the failure rates of cancer recurrence and the actual treatment effect; statistical power may be compromised similarly if such *dilution* impact is significant. Furthermore, if the treatment is only beneficial for younger patients as they can tolerate the associated toxicity, statistical inference based on composite endpoints may be prone to error as they miss the fine details of differential effects on different event types and the associated interaction with possible confounders. In either scenario, competing risks analysis may offer a potential remedy to help better utilize resources, understand the disease process, and assess the risk/benefit profile. On the other hand, the merit of composite endpoints sometimes outweighs the concerns mentioned earlier in certain settings, especially when the impacts of experimental intervention on the multiple event types are unclear and of interest as well. For example, unlike disease-specific survival (time to disease-specific death), overall survival eliminates the often-complex issue of cause of death ascertainment at both design and data collection stages and naturally corrects for both beneficial and deleterious effects of a given treatment.

10.2.2 Modeling Paradigm and Important Quantities

As a subdiscipline of survival analysis, competing risks deal with the situation when observed data constitute more than one event types. These event types can be mutually exclusive, for example, different cause of death, such that the occurrence of any type of event precludes the occurrence of any other types of events. They can also be observable in different sequences, such as local and distant cancer recurrence. However, in these cases, our primary interests may be the first failures only. Mathematically speaking, competing risks analysis considers a failure time T^*, and the cause of failure $D^* \in \{1, \ldots, K\}$, which takes the value k if it is a type k failure $(k = 1, \ldots, K)$. In presence of right censoring, let C be the censoring time, we observe $T = \min(T^*, C), \delta = I(T < C), D = I(\delta = 1)D^*$, where $I(\cdot)$ is an indicator function.

10.2.2.1 Competing Risks as a Multistate Model

Competing risks may be conveniently illustrated as a special case of the multistate model (Andersen et al., 2002; Putter et al., 2007), as shown in Figure 10.1.

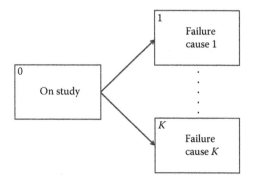

FIGURE 10.1
Schematic illustration of competing risks in the multistate model framework.

The basic model parameters in multistate models are transition probabilities between state h and state l ($h < l$), that is,

$$P_{hl}(s, t) = \Pr\left(\text{state } l \text{ at } t | \text{state } h \text{ at } s\right), \quad s < t.$$

10.2.2.2 Cause-Specific Hazard Functions

The transition intensities between the transient state 0 (*alive*) and absorbing states $k, k = 1, \ldots, K$ (*failure cause k*), as defined in Equation 10.1, represent the fundamental identifiable quantities in competing risks called CSH, or *crude* hazards as

$$\lambda_k(t) = \lim_{h \to 0} \frac{1}{h} \Pr\left(t \le T^* \le t + h, D^* = k | T^* \ge t\right). \tag{10.1}$$

Let us denote $\Lambda_k(t) = \int_0^t \lambda_k(u) du$ as the cumulative CSH and $\Lambda(t) = \sum_{k=1}^{K} \Lambda_k(t)$ as the cumulative hazard function corresponding to survival time T^*, which is uniquely related to the EFS function:

$$\Pr\left(T^* > t\right) = S(t) = \exp\left(-\Lambda(t)\right) = \exp\left(-\sum_{k=1}^{K} \int_0^t \lambda_k(u) \, du\right) = P_{00}(0, t). \tag{10.2}$$

Note that while it may be natural to consider $S_k(t) = \exp(-\Lambda_k(t))$, these quantities should not be interpreted as the corresponding *crude* survival function for event type k as they do not correspond to either state transition probabilities or state occupation probabilities. In fact, they have no simple probability interpretations without unverifiable assumptions (Gaynor et al.,

1993; Gooley et al., 1999). The interpretation of $S_k(t)$ is thus restricted to the hypothetical setting with unverifiable assumptions.

10.2.2.3 Cumulative Incidence Functions

Another important quantity in competing risks is *cumulative incidence function* (CIF), or cumulative crude failure probability. It is the cumulative probability of event type k having occurred by time t in the presence of other competing events and can be mathematically expressed as

$$F_k(t) = \Pr(T^* < t, D^* = k) = \int_0^t S(u)\lambda_k(u)\,du = P_{0k}(0,t). \qquad (10.3)$$

One may view the integrand as a product of the probability having no event occur up to time u (i.e., $S(u)$) and the instantaneous probability of event k failing at time u (i.e., $\lambda_k(u)$). Therefore, $F_k(t)$ may be heuristically considered as the expected proportion of patients with event k over a time interval t. The CIF is a nondecreasing function of time with $F_k(\infty) = \Pr(D^* = k) \leq 1$, and this is referred to as a subdistribution function. The CIF is also additive to the probability of failing from any event type, that is,

$$1 - S(t) = \sum_{k=1}^{K} F_k(t). \qquad (10.4)$$

Furthermore, Equation 10.3 implicitly implies that $F_k(t)$ depends on not only $\lambda_k(t)$ the CSH of cause k but also $\lambda_l(t)$ where $l \neq k$, the CSH of competing events by Equation 10.2. In other words, unlike classical univariate survival analysis where the hazard function has one-to-one correspondence with the survival function, in competing risks, the seeming mismatch between a CSH of interest and its subdistribution leads to some not-so-straightforward implications as we shall see in the subsequent sections.

10.2.2.4 Subdistribution Hazard Functions

Gray (1988) introduced the concept of the *subdistribution hazard* (SDH) function, which can be used for direct modeling of $F_j(t)$, as follows:

$$\tilde{\lambda}_k(t) = \lim_{h \to 0} \frac{1}{h} \Pr\left(t \leq T^* \leq t+h, D^* = k | \{T^* \geq t\} \cup \left(\{T^* \leq t\} \cap \{D^* \neq k\}\right)\right)$$

$$= \frac{dF_k(t)/dt}{1 - F_k(t)} = \frac{-d\log\{1 - F_k(t)\}}{dt}. \qquad (10.5)$$

We note that $\tilde{\lambda}_k(t)$ does not directly correspond to quantities interpretable within the multistate model. Based on Equation 10.5, one may heuristically view SDH as a classical hazard function of an artificial failure time $\tilde{T}$, defined as $\tilde{T} = T^*$ if $D^* = k$ and $\tilde{T} = \infty$ if $D^* \neq k$. As we delineate in Section 10.3, the definitions of SDH and CSH imply many fundamental differences.

10.2.2.5 Competing Risks as Latent Failure Times

Alternatively, competing risks may be formulated in the so-called latent or potential failure time framework, where there exist k latent failure times corresponding to the times to each distinct type of failure, $T_1^*, \ldots, T_k^*$, and the minimum of them or the first event time and type constitute the observed data, that is, $T^* = \min_k T_k^*, D^* = \arg\min_k T_k^*, k = 1, \ldots, K$. In presence of right-censoring, the joint survival distribution $Q(\cdot)$ becomes of interest, which is defined as

$$Q(t_1, \ldots, t_K) = \Pr\left(T_1^* \geq t_1, \ldots, T_K^* \geq t_K\right),$$

such that the EFS $S(t) = Q(t, t, \ldots, t)$, and the CSH for event type k as

$$\lambda_k(t) = \frac{-\partial \log Q(t_1, \ldots, t_K)}{\partial t_k}\Big|_{t_1 = \cdots = t_k = t}.$$

The corresponding marginal survival functions in the absence of competing events

$$S_k^*(t_k) = \Pr\left(T_k^* \geq t_k\right) = Q(0, \ldots, 0, t_k, 0, \ldots, 0),$$

and the corresponding hazard functions are often called as marginal *net* hazards for cause k, for example,

$$\lambda_k^*(t) = -\frac{\partial \log S_k^*(t)}{\partial t}.$$

It can be shown that given the data structure (T^*, D^*), neither the joint survival distribution $Q(\cdot)$ nor the marginal survival function $S_k^*(\cdot)$ (and the corresponding $\lambda_k^*(\cdot)$) is identifiable, in the sense that for any $Q(\cdot)$ with arbitrary dependence between latent failure times, it is always possible to find a different $Q^*(\cdot)$ with independence relationship between latent failure times, such that both have very same CSH (Tsiatis, 1975). A well-known counterexample was provided by Prentice et al. (1978). For mathematical traceability, one may assume the *independent competing risks* assumption (e.g., $Q(t_1, \ldots, t_K) = \prod_k S_k(t_k)$) to estimate relevant quantities of interest based on CSHs. However, such *independent competing risks* assumption is unverifiable based on the competing risks data structure. Therefore, despite the long history of the latent failure times approach, the related unverifiable assumptions seem to prevent their wide applications in practice for now.

10.3 Statistical Methods for Estimation and Inference

In this section, we review the estimation and statistical inference of the estimable quantities in competing risks. In particular, we focus on the two principal types of analysis methods based on CSH and CIF, respectively.

10.3.1 One-Sample Setting

Like other disciplines in medical research and epidemiology, in competing risks, we are frequently interested in *rates* and *risks* (Table 10.1). The average hazard rates are typically computed as the ratio of number of event type k during a specified follow-up period and the total follow-up times (either to any first event or censoring) over all individuals, that is, the person-time at risk. It can also be roughly viewed as a weighted average of the instantaneous *risk* (hazard) over a period of time for event type k and can be particularly informative if the underlying hazard is roughly time constant. Not surprisingly, the sum of average hazard rates for each event types equals the average hazard rate for any event.

The statistical counterparts of a risk is a *probability*. In competing risks, often the cumulative probability of observing event type k is of interest. While the cumulative incidence of failure cause k ($\hat{P}_{0k}(0, t)$ or $\hat{F}_k(t)$) renders the natural and appropriate probability interpretation in the presence of competing risks, it is not uncommon to see the practice of using the compliment of KM estimator ($1 - \hat{S}_k(t)$) for the same purpose, where $\hat{S}_k(t)$ is based only on failures from cause k and treating all other failure causes as censoring. Many authors have discussed the appropriateness of using these two estimators in the oncology context, for example, Gaynor et al. (1993) and Gooley et al. (1999).

TABLE 10.1

Commonly Used Statistical Methods for Competing Risks

Setting	CSH Based		CIF Based	
	Estimand	**Method**	**Estimand**	**Method**
1-sample	$1 - S_k(t)$	KM estimator	$F_k(t)$	Aalen–Johansen estimator
m-sample ($m \geq 2$)	$\lambda_k(t)$	Log-rank test	$\tilde{\lambda}_k(t)$	Gray's test
Regression	$\lambda_k(t)$	Cox PH model	$\tilde{\lambda}_k(t)$	F&G PH model

The numerical difference between $1 - S_k(t)$ and $F_k(t)$ is clear if we express

$$1 - S_k(t) = \int_0^t \lambda_k(u) S_k(u)\, du. \tag{10.6}$$

Comparing Equations 10.6 and 10.3, because $\lambda_k(t) \le \sum_{k=1}^K \lambda_k(t)$, and $S_k(t) = \exp\left(-\int_0^t \lambda_k(u)du\right)$ and $S(t) = \exp\left(-\int_0^t \sum_{k=1}^K \lambda_k(u)du\right)$, we have $S_k(t) \ge S(t)$ and consequently $1 - S_k(t) \ge F(t)$. The equality holds only if $K=1$, that is, when there are no competing risks.

In practice, both estimators may be viewed as step functions where jumps only occur at the observed event times. Let us assume there are n patients are under study, and each may experience one of the following outcomes: event type 1, event type 2, and independent censoring. Consider the ordered distinct event times $t_1 < t_2 < \cdots < t_m, m \le n$ and denote n_j as the number of patients who are known to be at risk of any failure *beyond* time t_j, and e_{1j}, e_{2j}, and c_j as the numbers of patients who fail from event type 1 and 2, and who are censored, respectively.

The KM estimator of $S_k(t)$ is

$$\hat{S}_k(t) = \prod_{j|t_j \le t} \left(1 - \frac{e_{1j}}{n_j}\right). \tag{10.7}$$

The corresponding Nelson–Aalen (NA) estimator for the cumulative hazard $\Lambda_k(t)$ is

$$\hat{\Lambda}(t) = \sum_{j|t_j \le t} \hat{\lambda}_k(t_j) = \sum_{j|t_j \le t} \frac{e_{1j}}{n_j}.$$

Similarly, the KM estimator of the EFS $S(t)$ (for $K = 2$) is

$$\hat{S}(t) = \prod_{j|t_j \le t} \left(1 - \sum_{k=1}^K \hat{\lambda}_k(t_j)\right) = \prod_{j|t_j \le t} \left(1 - \frac{e_{1j} + e_{2j}}{n_j}\right).$$

The cumulative incidence for the failure type k is estimated as

$$\hat{F}_k(t) = \sum_{j|t_j \le t} \hat{S}(t_{j-1}) \frac{e_{1j}}{n_j}, \tag{10.8}$$

which can be viewed as the cumulative product of $\hat{S}(\cdot)$ and $\hat{\lambda}_k(\cdot)$, and sometimes called the Aalen–Johansen estimator.

The underlying differences between $\hat{F}_k(t)$ and $1 - \hat{S}_k(t)$ center on how the presence of other events affects the *risk set* (patients who survive up to time t) of event of interest. For example, let us consider competing risks consisting of cancer recurrence (event type 1) and noncancer death (event type 2). If a patient dies from noncancer cause, then that person is no more at risk for cancer recurrence. That is, the computation of KM estimator involves removing patients who have experienced type 2 event (as well as the usual censored) prior to time t from the *risk set* of type 1 event, and assume those who are not censored nor experience other events prior to time t as being at risk for experiencing type 1 event afterward. On the other hand, the computation of CIF estimator would only remove patients who are censored but retain patients who have experienced type 2 event prior to time t *forever*.

Gooley et al. (1999) presented a relatively more intuitive way to distinguish these two estimators by representing the corresponding increments at each jump. The key is to rewrite the estimate of CIF or KM at time t with respect to event type 1 as $\sum_{j=1}^{s} J(t_j)e_{1j}$, a sum of individual increments (*jumps*). Then for some observed event time t_j, the increments of CIF and KM estimators with respect to event type 1, $J_{CIF}(t_j)$ and $J_{KM}(t_j)$, are

$$J_{CIF}(t_j) = J_{CIF}(t_{j-1})\left(1 + \frac{c_j}{n_j}\right); \quad J_{KM}(t_j) = J_{KM}(t_{j-1})\left(1 + \frac{c_j + e_{2j}}{n_j}\right).$$

Such representation clearly demonstrates why and how the CIF estimator is different from KM estimator. If we consider patients who fail from event type 2 (e.g., e_{2j}) as still possible to experience event type 1 beyond time t_j, we should then consider them as censored cases of event type 1 by removing e_{2j} patients from *risk set* when computing KM estimator or retaining e_{2j} patients when computing the complement of KM estimator. Therefore, $1 - \hat{S}_k(t) \geq \hat{F}_k(t)$ is always true, and the equality only holds up to the first competing event occurs. If there is no competing events, $1 - \hat{S}_k(t) = \hat{F}_k(t)$ all the time.

One of the variance estimators of CIF $\hat{F}_k(t_j)$ based on Delta method is as follows (Marubini and Valsecchi, 1995; Hosmer et al., 2008):

$$\text{var}\left\{\hat{F}_k(t_j)\right\} = \sum_{i=1}^{j}\left[\left\{\hat{F}_k(t_j) - \hat{F}_k(t_i)\right\}^2 \frac{e_{1i} + e_{2i}}{n_i(n_i - e_{1i} - e_{2i})}\right]$$

$$+ \sum_{i=1}^{j} \hat{S}(t_{i-1})^2 \left(\frac{n_i - e_{1i}}{n_i}\right)\left(\frac{e_{1i}}{n_i^2}\right)$$

$$- 2\sum_{i=1}^{j}\left\{\hat{F}_k(t_j) - \hat{F}_k(t_i)\right\}\hat{S}(t_{i-1})\frac{e_{1i}}{n_i}.$$

To calculate the point-wise confidence intervals of $\hat{F}_k(t_j)$, one of the commonly used transformations is complementary log–log (cloglog), for example,

$$\hat{F}_k(t)^{\exp\left[\frac{\pm z_{\alpha/2}\,\hat{\sigma}_k(t)}{\hat{F}_k(t)\log(\hat{F}_k(t))}\right]},$$

where

    $z_{\alpha/2}$ is the upper $\alpha/2$ percentile of the standard normal deviate

    $\hat{\sigma}_k(t)$ is the square root of var$\{\hat{F}_k(t)\}$ (Choudhury 2002, Equation 4)

Greenwood's formula may be used to obtain the variances and associated confidence intervals of $1 - \hat{S}_k(t)$.

10.3.1.1 Questions of Interest in Competing Risks

Kalbfleisch and Prentice (2002, Section 8.2) summarize *the problem* of competing risks as "the estimation of failure rates for certain types of failure given the *removal* of some or all other failure types." Alternatively, it is also of interest on the failure rates of certain event types observed in patients who are *subject to censoring* by the other types of events. We note that the two problems mentioned earlier are primarily *not* statistical problems. The associated statistical methods and models should be only utilized after we have a good understanding of the biological mechanisms and in the context of specific disease settings.

    A hypothetical example in the same line as Chappell (1996, 2012) may help to make the distinctions apparent. Consider a patient population where 60% of them may die due to noncancer cause (event type 2) by the end of first year, and suppose all survivors exhibit cancer recurrence (event type 1) by the end of second year. The 2-year cancer recurrence rate based on KM estimator (i.e., among patients who survive and in the absence of risks of death) is 100% as all survivors by 2-year experience recurrence and thus address *the problem* of competing risks. Alternatively, the crude rate of cancer recurrence at 2 years based on CIF (i.e., in presence of death) is only 40%, as the remaining 60% died too soon; this *real-life* estimate is affected by both risks of cancer recurrence and noncancer death. Therefore, it is fair to argue that both estimates simply provide different aspects of the disease process. The former (1-KM) characterizes the disease progression process when noncancer death may be viewed as nuisance and thus could be useful to compare the efficacy aspects of a same regimen used on patients with different risks of comorbidities. The latter (CIF), on the contrary, may be a more relevant quantity for patients and physicians to balance the benefits and risks of all possible outcomes, including death and complications, when receiving treatment.

10.3.2 Comparison between Two or More Groups

In the context of competing risks, often it is of interest to compare the failure histories of one or more particular event type(s) between two or more groups. To simplify our discussions, let us continue to consider competing risks comprising event types 1 and 2 and assume the primary interest is to compare whether the distributions of event type 1 are same between groups A and B. We denote groups A and B with subscripts a and b.

Following the lines of Freidlin and Korn (2005), the following null hypotheses based on estimable quantities may be of interest: (1) equality of CSH, $H_0^\lambda : \lambda_{1a}(t) = \lambda_{1b}(t)$; (2) equality of CIFs, $H_0^F : F_{1a}(t) = F_{1b}(t)$; (3) equality of SDHs, $H_0^{\tilde\lambda} : \tilde\lambda_{1a}(t) = \tilde\lambda_{1b}(t)$; and (4) equality of EFS, $H_0^{12} : S_a(t) = S_b(t)$. Within the framework of latent failure times, we may be also interested in the equality of (latent) marginal distributions, $H_0 : S_{1a}^*(t) = S_{1b}^*(t)$, and the equality of joint (latent) distribution $H_0^Q : Q_a(t_1, t_2) = Q_b(t_1, t_2)$, which is a global null hypothesis that no difference in either type 1 or 2 event between groups. Due to the nonidentifiability issues as discussed in Section 10.2.2, neither H_0 nor H_0^Q can be tested unless additional unverifiable assumptions are imposed. In these cases, one may choose to interpret test results based on estimable quantities in the context of latent failure times. In the sense of providing the same nominal levels, tests for H_0^λ are valid tests for H_0 under the assumption of independent latent failure times, and tests for H_0^F, $H_0^{\tilde\lambda}$ and H_0^{12} are also valid tests for H_0^Q.

The widely used log-rank test (Mantel, 1966) is known to compare the equality of survival probabilities between groups, and therefore, it can be used to test H_0^λ and H_0^{12}, as log-rank test in fact compares the underlying (cause-specific) hazards between groups. Gray (1988) proposed a class of G^ρ tests in m-sample settings ($m \geq 2$), which can be used to directly compare the equality of CIFs between groups (H_0^F) when $\rho = 0$. While some other tests were also proposed to test H_0^F, such as Pepe and Mori (1993); and Lin et al. (1997), we limit our attention to the log-rank test and Gray's test only to better contrast their differences in interpretations. Like the log-rank test, Gray's test compares the equality of CIFs between groups (H_0^F) through comparing the underlying SDH functions ($H_0^{\tilde\lambda}$).

Let us continue to consider the hypothetical example we refer in Section 10.3.1. Assume that in group A, 60% of patients may die due to noncancer cause (event type 2) by the end of first year, and all survivors exhibit cancer recurrence (event type 1) by the end of second year; in group B, 40% die from noncancer cause at end of first year, and the remaining experience cancer recurrence at the end of second year. The 2-year cancer recurrence rates in both groups based on KM estimators are 100%, as all survivors by 2-year experience recurrence, and log-rank test consequently concludes that the two groups are not different in the distributions of cancer recurrence, that is, accept H_0^λ. The cumulative crude probability of cancer recurrence at 2 years based on CIF are 40% in group A and 60% in group B, and if the sample sizes are sufficiently large, Gray's test may detect such *real-life* differences by rejecting H_0^F and $H_0^{\tilde\lambda}$ and conclude the cancer recurrence rate is higher in group B than group A. One may notice such absolute risk differences in cancer recurrence are the consequences of the differential failure rates of non-cancer deaths. Therefore, a log-rank test on event type 2 (which is identical to Gray's test) suggesting group A has a higher mortality rate than group B (60% vs. 40%) may be very useful to provide a complete story. A log-rank test for H_0^{12} on DFS fails to provide these fine details alone but may provide complementary information on H_0 or H_0^Q if additional unverifiable assumptions are made.

The mathematical distinctions between the log-rank test for H_0^λ and Gray's test for $H_0^{\tilde\lambda}$ essentially rely on those between CSH and SDH, namely, how to redistribute the occurred competing events in risks sets. Once a competing event occurs, that individual exits the risk set in log-rank test statistics, while remaining in the risk set forever with respect to the event of interest, as implied by viewing the artificial failure time for SDH $\tilde T = \infty$ in this case.

Recent literature has carefully reviewed the inference aspects of both tests through simulation studies of competing risks data with two event types (Freidlin and Korn, 2005; Williamson et al., 2007; Dignam and Kocherginsky, 2008). Particular attention was paid to evaluating the validity of the log-rank test for H_0^λ and Gray's test for H_0^F when H_0 holds but H_0^Q does not. In general, as log-rank test for H_0^λ correctly detects differences in CSHs, if one is interested in assessing the underlying causal relationship between treatment group and clinical benefit on event type 1, log-rank test for H_0^λ is preferred. It consistently provides correct inferences at desired nominal levels unless the correlation between competing events is high and positive (unverifiable from observed data though), as suggested by Freidlin and Korn (2005). If the (isolated) relative treatment effect on event type 2 is toward the null, log-rank test for H_0^{12} may provide relatively satisfactory inferential results at a slightly inflated type 1 error; if the treatment effect on event type 2 is in the same direction as event type 1 and only differs in magnitude, inferences based on log-rank test for H_0^{12} may be also acceptable with a reduced power. From this perspective, if the experimental treatment is only expected to impact-specific event type, study design using EFS-type endpoint is less efficient in the sense that same power may be achieved with less resources had competing risks methods been used. Meanwhile, as shown in Freidlin and Korn (2005) and Dignam and Kocherginsky (2008), because it is entirely possible to observe a beneficial effect of treatment detected by Gray's test for H_0^F under H_0, and the observed differences in CIF are simply driven by an increased incidence of the competing event, inferences based on Gray's test for H_0 should be considered with caution. Therefore, in clinical research, when treatment efficacy evaluation of specific event type (e.g., local failure) is of primary interest, especially in cancer research where independent or positive relation between competing events are speculated, we may find the log-rank test is more helpful than Gray's test to characterize the role of treatment by isolating the possibly auxiliary impacts of competing events.

On the other hand, Gray's test is a valid test for H_0^F as it correctly detects differences in CIFs, regardless of the cause of such differences, which could be due to differences in hazards between groups for the event of interest itself, *or/and* differences in hazards between groups for competing event. From patients' perspective, it is fully legitimate and important to know whether a planned treatment would reduce cancer recurrence rates without increased risks of systematic disease or comorbidity. That is, when risk–benefit assessment is the paramount goal, a good understanding on how the treatment cures the disease in the absence of competing events becomes a less relevant question to address, and thus, Gray's test for H_0^F may be preferred. Moreover, as shown by Williamson et al. (2007), it is also possible that Gray's test is more informative than the log-rank test even for H_0 if a negative relation between competing events is expected, such as in the context of epilepsy.

In summary, as Gray originally pointed out (Gray, 1988), inference based on log-rank test and Gray's test simply address different aspects of the event history of interest. No universal recommendations should be made unless we have good understanding on the context and disease, and the question of interest (i.e., null hypothesis) should drive the choice of tests.

10.3.3 Regression Models

We now extend the statistical models we have reviewed earlier to regression settings. Let $X = (X_1, X_2, \ldots)'$ be the vector of time-independent covariates and $Z(t) = (Z_1(t), Z_2(t), \ldots, Z_p(t))'$ be the vector of modeled covariates that are the functions of X and time, and to be used in models for (T^*, D^*). For the CSH function, it is straightforward to consider Cox PH model as

$$\lambda_k(t; X) = \lambda_{0k}\exp[Z'(t)\beta_k]. \tag{10.9}$$

As shown in Kalbfleisch and Prentice (2002), estimation and statistical inference may be obtained similarly in the univariate case based on partial likelihood. In addition, joint modeling of all failure types simultaneously is also allowed because a common filtration can be defined to yield a stratified partial likelihood. Of course, other commonly used regression models in univariate survival analysis, such as accelerated failure time (AFT) model or proportional odds model (Bennett, 1983), may be straightforwardly extended to CSH in competing risks.

Fine and Gray (1999) (F&G) proposed direct modeling of the CIF $F_k(t)$, or equivalently, the SDH $\tilde{\lambda}_k(t)$ following PH assumptions as

$$\tilde{\lambda}_k(t; X) = \tilde{\lambda}_{0k}(t) \exp\left[Z'\tilde{\beta}_k\right]. \tag{10.10}$$

where Z are time-independent covariates, and the given equation is equivalent to a transformation model for the subdistribution function,

$$\log\left\{-\log\left[1 - F_k(t; X)\right]\right\} = \log\left\{-\log\left[1 - F_{0k}(t; X)\right]\right\} + Z'\tilde{\beta}_k.$$

F&G showed that a partial likelihood approach can be applied for the estimation of $\tilde{\beta}$ when the censoring is either absent or always observed. Otherwise (e.g., the usual right censoring is present), additional technical complexity, such as using inverse probability of censoring weighting (IPCW) (Rotnitzky and Robins, 2005) to construct an unbiased estimating function, is involved to estimate the regression parameters. This is a consequence of the risk set definition used in SDH. In particular, if the potential censoring time is always known, for example, staggered entry with administrative censoring, the risk set at time t consists of two distinct groups: those who have neither failed nor been censored at time t and those who have previously failed from other causes *and* whose potential censoring time is greater than or equal to t. Therefore, when potential censoring time is unavailable, techniques like IPCW are required for unbiased estimation. Unlike CSH approach, separate F&G models need to be fitted for each event type of interest, and no simultaneous inference is allowed (nor reasonable, see the following text). Alternative regression methods, such as those based on *pseudovalues* proposed by Klein and Andersen (2005), are shown to have comparable performance to the F&G model, and not reviewed here.

As we repeatedly argue, the most important differences between the two PH models remain in their interpretations as they address different questions of interest and different aspects of the disease process. CSH model solely characterizes the effects of covariates on the risk of the event of interest, and the extent of how such isolated effects captured by CSH model depends on the interrelationship between failure causes. Dignam et al. (2012) showed that even under moderate dependence between failure types, the CSH model continues to reflect the isolated effects reasonably well. When multiple covariates are included in either regression model, the interpretation of the adjusted effects should be restricted to the corresponding model formulation, that is, either on the scale of CSH solely or on the scale of the absolute specific event cumulative probability (influenced by combined risks changes in all competing events).

In practice, it is not rare to see that F&G models are fitted for the CIFs of each failure caused separately. However, one should be aware that the proportionality assumption in SDHs *cannot* generally hold simultaneously for all causes. To see this, rewrite Equation 10.10 as

$$1 - F_k(t; Z) = (1 - F_{0k}(t))^{\exp[Z'\tilde{\beta}_k]},$$

and assume it holds for all k and $t > 0$. Let $\lim_{t \to \infty} F_{0k}(t) = p_{0k}$, where $p_{0k} > 0$, $k = 1, \ldots, K$ and $\sum_{k=1}^{K} p_{0k} = 1$. Therefore, given Z, it follows that

$$(1 - p_{01})^{\exp[Z'\tilde{\beta}_1]} + \cdots + (1 - p_{0K})^{\exp[Z'\tilde{\beta}_K]} = K - 1. \qquad (10.11)$$

Therefore $(1 - p_{0K})$ and $\tilde{\beta}_K$ are determined by all other parameters in the model, the SDH cannot be modeled independently any more due to the constraint (Equation 10.11).

Nonetheless, to have a better understanding of the competing risks, one still often tends to investigate the CIFs for all possible event types simultaneously by fitting separate F&G models. In this case, the practical implications are the proportionality assumptions may hold only for a limited duration of follow-up. As a result, approaches commonly used in univariate survival analysis to diagnose and take account of nonproportionality, such as including external time-dependent covariates (Kalbfleisch and Prentice, 2002), may be also considered in competing risks analysis. In other situations, internal time-dependent covariates, such as biomarkers, are also of interest in competing risks settings. For CSH PH models, one can easily incorporate them by carefully disseminating the at-risk set in partial likelihood (Kalbfleisch and Prentice, 2002). For SDH PH models, Beyersmann and Schumacher (2008) recently extended F&G models to incorporate internal time-dependent covariate in the SDH framework. However, it is noted that a prediction of CIF, either based on the relationship of Equation 10.3 or the proposal of Beyersmann and Schumacher (2008), cannot be done due to the nature of internal time-dependent covariates as they do not correspond to models for probabilities anymore (Andersen et al., 1993). Additionally, Prentice et al. (1978) suggested that one can use internal time-dependent covariates to assess the interrelationship between failure types based on CSH.

10.4 Application: Combination of Chemotherapy and Radiotherapy for Lung Cancer

Lung cancer is the leading cause of cancer death in the United States and worldwide. In 2013, there are about 228,190 new cases and 159,580 deaths related to lung cancer in the United States (Siegel et al., 2013). Approximately 80%–85% of lung cancers are non-small-cell lung cancer (NSCLC), and 40% of these are inoperable locally advanced (stage II/III) at diagnosis. Prognosis for these patients is generally poor, with 5-year overall survival of less than 25%.

Prior to the 1990s, thoracic radiotherapy (TRT) was the dominant and only treatment option for inoperable locally advanced non-small-cell lung cancer (LA-NSCLC). In the early 1990s, a combination of chemotherapy and TRT was found promising to improve patients' outcomes, and thus, efforts emerged to explore the optimal schedules of the combination delivery, that is, whether combining TRT with sequential or concurrent chemotherapy. A number of randomized phase III clinical trials from Japan (Furuse et al., 1999), Europe (Fournel et al., 2005), and North America (Curran et al., 2011), evaluated different schedules and regimens of combination therapy. The findings from these trials and subsequent meta-analysis (Auperin et al., 2006, 2010) collectively established the standard care for LA-NSCLC: concurrent delivery of chemotherapy and TRT.

One of these practice-changing clinical trials is the RTOG 9410 study (Curran et al., 2011). It was conducted in North America among adult patients with inoperable American Joint Committee on Cancer (AJCC) stage II, IIIA, or IIIB newly diagnosed NSCLC, and relatively good health status. RTOG 9410 compared three treatment delivery schedules: sequential treatment with cisplatin-vinblastine induction chemotherapy followed by TRT at a total dose of 63 Gy (arm 1); a concurrent schedule in which two cycles of cisplatin-vinblastine were administered during the same TRT schedule as in arm 1 (arm 2); and another concurrent schedule combining cisplatin-etoposide chemotherapy with bifractionated TRT at a total dose of 69.6 Gy (arm 3). Patients were stratified by performance status (KPS 70-80 vs. 90-100) and clinical stage (II vs. IIIA vs. IIIB) and randomly assigned to one of three arms. The primary trial endpoint was overall survival, and study results showed that concurrent cisplatin-vinblastine and radiation therapy statistically significantly increased overall survival compared with the sequential treatment, whereas there was no statistically significant difference between the two concurrent schedules (Curran et al., 2011).

In-field (local) tumor growth is the major cause of failures in the treatment of inoperable LA-NSCLC; therefore, improvements in local control have been the principal target of localized therapy like TRT. Meanwhile, competing risks naturally arise as following the treatment of primary tumor, patients may also encounter out-field failure (distant metastasis or second primary), or death resulting from noncancer causes. DFS, defined as the time from randomization to in-field failure (failure type 1), out-field failure (failure type 2), or death without disease (failure type 3), whichever occurs first, is also of interest. In the following, we illustrate the competing risks analytic methods reviewed earlier by using arm 1 (sequential) and arm 3 (concurrent) data of RTOG 9410. Among the 382 eligible patients, 195 and 187 were assigned to sequential therapy and concurrent therapy, respectively.

Figure 10.2a presents the estimators of cumulative in-field failure probability, based on (1) the complement of KM estimate $(1 - \hat{S}_1(t))$ and (2) CIF $(\hat{F}_1(t))$. It is clear to see $1 - \hat{S}_1(t)$ is systematically overestimates $\hat{F}_1(t)$. The difference

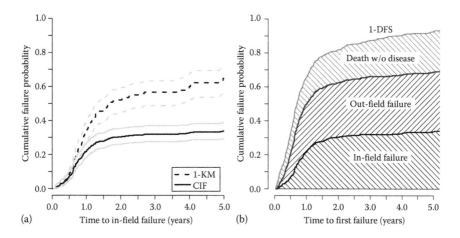

FIGURE 10.2

Cumulative incidence of failures in one-sample setting for LA-NSCLC patients in RTOG 9410: (a) The complement of KM estimate and the CIF estimate of in-field failure. Point-wise 95% confidence intervals are shown in gray. (b) Overlaid cumulative failure probabilities of first failures comprising DFS. Distinct shaded areas present the cumulative incidences of specific failure types.

is due to the patients who failed from out-field failure and death without disease. In $1 - \hat{S}_1(t)$, it is assumed patients who first experienced out-field failure or death without disease remain at the same likelihood of experiencing in-field failure; meanwhile, $\hat{F}_1(t)$ estimates the actual proportions that one may first experience in-field failure in the presence of the other competing events over time. As we discussed in Section 10.3.1, the two estimators simply answer questions under different contexts and provide complementary information about the development of local failures. For descriptive purposes, one may find estimates of cumulative incidences are more informative as they summarize the actual risks of experiencing certain events in presence of competing risks. The associated point-wise 95% confidence intervals based on Greenwood's formula and Delta method are also provided in gray.

Figure 10.2b provides an alternative illustration of CIF estimates that correspond to event types comprising DFS. The shaded areas (from bottom to top) represent $\hat{F}_1(t)$, $\hat{F}_2(t)$, and $\hat{F}_3(t)$, respectively. It is clear to see the relationship of Equation 10.4 by stacking the CIF estimates, as the top green curve also represents $1 - \hat{S}(t)$, where $\hat{S}(t)$ is the DFS probability.

Figure 10.3 compares whether distributions of each event type are same between the sequential (SEQ) delivery and concurrent (CON) delivery of chemotherapy and TRT. In addition to the estimates of cumulative incidences ($\hat{F}_k(t)$) and the complement of KM estimates ($1 - \hat{S}_k(t)$), the p-values from Gray's tests for CIF and log-rank tests for CSHs are provided accordingly.

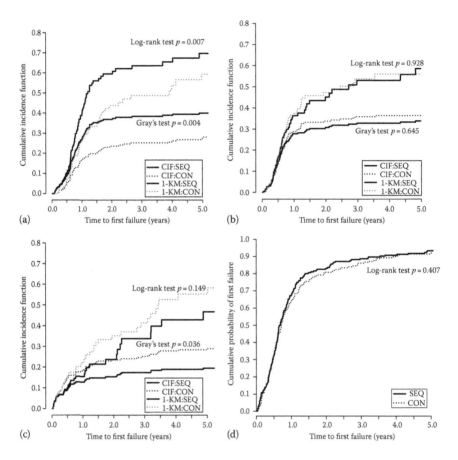

FIGURE 10.3
Comparing cumulative incidences of first failures for LA-NSCLC patients in RTOG 9410:
(a) In-field failures. (b) Out-field failures. (c) Deaths without disease. (d) Any failure types.

For in-field failure (Figure 10.3a), the hazard was significantly reduced among patients randomized to concurrent therapy (log-rank $p=0.007$), and cumulative incidence in concurrent therapy was also significantly lower than that in sequential therapy (Gray's test $p=0.004$). For out-field failure (Figure 10.3b), the hazard and cumulative incidence for patients assigned to concurrent therapy was nearly identical to their counterparts treated with sequential therapy, with log-rank $p=0.928$ and Gray's test $p=0.645$. Interestingly, for death without disease (Figure 10.3c), the log-rank test indicates there is no difference in CSHs ($p=0.149$), whereas Gray's test suggests concurrent therapy had a significantly higher cumulative incidence than sequential therapy ($p=0.036$). As there is no biologic explanation that supports differential risks of death without disease between treatments, the observed difference in cumulative incidences is likely a natural consequence of more patients having experienced in-field failure in the sequential therapy group, and thus, fewer patients remained at risk for death without disease. In other words, as out-field failure did not appear to be different between treatments, the fact that concurrent therapy significantly reduced in-field failure risk alone could possibly explain the increased incidence of death without disease because more patients were *exposed* to such risks. When combining all three types of failures together and comparing the distributions of DFSs (Figure 10.3d), log-rank test indicates no difference between treatments ($p = 0.407$). From the perspective of detecting a meaningful difference in (cause-specific) hazards of in-field failure, the observation of DFS here exemplifies the potential issue that such hazard difference in in-field failure is likely to be *diluted* when a composite endpoint (DFS) is used, unless a similar effect is also expected in other competing events.

To further evaluate the treatment effects of competing events after adjusting for other potential prognostic factors, we conduct Cox PH regression analysis based on CSHs, and F&G PH regression analysis based on SDHs. In general, cancer stage and histology are important prognostic factors for in-field failures and subsequent cancer deaths. Age at diagnosis is also important with respect to risks of noncancer deaths.

In summary, among the 382 eligible patients who participated in the trial and were randomized to either sequential or concurrent therapy, we evaluate treatment assignment (SEQ vs. CON), histology (nonsquamous vs. squamous), stage (IIIB vs. IIIA and II), and age at diagnosis (>60 vs. ≤60), with respect to the CSHs and SDHs of in-field failure, out-field failure, and death without disease. KPS (stratification variable) is not reported in the following analysis as it was not significantly associated with any of event types using either regression models (data not shown).

Table 10.2 summarizes the regression analysis results. Concurrent therapy reduced the hazard of in-field failure by 40% (HR = 0.598, $p = 0.004$); comparing with sequential therapy, the risk on the cumulative incidence scale (SDH) of in-field failure was reduced by 38% (HR = 0.621, $p = 0.007$) for concurrent therapy. For cancer stage, stage IIIB cancer significantly increased the hazard of in-field failure by 51% (HR = 1.514, $p = 0.02$) but only nominally increased the SDH of in-field failure (HR = 1.238, $p = 0.23$). Elderly patients at diagnosis (older than 60 years) were associated with a statistically nonsignificant 12% (HR = 0.876, $p = 0.454$) risk reduction of in-field failure, whereas had 29% (HR = 0.711, $p = 0.05$) lower risk of in-field cumulative incidence than younger patients.

For death without disease, concurrent therapy appears to only nominally increase the CSH (HR = 1.295, $p = 0.211$), whereas it significantly increases the cumulative incidence risk by 54% (HR = 1.538, $p = 0.036$). Furthermore, both Cox PH model based on CSH and F&G PH model based on SDH suggest that patients diagnosed with squamous cancer and older than 60 years were more likely to die without disease. Based on both regression models, none of the covariates (including treatment assignment) was found significantly

TABLE 10.2

Comparison of Competing Risks Regression Models Examining Treatment and Important Covariates for Competing Outcomes in LA-NSCLC (RTOG 9410)

Event Type	Cox CSH PH Model			Fine and Gray SDH PH Model		
	HR	95% CI	*p*-Value	HR	95% CI	*p*-Value
In-field failure						
CON vs. SEQ (RL)	0.598	(0.422, 0.849)	0.004	0.621	(0.439, 0.878)	0.007
Nonsquamous vs.						
squamous (RL)	0.843	(0.593, 1.199)	0.342	1.017	(0.715, 1.445)	0.930
Stage IIIB vs. IIIA/II (RL)	1.514	(1.067, 2.148)	0.020	1.238	(0.877, 1.748)	0.230
Age >60 vs. ≤60 (RL)	0.876	(0.619, 1.239)	0.454	0.711	(0.505, 1.000)	0.050
Out-field failure						
CON vs. SEQ (RL)	1.009	(0.720, 1.414)	0.959	1.084	(0.775, 1.516)	0.640
Nonsquamous vs.						
squamous (RL)	1.160	(0.811, 1.661)	0.417	1.374	(0.963, 1.961)	0.080
Stage IIIB vs. IIIA/II (RL)	1.143	(0.812, 1.608)	0.444	1.005	(0.714, 1.414)	0.980
Age >60 vs. ≤60 (RL)	1.097	(0.776, 1.549)	0.601	1.032	(0.729, 1.462)	0.860
Death without disease						
CON vs. SEQ (RL)	1.295	(0.863, 1.943)	0.211	1.538	(1.029, 2.300)	0.036
Nonsquamous vs.						
squamous (RL)	0.637	(0.428, 0.950)	0.027	0.646	(0.433, 0.964)	0.033
Stage IIIB vs. IIIA/II (RL)	1.022	(0.686, 1.523)	0.914	0.917	(0.618, 1.359)	0.660
Age >60 vs. ≤60 (RL)	2.153	(1.374, 3.374)	0.001	1.895	(1.230, 2.921)	0.004
DFS						
CON vs. SEQ (RL)	0.892	(0.726, 1.096)	0.276	—	—	—
Nonsquamous vs.						
squamous (RL)	0.875	(0.709, 1.080)	0.214	—	—	—
Stage IIIB vs. IIIA/II (RL)	1.232	(1.001, 1.517)	0.049	—	—	—
Age >60 vs. ≤60 (RL)	1.191	(0.963, 1.472)	0.107	—	—	—

Notes: CON, Concurrent chemotherapy and TRT; SEQ, Sequential chemotherapy and TRT; RL, Reference level.

associated with altered risks of out-field failure. The only exception is cancer histology, where nonsquamous cancer nominally increased the cumulative incidence risk (HR $= 1.374$, $p = 0.08$).

Cox PH model is also used to review the associations between covariates and DFS, the composite endpoint comprising the first occurrence of in-field failure, out-field failure, or any death. This approach avoids using competing risks and is still commonly used in oncology clinical trials. Concurrent therapy was not significantly associated with altered risks of DFS. Meanwhile, stage IIIB cancer was significantly associated with inferior DFS outcomes (HR $= 1.232$, $p = 0.049$).

Some reconciliation between the regression analysis results based on CSH, SDH, and DFS may be of interest. If concurrent therapy indeed reduced the risks of in-field failure and had no influence in out-field failure, as shown in the CSH and SDH regression analyses, then it is natural to expect an increased incidence of noncancer death simply because more patients were exposed to such risks. Therefore, from the perspective of determining the causal relationship between treatment and first failures, results based on CSH regression model may be more informative than SDH. For age at diagnosis, it is interesting to observe that elderly patients were less likely to experience in-field but more likely to die from noncancer causes with respect to cumulative incidences, that is, a *protective* effect of age with respect to in-field failure. This may be largely due to the tremendous risk of noncancer death as patients get older, such that the older the patient is, the less likely he/she would be alive to experience in-field failure simply because most of them already died from noncancer causes. Such phenomena is not rare in cancer studies, and frequently reported in competing risks literatures, including the original proposal by Fine and Gray (1999). Cancer stage appears to be prognostic with respect to DFS, as stage IIIB cancer was associated with inferior outcomes. CSH regression analysis indicates that such inferior outcomes (HR $= 1.232$, $p = 0.049$) were mainly through the increased hazard of in-field failure (HR $= 1.514$, $p = 0.02$). Meanwhile, cumulative incidences may be insensitive to identify the source of impacted failure type, as clinical stage was not significantly associated with the cumulative incidences of any failure types. Moreover, age at diagnosis had differential responses to noncancer death and cancer recurrence, and analysis based on DFS alone fails to identify such potentially important interactions, as discussed in Section 10.2.1.

10.5 Discussions

In classical univariate survival analysis, the fundamental quantity is the hazard function. Other important quantities, such as the subdistribution function (the complement of survival function in univariate settings), have one-to-one correspondence with the hazard alone. In competing risks, such one-to-one correspondence no longer holds anymore. For example, $F_k(t)$, the cumulative incidence of event type k, not only depends on $\lambda_k(t)$, the CSH for event type k, but the ones for other causes. Among the key identifiable quantities in competing risks, CSHs continue to play a pivotal role, and the CIFs may be viewed as functions of CSHs from all event types. Such feature becomes apparent when viewing competing risks and univariate survival analysis from the perspective of a multistate model, where the occurrence probabilities of specific absorbing states naturally depend on all preceding transitions between states (Andersen et al., 2002). A number of implications are rooted from such feature in competing risks.

The CSHs are instantaneous transition rates between adjacent states and are well-defined and valid by definition. Statistical inferences based on CSHs, such hypothesis testing for equality between two or more groups, and associated regression analysis are therefore also built upon minimal assumptions and may be comprehended in a similar fashion as univariate survival analysis. On the other hand, the *crude* event-specific survival probability $S_k(t) = \exp\left(-\int_0^t \lambda_k(u)\,du\right)$, which depends only on the CSH $\lambda_k(t)$, does not have a direct probability interpretation in the multistate model where both event of interest and competing events are operating, because it cannot correspond to any state occupation probability. Therefore, $\hat{S}_k(t)$ or $1 - \hat{S}_k(t)$ is meaningful only under the hypothetical scenario where the competing events are absent. Such hypothetical scenario assessment is not rare in practice; for example, a radiation biologist may be interested in characterizing the treatment-related toxicity process by treating death as nuisance, and policymakers may be interested in the social burden of prostate cancer-specific death had risks of comorbidity been removed. Indeed, one may view the hypothetical assessments in the context of *sensitivity analysis*, and make appropriate interpretations accordingly.

The CIFs measure the absolute risks of specific event occurrences over time. From the perspective of a multistate model, they simply represent the corresponding state occupation probabilities indexed by time. Therefore, they are the ideal estimands when quantifying the changes in absolute risks and perform risk/benefit analysis of a treatment. Meanwhile, the construction of the associated SDH functions is artificial in the sense that they do not correspond to actual instantaneous transition rates in the multistate model. Instead, $\tilde{\lambda}_k(t)$ can be heuristically viewed as the hazard of event type k at time t from a population who are either event-free or experienced any event other than type k. Therefore, the interpretation of hazard ratios from F&G PH model is neither straightforward nor intuitive in some cases. Nonetheless, the direct modeling CIF through SDHs, such as F&G PH model, still provides a much needed approach to assess the associations between covariates and CIF and may be useful to identify subgroups who are most likely to benefit from a treatment when all possible outcomes are considered.

We have reviewed the most commonly used analytic approaches in the presence of competing risks by characterizing them into two principal types of competing risks analysis: the ones based on CSH and CIF, respectively. We have attempted to illustrate the sources of differences from both mathematical formulation and interpretation and argued the choice of estimands and methods should be driven by the question of interest and context-dependent. An in-depth competing risks analysis is provided in Section 10.4 to illustrate these issues. Another competing risks example in oncology clinical trials can be found in Section 9.3.2.2. We are not able to systematically review analytic methods within latent failure time framework. As the observed data structure is not informative enough to identify the correlations between latent failure times, the joint distribution $Q(\cdot)$ is not estimable except under additional verifiable assumptions, for example, a known copula class that generates $Q(\cdot)$ or independence. Therefore, many (Andersen and Keiding, 2012 and references therein) have argued that models for latent failure times are not very useful in practice. We agree and believe that methods based on latent failure times, if used in practice, should be only considered for secondary exploratory analysis and viewed as another type of *sensitivity analysis* under assumed extreme cases.

Analysis based on an EFS-type endpoint is still frequently used in cancer clinical trials due to its clinical relevance and methodological simplicity. In this case, the key quantity in estimation is the sum of CSHs $\lambda(t) = \sum_{k=1}^{K} \lambda_k(t)$, with one-to-one correspondence with survival and CIF. Intuitively, using the sum of CSHs is subject to loss of information from individual CSH. Therefore, in practice, using an EFS endpoint may be most informative if the intervention is expected to have similar effects (both in sign and magnitude) on all event types. Otherwise, as we have seen in Section 10.4, solely using DFS may fail to identify the differential effects of treatment and age with respect to different failure types. As a result, analysis of EFS endpoints should always be accompanied with the analysis of CSH of primary and competing events, such that stakeholders can have a more complete picture of the disease process and make informed decisions accordingly.

Grant Support

This work was supported by U10 CA21661 (RTOG) and U10 CA180822 (NRG Oncology) from the National Cancer Institute, National Institutes of Health, Department of Health and Human Services.

References

Andersen, P. K., Abildstrom, S. Z., and Rosthøj, S. (2002). Competing risks as a multistate model. *Statistical Methods in Medical Research*, 11(2):203–215.

Andersen, P. K., Borgan, Ø., Gill, R. D., and Keiding, N. (1993). *Statistical Models Based on Counting Processes*. Springer, New York.

Andersen, P. K. and Keiding, N. (2012). Interpretability and importance of functionals in competing risks and multistate models. *Statistics in Medicine*, 31(11–12): 1074–1088.

Auperin, A. et al. (2006). Concomitant radio-chemotherapy based on platin compounds in patients with locally advanced non-small cell lung cancer (NSCLC): A meta-analysis of individual data from 1764 patients. *Annals of Oncology*, 17(3):473–483.

Auperin, A. et al. (2010). Meta-analysis of concomitant versus sequential radio-chemotherapy in locally advanced non-small-cell lung cancer. *Journal of Clinical Oncology*, 28(13):2181–2190.

Bennett, S. (1983). Analysis of survival data by the proportional odds model. *Statistics in Medicine*, 2(2):273–277.

Beyersmann, J. and Schumacher, M. (2008). Time-dependent covariates in the proportional subdistribution hazards model for competing risks. *Biostatistics*, 9(4):765–776.

Chappell, R. (1996). Re: Caplan et al. *IJROBP* 29: 1183–1186; 1994, and Bentzen et al. *IJROBP* 32: 1531–1534; 1995. *International Journal of Radiation Oncology Biology Physics*, 36(4):988–989.

Chappell, R. (2012). Competing risk analyses: How are they different and why should you care? *Clinical Cancer Research*, 18(8):2127–2129.

Choudhury, J. B. (2002). Nonparametric confidence interval estimation for competing risks analysis: Application to contraceptive data. *Statistics in Medicine* 21:1129–1140.

Curran, W. J. et al. (2011). Sequential vs concurrent chemoradiation for stage iii non-small cell lung cancer: Randomized phase iii trial rtog 9410. *Journal of the National Cancer Institute*, 103(19):1452–1460.

Dignam, J. J. and Kocherginsky, M. N. (2008). Choice and interpretation of statistical tests used when competing risks are present. *Journal of Clinical Oncology*, 26(24):4027–4034.

Dignam, J. J., Zhang, Q., and Kocherginsky, M. (2012). The use and interpretation of competing risks regression models. *Clinical Cancer Research*, 18(8):2301–2308.

Fine, J. P. and Gray, R. J. (1999). A proportional hazards model for the subdistribution of a competing risk. *Journal of the American Statistical Association*, 94(446):496–509.

Fournel, P. et al. (2005). Randomized phase iii trial of sequential chemoradiotherapy compared with concurrent chemoradiotherapy in locally advanced non-small-cell lung cancer: Groupe lyon-saint-etienne d'oncologie thoracique–groupe français de pneumo-cancérologie npc 95-01 study. *Journal of Clinical Oncology*, 23(25):5910–5917.

Freidlin, B. and Korn, E. L. (2005). Testing treatment effects in the presence of competing risks. *Statistics in Medicine*, 24(11):1703–1712.

Furuse, K. et al. (1999). Phase III study of concurrent versus sequential thoracic radiotherapy in combination with mitomycin, vindesine, and cisplatin in unresectable stage iii non-small-cell lung cancer. *Journal of Clinical Oncology*, 17(9):2692–2692.

Gaynor, J. J., Feuer, E. J., Tan, C. C., Wu, D. H., Little, C. R., Straus, D. J., Clarkson, B. D., and Brennan, M. F. (1993). On the use of cause-specific failure and conditional failure probabilities: Examples from clinical oncology data. *Journal of the American Statistical Association*, 88(422):400–409.

Gooley, T. A. et al. (1999). Estimation of failure probabilities in the presence of competing risks: New representations of old estimators. *Statistics in Medicine*, 18(6):695–706.

Gray, R. J. (1988). A class of k-sample tests for comparing the cumulative incidence of a competing risk. *The Annals of Statistics*, 16:1141–1154.

Hosmer, D. W., May, S., and Lemeshow, S. (2008). *Applied Survival Analysis*. Wiley-Interscience, Hoboken, NJ.

Kalbfleisch, J. D. and Prentice, R. L. (2002). *The Statistical Analysis of Failure Time Data*, Vol. 360, 2nd edn. Wiley-Interscience, Hoboken, NJ.

Kaplan, E. L. and Meier, P. (1958). Nonparametric estimation from incomplete observations. *Journal of the American Statistical Association*, 53(282):457–481.

Klein, J. P. and Andersen, P. K. (2005). Regression modeling of competing risks data based on pseudovalues of the cumulative incidence function. *Biometrics*, 61(1):223–229.

Lin, D. et al. (1997). Non-parametric inference for cumulative incidence functions in competing risks studies. *Statistics in Medicine*, 16(8):901–910.

Mantel, N. (1966). Evaluation of survival data and two new rank order statistics arising in its consideration. *Cancer Chemotherapy Reports*, 50:163–170.

Marubini, E. and Valsecchi, M. G. (1995). *Analysing Survival Data from Clinical Trials and Observational Studies*. John Wiley, Chichester, U.K.

Mell, L. K. and Jeong, J.-H. (2010). Pitfalls of using composite primary end points in the presence of competing risks. *Journal of Clinical Oncology*, 28(28):4297–4299.

Pepe, M. S. and Mori, M. (1993). Kaplan-meier, marginal or conditional probability curves in summarizing competing risks failure time data? *Statistics in Medicine*, 12(8):737–751.

Prentice, R. L., Kalbfleisch, J. D., Peterson Jr., A. V., Flournoy, N., Farewell, V., and Breslow, N. (1978). The analysis of failure times in the presence of competing risks. *Biometrics*, 34:541–554.

Putter, H., Fiocco, M., and Geskus, R. (2007). Tutorial in biostatistics: Competing risks and multi-state models. *Statistics in Medicine*, 26(11):2389–2430.

Rotnitzky, A. and Robins, J. M. (2005). Inverse probability weighting in survival analysis. In *Encyclopedia of Biostatistics*, P. Armitage and T. Cotton, eds. John Wiley, Chichester, U.K.

Siegel, R., Naishadham, D., and Jemal, A. (2013). Cancer statistics, 2013. *CA: A Cancer Journal for Clinicians*, 63(1):11–30.

Tsiatis, A. (1975). A nonidentifiability aspect of the problem of competing risks. *Proceedings of the National Academy of Sciences*, 72(1):20–22.

Williamson, P., Kolamunnage-Dona, R., and Smith, C. T. (2007). The influence of competing-risks setting on the choice of hypothesis test for treatment effect. *Biostatistics*, 8(4):689–694.

11

Dose Finding with Escalation with Overdose Control in Cancer Clinical Trials

André Rogatko and Mourad Tighiouart

CONTENTS

11.1 Introduction

Clinical trials of new anticancer therapies are widespread, critically important tools in the search for more effective cancer treatments. Cancer trials
typically proceed through several distinct phases. The major objective in
dose-finding (phase I) trials is to identify a working dose for subsequent
studies, whereas the major endpoint in phase II and III trials is treatment efficacy. Dose-finding trials represent the first testing of an investigational agent
in humans and act as a point of translation of years of laboratory research
into the clinic. Whereas dose-finding trials in other areas of medicine enroll
healthy participants, oncology dose finding trials typically enroll patients
who have cancer and who have exhausted standard treatment options
(Roberts et al. 2004).

Dose finding is the crucial first step in the process of scrutinizing whether a
new agent will help cancer patients. The fundamental conflict underlying the
design of cancer dose-finding trials is that increasing the dose slowly to avoid
unacceptable toxic events must be balanced against treating many patients at
suboptimal or nontherapeutic doses. Ideally, from a therapeutic perspective,
dose-finding trials should be designed to maximize the number of patients
receiving an optimal dose. Consequently, more patients would be treated
with therapeutic doses of promising new agents, and fewer patients would
have to suffer the deleterious effects of toxic doses.

Evidence of treatment benefit, usually expressed as a reduction in tumor
size or an increase in survival, requires months (if not years) of observation
and is therefore unlikely to occur during the relatively short-time course of
a dose-finding trial (O'Quigley et al. 1990, Whitehead 1997). Consequently,
the target dose is usually defined in terms of the prevalence of treatment side
effects without direct regard for treatment efficacy. Thus, it can be defined
that the primary objective of a dose-finding trial is to determine the safe
dose of a new drug or a combination of drugs for subsequent clinical evaluation of efficacy. The dose sought is typically referred to as the maximum
tolerated dose (MTD), working dose, target dose, or phase II dose; and its
definition depends on the treatment under investigation, the severity and
reversibility of its side effects, and clinical attributes of the target patient
population. Specifically, the MTD, γ, is defined as the dose expected to

produce some degree of medically unacceptable, dose-limiting toxicity (DLT) in a prespecified proportion θ of patients (Gatsonis and Greenhouse 1992),

$$P(\text{DLT}|\text{Dose} = \gamma) = \theta \qquad (11.1)$$

In other words, if a population of patients is treated at the MTD, a proportion θ of them is expected to manifest DLT. The value chosen for the target probability θ would depend on the nature and consequences of the DLT; it would set relatively high when the DLT is a transient, correctable, or nonfatal condition and low when it is life threatening or lethal (Babb et al. 1998).

Several statistical methodologies have been proposed in the literature to select the MTD in cancer phase I trials, and many of them have been reviewed (Rosenberger and Haines 2002, Ting 2006). In particular, the continual reassessment method (CRM) (O'Quigley et al. 1990) and its modifications (Faries 1994, Goodman et al. 1995, Moller 1995, Piantadosi et al. 1998, Storer 2001), and escalation with overdose control (EWOC) (Babb et al. 1998, Zacks et al. 1998, Babb and Rogatko 2001, Tighiouart et al. 2005, Rogatko et al. 2008) are Bayesian adaptive designs that produce consistent sequences of doses. CRM and EWOC can be easily implemented in practice using published tutorials and free interactive software (Zohar et al. 2003, Garrett-Mayer 2006, Xu et al. 2007, Rogatko et al. 2012a,b).

The focus of the present chapter is on EWOC, a Bayesian adaptive design. The characteristics of a typical dose-finding trial are very amenable to a Bayesian approach. At the onset of the trial, there is wide uncertainty and no experimental information, and evidence is accumulated sequentially. Thus, the design and conduct of dose-finding trials would benefit from methods that can incorporate information from preclinical studies and sources outside the trial. At the same time, the adopted methodology should be able to coherently combine this preclinical information and evidence accrued during the trial. Furthermore, both the investigator and the patient might benefit if updated assessments of the risk of toxicity were available during the trial. Both of these needs can be addressed within a Bayesian framework.

EWOC is primarily intended to be used in cancer clinical trials involving agents whose side effects at dose levels above the MTD might be quite severe. Cancer dose-finding trials typically are performed with patients for whom known indicated treatments have failed or do not exist. To these terminally ill patients, a dose-finding trial offers an experimental treatment with some probability of success. Horstmann et al. (2005) reviewed all nonpediatric dose-finding oncology trials sponsored by the Cancer Therapy Evaluation Program at the National Cancer Institute (NCI) between 1991 and 2002. They analyzed 460 trials involving 11,935 participants, all of whom were assessed for toxicity and 10,402 of whom were assessed for a response to therapy. The overall response rate (i.e., for both complete and partial responses) was 10.6%, with considerable variation among the trials. These results demonstrate that it is reasonable to expect a therapeutic intent from a dose-finding trial.

Consequently, since there is an ethical motivation to not harm the patient, controlling this risk is a desirable goal for physicians and patients. A dose-searching algorithm that conforms to such an ethical constraint ideally should converge to the target dose from below, that is, from lower doses. EWOC was the first dose-finding procedure to directly incorporate the ethical constraint of minimizing the chance of treating patients at unacceptably high doses. Its defining property is that the expected proportion of patients treated at doses above the MTD is equal to a specified value α, the *feasibility bound*. This value is selected by the clinician and reflects his/her level of concern about overdosing. Zacks and colleagues (1998) showed that among designs with this defining property, EWOC minimizes the average amount by which patients are underdosed. This means that EWOC approaches the MTD as rapidly as possible, while keeping the expected proportion of patients overdosed less than the value α. They also showed that, as a trial progresses, the dose sequence defined by EWOC approaches the MTD (i.e., the sequence of recommended doses converges in probability to the MTD). Eventually, all patients beyond a certain time would be treated at doses sufficiently close to the MTD.

11.2 Escalation with Overdose Control

Denote by Y the binary indicator of DLT. Assume that there exists x^* and x^{**}, $x^* < x^{**}$ such that

$$P(Y = 1|x = x^*) = 0$$
$$P(Y = 1|x = x^{**}) = 1 - \varepsilon \tag{11.2}$$

where $0 < \varepsilon < 1$ is known and $\theta < 1 - \varepsilon$.

Let $F(z)$ be a strictly increasing cumulative distribution function (CDF) having probability density function $f(z)$. We consider a dose–toxicity relationship of the form

$$P(Y = 1|x) = F\left(F^{-1}(1 - \varepsilon) + \beta \log\left(\frac{x - x^*}{x^{**} - x^*}\right)\right) \tag{11.3}$$

where β is unknown, and $0 < \beta^* < \beta < \beta^{**}$ for some positive real numbers β^* and β^{**}. It is easy to verify that model (Equation 11.3) satisfies the constraints (Equation 11.2). The condition $\beta > 0$ implies that the probability of DLT is an increasing function of dose. Assume that the MTD γ defined in

(Equation 11.1) belongs to the interval $[x^*, x^{**}]$. Using (Equation 11.3), it can be shown that

$$\gamma = x^* + (x^{**} - x^*)e^{\gamma'} \tag{11.4}$$

where

$$\gamma' = \log\left(\frac{\gamma - x^*}{x^{**} - x^*}\right) = \frac{F^{-1}(\theta) - F^{-1}(1 - \varepsilon)}{\beta} = -\frac{\phi}{\beta} \tag{11.5}$$

is the MTD on the log-standardized scale and $\Phi = F^{-1}(1 - \varepsilon) - F^{-1}(\theta)$.

11.2.1 Dose Escalation Based on Bayesian Estimates

Let $G(\beta x) = F(F^{-1}(\theta) + \Phi + \beta x)$, $g(x) = G'(x)$, and $z_1 = -\Phi/\beta^*$ be the level assigned to the first patient. This log-standardized dose is safe in the sense that $G(\beta z_1) \leq \theta$. Let $D_n = \{(z_i, Y_i), i = 1, \ldots, n\}$ be the data after enrolling n patients to the trial where Y_i is the observed DLT status of the patient getting level z_i, $z_i \in L^* = [-\Phi/\beta^*, -\Phi/\beta^{**}]$.

Let $h(\beta)$ be a prior density function for the parameter β on $[\beta^*, \beta^{**}]$ and $\Pi_n(\beta) = \Pi(\beta|D_n)$ the posterior CDF given the data D_n. For $0 < \alpha < 1$, let

$$\xi_n^{(\alpha)} = -\frac{\phi}{\Pi_{n-1}^{-1}(\alpha)}, \quad n \geq 1 \tag{11.6}$$

Then, it is easy to verify that for all $n \geq 1$,

$$P\left(\xi_n^{(\alpha)} \leq -\frac{\phi}{\beta|D_{n-1}}\right) \geq 1 - \alpha \tag{11.7}$$

Any sequence of levels that satisfy Equation 11.7 is called Bayesian feasible at level $(1 - \alpha)$ (Zacks et al. 1998). The choice of $\xi_n^{(\alpha)}$ as the log-standardized dose levels in the trial implies that the posterior probability of exceeding the MTD is equal to the feasibility bound α. Let $\mathcal{F}_n = \sigma(D_n)$ be the sigma-field generated by D_n and $\psi^{(\alpha)}$ be the class of all Bayesian-feasible sequences $z_n \in \mathcal{F}_n$ of level $(1 - \alpha)$.

Definition 11.1 A sequence of levels $\{z_n^*, n \geq 1\} \in \psi^{(\alpha)}$ is called optimal Bayesian feasible at level $(1 - \alpha)$, if for all $N \geq 1$,

$$\sum_{n=1}^{N} E_h\left\{(\gamma' - z_n^*)^+\right\} = \inf_{\{z_n\} \in \psi^{(\alpha)}} \sum_{n=1}^{N} E_h\left\{(\gamma' - z_n)^+\right\} \tag{11.8}$$

where $z^+ = zI(z > 0)$ denotes the positive part of a random variable. This means that z_n^* minimizes the average amount by which patients are underdosed. Using the law of total expectation, Zacks and colleagues (1998) showed that $\xi_n^{(\alpha)}$ is optimal Bayesian feasible. Conditions under which this sequence converges to the true MTD in probability are stated in the next theorem.

Theorem 11.1   Suppose that for $\beta_0 \in [\beta^*, \beta^{**}]$,

(i) $0 < \varepsilon_1 < G(-\beta_0 \Phi/\beta^*) \le G(-\beta_0 \Phi/\beta^{**}) \le 1 - \varepsilon$
(ii) $0 < \varepsilon_2 < \inf\{g(\beta_0 x): x \in L^*\} \le \sup\{g(\beta_0 x): x \in L^*\} \le g^*$
(iii) $g(x)$ is continuously differentiable
(iv) $-\infty < \inf\{g'(\beta_0 x): x \in L^*\} \le \sup\{g'(\beta_0 x): x \in L^*\} < \infty$.
(v) $h(\beta)$ is uniform on $[\beta^*, \beta^{**}]$.

Then,

$$\xi_n^{(\alpha)} \xrightarrow{p} -\frac{\Phi}{\beta_0} \text{ as } n \to \infty$$

Proof.   See Zacks and colleagues (1998).

11.2.2 Coherence of EWOC

Due to ethical concerns, the dose of a cytotoxic agent for the next patient in a trial should not be higher than the current allocated dose if the current patient exhibits DLT. Likewise, the dose for the next patient should not be lower than the current one if the current patient does not exhibit DLT. This desirable property is known as coherence, and Cheung (Cheung 2005), who proposed this concept, also showed that CRM is coherent. The author also showed how the coherence property can be lost when ad hoc modifications are introduced to CRM. In this section, we show that EWOC, as described in Section 11.2.1, is coherent.

Let $F(x, \gamma) = P(Y = 1|x)$ be the model given in Equation 11.3 reparameterized in terms of the MTD. Let $D_n = \{(x_1, Y_1), \ldots, (x_n, Y_n)\}$ be the data generated using the EWOC scheme described in Section 11.2.1. This design is said to be coherent in escalation if for all $n \ge 2, x_n \ge x_{n+1}$ whenever $Y_{n-1} = 0$. The design is said to be coherent in de-escalation if for all $n \ge 2, x_n \le x_{n+1}$ whenever $Y_{n-1} = 1$. The design is said to be coherent if it is coherent in both escalation and de-escalation.

Theorem 11.2   Suppose that $F(x, \gamma)$ is nonincreasing in γ for fixed dose x. Then the EWOC scheme described in Section 11.2.1 is coherent.

The proof of Theorem 11.2 is given in Tighiouart and Rogatko (2010). It is easy to verify that the monotonicity condition on $F(x, \gamma)$ is satisfied by Equation 11.3, and in particular, the logistic function.

11.2.3 Two-Parameter Logistic Model

Denote by $X_{\min}$ and $X_{\max}$ the minimum and maximum dose levels available for use in the trial. One chooses these levels in the belief that $X_{\min}$ is safe when administered to humans. Babb and colleagues (1998) considered a two-parameter logistic model for the dose–toxicity relationship:

$$P(Y = 1|\text{Dose} = x) = \frac{\exp(\beta_0 + \beta_1 x)}{1 + \exp(\beta_0 + \beta_1 x)} \tag{11.9}$$

where we assume that $\beta_1 > 0$ so that the probability of DLT is a monotonic increasing function of dose. Equation 11.9 is reparameterized in terms of the MTD γ and the probability of DLT at the starting dose ρ_0, parameters that clinicians can easily interpret (Figure 11.1). This might be advantageous since γ is the parameter of interest and one often conducts preliminary studies at or near the starting dose so that one can select a meaningful informative prior for ρ_0. Using the definition of the MTD in Equations 11.1 and 11.9, it can be shown that

$$\beta_0 = \frac{X_{\min}\text{logit}(\theta) - \gamma\,\text{logit}(\rho_0)}{X_{\min} - \gamma}$$

$$\beta_1 = \frac{\text{logit}(\rho_0) - \text{logit}(\theta)}{X_{\min} - \gamma} \tag{11.10}$$

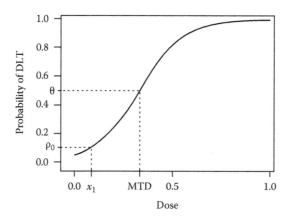

FIGURE 11.1
Example of the logistic tolerance distribution used to model dose–toxicity relationship. Dose has been standardized to the unit interval. The minimum dose is taken to be $x_1 = 0.1$, and the probability of DLT at x_1 is denoted by ρ_0. MTD has arbitrarily been defined as the standardized dose for which the probability of DLT is equal to 0.5.

The second equation in Equation 11.10 shows that the assumption that $\beta_1 > 0$ implies $0 < \rho_0 < \theta$. Thus, Equation 11.9 becomes

$$P(Y = 1|\text{Dose} = x) = p(\rho_0, \gamma, x) = \frac{\exp\left\{\ln\left[\frac{\rho_0}{1-\rho_0}\right] + \ln\left[\frac{\theta(1-\rho_0)}{\rho_0(1-\theta)}\right]\frac{x}{\gamma}\right\}}{1 + \exp\left\{\ln\left[\frac{\rho_0}{1-\rho_0}\right] + \ln\left[\frac{\theta(1-\rho_0)}{\rho_0(1-\theta)}\right]\frac{x}{\gamma}\right\}},$$

$$(11.11)$$

11.2.3.1 Trial Design

After specifying a prior distribution $h(\rho_0, \gamma)$ for (ρ_0, γ), denote by $\Pi_n(\gamma)$ the marginal posterior CDF of γ given D_n. EWOC can be described as follows. The first patient receives the dose $x_1 = X_{\min}$, and conditional on the event $\{y_1 = 0\}$, the $(n+1)$st patient receives the dose $x_{n+1} = \Pi_n^{-1}(\alpha)$ so that the posterior probability of exceeding the MTD is equal to the feasibility bound α. If $y_1 = 1$, we recommend that the clinician stop the trial. Calculation of the marginal posterior distribution of γ is performed using numerical integration (Babb et al. 1998). Often in practice, phase I clinical trials are typically based on a small number of prespecified dose levels $d_1, \dots, d_r$. In this case, the $(n+1)$st patient receives the dose

$$\widehat{d}_{n+1} = \max\left\{d_1, \dots, d_r : d_i - x_{n+1} \leq T_1 \text{ and } \Pi_n(x_{n+1}) - \alpha \leq T_2\right\} \quad (11.12)$$

where T_1, T_2 are nonnegative numbers we refer to as tolerances. We note that this design scheme does not require that we know all patient responses before we can treat a newly accrued patient. Instead, we can select the dose for the new patient on the basis of the data currently available.

At the conclusion of the trial, the MTD is estimated by minimizing the posterior expected loss with respect to some suitable loss function l. One should consider asymmetric loss functions since underestimation and overestimation have very different consequences. Indeed, the dose x_n selected by EWOC for the nth patient corresponds to the estimate of γ having minimal risk with respect to the asymmetric loss function

$$l_\alpha(x, y) = \begin{cases} \alpha(\gamma - x) & \text{if } x \leq \gamma, \text{ that is, if } x \text{ is an underdose} \\ (1 - \alpha)(x - \gamma) & \text{if } x > \gamma, \text{ that is, if } x \text{ is an overdose} \end{cases} \quad (11.13)$$

Note that the loss function l_α implies that for any $\delta > 0$, the loss incurred by treating a patient at δ units above the MTD is $(1 - \alpha)/\alpha$ times greater than the loss associated with treating the patient at δ units below the MTD. This interpretation might provide a meaningful basis for the selection of the feasibility bound.

11.2.3.2 Design Operating Characteristics

Operating characteristics of this design and its comparison with seven phase I dose escalation schemes consisting of four up-and-down (UD) designs (Storer 1989), two stochastic approximation (SA) methods (Anbar 1978, Wu 1985), and CRM (O'Quigley 1990) were studied by Babb and colleagues (1998) using extensive simulations.

These methods were compared with respect to proportion of patients assigned dose levels above and below the MTD and mean squared error observed for the estimator of the MTD. In all simulations, the MTD was taken to be the dose for which the probability of DLT is $1/3$. EWOC was implemented with vague priors for the parameters (ρ_0, γ) in the simulations. Specifically, it was assumed that $h(\rho_0, \gamma)$ is uniform on $[0, \theta] \times [X_{min}, X_{max}]$ with ρ_0 and γ independent. EWOC was set up so that the proportion of patients overdosed (treated at dose levels above the MTD) was expected to be less than $\alpha = 0.25$. Since the observed proportion of patients overdosed by EWOC was 0.193, EWOC did provide effective overdose control. When compared to the other methods, EWOC was observed to overdose fewer patients than CRM but tended to overdose more patients than the nonparametric designs (UD, SA). However, it should be noted that, by choosing smaller values of α, it is possible to reduce the proportion of patients overdosed by EWOC to levels that are comparable to those observed for any of the nonparametric designs. Furthermore, the latter schemes were seen to be overly conservative, treating a large proportion of patients at extremely low dose levels.

Methods were compared with respect to the proportion of patients treated at subtherapeutic doses, defined as doses for which the probability of DLT was less than 0.2. Results indicate that EWOC treated fewer patients at low, possibly nontherapeutic, dose levels than did any of the nonparametric schemes. However, EWOC was observed to treat more patients at low dose levels than did CRM. This was expected since EWOC attempts to protect patients from being overdosed and so will tend to treat patients at lower dose levels than CRM does. Babb and colleagues (Babb et al. 1998) also compared these methods in terms of optimal dose levels defined as doses for which the probability of DLT was less than $1/3$ and greater than 0.2, that is, doses near but not above the MTD. Overall, the proportion of patients treated at optimal dose levels was higher for EWOC than it was for any of the other dose allocation schemes. All of the nonparametric methods tended to provide significantly biased estimates of the MTD while the overall mean bias associated with both Bayesian schemes was near zero. With respect to both bias and mean squared error, EWOC provided more efficient estimates of the MTD than any of the nonparametric methods. However, CRM estimated the MTD with smaller mean squared error than did EWOC. Thus, a slight decrease in the accuracy of the MTD estimate is the price paid for incorporating protection from overdosing into EWOC. Overall, simulations showed EWOC to be effective in controlling the frequency of overdosing in a phase I trial. Relative to CRM, EWOC overdosed a smaller proportion of patients and estimated the MTD with comparable accuracy. When compared to the nonparametric schemes, EWOC assigned fewer patients to either subtherapeutic or severely toxic dose levels, treated more patients at optimal dose levels and estimated the MTD with smaller average bias and mean squared error.

Compared to UD methods, EWOC assigned fewer patients to either subtherapeutic or severely toxic dose levels, and estimated the MTD with smaller average bias and mean squared error than UD methods. Furthermore, EWOC treated 55% of the patients at optimal levels compared to 35% for UD designs on the average.

11.2.3.3 Correlated Priors on ρ_0 and γ

In models (Equations 11.3 and 11.9), we assumed that the support of the MTD was strictly contained in $[x^*, x^{**}]$ and $[X_{min}, X_{max}]$, respectively. The assumption that γ has an upper bound may be too restrictive. In the absence of toxicity, this assumption causes the dose escalation rate to slow down, and in general, the target MTD will never be achieved if it lies outside the support of γ. Furthermore, since the support of the probability of DLT at the initial dose ρ_0 is $[0, \theta]$ and γ is a function of θ, the assumption of prior independence between ρ_0 and γ may not be realistic. Intuitively, the closer is ρ_0 to θ, the closer the MTD is to X_{min}. Tighiouart and colleagues (2005) introduced a class of correlated priors for $h(\rho_0, \gamma)$ on $[0, \theta] \times [X_{min}, \infty)$ using truncated normal distributions for the parameter γ. They showed that a candidate joint prior for (ρ_0, γ) with negative a priori correlation structure results in a safer trial than the one that assumes independent priors for these two parameters while keeping the efficiency of the estimate of the MTD essentially unchanged.

11.2.3.4 EWOC with Varying Feasibility Bound

Almost all the phase I cancer clinical trials we designed at Fox Chase Cancer Center, Winship Cancer Institute, and Samuel Oschin Comprehensive Cancer Institute used a variable feasibility bound α (Babb and Rogatko 2001, 2004, Cheng et al. 2004, Tighiouart and Rogatko 2006a,b). The rationale behind this approach is that uncertainty about the MTD is high at the onset of the trial and a small value of α offers protection against the possibility of administering dose levels much greater than the MTD. As the trial progresses, uncertainty about the MTD declines and the likelihood of selecting a dose level significantly above the MTD becomes significantly smaller. However, design operating characteristics were not studied. Chu et al. (2009) showed how a version of CRM can be viewed as a special case of EWOC and conducted extensive simulations to compare the performance of EWOC with varying feasibility bound α to EWOC with constant α and four different versions of CRM. As expected, it was found that in general, EWOC with varying feasibility bound had a faster convergence rate than EWOC with fixed α and a better overdose protection than CRM.

11.2.3.5 Cohort Size

Denote by ^mP a design that treats patients in successive cohorts of size m simultaneously at the same dose level. For a given fixed number of patients in the trial, an advantage of an ^mP design with $m > 1$ over a ^1P design is a shorter time of completion of the trial. However, it is not clear how the two designs compare with respect to safety of the trial and efficiency of the estimate of the MTD using EWOC. Goodman and colleagues (1995) argue for the use of more than one patient per dose level in a modified version of the CRM to reduce the duration of the trial and toxicity incidence associated with the original CRM. Tighiouart and Rogatko (2012) compared a ^3P design with a ^1P design in terms of the number of patients given therapeutic doses, that is, doses in a neighborhood of the *true* MTD. They showed through simulations that on the average, design ^3P is practically no better than design ^1P in terms of assigning therapeutic doses to patients and safety of the trial and efficiency of the estimate of the MTD are essentially the same under the two designs.

For a given sample size n, Tighiouart and Rogatko (2012) compared the performance of a ^1P with a ^3P design by estimating the percent of patients treated within a neighborhood of the true MTD. Safety and efficiency of the estimate of the MTD under the two designs were also compared.

11.2.3.6 Comparing ^3P and ^1P Designs: Operating Characteristics

Simulation results presented in this section assume that the feasibility bound $\alpha = 0.25$ and that the dose levels are standardized so that the starting dose for each trial is $x_1 = 0$ and all subsequent dose levels are selected from the unit interval. The target probability of DLT is taken as $\theta = 0.3$. Independent vague prior distributions were put on the parameters ρ_0 and γ on the intervals $[0, \theta]$, $[0, 1]$, respectively. The dose–toxicity relationship was modeled using the logistic function, as in Section 11.2.3. Trials were simulated under different scenarios corresponding to different values of ρ_0 and γ. We considered 12 scenarios corresponding to combinations of three values of ρ_0, $\{\theta/4, \theta/2, 3\theta/4\}$ with four values of γ, 0.2, 0.4, 0.6, and 0.8.

For each design, each sample size $n = 12$, 30, and each combination of (ρ_0, γ), 5000 trials were simulated, and the proportions of patients given therapeutic doses, that is, doses in an ε-neighborhood of the true MTD, for $\varepsilon = 0.05, 0.1, 0.15, 0.2$ were calculated.

TABLE 11.1

Estimated Proportions of Patients Given Doses in an ε-Neighborhood of the True MTD under Designs ^1P and ^3P and Differences between These Proportions on the Average (Diff.)

ε		Sample Size n	
		12	30
	^1P	0.1436	0.1920
0.05	^3P	0.0924	0.1644
	Diff.	0.0512	0.0276
	^1P	0.2956	0.3587
0.1	^3P	0.2860	0.3497
	Diff.	0.0096	0.0090
	^1P	0.4115	0.4839
0.15	^3P	0.3888	0.4714
	Diff.	0.0227	0.0125
	^1P	0.4988	0.5801
0.2	^3P	0.4517	0.5564
	Diff.	0.0471	0.0237

For each scenario, the estimated proportions of patients given doses in an ε-neighborhood of the true MTD under designs ^1P and ^3P and the difference in these proportions between the two designs were calculated. Table 11.1 provides the average of these estimates across the 12 scenarios. The estimated difference in the proportions of patients given doses in an ε-neighborhood of the true MTD between the ^1P design and the ^3P design averaged across the 12 entertained scenarios for (ρ_0, γ) for different sample sizes shows that the proportion of patients given therapeutic doses under design ^1P is always greater than the corresponding proportion under design ^3P, the largest of these differences being about 5%. The practical impact of this difference is unimportant because of the relatively small number of patients involved in phase I cancer clinical trials. Table 11.2 gives the average values of the statistics: (1) proportions of patients exhibiting DLT, (2) proportions of patients given doses above the *true* MTD, (3) bias, and (4) mean squared error between the ^1P and ^3P designs, averaged across the 12 entertained scenarios for (ρ_0, γ). The results from (1) and (2) indicate that the two designs are equally safe, and those from (3) and (4) indicate that no practical gain is achieved in terms of the efficiency of the estimate of the MTD. From an ethical point of view, we recommend the ^1P design prevent the occurrence of three simultaneous DLTs if we were to use the ^3P design. This should be discussed with the clinician after assessing the importance of the length of the trial.

TABLE 11.2

Estimated Proportions of Patients Exhibiting DLTs, Treated above the MTD, MSE, and Bias of the MTD under Designs ^1P and ^3P and Differences between These Proportions on the Average (Diff.)

		Sample Size n	
		12	30
Proportion of DLTs	^1P	0.2546	0.2616
	^3P	0.2444	0.2595
	Diff.	0.0102	0.0021
Proportion above the MTD	^1P	0.1895	0.2067
	^3P	0.1685	0.2029
	Diff.	0.0210	0.0038
MSE	^1P	0.0427	0.0351
	^3P	0.0429	0.0344
	Diff.	−0.0002	0.0007
Bias	^1P	0.0186	0.0193
	^3P	0.0271	0.0244
	Diff.	−0.0085	−0.0051

11.2.3.7 Sample Size Determination

An important design consideration for a cancer phase I clinical trial is the choice of number of patients to enroll. Most sample size recommendations in the literature are based on prespecified stopping rules; see, for example, Zohar and Chevret (2001) on selecting the number of patients by considering different stopping rules using the CRM. Lin and Shih (2001) and Ivanova (2006) describe sample size recommendations based on the expected number of patients allocated to each dose selected from a set of prespecified dose levels. However, these methods apply to a prespecified set of discrete doses, and it is not clear how they can be applied to continuous doses. Unlike the frequentist approach, there is no consensus on a specific Bayesian method for the sample size determination problem; see Adcock (1997) for a review of Bayesian approaches. Tighiouart and Rogatko (2012) conducted extensive simulation studies in order to estimate the sample size based on a desired accuracy of the Bayes estimate on the average. Specifically, they determined the minimum number of patients so that the posterior variance of the MTD on the average over all possible trials is no more than a specified margin. Table 11.3 shows that with six patients, we can estimate the MTD with an average posterior standard deviation equal to 25% of the range of the dose and that a 17% decrease in the average posterior standard deviation is achieved when increasing the sample size from 6 to 40 patients. Similarly, the average length of the 90% highest posterior density (HPD)

TABLE 11.3

Average Posterior Standard Deviation and Average
Length of HPD of the Posterior Distribution of the MTD
That Are Achieved for a Given Sample Size for $\theta = 0.3$

N	Mean SD	Length of 90% HPD	Length of 95% HPD
10	0.2351	0.7111	0.7925
20	0.2197	0.6673	0.7546
30	0.2102	0.6410	0.7313
40	0.2036	0.6200	0.7123

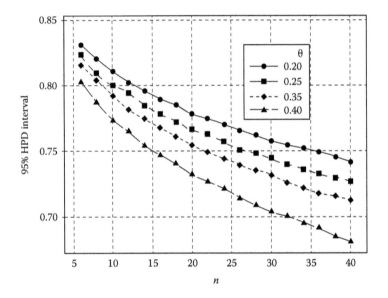

FIGURE 11.2
Estimated mean length of HPD of the posterior distribution of the MTD as a function of the number of patients accrued to the trial for different target probabilities of DLT θ.

interval is 74% of the dose range when six patients are enrolled in the phase I trial, and a reduction of 16% of this length is achieved when increasing the number of patients from 6 to 40. Figure 11.2 shows the 95% HPD intervals as functions of the sample size n and target probability of DLT θ.

11.2.4 R115777 Trial

EWOC was used to design a phase I clinical trial that involved the R115777 drug at Fox Chase Cancer Center in Philadelphia, United States, in 1999. R115777 is a selective nonpeptidomimetic inhibitor of farnesyltransferase

(FTase), one of several enzymes responsible for posttranslational modification that is required for the function of p21(ras) and other proteins. This was a repeated dose, single center trial designed to determine the MTD of R115777 in patients with advanced incurable cancer.

The dose-escalation scheme was designed to determine the MTD of R115777 when drug is administered orally for 12 h during 21 days followed by a 7-day rest. This constitutes one cycle of therapy. Toxicity was assessed by the NCI Common Toxicity Criteria. DLT was determined by week 3 of cycle 1, as defined by Grade III nonhematological toxicity (with the exception of alopecia or nausea/vomiting) or hematological Grade IV toxicity with a possible, probable, or likely causal relationship to the administration of R115777. Dosing continued until there was evidence of tumor progression or DLT leading to permanent discontinuation. The value chosen for the targeted probability of DLT was $\theta = 1/3$. The initial dose judged to be safe by the clinician for this study was $X_{min} = 60 \, mg/m^2$, and the maximum allowable dose was $X_{max} = 600 \, mg/m^2$ (Tighiouart and Rogatko 2006a,b). Assuming vague priors for ρ_0 on $[0, \theta]$ and γ on $[60, 600]$, the prior probability density of (ρ_0, γ) is

$$h(\rho_0, \gamma) = \begin{cases} \dfrac{1}{180} & \text{if } (\rho_0, \gamma) \in [0, 1/3] \times [60, 600] \\ 0 & \text{otherwise} \end{cases} \tag{11.14}$$

Thus, ρ_0 and γ are independent a priori, uniformly distributed over their corresponding interval. Figure 11.3 shows the posterior distributions of the MTD as the trial progressed. Figure 11.4 displays, in detail, the posterior density of the MTD after 33 patients have been treated. The posterior mode is 323, which correspond to the 47th percentile of the distribution. In this trail, we used a variable feasibility bound α, starting with $\alpha = 0.3$, this value being a compromise between the therapeutic aspect of the agent and its toxic side effects. As the trial progressed, α increased in small increments until $\alpha = 0.5$ so that, by the end of the trial, the given dose corresponds to the 50th percentile, that is, the median of the marginal posterior probability density function. Thus, the dose to be given to the 34th patient is 328. The 95% highest posterior density interval is [160.5, 536.1].

11.3 Use of Covariate in Prospective Clinical Trial

A key assumption implied by the definition of the phase I target dose (MTD) is that every subgroup of the patient population has the same MTD. That is, it is assumed that the patient population is homogeneous in terms of treatment tolerance and every patient should be treated at the same dose. As a result,

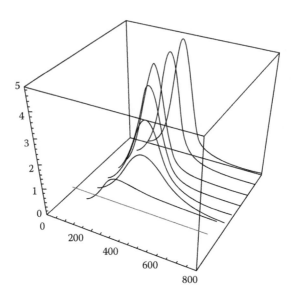

FIGURE 11.3
Posterior density of the MTD when the number of treated patients (from bottom to top) is 1, 5, 10, 15, 20, 25, 30, 33.

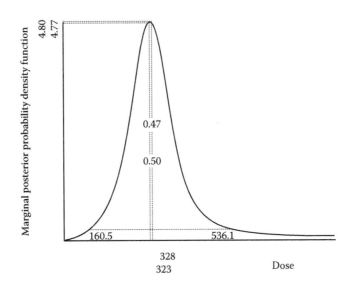

FIGURE 11.4
Posterior density of the MTD after 33 patients have been treated. The posterior mode is 323 (47th percentile), and the median and dose to be given to the 34th patient is 328. The 95% HPD interval is [160.5, 536.1].

no allowance is made for individual patient differences in susceptibility to treatment (Dillman and Koziol 1992).

Babb and Rogatko (2001) extended EWOC to allow the incorporation of information concerning individual patient differences in susceptibility to treatment. The method adjusts doses according to patient-specific characteristics while safeguarding against overdosing.

11.3.1 Model

In this section, we describe a Bayesian logistic model that accounts for patient heterogeneity thought to be related to treatment susceptibility. Denote by Z the observable binary covariates taking values 0 or 1 and let

$$P(Y = 1 | \text{Dose} = x, Z = z) = \frac{\exp\{\beta_0 + \beta_1 x + \beta_2 z\}}{1 + \exp\{\beta_0 + \beta_1 x + \beta_2 z\}} \qquad (11.15)$$

As in Section 11.2, we will assume that $\beta_1 > 0$ so that the probability of DLT is an increasing function of dose x for fixed z. The MTD for patients with covariate value z is defined as the dose $\gamma(z)$ that results in a probability equal to $\theta(z)$ that a DLT will manifest. We will assume that $\theta(z)$ is constant in z although the methodology can be adapted to different target probabilities of toxicities. It follows from Equation 11.15 that

$$\gamma(z) = \frac{1}{\beta_1} [\text{logit}(\theta) - \beta_0 - \beta_1 z] \qquad (11.16)$$

We reparametrize model (Equation 11.15) in terms of $\gamma_0 = \gamma(0)$, $\gamma_1 = \gamma(1)$, and $\rho_{0,0}$, the probability of DLT at the initial dose for patients with covariate value $Z = 0$. We chose this reparameterization because the MTDs for each group are the parameters of interest. Other parameterizations such as difference between the MTDs in both groups are possible. An advantage of this reparameterization is the natural specification of vague but proper prior densities for the model parameters. Indeed, under the assumption that γ_0, γ_1 belong to $[X_{min}, X_{max}]$ with prior probability 1 and no prior assumptions on whether one group can tolerate higher doses better than the other, we can take $(\gamma_0, \gamma_1) \sim$ Uniform $[X_{min}, X_{max}]^2$ and γ_0 independent of γ_1. If on the other hand, we have a priori belief that one group can tolerate higher doses better than the other group, for example, then (γ_0, γ_1) can be taken to be uniform on the triangle $X_{min} < \gamma_0 < \gamma_1 < X_{max}$. The prior distribution for $\rho_{0,0}$ is taken as a uniform in $[0, \theta]$, which reflects a lack of prior knowledge regarding the probability of DLT at the initial dose.

11.3.1.1 Trial Design

Denote by A and B the two groups of patients corresponding to covariate values 0 and 1, respectively. We note that if the prior distribution $\pi(\gamma_1)$ is independent of the joint prior distribution of $(\rho_{0,0}, \gamma_0)$, then $\pi(\gamma_1)$ is never updated unless a patient in group B is enrolled in the trial. In the case of such priors, the trial proceeds as follows:

The first patient in either group receives the dose $x_1 = X_{\min}$. Let $\Pi_{z,1}$ be the marginal posterior CDF of the MTD γ_z, $z = 0, 1$. Suppose that the first patient belongs to group A. If the second patient belongs to group A, then he or she will receive the dose $x_{0,2} = \Pi_{0,1}^{-1}(\alpha)$ so that the posterior probability of exceeding the MTD γ_0 is equal to the feasibility bound α. If the second patient belongs to group B, then he or she will receive the dose $x_1 = X_{\min}$. In general, the first time a patient is assigned to a given group always receives $x_1 = X_{\min}$ no matter how many patients have been enrolled in the other group. Once l patients have been enrolled in the trial with at least one patient treated in each group, the $(l+1)$st patient with covariate value z receives the dose $x_{z,l+1} = \Pi_{z,l}^{-1}(\alpha)$. The trial proceeds until a total of n patients have been accrued.

11.3.1.2 Operating Characteristics

Tighiouart and colleagues (2007, 2012a) carried out extensive simulations to compare the performance of the following designs: (1) Design using a covariate; patients are accrued to the trial sequentially and the dose given to a patient is calculated assuming model (Section 11.3.1), (2) design ignoring the covariate; patients are accrued to the trial sequentially and the dose given to a patient is calculated assuming a logistic model (Equation 11.3) without the covariate, that is, as in the original EWOC, and (3) design using separate trials; in each group, patients are accrued to the trial sequentially and EWOC is implemented in each group. O'Quigley and colleagues (1999) investigated the performance of a two-stage CRM using a binary covariate. They considered three different models for the dose–toxicity relationship, and maximum likelihood method was used to estimate the model parameters. This required starting the escalation scheme using some ad hoc mechanism until the first toxicity is observed. They found that significant gains can be made using the two-sample CRM when there are group imbalances. However, there may not be enough patients in one group to detect that effect.

We considered several scenarios for the *true* values of γ_1 and γ_2 and $\rho_{0,0}$, sample sizes n_1 and n_2 for each group. We have found that if the two MTDs are different and the design does not adjust for this heterogeneity, then the trial will result in more patients being overdosed; the percent of patients being overdosed can be two times higher than with a design that uses covariate. If the two MTDs are different and parallel trials are used, then the estimates of the MTDs are less efficient, and more patients may be over-dosed for low values of the MTDs. If the two MTDs are the same and the design adjusts for patients' heterogeneity, then it can happen that few more patients will be overdosed if the true MTD is low relative to a design with no covariate but the difference is not practically important. Thus, we stand to lose little if we do include a statistically nonsignificant covariate in the model. Incidentally, this conclusion is in agreement with the findings in O'Quigley and colleagues (1999). More details on the simulation set-up and results can be found in Tighiouart and colleagues (2007).

11.3.2 PNU Trial

In this example, we describe the use of EWOC with a continuous base-line covariate in a phase I study of PNU-214565 (PNU) involving patients with advanced adenocarcinomas of gastrointestinal origin. Preclinical studies demonstrated that the action of PNU is moderated by the neutralizing capac-ity of anti-SEA anÂntibodies. Based on this, the MTD for patients with pre-treatment anti-SEA concentration c was defined as the dose $\gamma(c)$ that results in a probability equal to $\theta = 0.1$ that a DLT will be manifest within 28 days. The small value chosen for θ reflects the severity of treatment-attributable toxicities (e.g., myelosuppression) observed in previous studies.

Let $\rho_x(c)$ be the probability of DLT for a patient with baseline covari-ate value c treated with dose level x. We assume that $\beta_1 > 0$ and $\beta_2 < 0$ in a logistic model (Equation 11.15) so that the probability of DLT is (1) an increasing function of dose for fixed anti-SEA and (2) a decreasing func-tion of anti-SEA for fixed dose since anti-SEA has a neutralizing effect on PNU. The model is reparameterized in terms of $\gamma_{\max} = \gamma(c_2)$, $\rho_1 = \rho_{0.5}(c_1)$, and $\rho_2 = \rho_{0.5}(c_2)$ for values $c_1 = 0.01$ and $c_2 = 1800$ selected to span the range of anti-SEA concentrations expected in the trial and $x = 0.5$ ng/kg is the minimum dose of PNU allowed in the trial. Since the probability of DLT at a given dose is a decreasing function of anti-SEA, we have $\rho_2 < \rho_1$. Fur-thermore, since the MTD was assumed to be greater than 0.5 ng/kg for all values of anti-SEA, we have $\rho_1 < \theta$. The prior distribution of $(\gamma_{\max}, \rho_1, \rho_2)$ was then specified by assuming that $\gamma_{\max}$ and (ρ_1, ρ_2) are independent a priori, with (ρ_1, ρ_2) uniformly distributed on $\Omega = \{(x, y): 0 \leq y \leq x \leq \theta\}$ and $\ln(\gamma_{\max})$ uniformly distributed on the interval $[\ln(3.5), \ln(1000)]$ (see Babb and Rogatko (2001) for the rationale behind the choice of the support of the MTD $\gamma_{\max}$).

The PNU trial was designed according to the scheme described in Section 11.3.1.1 except that Z is replaced by the continuous covariate representing the baseline anti-SEA. A total of 56 patients were treated in the phase I trial of which 3 (5.4%) experienced DLT. Patients were observed to tolerate doses of PNU as high as 44% of their anti-SEA concentration without significant toxicity. None of the 96 patients treated at a dose less than 7% of their anti-SEA concentration exhibited DLT. Of the 63 patients treated with a dose greater

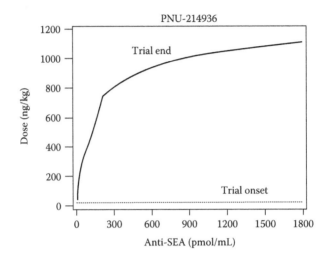

FIGURE 11.5
Recommended dose of PNU as a function of anti-SEA concentration at both the onset and the conclusion of the phase I trial.

than their anti-SEA/30 (the lowest permissible dose during the phase I trial), 7 patients (11.1%) manifest DLT, a rate of toxicity not far above the targeted proportion $\theta = 0.1$.

Figure 11.5 shows the recommended dose level as a function of anti-SEA at both the start and the conclusion of the trial. The latter (uppermost) curve corresponds to the dose levels recommended for phase II evaluation. At trial onset, the recommended dose curve was nearly horizontal beyond an anti-SEA concentration of 100 pmol/mL. In other words, nearly the same dose was recommended for all patients with sufficiently high anti-SEA concentration. Essentially, this was a reflection of the fact that data from only 36 patients with anti-SEA greater than 100 were available at the start of the trial. Since all of these patients received a dose less than 2.6% of their anti-SEA concentration (no dose exceeded 4 ng/kg) and none experienced DLT, little was initially known about the effect of high anti-SEA concentrations on treatment response.

11.4 Use of Ordinal Toxicity Grades

The majority of the statistical designs that were proposed in the last two decades allocate future doses based on a binary outcome of DLT of previously treated patients. Such designs may not be efficient in the sense that the dose recommended for the next patient is the same regardless whether the previously treated patient had no toxicity or had intermediate grade 2 toxicity. Tighiouart and colleagues (2012b) extended EWOC by introducing an intermediate grade 2 toxicity when assessing DLT. Under the proportional odds model assumption of dose–toxicity relationship, they proved that in the absence of DLT, the dose allocated to the next patient given that the previously treated patient had a maximum of grade 2 toxicity is lower than the dose given to the next patient had the previously treated patient exhibited grade 0 or 1 toxicity at the most. They also proved that the coherence properties of EWOC are preserved. Simulation results indicated that the safety of the trial was not compromised and the efficiency of the estimate of the MTD was maintained relative to EWOC treating DLT as a binary outcome and that fewer patients are overdosed using this design when the true MTD is close to the minimum dose.

11.4.1 Model

Let $G = 0, 1, \ldots, 4$ be the maximum grade of toxicity experienced by a patient by the end of one cycle of therapy and define DLT as a maximum of grade 3 or 4 toxicity. Let

$$Y = \begin{cases} 0 & \text{if } G = 0 \text{ or } 1 \\ 1 & \text{if } G = 2 \\ 2 & \text{if } G = 3 \text{ or } 4 \end{cases} \tag{11.17}$$

We model the dose–toxicity relationship by assuming that

$$P(Y \geq j|x) = F(\alpha_j + \beta x), \quad j = 1, 2 \tag{11.18}$$

where $F(\cdot)$ is a known strictly increasing CDF. This implies that $\alpha_2 \leq \alpha_1$. We assume that $\beta > 0$ so that the probability of DLT is an increasing function of dose. The MTD, γ, is defined as the dose that is expected to produce DLT in a specified proportion θ of patients:

$$P(Y = 2|x = \gamma) = F(\alpha_2 + \beta\gamma) = \theta \tag{11.19}$$

Suppose that dose levels in the trial are selected in the interval $[X_{\min}, X_{\max}]$. Let $D_n = \{(x_i, Y_i), i = 1, \ldots, n\}$ be the data after enrolling n patients to the trial. The likelihood function for the parameters α_1, α_2, and β is

$$L(\alpha_1, \alpha_2, \beta|D_n) = \prod_{i=1}^{n} [1 - F(\alpha_1 + \beta x_i)]^{I(Y_i=0)} [F(\alpha_1 + \beta x_i)$$
$$- F(\alpha_2 + \beta x_i)]^{I(Y_i=1)} [F(\alpha_2 + \beta x_i)]^{I(Y_i=2)} \tag{11.20}$$

where $I(\cdot)$ is the indicator function.

We reparameterize model (11.18) in terms of $\rho_0 = P(Y = 2 | x = X_{min})$, the probability that a DLT manifests within the first cycle of therapy for a patient given dose $x = X_{min}$, $\rho_1 = P(Y \geq 1 | x = X_{min})$, the probability that a grade 2 or more toxicity manifests within the first cycle of therapy for a patient given dose $x = X_{min}$, and the MTD γ. This reparameterization is convenient to clinicians since γ is the parameter of interest. Assuming that the dose is standardized to be in the interval [0, 1], it can be shown that

$$\alpha_1 = F^{-1}(\rho_1), \ \alpha_2 = F^{-1}(\rho_0)$$

$$\beta = \frac{1}{\gamma} \left(F^{-1}(\theta) - F^{-1}(\rho_0) \right)$$

(11.21)

The conditions $\alpha_2 \leq \alpha_1$, $\beta > 0$, and (11.18) imply that $0 \leq \rho_0 \leq \rho_1$ and $0 \leq \rho_0 \leq \theta$. Define

$$F_1(\rho_0, \rho_1, \gamma; x) = F \left(F^{-1}(\rho_1) + \frac{\left(F^{-1}(\theta) - F^{-1}(\rho_0) \right) x}{\gamma} \right)$$

$$F_2(\rho_0, \rho_1, \gamma; x) = F \left(F^{-1}(\rho_0) + \frac{\left(F^{-1}(\theta) - F^{-1}(\rho_0) \right) x}{\gamma} \right)$$

(11.22)

Using (11.20 through 11.22), the likelihood of the reparameterized model is

$$L(\rho_0, \rho_1, \gamma | D_n) = \prod_{i=1}^{n} [1 - F_1 (\rho_0, \rho_1, \gamma; x_i)]^{I(Y_i=0)}$$

$$\times [F_1 (\rho_0, \rho_1, \gamma; x_i) - F_2 (\rho_0, \rho_1, \gamma; x_i)]^{I(Y_i=1)}$$

$$\times [F_2 (\rho_0, \rho_1, \gamma; x_i)]^{I(Y_i=2)}$$

(11.23)

Let $g(\rho_0, \rho_1, \gamma)$ be the prior distribution on Ω, where $\Omega = \{(x, y, z): 0 \leq x \leq \theta, x \leq y \leq 1, X_{min} \leq z \leq X_{max}\}$. Using Bayes rule, the posterior distribution of the model parameters is proportional to the product of the likelihood and prior distribution

$$\pi(\rho_0, \rho_1, \gamma | D_n) \propto L(\rho_0, \rho_1, \gamma | D_n) \times g(\rho_0, \rho_1, \gamma)$$

(11.24)

WinBUGS (Lunn et al. 2000) can be used to estimate features of the posterior distribution of the MTD and design a trial. In the absence of prior information about the MTD and probability of DLT at X_{min}, we specify vague priors for the model parameters as follows:

$$\gamma \sim \text{Unif}\,[X_{min}, X_{max}]$$
$$\rho_0 \sim \text{Unif}[0, \theta]$$
$$\rho_1|\rho_0 \sim \text{Unif}\,[\rho_0, 1] \qquad\qquad (11.25)$$

The trial design follows the same procedures described in Section 11.2.3.1.

11.5 Dose Finding beyond Phase I

The impact of anti-SEA levels on susceptibility to PNU treatment and the possibility to adjust doses according to patient-specific characteristics begged the question: Are there any other treatments where covariates should be used in dose adjustment?

11.5.1 Patient-Specific Characteristics Are Important Predictors of Toxicity

Various studies have examined prognostic factors for survival and antitumor activity as an endpoint in early-phase cancer clinical trials such as that of Yamamoto et al. (1999) and Bachelot et al. (2000). However, we found no systematic study of prognostic factors for the endpoint of toxicity. The lack of studies on toxicity predictors led us to propose a retrospective study to identify readily available patient characteristics that may constitute risk factors or markers of susceptibility to adverse treatment effects in cancer phase I and II clinical trials. We studied 459 patients enrolled in 23 investigator-initiated therapeutic phase I and II studies at Fox Chase Cancer Center during the time period 1991–1999. Our purpose was to identify patient characteristics that may be risk factors or markers of susceptibility to adverse treatment effects in cancer phase I and II clinical trials.

We found that patient characteristics compete with dose as predictors of the toxic response to several chemotherapeutic agents. Furthermore, pretreatment characteristics, even within the normal range, may be predictors of treatment toxicity. Seventeen pretreatment factors, including performance status, alkaline phosphatase, total bilirubin, serum creatinine, and tobacco use, emerged as significant predictors of toxicity. Unexpectedly, dose was not always a predictor of toxicity. Even for values within the normal range, serum bilirubin and alkaline phosphatase were identified as predictors of toxicity following treatment with docetaxel and alkaline phosphatase as a predictor for toxicity following irinotecan (Rogatko et al. 2004). Since the factors considered in this study can be observed prior to treatment, they may form the foundation for customized dosing regimens wherein drug doses are adjusted according to individual patient susceptibilities.

Phase I studies assume that dose is the most significant determinant of toxicity. Our analysis of multiple phase I and early phase II trials revealed that dose is not always a significant predictor of toxicity (Rogatko et al. 2004). Even with conventional patient selection criteria that included the requirement for normal or near-normal hepatic and renal function, patient characteristics had greater predictive value than dose for the toxicity of several agents. Thus, the current eligibility criteria for most phase I and II trials do not provide populations that are uniform enough to conclude that differences in toxic response are primarily dose-related.

11.5.2 Is the Phase I MTD the Appropriate Dose for Phases II and III?

While the major endpoint in phase II trials is efficacy, the search for the optimal dose is usually restricted to the phase I setting. In this case, the phase II dose determination may be based on the data accumulated from a patient population with very different characteristics than the target phase II population. The implicit assumption that the dose selected in one patient population will be optimal for other populations may be incorrect. In a phase II trial, the frequent occurrence of toxicity leads generally to dose reduction. In contrast, dose increase due to lack of toxicity is much less common, since such treatments are generally considered *well tolerated*. Since maximizing dose-intensity is still regarded as an important condition to achieve an optimal therapeutic effect, failure to increase the dose in the absence of toxicity may result in patients being treated at subtherapeutic dose levels. As an illustration of this, EWOC was applied retrospectively to data from two trials completed at the Fox Chase Cancer Center (Hudes, personal communication). In the phase I study, 26 patients with a wide variety of malignancies were treated with paclitaxel and estramustine. The phase II dose was chosen to be $120\,mg/m^2$. In the phase II trial, 34 patients with prostate cancer received the recommended phase II dose, following the same regimen. Figure 11.6 shows, for each trial, the EWOC recommended dose for selected values of the targeted probability of DLT. One sees that different doses are required in the two patient populations to obtain the same targeted probability of DLT. This example shows how EWOC can be used to aid the clinician in selecting a working dose for a cytotoxic treatment. Once the clinician decides upon a target probability of DLT, θ, the phase II (or III) dose can be determined.

11.5.3 Dose Finding through All Phases

According to the current paradigm for the clinical evaluation of new cancer therapies, (A) the dose of a therapeutic agent is not adjusted to accommodate individual patient differences, and (B) the identification of a working

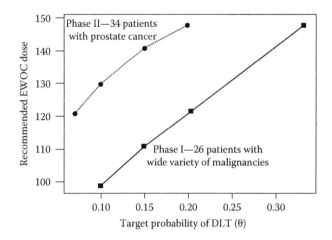

FIGURE 11.6
EWOC recommended dose for selected values of the targeted probability of DLT in two distinct trials. Paclitaxel and estramustine were used to treat 26 patients in the phase I study with a wide variety of malignancies. The phase II dose was chosen to be 120 mg/m^2 of estramustine. Thirty-four patients with prostate cancer received the recommended phase II dose in the phase II trial. One sees that different doses are required in the two patient populations to obtain the same targeted probability of DLT.

dose of new cancer therapies is mainly restricted to phase I trials. Rogatko and colleagues (2005) proposed that (A') the dose should be fine-tuned using patient-specific attributes, and (B') the search for the optimal dose should be extended beyond phase I and into phases II and III. This paper also provided examples of how phase I design methods can be used to update the working dose for phases II and III and how fine-tuning the dose may involve the utilization of patient-specific attributes to obtain a personalized treatment regimen. The standard paradigm of clinical evaluation of new cancer therapies restricts the dose determination to the initial phase of the process (Figure 11.7a). At the same time, it would be unreasonable to design phase I trials with sufficient power to distinguish the important patient-specific characteristics for a given therapy.

One solution is to continue the quest for determining the best dose throughout phases II and III (Figure 11.7b). Hence, clinical trials might progress as follows. First, a phase I trial is conducted to characterize the toxicity profile of the treatment and determine a starting dose for phase II investigation. Subsequently, the phase II and III trials can be designed in stages with the data from each stage used to determine if and what adjustment of the dose is needed. Dose modification would continue until either a specific number of patients have been treated or the dose has converged to the MTD according to some criterion (such as the posterior variance of the estimated MTD).

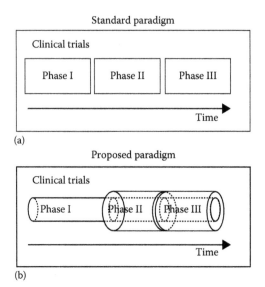

FIGURE 11.7

Standard paradigm of clinical evaluation of new cancer therapies restricts the determination of dose to the initial phase of the process (a). The proposed paradigm entails extending the search for the optimal dose beyond phase I and into phases II and III (b).

11.6 Designing a Trial with EWOC

A large proportion of cancer dose-finding trials employ the UD design (Section 11.2.3.2), which requires little or no participation of a biostatistician in their design. Conversely, designing a trial with EWOC will involve a higher level of interaction among the team of biostatisticians and medical researchers.

11.6.1 Probability of Dose-Limiting Toxicity—θ

The first parameter value to be decided upon by the team is the probability of DLT (θ). That is, the proportion of patients expected to experience a medically unacceptable DLT if administered the target dose (or MTD). Its value, generally between 0.1 and 0.5, depends on the nature of the DLT. It would be set relatively high when the DLT is a transient, reversible, correctable, or nonfatal condition and low when it is lethal or life threatening. In Section 11.2.4, the targeted probability of DLT for R115777 was selected as $\theta = 1/3$ (grade 3–4 neutropenia and thrombocytopenia were principal DLTs), whereas in PNU trial (Section 11.3.2) it was $\theta = 0.1$ since the DLT event was life threatening. The possibility of explicitly choosing θ as a design consideration in a dose-finding trial allows the incorporation of the severity of the DLT into the design

TABLE 11.4

Number of Patients Accrued to the Trial
(N) to Achieve an Estimated Mean Length
of 0.74 for the 95% HPD Interval of the
Posterior Distribution of the MTD for
Selected Target Probabilities of DLT(θ)

θ	N
0.2	40
0.25	30
0.3	24
0.4	18

of the trial. This capability enriches the design process and empowers the trial
designer to develop customized and ethical studies.

Although the choice of θ is to be made solely on ethical grounds, it will
have important consequences regarding the achievable precision of the MTD
estimate for a given sample size. Table 11.4 shows that to achieve an estimated
mean length of 0.74 for the 95% HPD interval of the marginal posterior distribution of the MTD, 40 patients are required for $\theta = 0.2$, whereas only 18 are
needed for $\theta = 0.4$ (Tighiouart and Rogatko 2012).

11.6.2 Probability of Exceeding the Target Dose—α

The second parameter to be chosen is the probability of exceeding the target
dose (α), that is, the probability that the dose selected by EWOC is higher than
the target dose. This parameter allows EWOC to exert overdose control and is
its distinctive feature. Low values make the escalation cautious; high values
cause larger steps. In the beginning of a trial, there is a higher level of uncertainty about the target dose. Consequently, at the onset of a dose-finding trial,
the probability of exceeding the target dose is typically set to a low value (e.g.,
0.2) in order to minimize the possibility of harming patients by administering doses much greater than the target dose. As the trial progresses, both the
uncertainty about the target dose and the likelihood of administering a dose
considerably higher than the target dose decrease. Thus, one should consider
gradually increasing alpha during the course of the trial. When the probability of exceeding the target dose is set to 0.5, it implies that underdosing a
patient (treating with a dose lower than the target dose) is as likely as overdosing. Note that the recommended dose when $\alpha = 0.5$ is the median of the
marginal posterior distribution of γ.

11.6.3 Cohort and Sample Sizes

The next two parameters necessary to design an EWOC trial are cohort and
sample sizes. Both topics have been discussed in detail in Sections 11.2.3.6

and 11.2.3.7. In these sections, we provided a rational for the choice of cohort sizes and the number of patients to accrue in a dose-finding cancer clinical trial when the Bayesian adaptive design EWOC is used. We have shown through simulations that the two designs are equally safe and that no practical gain is achieved in terms of the efficiency of the estimate of the MTD. Depending on how important the length of the trial is to the clinician and the institution, we recommend using one patient per dose level to avoid seeing simultaneous toxic events when a group of patients are treated at the same dose level as was the case in a recent dose-finding trial of the drug TGN1412 (Gitlin 2006). In that trial, six volunteers were given what was believed to be a safe dose of an anti-inflammatory drug TGN1412. Shortly after, all six were admitted into intensive care due to severe reactions, including swelling of the head and neck.

Choosing one patient per cohort gives more flexibility to accrue patients into the trial. Because it may take many weeks to resolve toxicity, a patient may be accrued to the trial before the responses of all previously treated patients have been determined. It will be at the investigator's discretion whether to treat the newly accrued patient at the dose level determined on the basis of the currently available data or to wait until one or more toxicities are resolved. In this case, one can specify in the protocol that no more than some predetermined number of patients will be treated at the same dose level.

EWOC trials allow the general philosophy of "always give to the patient the best dose at the present time" to be followed. If several patients are eligible to start treatment at a given time, the best known dose can be given to them. As new toxicity information becomes available, it is incorporated to the model and an updated best dose is calculated.

We addressed the sample size determination problem by giving tabulated values of the number of patients to accrue in a cancer dose-finding clinical trial as a function of the posterior standard deviation and length of the HPD interval of the MTD on the average over all possible trials. Although this aspect of the trial never received much emphasis in the literature due to the relatively small number of patients and logistical issues associated with such trials, we felt that providing a measure of the accuracy of the estimate of the MTD that can be achieved for a given sample size would help the clinicians understand what can and cannot be achieved during this phase of the trial. Our results show that in general, there is 17% decrease in the average posterior standard deviation of the MTD when the sample size increases from 6 to 40 patients and that for a sample size of 20 patients, the average posterior standard deviation of the MTD is about one-fifth the range of the dose levels.

11.6.4 Minimum and Maximum Doses and Minimum Dose Increment

There are three dose-related quantities that need to be specified: minimum and maximum doses and minimum dose increment. EWOC will never assign doses below the minimum dose nor above the maximum dose. It is highly desirable that the first dose administered to a patient in the trial, the initial dose, is a safe dose. That is, one expects that the probability of DLT of a patient receiving the initial dose to be zero. However, it may be a good strategy to let the initial dose be higher than the minimum dose. In case that the first patient exhibits DLT, the second patient can receive a lower dose instead of terminating the trial. With respect to the choice of the maximum dose, Eichhorn and Zacks (1981) have shown that the only necessary assumption is that the maximum dose must be greater than the MTD. In their formulation, this assumption was satisfied by setting the maximum dose equal to a dose level sufficiently high that almost surely all patients administered this dose will experience DLT. In the absence of prior information regarding the maximum dose, we recommend the same approach when using EWOC. The elicitation of the maximum dose is not difficult in trials involving established agents in novel combinations. This is because the agents are believed to be more toxic in combination. Therefore, each drug's known MTD as a single agent constitutes a suitable maximum for the drug in the combination. However, in the case of a new agent, the lack of toxicity information from humans makes it more difficult to provide a reasonable guess for the maximum dose. When this is the case, specification of the maximum dose can be delayed until information about the toxicity is acquired. For example, a clinician could initially escalate doses by 50% or 100% between consecutive patients until some toxicity is seen. At this point, based on the nature of the observed toxicity and the dose at which it occurred, a maximum dose can be specified. EWOC can then be used to provide safeguards against subsequent overdosing. An alternative approach is to use truncated normal distributions for the parameter γ (Tighiouart et al. 2005). Regarding the minimum dose increment, it depends on how the chemotherapeutic agent is delivered to the patient. If the agent is administered by infusion, the dose support can be considered as continuous in the interval $[X_{min}, X_{max}]$. Same applies to radiation therapy. If the agent is delivered orally, such as pills or capsules, doses may be constrained by pill sizes.

11.6.5 Software, Tutorial, and Protocol Template

A website (http://biostatistics.csmc.edu/ewoc/index.php) was established to facilitate access to software applications developed by our group to design and conduct dose-finding clinical trials in cancer using the EWOC method: Standalone-EWOC application available for download and Web-EWOC, a web-based calculator that includes documentation, storage, and forum. In addition, information about both the software and the method itself is available at that website to facilitate the implementation of EWOC in cancer trials.

Also available at the same location is a tutorial that provides step-by-step instructions on how to design a trial using EWOC. It provides guidance on how to select values for the probability of DLT (θ), dose range (minimum and maximum doses), minimum dose increment, probability of exceeding the target dose (α), cohort size, prior distributions for target dose γ, and probability of DLT at the initial dose ρ_0. It also shows, using a hypothetical dose-finding trial, how to adjust these parameters to generate a dose escalation with desired characteristics. In addition, a simulation module allows operating characteristics of a particular design to be generated. The last section of the tutorial illustrates how all the information generated can be incorporated into a template to produce the statistical section of the dose-finding protocol, ready to be submitted to a Protocol Review Committee and Institutional Review Board.

11.7 Final Remarks

In this chapter, we described EWOC, a Bayesian dose-finding design for cancer dose-finding clinical trials. The method is flexible enough to allow prior information about the drug from laboratory or animal studies to be incorporated in the model, makes use of all the information available at the time of each dose assignment, controls the probability of overdosing patients at each stage, allows the estimation of the precision of the MTD, is optimal Bayesian feasible, produces a sequence of doses that converges in probability to the MTD, is coherent, and accounts for patients' pretreatment characteristics. EWOC can be implemented with the user-friendly software EWOC (Rogatko et al. 2012a,b) or WinBUGS (Lunn et al. 2000) for general class of prior distributions, covariates, or ordinal toxicity grades. EWOC allows flexible patient enrollment and conforms with the ethical goal of maximizing the number of patients receiving optimal doses. At the time this chapter was written, we were aware of 20 peer-reviewed articles describing trials designed with EWOC. Currently, active research includes extensions allowing time to toxicity and drug combinations.

References

Adcock, C. J. (1997). Sample size determination: A review. *Statistician* 46 (2):261–283.
Anbar, D. (1978). Stochastic approximation methods and their use in bioassay and phase I clinical trials. *Communications in Statistics* 13 (19):2451–2467.
Babb, J. and Rogatko, A. (2004). Bayesian methods for cancer phase I clinical trials. In *Contemporary Biostatistical Methods in Clinical Trial*, N. Geller (ed.), pp. 1–39. New York: Marcel Dekker.
Babb, J., Rogatko, A., and Zacks, S. (1998). Cancer phase I clinical trials: Efficient dose escalation with overdose control. *Statistics in Medicine* 17:1103–1120.

Babb, J. S. and Rogatko, A. (2001). Patient specific dosing in a cancer phase I clinical trial. *Statistics in Medicine* 20 (14):2079–2090.

Bachelot, T., Ray-Coquard, I., Catimel, G., Ardiet, C., Guastalla, J. P., Dumortier, A., Chauvin, F., Droz, J. P., Philip, T., and Clavel, M. (2000). Multivariable analysis of prognostic factors for toxicity and survival for patients enrolled in phase I clinical trials. *Annals of Oncology* 11 (2):151–156.

Cheng, J. D., Babb, J. S., Langer, C., Aamdal, S., Robert, F., Engelhardt, L. R., Fernberg, O. et al. (2004). Individualized patient dosing in phase I clinical trials: The role of EWOC in PNU-214936. *Journal of Clinical Oncology* 22 (4):602–609.

Cheung, Y. K. (2005). Coherence principles in dose-finding studies. *Biometrika* 92 (4): 863–873.

Chu, P. L., Lin, Y., and Shih, W. J. (2009). Unifying CRM and EWOC designs for phase I cancer clinical trials. *Journal of Statistical Planning and Inference* 139 (3):1146–1163. doi: 10.1016/j.jspi.2008.07.005.

Dillman, R. O. and Koziol, J. A. (1992). Phase I cancer trials: Limitations and implications. *Molecular Biotherapy* 4 (3):117–121.

Eichhorn, B. H. and Zacks, S. (1981). Bayes sequential search of an optimal dosage: Linear regression with both parameters unknown. *Communications in Statistics: Theory and Methods* 10:931–953.

Faries, D. (1994). Practical modifications of the continual reassessment method for phase I cancer clinical trials. *Journal of Biopharmaceutical Statistics* 4 (2): 147–164.

Garrett-Mayer, E. (2006). The continual reassessment method for dose-finding studies: A tutorial. *Clinical Trials* 3 (1):57–71.

Gatsonis, C. and Greenhouse, J. B. (1992). Bayesian methods for phase I clinical trials. *Statistics in Medicine* 11 (10):1377–1389.

Gitlin, J. M. Phase I trial gone awry: Follow up (2006). Available from http:arstechnica.com/journals/science.ars/2006/8/1/4840.

Goodman, S. N., Zahurak, M. L., and Piantadosi, S. (1995). Some practical improvements in the continual reassessment method for phase I studies. *Statistics in Medicine* 14 (11):1149–1161.

Horstmann, E., McCabe, M. S., Grochow, L., Yamamoto, S., Rubinstein, L., Budd, T., Shoemaker, D., Emanuel, E. J., and Grady, C. (2005). Risks and benefits of phase 1 oncology trials, 1991 through 2002. *New England Journal of Medicine* 352 (9): 895–904.

Ivanova, A. (2006). Escalation, group and A+B designs for dose-finding trials. *Statistics in Medicine* 25:3668–3678.

Lin, Y. and Shih, W. J. (2001). Statistical properties of the traditional algorithm-based designs for phase I cancer clinical trials. *Biostatistics* 2 (2):203–215.

Lunn, D. J., Thomas, A., Best, N., and Spiegelhalter, D. (2000). WinBUGS—A Bayesian modelling framework: Concepts, structure, and extensibility. *Statistics and Computing* 10 (4):325–337.

Moller, S. (1995). An extension of the continual reassessment methods using a preliminary up-and-down design in a dose finding study in cancer patients, in order to investigate a greater range of doses. *Statistics in Medicine* 14 (9–10):911–922; discussion 923.

O'Quigley, J. (1990). Sequential design and analysis of dose finding studies in patients with life threatening disease. *Fundamental & Clinical Pharmacology* 4 (Suppl 2): 81s–91s.

O'Quigley, J., Pepe, M., and Fisher, L. (1990). Continual reassessment method: A practical design for phase 1 clinical trials in cancer. *Biometrics* 46 (1):33–48.

O'Quigley, J., Shen, L. Z., and Gamst, A. (1999). Two-sample continual reassessment method. *Journal of Biopharmaceutical Statistics* 9 (1):17–44.

Piantadosi, S., Fisher, J. D., and Grossman, S. (1998). Practical implementation of a modified continual reassessment method for dose-finding trials. *Cancer Chemotherapy and Pharmacology* 41 (6):429–436.

Roberts, T. G. Jr., Goulart, B. H., Squitieri, L., Stallings, S. C., Halpern, E. F., Chabner, B. A., Gazelle, G. S., Finkelstein, S. N., and Clark, J. W. (2004). Trends in the risks and benefits to patients with cancer participating in phase 1 clinical trials. *JAMA* 292 (17):2130–2140.

Rogatko, A., Babb, J. S., Tighiouart, M., Khuri, F. R., and Hudes, G. (2005). New paradigm in dose-finding trials: Patient-specific dosing and beyond phase I. *Clinical Cancer Research* 11 (15):5342–5346.

Rogatko, A., Babb, J. S., Wang, H., Slifker, M. J., and Hudes, G. R. (2004). Patient characteristics compete with dose as predictors of acute treatment toxicity in early phase clinical trials. *Clinical Cancer Research* 10 (14):4645–4651.

Rogatko, A., Ghosh, P., Vidakovic, B., and Tighiouart, M. (2008). Patient-specific dose adjustment in the cancer clinical trial setting. *Pharmaceutical Medicine* 22 (6): 345–350.

Rogatko, A., Tighiouart, M., and Bresee, C. (2012a). Escalation with overdose control. *User Guide Web-Based Portal.* http://biostatistics.csmc.edu/ewoc/.

Rogatko, A., Tighiouart, M., and Cook-Wiens, G. (2012b). Escalation with overdose control. *User's Guide.* Version 3.1., http://biostatistics.csmc.edu/ewoc/.

Rosenberger, W. F. and Haines, L. M. (2002). Competing designs for phase I clinical trials: A review. *Statistics in Medicine* 21 (18):2757–2770.

Storer, B. E. (1989). Design and analysis of phase I clinical trials. *Biometrics* 45 (3): 925–937.

Storer, B. E. (2001). An evaluation of phase I clinical trial designs in the continuous dose-response setting. *Statistics in Medicine* 20 (16):2399–2408.

Tighiouart, M., Cook-Wiens, G., and Rogatko, A. (2012a). Incorporating a patient dichotomous characteristic in cancer phase I clinical trials using escalation with overdose control. *Journal of Probability and Statistics* 2012:10. doi: 10.1155/2012/567819.

Tighiouart, M., Cook-Wiens, G., and Rogatko, A. (2012b). Escalation with overdose control using ordinal toxicity grades for cancer phase I clinical trials. *Journal of Probability and Statistics* 2012:1–18.

Tighiouart, M. and Rogatko, A. (2006a). Dose finding in oncology—Parametric methods. In *Dose Finding in Drug Development*, N. Ting (ed.), pp. 59–72. New York: Springer.

Tighiouart, M. and Rogatko, A. (2006b). Dose escalation with overdose control. In *Statistical Methods for Dose-Finding Experiments*, S. Chevret (ed.). New York: John Wiley and Sons.

Tighiouart, M. and Rogatko, A. (2010). Dose finding with escalation with overdose control (EWOC) in cancer clinical trials. *Statistical Science* 25 (2):217–226.

Tighiouart, M. and Rogatko, A. (2012). Number of patients per cohort and sample size considerations using dose escalation with overdose control. *Journal of Probability and Statistics* 2012:16. doi: 10.1155/2012/692725.

Tighiouart, M., Rogatko, A., and Babb, J. S. (2005). Flexible Bayesian methods for cancer phase I clinical trials. Dose escalation with overdose control. *Statistics in Medicine* 24 (14):2183–2196.

Tighiouart, M., Rogatko, A., and Xu, Z. (2007). Incorporating patient's characteristics in cancer phase I clinical trials using escalation with overdose control. *Paper Read at Joint Statistical Meetings*, Salt Lake City, UT.

Ting, N. (2006). *Dose Finding in Drug Development*, 1st edn. New York: Springer.

Whitehead, J. (1997). Bayesian decision procedures with application to dose-finding studies. *International Journal of Pharmaceutical Medicine* 11:201–208.

Wu, C. F. J. (1985). Efficient sequential designs with binary data. *Journal of the American Statistical Association* 80 (392):974–984.

Xu, Z. H., Tighiouart, M., and Rogatko, A. (2007). EWOC 2.0: Interactive software for dose escalation in cancer phase I clinical trials. *Drug Information Journal* 41 (2):221–228.

Yamamoto, N., Tamura, T., Fukuoka, M., and Saijo, N. (1999). Survival and prognostic factors in lung cancer patients treated in phase I trials: Japanese experience. *International Journal of Oncology* 15 (4):737–741.

Zacks, S., Rogatko, A., and Babb, J. (1998). Optimal Bayesian-feasible dose escalation for cancer phase I trials. *Statistics and Probability Letters* 38:215–220.

Zohar, S. and Chevret, S. (2001). The continual reassessment method: Comparison of Bayesian stopping rules for dose-ranging studies. *Statistics in Medicine* 20 (19): 2827–2843.

Zohar, S., Latouche, A. Taconnet, M. and Chevret, S. (2003). Software to compute and conduct sequential Bayesian phase I or II dose-ranging clinical trials with stopping rules. *Computer Methods and Programs in Biomedicine* 72 (2):117–125.

12

Interval-Censored Time-to-Event Data and Their Applications in Clinical Trials

Ling Ma, Yanqin Feng, Ding-Geng (Din) Chen, and Jianguo Sun

CONTENTS

12.1 Introduction

Interval-censored time-to-event or failure time data occur in many areas especially in cancer oncology and biopharmaceutics. Their analysis has attracted a great deal of attention over last two decades or so. For example, two books have been published on the topic (Sun, 2006; Chen et al., 2012). In addition, a couple of review papers have been published including Gómez et al. (2009) and Zhang and Sun (2010). In the context of failure time data, interval censoring means that the failure time variable of interest is observed or known only to lie within some intervals or windows instead of being observed exactly (Finkelstein, 1986; Kalbfleisch and Prentice, 2002; Sun, 2006). If the interval includes only or reduces to a single time point, one obtains the exact failure time.

One field that often produces interval-censored failure time data is clinical trials and the same is true in medical or health studies that entail periodic follow-up. In these studies, an individual due for the prescheduled observations for a clinically observable change in disease or health status may miss some observations and return with a changed status. As a consequence, we only know that the true event time is greater than the last observation time at which the change has not occurred and less than or equal to the first observation time at which the change has been observed to occur. That is, we only have an interval that contains the real (but unobserved) time of occurrence of the change. A typical and well-known example of interval-censored data is given in Finkelstein (1986) on comparing *Radiotherapy Only* treatment to *Radiation + chemotherapy* treatment. The study consists of 94 early breast cancer patients and they were supposed to be monitored every 4–6 months. Among them, 46 patients received radiation only and 48 received radiation plus chemotherapy. For the study, the event of interest is the time to first appearance of moderate or severe breast retraction and for which only interval-censored data are available. More details on the study are given in the following. One can find similar data in the acquired immune deficiency syndrome (AIDS) trials too (De Gruttola and Lagakos, 1989). In these cases, one may be interested in times to AIDS for human immunodeficiency virus (HIV) infected subjects. The determination of AIDS onset is usually based on blood testing, which can be performed obviously only periodically but not continuously. In consequence, only interval-censored data may be available for AIDS diagnosis times. A similar case is for studies on HIV infection times. If a patient is HIV positive at the beginning of a study, then the HIV infection time is usually determined by a retrospective analysis of his or her medical history. Therefore, we are only able to obtain an interval given by the last HIV negative test date and the first HIV positive test date for the HIV infection time.

In reality, interval censoring can occur in different forms and each form represents one type of interval-censored failure time data. Among them, an important type of interval-censored failure time data is the so-called current status data (Sun and Kalbfleisch, 1993; Jewell and van der Laan, 1995). This type of interval censoring means that each subject is observed only once for the status of the occurrence of the failure event of interest. In other words, one does not directly observe the occurrence of the failure event of interest but instead only knows the observation time and whether or not the event has occurred at the time. In consequence, the failure time is either left or right censored. One type of studies that usually produces current status data is cross-sectional studies on failure events (Keiding, 1991). Another type is tumorigenicity studies and in this situation, the time to tumor onset is usually of interest but not directly observable (Dinse and Lagakos, 1983). In these cases, one only knows or observes the exact value of the observation time, which is usually the death or sacrifice time of the subject. Note that for the first example, current status data occur due to the study design, while for the second case, they are often observed because of the inability of measuring the failure variable directly and exactly. Sometimes we also refer current status data to as case I interval censored data and the general case as case II interval-censored data (Groeneboom and Wellner, 1992).

Another type of interval-censored data is the so-called doubly censored data (De Gruttola and Lagakos, 1989; Sun, 2004). By this, we mean that the failure time of interest is defined as or represents the time between two related events and the observed data on the times to the occurrences of both events are interval-censored. In contrast, the interval censored data discussed earlier can be regarded as a special case of such doubly censored data in which one observes the times to the first event exactly and thus can treat them being zero for simplicity. An example of doubly censored data is provided by the AIDS studies discussed earlier when the variable of interest is AIDS incubation time (De Gruttola and Lagakos, 1989), the time from HIV infection to AIDS diagnosis, with both the HIV infection time and the AIDS diagnosis time being right or interval censored. Grouped failure time data are another special case of interval-censored data and often arise in, for example, large animal studies. By grouped failure time data, we usually mean that the intervals for any two subjects either are completely identical or have no overlapping. It is easy to see that for the analysis of such data, one could readily employ the methods available for right-censored data and no new methods are needed in theory. In other words, the statistical inference about grouped failure time data is relatively straightforward.

This chapter will provide a review of the recent development on the topic with the focus on the applications of the existing methodology in clinical trials and available software packages. More specifically, we will first discuss several common issues in the analysis of interval-censored data including the analysis of current status data, the analysis of univariate data, the analysis of multivariate data, and the analysis of competing risks data as well as informatively censored data. The chapter will conclude with the discussion of available software packages for interval-censored data, an illustrative example, and some concluding remarks. We remark that this chapter is a modified and updated version of Sun and Li (2013).

12.2 Notation and Likelihood Function

To describe the likelihood function, we will first define some notation. In the following, we will use T to denote the failure time of interest. By saying that T is interval censored, we mean that only available information for T is an interval denoted by $I = (L, R]$ such that $T \in I$. Using this notation, we see that current status data correspond to the situation where either $L = 0$ or $R = \infty$. Interval-censored data reduce to right-censored data if $L = R$ or $R = \infty$ for all subjects in the study. Note that a more general way to describe interval-censored data is to assume that the observation on T is given by a group of intervals (Turnbull, 1976). However, we will not discuss this general representation as the resulting likelihood functions in both cases have essentially the same structure.

Now suppose that there is a failure time study consisting of n independent subjects and let T_i and $I_i = (L_i, R_i]$ be defined as aforementioned but associated with subject i. Define $F(t) = P(T \leq t)$, the cumulative distribution function of T, and $S(t) = 1 - F(t)$, the survival function of T. Let $0 = t_0 < t_1 < \cdots < t_m < t_{m+1} = \infty$ denote the unique ordered elements of $\{0, \{L_i\}_{i=1}^n, \{R_i\}_{i=1}^n, \infty\}$, α_{ij} the indicator of the event $(t_{j-1}, t_j] \subseteq I_i$, and $p_j = S(t_{j-1}) - S(t_j)$. Then for inference about S or for $p = (p_1, \ldots, p_{m+1})'$, the likelihood function that is commonly used has the form

$$L_s(p) = \prod_{i=1}^n [S(L_i) - S(R_i)] = \prod_{i=1}^n \sum_{j=1}^{m+1} \alpha_{ij} p_j. \tag{12.1}$$

In reality, there may exist some covariates denoted by Z and in this case, the likelihood function aforementioned becomes

$$L_s(p|Z_i's) = \prod_{i=1}^n [S(L_i|Z_i) - S(R_i|Z_i)] = \prod_{i=1}^n \sum_{j=1}^{m+1} \alpha_{ij} p_j(Z_i). \tag{12.2}$$

Of course, the likelihood functions given previously come with some assumptions. The most fundamental and important one is perhaps the so-called noninformative interval censoring, which can be described by the following equality (Oller et al., 2004; Sun, 2006)

$$P(T \leq t|L = l, R = r, L < T \leq R) = P(T \leq t|l < T \leq r). \tag{12.3}$$

The earlier assumption essentially says that, except for the fact that T lies between l and r, which are the realizations of L and R, the interval $(L, R]$ (or equivalently its endpoints L and R) does not provide any extra information for T. In other words, the probabilistic behavior of T remains the same except that the original sample space $T \geq 0$ is now reduced to $l = L < T \leq R = r$. In the existence of covariates, the assumption (12.3) becomes

$$P(T \leq t|L = l, R = r, L < T \leq R, Z = z) = P(T \leq t|l < T \leq r, Z = z). \tag{12.4}$$

One could also employ different ways to characterize the noninformative interval censoring assumption. For example, one can use a stochastic process to describe the underlying interval censoring mechanism by assuming that there exists a sequence of observation times or an observation process. Then the noninformative assumption means that the process is independent of the failure time or process of interest (Groeneboom and Wellner, 1992; Oller et al., 2004). In practice, one question of interest is the conditions under which the assumption (12.3) or (12.4) holds and for this, the readers are referred to the discussions given in Oller et al. (2004, 2007) and Betensky (2000).

In the following, all discussions will be based on the assumption (12.3) or (12.4) unless specified otherwise.

It is worth noting that in the case of right-censored failure time data, the noninformative censoring can be described in a much simpler format. In this case, it means that the censoring time or variable is independent of the failure time of interest completely or conditionally given covariates. It is clear that the two censoring mechanisms are quite different as only one variable is involved or needed with respect to right censoring. In the case of interval censoring, two variables L and R are needed and furthermore, they together with T have a natural relationship $L < T \leq R$.

To help understanding and illustrate the concepts and discussion said earlier, we now consider a specific example of interval-censored data arising from an AIDS clinical trial, AIDS Clinical Trial Group 181, on HIV-infected individuals. The study is a natural history substudy of a comparative clinical trial of three antipneumocystis drugs and concerns the opportunistic infection cytomegalovirus (CMV). During the study, among other activities, blood and urine samples were collected from the patients at their clinical visits and tested for the presence of CMV, which is also commonly referred to as shedding of the virus. These samples and tests provide observed information on the two variables of interest, the times to CMV shedding in blood and in urine.

The observed data set is given in data set I of Appendix A in Sun (2006) and contains the observed intervals for the times to CMV shedding in blood and urine from 204 patients who provided at least one urine and blood samples during the study. More specifically, for the two failure times of interest, only interval-censored data are available with the intervals given by the last clinical visit time at which the shedding had not happened and the first clinical visit time at which the shedding had already occurred. Note that in this case, we actually have bivariate interval-censored data. If it is reasonable to assume that the clinical visit times of the patients have nothing to do with their disease status, then we would have noninformative interval censoring. Otherwise, one may have to consider the existence of informative interval censoring and this latter situation could be the case if, for example, the patients paid clinical visits because they felt their disease got worse. Among others, Goggins and Finkelstein (2000) also discussed this data set.

12.3 Statistical Analysis of Current Status Data

In this section, we will briefly discuss statistical analysis of current status data. Let $T, F(t)$ and $S(t)$ be defined as earlier and suppose that one observes only current status data on T, which are usually denoted by $\{C_i, \delta_i = I(T_i \leq C_i)\}_{i=1}^{n}$, where C_i represents the observation time on subject i.

Note that in this case, the noninformative censoring means that T_i and C_i are independent completely or given covariates. For the situation, it is easy to see that the likelihood function given in (12.1) reduces to

$$L_S(p) = \prod_{i=1}^{n} [1 - S(C_i)]^{\delta_i} [S(C_i)]^{1-\delta_i}.$$

To find the nonparametric maximum likelihood estimator (NPMLE) of S or maximize the previous likelihood function, let the t_j's denote the ordered observation times as defined earlier and Q_j the set of subjects who are observed at t_j, $j = 1, \ldots, m$. Define $d_j = \sum_{i \in Q_j} I(T_i \leq t_j)$ and let n_j denote the number of elements in Q_j. Then the NPMLE of S can be shown (Sun, 2006) to be equal to the isotonic regression of $\{d_1/n_1, \ldots, d_m/n_m\}$ with weights $\{n_1, \ldots, n_m\}$. Using the max-min formula for the isotonic regression (Barlow et al., 1972; Silvapulle and Sen, 2005), one can derive the NPMLE of S as

$$\hat{S}(t_j) = 1 - \max_{u \leq j} \min_{v \geq j} \left(\frac{\sum_{l=u}^{v} d_j}{\sum_{l=u}^{v} n_j} \right).$$

One question of both practical and theoretical interest about the NPMLE of S is its asymptotic properties such as the convergence rate. For this, many studies have been performed including a sequence of papers by Groeneboom and his collaborators (Groeneboom and Wellner, 1992; Groeneboom et al., 2010, 2012a). In particular, by viewing the problem as an inverse statistical model, Groeneboom et al. (2012b) developed some smooth plug-in inverse estimators and Banerjee (2012) provided a review and discussion of recent literature on the problem.

With respect to nonparametric comparison of survival functions based on current status data, several procedures have been proposed including the ones given in Sun and Kalbfleisch (1993), Andersen and Ronn (1995), Sun (1999), and Groeneboom (2012). Here we remark that most of the existing procedures including the ones for general interval-censored data assume that the censoring mechanism is the same for different treatments. More specifically for current status data, this means that the observation times C_i's follow the same distributions for subjects in different arms. One exception is the test procedure given in Sun (1999), which allows the distributions of the C_i's to depend on the treatment arms.

One needs to perform regression analysis if there exist covariates and one is interested in, for example, quantifying the effect of some covariates on the failure time of interest or predicting survival probabilities for new individuals. For this, the first step is usually to specify an appropriate regression model. For failure time data, the most commonly used model is perhaps the proportional hazards model (Cox, 1972) given by

$$\lambda(t|Z = z) = \lambda_0(t)e^{\beta'z} \tag{12.5}$$

with respect to the hazard function of T given covariates $Z = z$. Here $\lambda_0(t)$ denotes the unknown baseline hazard function and β the vector of unknown regression parameters. To fit current status data to the model, among others, Huang (1996) provided iterative convex minorant (ICM) type algorithm for estimation of unknown parameters in addition to investigating the properties of the resulting estimates of unknown parameters. In addition, one could employ the Newton–Raphson algorithm (Sun, 2006). More recently, Cai et al. (2011) developed a Bayesian approach by modeling the baseline cumulative hazard function with monotone splines and Zhang (2012) provided some general discussion.

In addition to the proportional hazards model (12.5), of course, there exist many other models that may be used. For example, another attractive semiparametric regression model often used in practice is the additive hazards model given by

$$\lambda(t|Z = z) = \lambda_0(t) + \beta'z, \tag{12.6}$$

again with respect to the hazard function of T given $Z = z$. It specifies that the effects of the covariates are additive rather than multiplicative as in model (12.5). For inference about this model based on current status data, among others, Lin et al. (1998) and Martinussen and Scheike (2002) developed some estimating equation approaches and Chen and Sun (2009) presented a multiple imputation procedure. More recently, Zhang (2012) discussed the fitting of both the proportional odds model and the linear transformation model described in the following to current status data. More recent references on the proportional odds model for current status data include McMahan et al. (2013), Wang and Dunson (2011), and Wen and Chen (2012, 2013). In particular, McMahan et al. (2013) proposed some expectation–maximization (EM) algorithms and provided closed form variance estimates, while Wang and Dunson (2011) discussed the problem in the Bayesian framework. We remark that unlike most methods developed for right-censored data, estimating regression parameters under interval censoring usually involves estimation of both the parametric and nonparametric parts. In other words, for interval-censored data, one has to deal with estimation of some unknown baseline functions in order to estimate regression parameters.

There are many other issues in analyzing current status failure time data. One is that in dealing with failure time data, a basic assumption that is usually not explicitly described is that the failure is assumed to occur. In some situations, this may not be the case and the so-called cure model is often employed. Among others, Ma (2009, 2011) recently studied the fitting of the cure model to current status data. Another issue of practical interest is the misclassification problem, and McKeown and Jewell (2010) and Saly Rosas and Hughes (2011) discussed it in the context of current status data.

12.4 Statistical Analysis of Univariate Interval-Censored Data

Now we will consider statistical analysis of general univariate interval-censored failure time data. As mentioned earlier, we will mainly discuss the three basic topics, nonparametric estimation of a survival function, nonparametric comparison of survival functions, and regression analysis with the focus on some of the recent advances.

For nonparametric estimation, one can easily see that given general interval-censored data, the problem of finding the NPMLE of S becomes that of maximizing $L_S(p)$ given in (12.1) under the constraint that $\sum_{j=1}^{m+1} p_j = 1$ and $p_j \geq 0, j = 1, \ldots, m + 1$ (Li et al., 1997). Obviously, the likelihood function L_S depends on S only through the values $\{S(t_j)\}_{j=1}^{m}$. Thus, the NPMLE $\hat{S}$ of S can be uniquely determined only over the observed intervals $(t_{j-1}, t_j]$ and the behavior of S within these intervals will be unknown. Several methods have been proposed for maximizing $L_S(p)$ with respect to p. The first and simplest one is perhaps the self-consistency algorithm given by Turnbull (1976). One drawback of the algorithm is that the convergence can be slow. Corresponding to this, Groeneboom and Wellner (1992) developed an ICM algorithm, which can be seen as an optimized version of the well known pool-adjacent-violator algorithm for isotonic regression (Robertson et al., 1988). Another faster algorithm, a hybrid one that combines the two approaches aforementioned, is the EM-ICM algorithm given in Wellner and Zhan (1997). The other authors who recently investigated the same problem include Yavuz and Lambert (2011) and Dehghan and Duchesne (2011). The former employed Bayesian penalized B-splines to obtain smooth estimation, while the latter generalized the self-consistency algorithm to the case where there exists a continuous covariate.

It is worth to emphasize that all algorithms mentioned earlier are iterative and for general interval-censored data, there is no closed form for the NPMLE of S unlike with current status data. Also it should be noted that for general interval-censored data, the NPMLE may not be unique and the solutions derived from the algorithms mentioned earlier may not be the NPMLE. One sufficient condition for the uniqueness of the NPMLE is that the log likelihood is strictly concave. Another important but difficult issue related to the NPMLE of S is the consistent variance estimation as the standard Fisher information matrix approach does not work here. For this, one way is to employ the profile likelihood approach (Murphy and van der Vaart, 1999, 2000), which is often computationally intensive. Recently

Huang et al. (2012) proposed a least-squares approach based on the efficient score function. For detailed discussion on these issues and the asymptotic properties as well as the differences between right-censored and interval-censored data, the readers are referred to Groeneboom and Wellner (1992), Maathuis and Wellner (2008), and Zhang and Sun (2010), among others.

The comparison of different treatments or survival functions is another primary objective in most medical or clinical studies. To formalize the problem, suppose that there are K treatment arms in a clinical study and let $S^{(k)}(t)$ denote the survival function of the kth arm with $k = 1, \ldots, K$. Then the problem becomes testing the null hypothesis

$$H_0 : S^{(1)}(t) = S^{(2)}(t) = \cdots = S^{(K)}(t) \quad \text{for all } t.$$

In the case of right-censored data, many nonparametric test procedures have been developed and most of them can be classified into two categories: rank-based tests and survival-based tests. The fundamental difference between them is that the former relies on the differences between the estimated hazard functions, while the latter bases the comparison on the differences between the estimated survival functions. Among them, the log-rank test is perhaps the most widely used method and a few of them have been generalized to the case of interval-censored data (Sun, 2006). For example, recently Huang et al. (2008) presented an imputation-based generalized log-rank test, and also Fay and Shih (2012) and Oller and Gómez (2012) gave some generalized rank-based and survival-based test procedures, respectively, for interval-censored data. To give a representative of such procedures, in the following, we describe the one proposed by Zhao et al. (2008).

Consider a survival study consisting of n independent subjects with n_k subjects from treatment arm k, $k = 1, \ldots, K$. Let the T_i's and I_i's be defined as before and D_{k1} and D_{k2} denote the sets of indices of the subjects in treatment arm k whose failure times are observed exactly and interval censored, respectively. Define $n_{k1} = |D_{k1}|, n_{k2} = |D_{k2}|, n_1 = \sum_{k=1}^{K} n_{k1}$ and $n_2 = \sum_{k=1}^{K} n_{k2}$, and let $\hat{S}$ denote the NPMLE of the common survival function under the hypothesis H_0. To test the hypothesis H_0, Zhao et al. (2008) proposed to use the test statistic $U_\xi = (U_\xi^{(1)}, \ldots, U_\xi^{(K)})'$, where

$$U_\xi^{(k)} = \frac{n_1}{n_{k1}} \sum_{i \in D_{k1}} \frac{\xi(\hat{S}(T_i-)) - \xi(\hat{S}(T_i))}{\hat{S}(T_i-) - \hat{S}(T_i)} - \frac{n_2}{n_{k2}} \sum_{i \in D_{k2}} \frac{\xi(\hat{S}(L_i)) - \xi(\hat{S}(R_i))}{\hat{S}(L_i-) - \hat{S}(R_i)},$$

and ξ is a positive known function over $(0, 1)$. In practice, different ξ can be used and will yield different test statistics. The aforementioned statistics were motivated by that given in Peto and Peto (1972), who discussed similar test statistics with $\xi(x) = x \log(x)$ for right-censored data. If $n_1 = 0$, that is, only interval-censored failure time data are available, the statistics U_ξ reduces to that proposed in Sun et al. (2005), a generalization of the log-rank test discussed in Peto and Peto (1972). Under some regularity conditions and H_0, Zhao et al. (2008) showed that as $n \to \infty, n_{k1}/n \to p_{k1}, n_{k2}/n \to p_{k2}$ with $0 < p_{kj} < 1, U_\xi/\sqrt{n}$ has an asymptotic normal distribution with mean zero. They also gave a consistent estimate of the asymptotic covariance matrix.

With respect to regression analysis of general interval-censored data, as with other types of failure time data, the proportional hazards model (12.5) and the additive hazard model (12.6) are commonly used. Among recent work on this, for example, Zhang et al. (2010) and Heller (2011) investigated the fitting of model (12.5) to interval-censored data. The former studied the spline-based maximum likelihood approach in which monotone B-spline were used to approximate the baseline cumulative hazard function, while the latter gave a weighted estimating equation method. To fit model (12.6) to interval-censored data, Chen and Sun (2010) and Wang et al. (2010) developed a multiple imputation procedure and an estimating equation approach, respectively.

Many other semiparametric models have been employed for regression analysis of interval-censored data. One such model is the proportional odds model (Sun, 2006; Sun et al., 2007) expressed as

$$\log \left(\frac{F(t|z)}{1 - F(t|z)} \right) = h(t) + \beta'z \qquad (12.7)$$

with respect to the CDF $F(t|z)$ of the failure time T of interest given $Z = z$. Here $h(t)$ is an unknown monotone-increasing function, also referred to as the baseline log odds, and β represents a vector of regression parameters as in models (12.5) and (12.6). The accelerated failure time model (Sun, 2006) has also been considered for analyzing interval-censored data, and it assumes that T and Z have the following relationship

$$\log(T) = \beta'Z + \varepsilon. \qquad (12.8)$$

In the previous equation, β is defined as before and ε is an error term whose distribution is usually unspecified.

It is easy to see that the four semiparametric models (12.5) through (12.8) are all specific models in terms of the functional form of the effects of covariates. Sometimes one may prefer a model that gives more flexibility. One such model that has been investigated for interval-censored data is the linear transformation model that specifies the relationship between the failure time T and the covariate Z as

$$h(T) = \beta'Z + \varepsilon. \qquad (12.9)$$

Here $h : \mathcal{R}^+ \to \mathcal{R}$ ($\mathcal{R}$ denotes the real line and $\mathcal{R}^+$ the positive half real line) is an unknown strictly increasing function and the distribution of ε is assumed to be known. It is apparent that the model discussed earlier gives different models depending on the specification of the distribution of ε and especially, it includes models (12.5) and (12.7) as special cases. Among the authors who recently discussed this model for regression analysis of interval-censored data, Chen and Sun (2010) gave a multiple imputation approach and Feng et al. (2013) and Zhang and Zhao (2013) developed some empirical likelihood-based methods.

For a practical problem, of course, one may also employ some parametric models such as piecewise exponential models (Zhang and Sun, 2010) if there exists some prior information about the appropriateness of the model. A main advantage of adopting a parametric model is that one can readily apply the maximum likelihood approach for inference. On the other hand, it is well known that parametric models usually may be too difficult to be verified and are less flexible than semiparametric models.

Other recent work on regression analysis of interval-censored data includes Kim and Jhun (2008), Liu and Shen (2009), Li and Ma (2010), and Lam et al. (2013), who considered the fitting of a cure model to interval-censored data, and Wang et al. (2012a,b, 2013), who discussed the application of Bayesian approaches to the problem. In addition, Zhu et al. (2008) proposed a transformation approach, Lin and Wang (2010) studied the fitting of a probit model to interval censored, and Lee et al. (2011) examined the use of three imputation approaches for the problem. Furthermore, Zhang and Davidian (2008) proposed a group of smooth semiparametric regression models, Zhang (2009) discussed the survival prediction under model (12.9), Pawel and Leon (2012) presented a linear regression method based on a convex piecewise linear criterion function, and MacKenzie and Peng (2013) developed some likelihood methods under parametric proportional and nonproportional hazards regression models for the longitudinal study.

12.5 Statistical Analysis of Multivariate Interval-Censored Data

Multivariate interval-censored data arise if a failure time study involves several related failure time variables of interest and each of them suffers interval censoring. It is apparent that the analysis would be straightforward if the variables are independent and when they are not independent of each other as usually the case, one needs and should employ the inference procedures that can take into account the correlation among the failure time variables. Another difference between univariate and multivariate interval-censored data is that for the latter, a new and unique issue that does not exist for the former is to make inference about the association between the failure time variables. In the following, we will first discuss a couple of basic issues related to the analysis of multivariate interval-censored data and then some recent advances with the focus on bivariate interval-censored data.

As with univariate interval-censored data, nonparametric estimation of a survival function is also one of the primary objectives of interest in practice for multivariate interval-censored data. To discuss this, consider a survival study that involves n independent subjects from a homogeneous population with each subject giving rise to two failure times denoted by T_{1i} and $T_{2i}, i = 1, \ldots, n$. Let $F(t_1, t_2) = P(T_{1i} \leq t_1, T_{2i} \leq t_2)$ denote their joint cumulative distribution function and suppose that only interval-censored failure time data in the form $\{U_i = (L_{1i}, R_{1i}] \times (L_{2i}, R_{2i}], i = 1, \ldots, n\}$ are available, where $(L_{1i}, R_{1i}]$ and $(L_{2i}, R_{2i}]$ represent the intervals to which T_{1i} and T_{2i} belong, respectively. It is easy to see that the observation on each subject could be a point, line segment, or rectangle. If one treats points as rectangles that are degenerate in both dimensions and line segments as rectangles that are degenerate in one dimension, then the observed data consist of a collection of n rectangles.

For the determination of the NPMLE of F, note that as with univariate data, it will be a step function. Let $H = \{H_j = (r_{1j}, s_{1j}] \times (r_{2j}, s_{2j}], j = 1, \ldots, m\}$ denote the disjoint rectangles that constitute the regions of possible support of the NPMLE of F and define $\alpha_{ij} = I(H_j \subseteq (L_{1i}, R_{1i}] \times (L_{2i}, R_{2i}])$ and $p_j = F(H_j) = F(s_{1j}, s_{2j}) - F(r_{1j}, s_{2j}) - F(s_{1j}, r_{2j}) + F(r_{1j}, r_{2j}), i = 1, \ldots, n; j = 1, \ldots, m$. Then the likelihood function has the form

$$L_F(p) = \prod_{i=1}^{n} F(U_i) = \prod_{i=1}^{n} \sum_{j=1}^{m} \alpha_{ij} p_j \tag{12.10}$$

with $p = (p_1, \ldots, p_m)'$, and the NPMLE of F can be determined by maximizing (12.10) over the p_j's subject to $p_j \geq 0$ and $\sum_{j=1}^{m} p_j = 1$. One can easily see that the likelihood functions given (12.1) and (12.10) actually have the same structure, and thus, the algorithms developed for the maximization of (12.1) could be employed here for the maximization of (12.1) assuming that H is known. However, for a given data set, the determination of H is actually not straightforward or difficult and for this, some algorithms have been developed and need to be used (Sun, 2006). It is apparent that the same holds for general multivariate interval-censored data.

As mentioned earlier, the estimation of the association between related failure time variables is often of interest for multivariate failure time data. To discuss this in the context of bivariate data, let $S_1(t)$ and $S_2(t)$ denote the marginal survival functions of possibly related T_1 and T_2, respectively, and $S(t_1, t_2) = P(T_1 > t_1, T_2 > t_2)$ their joint survival function. For the problem, one common way is to assume that $S(t_1, t_2)$ can be expressed by a copula model as $S(t_1, t_2) = C_\alpha(S_1(t_1), S_2(t_2))$, where C_α is a distribution function on the unit square and $\alpha \in R$ is a global association parameter. One attractive feature of the aforementioned expression is its flexibility as it includes as special cases many useful bivariate failure time models such as the Archimedean copula family. Another attractive feature is that under this expression, the marginal distributions do not depend on the choice of the association structure and thus, one can model the marginal distributions and the association separately. Among others, Wang and Ding (2000) and Sun et al. (2006) discussed this approach for bivariate interval-censored data. Of course, by using high order copula models instead of the bivariate model, one can apply the same approach to general multivariate interval-censored data.

The copula model approach discussed earlier can be employed for regression analysis of multivariate interval-censored data too. For example, Wang et al. (2008) and Zhang et al. (2009) recently applied it to the fitting of models (12.5) and (12.7), respectively, to bivariate current status data and developed efficient estimates of regression parameters. To describe the relationship among correlated failure time variables, instead of using the copula model, another commonly used approach is to employ frailty or latent variable models in which some latent variables are used to characterize the correlation. Among others, Jonker et al. (2009) and Hens et al. (2009) considered this approach for regression analysis of bivariate interval-censored data, while Chen et al. (2009) and Nielsen and Parner (2010) applied it to regression analysis of general multivariate interval-censored data. A third approach that is often adopted for multivariate failure time data is the marginal approach that leaves the correlation arbitrary. For example, Chen et al. (2007) and Tong et al. (2008) developed such approaches for fitting models (12.7) and (12.6), respectively, to general multivariate interval-censored data. For all three approaches, of course, one could put them in the Bayesian framework (Komàrek and Lesaffre, 2007).

It is well known that multivariate failure time data can be seen as a special case of clustered failure time data and a key feature of the latter is that the cluster size, the number of correlated failure time variables, can differ from cluster to cluster. Thus, the existing methods for multivariate interval-censored data cannot be directly applied to clustered interval-censored data. Some recent references on the latter include Chen et al. (2009), Zhang and Sun (2010), Xiang et al. (2011), and Kor et al. (2013). The first three discussed the fitting of the proportional hazards model (12.5), while the last considered the inference about a cure model. Compared to other types of interval-censored data discussed earlier, however, there exists very limited literature on clustered interval-censored data.

12.6 Statistical Analysis of Competing Risks and Informatively Interval-Censored Data

Competing risks failure time data arise when there exist several related failure causes or types. The underlying structure behind them is actually similar to that behind multivariate failure time data. A key difference between the two types of data, also a key feature of the former, is that one observes only one failure type or one of several related underlying failure time variables. For a patient with several diseases, for example, one can observe the death time caused by one disease, but not the times due to other diseases. In the following content, we will first discuss the analysis of competing risks interval-censored data, and then discuss the analysis of informatively interval-censored data with the focus on nonparametric estimation and regression analysis.

Consider a competing risks study that involves m different types of failures. Let T denote the real or observable failure time and $J \in \{1, \ldots, m\}$ the cause or type of failure. In this case, instead of the overall survival function, one is usually interested in estimating the cumulative incidence function defined as $F_j(t) = P(T \leq t, J = j)$ (Fine and Gray, 1999) and the cause-specific hazard function defined as $\lambda_j(t) = dF_j / (1 - F_+(j))$ (Larson, 1984) for failure type j, where $F_+(t) = \sum_{j=1}^{B} F_j(t) = P(T \leq t)$. Among others, one of the early works was given by Jewell et al. (2003), who discussed two nonparametric estimators of the cumulative incidence function, NPMLE and a simpler naive NPMLE, based on reduced current status data. Also Groeneboom et al. (2008a,b) rigorously derived the large sample properties of the two nonparametric estimators and showed that they both converge at a cube root rate. Recently, Li and Fine (2013) investigated the same estimation problem under some smoothness assumption and showed that the smoothing of the naive estimator can yield an improved convergence rate relative to the unsmoothed NPMLE and the unsmoothed naive estimator. Maathuis and Hudgens (2011) also investigated the same problem. Note that all references mentioned earlier are for competing risks current status data only and there does not seem to exist similar work for the same problems for general competing risks interval-censored data except Frydman and Liu (2013), who recently discussed the nonparametric estimation of the cumulative intensity for general situations.

In addition to the notation aforementioned, suppose that there also exists a vector of covariates Z and one is interested in regression analysis. In this case, one can write the conditional cumulative incidence function $F_j(t|Z) = P(T \leq t; J = j|Z)$ as $F_j(t|Z) = P(T \leq t|J = j, Z)$, $j = 1, \ldots, m$. To conduct regression analysis, one way is to directly model $F_j(t|Z)$ or the conditional distribution function $P(T \leq t|J = j, Z)$ along with $P(J = j|Z)$. Note that the regression parameters in these two different modeling approaches will have different meanings and one may base the selection on the regression parameter that is preferred. Assume that one would like to take the second approach and T is continuous. For this, a common approach is to assume that the cause-specific hazard function corresponding to $P(T \leq t|J = j, Z)$ and $P(J = j|Z)$ follow, for example, the proportional hazards model $\lambda_j(t|Z) = \lambda_{0j}(t)e^{\beta_j'Z}$ and the parametric model $g_j(\gamma, Z)$. Here $\lambda_{0j}(t)$ is an unknown baseline hazard function, β_j and γ are regression parameters, and g_j is a known and positive function satisfying $\sum_{j=1}^{m} g_j(\gamma, Z) = 1$. For each j, let $\Lambda_j(t) = \int_0^t \lambda_{0j}(t)$. Then we have $F_j(t|Z) = g_j(\gamma, Z) \left(1 - \exp\left\{-\Lambda_j(t)\exp\left(\beta_j' Z\right)\right\}\right)$.

For inference, suppose that one observes only current status data discussed in Section 12.3. For each j, define $\delta_j = I\{T \leq C, J = j\}, j = 1, \ldots, m$, $\delta_{m+1} = I\{T > C\}$ and $\Delta = (\delta_1, \ldots, \delta_m, \delta_{m+1})'$. Let $\theta = (\beta_1', \ldots, \beta_m', \gamma')'$ and $\Lambda = (\Lambda_1, \ldots, \Lambda_m)'$ and suppose that the joint distribution function of (C, Z) does not involve θ and Λ. Then the likelihood contribution from a single observation $X = (C, \Delta, Z)$ has the form

$$
\begin{aligned}
L(\theta, \Lambda; X) &= \prod_{j=1}^{m} F_j(C|Z)^{\delta_j} \left(1 - \sum_{k=1}^{m} F_k(C|Z)\right)^{\delta_{m+1}} \\
&= \prod_{j=1}^{m} \left[g_j(\gamma, Z)\left(1 - e^{-\Lambda_j(C)e^{\beta_j' Z}}\right)\right]^{\delta_j} \\
&\quad \times \left[\sum_{k=1}^{m} g_k(\gamma, Z)\left(1 - e^{-\Lambda_k(C)e^{\beta_k' Z}}\right)\right]^{\delta_{m+1}}
\end{aligned}
$$

and the likelihood function from n independent, identically distributed (i.i.d.) copies of X is given by the product of $L(\theta, \Lambda; X)$. Sun and Shen (2009) investigated this and established the asymptotic properties of the resulting maximum likelihood estimates.

Now we discuss the analysis of informatively interval-censored data and for this, we will first consider current status data. As before, let T and C denote the failure time of interest and the observation time, respectively. By informative censoring, it means that T and C are correlated and thus, the underlying probability space is similar to that for $m = 2$ competing risks problems without censoring. On the other hand, the data structures behind them are different as for current status data, C is always observable but not T, while both T and C could be observed for exact or right-censored competing risks data. As with informatively right-censored failure time data, the survival function of T is generally unidentifiable based on informatively current status data without some assumptions. Wang et al. (2012) discussed this and proposed two estimates of the survival function under the copula model framework. Following them, Titman (2014) also investigated the same problem and gave a pool-adjacent-violators algorithm.

As mentioned earlier, one area that often produces current status data is tumorigenicity experiments and in this case, T and C represent the tumor onset time and the death time, respectively. In addition, sacrifice is often used in these studies and in this situation, the death will not be observed. In other words, there may exist right censoring on the observation time C. Also often used in tumorigenicity studies, one way to analyze informatively current status data with right censoring is to formulate the problem using the illness-death or three-state model consisting of health, illness (tumor in the tumorigenicity study), and death. Here the illness and death correspond to the failure event of interest and observation event, respectively, and actually an extensive literature has been established on this in the context of tumorigenicity studies. Recent references on this for informatively current status data include Frydman and Szarek (2010) and Kim et al. (2012). The first developed the nonparametric maximum likelihood approach and the second reference discussed the regression analysis problem under the proportional hazards model. In addition, Chen et al. (2012) presented a class of semiparametric transformation models with log-normal frailty and developed an EM algorithm for the sieve parameter estimation.

For the analysis of general informatively interval-censored data, first note that as discussed earlier, the censoring mechanism behind interval censoring is usually much more complicated than that behind both right censoring and current status data case. In consequence, it is usually difficult or impossible to generalize the methods developed for informatively right-censored data or informative current status data to them. To further see this, note that one can write the likelihood contribution from a single interval-censored observation as

$$P(L \leq T \leq R) = P(l \leq T \leq r | L = l, R = r)P(L = l, R = r).$$

This indicates that to conduct regression analysis, one would have to specify some joint models for $P(L \leq T \leq R)$ or for both $P(l \leq T \leq r | L = l, R = r)$ and $P(L = l, R = r)$. This is quite different from the case of right-censored data or current status data, and usually not easy. The authors who recently discussed this joint modeling approach include Zhang et al. (2007) and Wang et al. (2010). They considered the cases where T follows models (12.5) and (12.6), respectively, marginally and in both cases, model (12.5) was also used to model the censoring variables.

12.7 Statistical Analysis of Other Interval-Censored Data

In the previous sections, we have discussed several types of interval-censored failure time data that one commonly faces in practice. A few other types of interval-censored data that were not touched earlier may occur too in failure time studies. They include doubly censored data, interval-censored data from multistate models, and interval-censored data with missing or mismeasured covariates. In this section, we will briefly discuss them along with some others.

By failure time variable or interval-censored failure time variable, one generally means a variable measuring the time from zero to the occurrence of an event of interest. A generalization of this is the variable that measures the elapse time between two successive events such as the onset of a disease and the death due to the disease. Given such a variable of interest plus interval censoring, one will have a doubly censored data analysis problem as discussed earlier. In other words, the interval-censored data discussed earlier can be regarded as a special case of doubly censored data. Sun (2006) devoted one chapter for the analysis of doubly censored data. Recent references on doubly censored data include Komàrek and Lesaffre (2008), Deng et al. (2009), and Ji et al. (2012), which discussed the Bayesian approach, the nonparametric estimation problem and the quantile regression problem, respectively.

Note that one way to formulate doubly censored data is to employ a three-state model similar to the one discussed in the previous section. Corresponding to this, a natural question of interest will be the analysis of interval-censored data arising from a multistate model, naturally and commonly used in, for example, epidemiological or disease progression studies. Several authors actually have recently considered this problem including Barrett et al. (2011), Chen et al. (2010), Cook et al. (2008), Joly et al. (2012), Kim (2013), McKeown and Jewell (2011), and Yang and Nair (2011). In particular, Chen et al. (2010) investigated the maximum likelihood approach, and Yang and Nair (2011) compared the multistate-based analysis and the simple failure time data analysis in the presence of interval censoring.

Missing covariates often occur in regression analysis as well as covariate mismeasurement and other related problems. The same can happen in failure time studies in the presence of interval censoring. For example, Wen and Lin (2011) gave a set of current status data with missing covariates arising from the 2005 Taiwan National Health Survey and developed a semiparametric maximum likelihood estimation procedure under the proportional hazards model (12.5). Wen (2012) and Wen et al. (2011) discussed regression analysis of interval-censored data under model (12.5) in the presence of measurement errors of covariates. Also Li and Nan (2011) considered current status data arising from case-cohort studies. In all discussions so far, it is supposed that interval censoring occurs on the failure time variable of interest and in reality, it could occur on covariates. In other words, covariates may have missing values in the form of censoring and Schoenfeld et al. (2011) provided some discussion on such interval-censored data.

12.8 Software: An Example and Concluding Remarks

It is definitely important to implement the available inference procedures numerically for practitioners. Unfortunately and surprisingly, there is no commercially available statistical software yet that provides an extensive coverage for interval-censored data. This is perhaps due to the complexity of both the algorithms and the theory behind it. One can, however, find some simple functions in SPLUS, R, and SAS that can be applied to interval-censored data.

In SPLUS, the function *kaplanMeier* can be used to compute the Turnbull estimator (1976). In R, the packages *interval* and *Icens* can be used to find the NPMLE of the survival and distribution function, while *interval* package can be used to perform two-sample weighted log-rank tests and k-sample tests. In addition, parametric methods are provided in the *survival* package using the *survreg* function and for semiparametric estimation, one can use *Icens* function in the *Epi* package. The *intcox* package can also be used for semiparametric estimation under the proportional hazards model using the method proposed by Pan (1999). However, the package does not directly provide standard error estimation for the estimated regression parameters and the bootstrapping approach is one of the remedies for this drawback. One can find some details and examples on the use of R for the analysis of interval-censored data in Fay and Shaw (2010) and Gómez et al. (2009).

With respect to SAS, one can employ the *LIFEREG* and *RELIABILITY* procedures to fit some popular parametric lifetime distributions to interval-censored data by maximum likelihood estimation and the macro %*EMICM* for the determination of the NPMLE of a survival function based on the EM-ICM algorithm. For treatment comparison, the macros %*ICSTEST* and %*ICE* can be used to compute the generalized log-rank test I of Zhao and Sun (2004) and generalized log-rank test II of Sun et al. (2005), and So et al. (2010) gave some illustrations and examples about the use of these macros. Other useful macros for interval-censored data analysis include %*INTCEN_SURV_P* and %*TURNBULL_INT*. The former can be used for median estimation, and the latter can be used to compute the Turnbull intervals (1976).

To illustrate the software aforementioned, we now return to the breast cancer clinical trial discussed earlier. As mentioned before, the trial consists of two treatments, *radiotherapy only* ($x = 0$) and *radiation + chemotherapy* ($x = 1$), given to 94 early breast cancer patients. Forty-six (46) patients received radiation only and 48 received radiation plus chemotherapy. Patients were monitored initially every 4–6 months. For this study, the failure time of interest is the time to first appearance of moderate or severe breast retraction and observed to occur only within intervals between visits. Therefore, the data are interval censored. Among the 94 patients, 56 were interval censored (*cens* = 1) with specific values of t_L and t_R, and 38 were right-censored (*cens* = 0) observations for patients who had not achieved moderate or severe breast retraction until the last visit and $t_R = NA$. The data, reproduced from Sun (2006), are given in Table 12.1.

For the analysis, we first apply the Turnbull's (1976) estimator (as a reference) for interval-censored data to estimate the survival functions and then apply the *intcox* to this data. The implementation of *intcox* is very straightforward with just a function call as follows:

TABLE 12.1

Interval-Censored Breast Cancer Data

Treatment	Time to First Appearance of Moderate or Severe Retraction (94 Patients)
RT Total 46	(45,],(37,],(46,],(33,],(15,],(18,],(46,],(46,],(46,]
	(24,],(46,],(36,],(37,],(22,],(38,],(34,],(17,],(46,]
	(46,],(36,],(46,],(37,],(24,],(40,],(33,],(25,37],(4,11]
	(17,25],(6,10],(0,5],(0,7],(26,40],(19,26],(11,15],(11,18]
	(27,34],(7,16],(36,44],(5,12],(19,35],(5,12],(9,14]
	(36,48],(17,25],(37,44]
RCT Total 48	(13,],(13,],(11,],(35,],(33,],(31,],(11,],(23,],(13,]
	(22,],(34,],(34,],(48,],(8,12],(0,5],(30,34],(16,20],(0,22]
	(5,8],(30,36],(18,25],(24,31],(12,20],(10,17],(17,24],(18,24]
	(17,27],(8,21],(17,26],(17,23],(33,40],(4,9],(16,60],(24,30]
	(15,22],(35,39],(16,24],(13,39],(15,19],(11,17],(19,32],(4,8]
	(44,48],(11,13],(22,32],(11,20],(14,17],(10,35]

```
# load the library
library(intcox)
# Call function "intcox" to fit the intcox method
fit = intcox(Surv(left,right,type="interval2"~treatment,
             data="data name")
# print the summary
summary(fit)
```

The estimated survival functions from both the Turnbull's estimator and the *intcox* appear in Figure 12.1. In this figure, the lower two dotted-lines denote *radiation + chemotherapy* treatment and the upper two-solid lines denote *radiation only*. In each treatment, the thin line depicts Turnbull's estimator and the thick line depicts *intcox*.

From this figure, we note that the estimated survival (without breast retraction) functions do not differ noticeably in the early stage, up to 20 months. After 20 months, there is a sharp decline in the curves, particularly for patients in the *radiation + chemotherapy* treatment group. For example, at *time* = 44 months, 11.06% of patients are estimated to be free of breast retraction in the *radiation + chemotherapy* group, as compared to 46.78% of patients in the *radiation only* treatment.

Fitting the *intcox* approach, the estimated treatment difference is 0.776 with Akaike information criterion (AIC) = 334.73. The standard error for this treatment parameter cannot be obtained directly from *intcox* but be estimated using standard bootstrap methods. We obtain random samples of the observed data with replacement 1000 times and fit the *intcox* for the resulting bootstrap samples. The bootstrap distribution can then be constructed

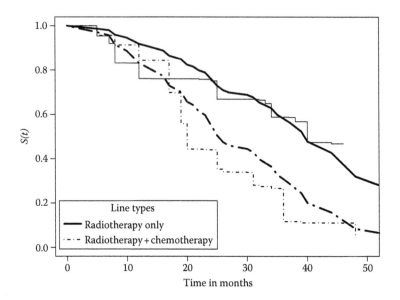

FIGURE 12.1
Turnbull's nonparametric estimator (the thin lines) overlaid with *intcox* estimator (the thick lines).

and used for statistical inference. From the bootstrap distribution, the 95% confidence interval for treatment effect is $(0.237, 1.412)$ with estimated regression parameter $\hat{\beta} = 0.776$, which again confirms the statistical significance of treatment effect. More detailed analysis for this data can be found in Chen et al. (2011, 2013), among others.

Methodologically, there are still many open questions in the analysis of interval-censored data. Examples include but are not limited to model checking techniques and joint modeling of longitudinal and interval-censored data. Some of the methods discussed in the previous sections also need proper theoretical justification. The major difficulty is that there are no basic tools that are as simple and elegant as the partial likelihood and the martingale theory for right-censored data. The works by Groeneboom and Wellner (1992) and Huang and Wellner (1997) are perhaps the most comprehensive studies for interval censoring, which mainly rely on complicated empirical processes and the optimization theory and are difficult to generalize.

References

Andersen, P.K. and Ronn, B.B. 1995. A nonparametric test for comparing two samples where all observations are either left- or right-censored. *Biometrics*, 51:323–329.

Banerjee, M. 2012. Current status data in the 21st century. In Chen, D., Sun, J. and Peace, K. (Eds.) *Interval Censored Time-to-Event Data: Methods and Applications* (pp. 45–88). Chapman & Hall/CRC, Boca Raton, FL.

Barlow, R.E., Bartholomew, D.J, Bremner, J.M., and Brunk, H.D. 1972. *Statistical Inference under Order Restrictions*. Wiley, New York.

Barrett, J.K., Siannis, F., and Farewell, V.T. 2011. A semi-competing risks model for data with interval-censoring and informative observation: An application to the MRC cognitive function and aging study. *Statistics in Medicine*, 30:1–10.

Betensky, R.A. 2000. On nonidentifiability and noninformative censoring for current status data. *Biometrika*, 87:218–221.

Cai, B., Lin, X., and Wang, L. 2011. Bayesian proportional hazards model for current status data with monotone splines. *Computational Statistics & Data Analysis*, 55:2644–2651.

Chen, B., Yi, G.Y., and Cook, R.J. 2010. Analysis of interval-censored disease progression data via multi-state models under a nonignorable inspection process. *Statistics in Medicine*, 29:1175–1189.

Chen, C.M., Lu, T.F., Chen, M.H., and Hsu, C.M. 2012. Semiparametric transformation models for current status data with informative censoring. *Biometrical Journal*, 54:641–656.

Chen, D.G. and Peace, K.E. 2011. *Clinical Trial Data Analysis Using R*. Chapman & Hall/CRC, Boca Raton, FL.

Chen, D.G., Sun, J., and Peace, K.E. 2012. *Interval-Censored Time-to-Event Data: Methods and Applications*. Chapman & Hall/CRC, Boca Raton, FL.

Chen, D.G., Yu, L. Peace, K.E. Lio, Y.L., and Wang, Y. 2013. Approximating the baseline hazard function by Taylor series for interval-censored time-to-event data. *Journal of Biopharmaceutical Statistics*, 23(3):695–708.

Chen, L. and Sun, J. 2009. A multiple imputation approach to the analysis of current status data with the additive hazards model. *Communications in Statistics: Theory and Methods*, 38:1009–1018.

Chen, L. and Sun, J. 2010. A multiple imputation approach to the analysis of interval-censored failure time data with the additive hazards model. *Computational Statistics & Data Analysis*, 54:1109–1116.

Chen, M., Tong, X., and Sun J. 2007. The proportional odds model for multivariate interval-censored failure time data. *Statistics in Medicine*, 26:5147–5161.

Chen, M., Tong, X., and Sun J. 2009. A frailty model approach for regression analysis of multivariate current status data. *Statistics in Medicine*, 28:3424–3436.

Cook, R.J., Zeng, L., and Lee, K.-A. 2008. A multistate model for bivariate interval-censored failure time data. *Biometrics*, 64:1100–1109.

Cox, D.R. 1972. Regression models and life-tables (with discussion). *Journal of the Royal Statistical Society, Series B*, 34:187–220.

De Gruttola, V.G. and Lagakos, S.W. 1989. Analysis of doubly-censored survival data, with application to AIDS. *Biometrics*, 45:1–11.

Dehghan, M.H. and Duchesne, T. 2011. A generalization of Turnbull's estimator for nonparametric estimation of the conditional survival function with interval-censored data. *Lifetime Data Analysis*, 17:234–255.

Deng, D., Fang, H., and Sun, J. 2009. Nonparametric estimation for doubly censored failure time data. *Journal of Nonparametric Statistics*, 21:801–814.

Dinse, G.E. and Lagakos, S.W. 1983. Regression analysis of tumor prevalence data. *Applied Statistics*, 32:236–248.

Fay, M.P. and Shaw, P.A. 2010. Exact and asymptotic weighted logrank tests for interval censored data: The interval R package. *Journal of Statistical Software*, 36:1–34.

Fay, M.P. and Shih, J.H. 2012. Weighted logrank tests for interval censored data when assessment times depend on treatment. *Statistics in Medicine*, 31:3760–3772.

Feng, Y., Ma, L., and Sun, J. 2013. Empirical analysis of interval-censored failure time data with linear transformation models. *Communications in Statistics–Theory and Methods*. In press.

Fine, J.P. and Gray, R.J. 1999. A proportional hazards model for the subdistribution of a competing risk. *Journal of the American Statistical Association*, 94:496–509.

Finkelstein, D.M. 1986. A proportional hazards model for interval-censored failure time data. *Biometrics*, 42:845–854.

Frydman, H. and Liu, J. 2013. Nonparametric estimation of the cumulative intensities in an interval censored competing risks model. *Lifetime Data Analysis*, 19: 79–99.

Frydman, H. and Szarek, M. 2010. Estimation of overall survival in an 'illness-death' model with application to the vertical transmission of HIV-1. *Statistics in Medicine*, 29:2045–2054.

Goggins, W.B. and Finkelstein, D.M. 2000. A proportional hazards model for multivariate interval-censored failure time data. *Biometrics*, 56:940–943.

Gómez, G., Oller, R., Calle, M.L., and Langohr, K. 2009. Tutorial on methods for interval-censored data and their implementation in R. *Statistical Modelling*, 9:259–297.

Groeneboom, P. 2012. Likelihood ratio type two-sample tests for current status data. *Scandinavian Journal of Statistics—Theory and Applications*, 39:645–662.

Groeneboom, P., Jongbloed, G., and Witte, B.I. 2010. Maximum smoothed likelihood estimation and smoothed maximum likelihood estimation in the current status model. *Annals of Statistics*, 38:352–387.

Groeneboom, P., Jongbloed, G., and Witte, B.I. 2012a. A maximum smoothed likelihood estimator in the current status continuous mark model. *Journal of Nonparametric Statistics*, 24:85–101.

Groeneboom, P., Jongbloed, G., and Witte, B. 2012b. Smooth plug-in inverse estimators in the current status continuous mark model. *Scandinavian Journal of Statistics*, 39:15–33.

Groeneboom, P., Maathuis, M.H., and Wellner, J.A. 2008a. Current status data with competing risks: Consistency and rates of convergence of the MLE. *The Annals of Statistics*, 36:1031–1063.

Groeneboom, P. Maathuis, M.H., and Wellner, J.A. 2008b. Current status data with competing risks: Limiting distribution of the MLE. *The Annals of Statistics*, 36:1064–1089.

Groeneboom, P. and Wellner, J.A. 1992. *Information Bounds and Nonparametric Maximum Likelihood Estimation*. DMV Seminar, Band 19, Birkhauser, New York.

Heller, G. 2011. Proportional hazards regression with interval censored data using an inverse probability weight. *Lifetime Data Analysis*, 17:373–385.

Hens, N., Wienke, A., Aerts, M., and Molenberghs, G. 2009. The correlated and shared gamma frailty model for bivariate current status data: An illustration for cross-sectional serological data. *Statistics in Medicine*, 28:2785–2800.

Huang, J. 1996. Efficient estimation for the proportional hazards model with interval censoring. *The Annals of Statistics*, 24:540–568.

Huang, J., Lee, C., and Yu, Q. 2008. A generalized log-rank test for interval-censored failure time data via multiple imputation. *Statistics in Medicine*, 27:3217–3226.

Huang, J. and Wellner, J.A. 1997. Interval censored survival data: A review of recent progress. In Lin, D. and Fleming, T. (Eds.) *Proceedings of the First Seattle Symposium in Biostatistics: Survival Analysis*. Springer-Verlag, New York.

Huang, J., Zhang, Y., and Hua, L. 2012. Consistent variance estimation in interval-censored data. In Chen, D., Sun, J., and Peace, K. (Eds.) *Interval Censored Time-to-Event Data: Methods and Applications* (pp. 233–268). Chapman & Hall/CRC, Boca Raton, FL.

Jewell, N.P. and van der Laan, M.J. 1995. Generalizations of current status data with applications. *Lifetime Data Analysis*, 1:101–110.

Jewell, N.P., van der Laan, M., and Henneman, T. 2003. Nonparametric estimation from current status data with competing risks. *Biometrika*, 90:183–197.

Ji, S., Peng, L., Cheng, Y., and Lai, H. 2012. Quantile regression for doubly censored data. *Biometrics*, 68:101–112.

Joly, P., Gerds, T.A., Qvist, V., Commenges, D., and Keiding, N. 2012. Estimating survival of dental fillings on the basis of interval-censored data and multi-state models. *Statistics in Medicine*, 31:1139–1149.

Jonker, M.A., Bhulai, S., Boomsma, D.I., Ligthart, R.S.L., Posthuma, D., and van der Vaart, A.W. 2009. Gamma frailty model for linkage analysis with application to interval censored migraine data. *Biostatistics*, 10:187–200.

Kalbfleisch, J.D. and Prentice, R.L. 2002. *The Statistical Analysis of Failure Time Data*. Wiley, New York.

Keiding, N. 1991. Age-specific incidence and prevalence: A statistical perspective (with discussion). *Journal of the Royal Statistical Society, Series A*, 154:371–412.

Kim, Y. and Jhun, M. 2008. Cure rate model with interval censored data. *Statistics in Medicine*, 27:3–14.

Kim, Y.J. 2013. Regression analysis of bivariate current status data using a multi-state model. *Communications in Statistics—Simulation and Computation*, DOI:10.1080/03610918.2012.705937.

Kim, Y.J., Kim, J., Nam, C.M., and Kim, Y.N. 2012. Statistical analysis of dependent current status data. In Chen, D., Sun, J., and Peace, K. (Eds.) *Interval Censored Time-to-Event Data: Methods and Applications* (pp. 113–148). Chapman & Hall/CRC, Boca Raton, FL.

Komàrek, A. and Lesaffre, E. 2007. Bayesian accelerated failure time model for correlated interval-censored data with a normal mixture as error distribution. *Statistica Sinica*, 17:549–569.

Komàrek, A. and Lesaffre, E. 2008. Bayesian accelerated failure time model with multivariate doubly interval-censored data and flexible distributional assumptions. *Journal of the American Statistical Association*, 103:523–533.

Kor, C.T., Cheng, K.F., and Chen, Y.H. 2013. A method for analyzing clustered interval-censored data based on Cox's model. *Statistics in Medicine*, 32:822–832.

Lam, K.F., Wong, K.Y., and Zhou, F. 2013. A semiparametric cure model for interval-censored data. *Biometrical Journal*, DOI: 10.1002/bimj.201300004.

Larson, M.G. 1984. Covariate analysis of competing risks models with log-linear models. *Biometrics*, 40:459–469.

Lee, T.C.K., Zeng, L., Thompson, D.J.S., and Dean, C.B. 2011. Comparison of imputation methods for interval censored time-to-event data in joint modelling of tree growth and mortality. *Canadian Journal of Statistics*, 39:438–457.

Li, C. and Fine, J.P. 2013. Smoothed nonparametric estimation for current status competing risks data. *Biometrika*, 100:173–187.

Li, J. and Ma, S. 2010. Interval-censored data with repeated measurements and a cured subgroup. *Journal of the Royal Statistical Society, Series C*, 59:693–705.

Li, L., Watkins, T., and Yu, Q. 1997. An EM algorithm for estimating survival functions with interval-censored data. *Scandinavian Journal of Statistics*, 24:531–542.

Li, Z. and Nan, B. 2011. Relative risk regression for current status data in case-cohort studies. *Canadian Journal of Statistics*, 39:557–577.

Lin, D.Y., Oakes, D., and Ying, Z. 1998. Additive hazards regression with current status data. *Biometrika*, 85:289–298.

Lin, X. and Wang, L. 2010. A semiparameteric probit model for case 2 interval-censored failure time data. *Statistics in Medicine*, 29:972–981.

Liu, H. and Shen, Y. 2009. A semiparametric regression cure model for interval-censored data. *Journal of the American Statistical Association*, 104:1168–1178.

Ma, S. 2009. Cure model with current status data. *Statistica Sinica*, 19:233–249.

Ma, S. 2011. Additive risk model for current status data with a cured subgroup. *Annals of the Institute of Statistical Mathematics*, 63:117–134.

Maathuis, M. and Hudgens, M. 2011. Nonparametric inference for competing risks current status data with continuous, discrete or grouped observation times. *Biometrika* 98:325–340.

Maathuis, M.H. and Wellner, J.A. 2008. Inconsistency of the MLE for the joint distribution of interval censored survival times and continuous marks. *Scandinavian Journal of Statistics*, 35:83–103.

MacKenzie, K. and Peng, D. 2013. Interval-censored parametric regression survival models and the analysis of longitudinal trials. *Statistics in Medicine*, 32:2804–2822.

Martinussen, T. and Scheike, T.H. 2002. Efficient estimation in additive hazards regression with current status data. *Biometrika*, 89:649–658.

McKeown, K. and Jewell, N.P. 2010. Misclassification of current status data. *Lifetime Data Analysis*, 16:215–230.

McKeown, K. and Jewell, N.P. 2011. Current status observation of a three-state counting process with application to simultaneous accurate and diluted HIV test data. *The Canadian Journal of Statistics*, 39:475–487.

McMahan, C.S., Wang, L., and Tebbs, J.M. 2013. Regression analysis for current status data using the EM algorithm. *Statistics in Medicine*, DOI: 10.1002/sim.5863.

Murphy, S.A. and van der Vaart, A.W. 1999. Observed information in semi- parametric models. *Bernoulli*, 5:381–412.

Murphy, S.A. and van der Vaart, A.W. 2000. On profile likelihood. *Journal of the American Statistical Association*, 95:449–465.

Nielsen, J. and Parner, E.T. 2010. Analyzing multivariate survival data using composite likelihood and flexible parametric modeling of the hazard functions. *Statistics in Medicine*, 29:2126–2136.

Oller, R. and Gómez, G. 2012. A generalized Fleming and Harringtons class of tests for interval-censored data. *The Canadian Journal of Statistics*, 40:501–516.

Oller, R., Gómez, G., and Calle, M.L. 2004. Interval censoring: Model characterizations for the validity of the simplified likelihood. *The Canadian Journal of Statistics*, 32:315–326.

Oller, R. Gómez, G., and Calle, M.L. 2007. Interval censoring: Identifiability and the constant-sum property. *Biometrika*, 94:61–70.

Pan, W. 1999. Extending the iterative convex minorant algorithm to the Cox model for interval-censored data. *Journal of Computational and Graphical Statistics*, 8:109–120.

Pawel, K. and Leon, B. 2012. Linear regression modeling of interval-censored survival times based on a convex piecewise-linear criterion function. *Biocybernetics and Biomedical Engineering*, 32:69–78.

Peto, R. and Peto, J. 1972. Asymptotically efficient rank invariant test procedures. *Journal of the Royal Statistical Society, Series A*, 135:185–207.

Robertson, T., Wright, F.T., and Dykstra, R. 1988. *Order Restrict Statistical Inference*. John Wiley, New York.

Saly Rosas, V.G. and Hughes, J.P. 2011. Nonparametric and semiparametric analysis of current status data subject to outcome misclassification. *Statistical Communications in Infectious Diseases*, 3:Article 7.

Schoenfeld, D.A., Rajicic, N., Ficociello, L.H., and Finkelstein, D.M. 2011. A test for the relationship between a time-varying marker and both recovery and progression with missing data. *Statistics in Medicine*, 30:718–724.

Silvapulle, M.J. and Sen, P.K. 2005. *Constrained Statistical Inference*. Wiley, New York.

So, Y., Johnston, G., and Kim, S.H. 2010. Analyzing interval-censored survival data with SAS software. In *Proceedings of the SAS Global Forum 2010 Conference*, Seattle, WA. SAS Institute Inc., Cary, NC.

Sun, J. 1999. A nonparametric test for current status data with unequal censoring. *Journal of the Royal Statistical Society, Series B*, 61:243–250.

Sun, J. 2004. Statistical analysis of doubly interval-censored failure time data. In Balakrishnan, N. and Rao, C.R. (Eds.) *Handbook of Statistics: Advances in Survival Analysis*. Elsevier, North Holland.

Sun, J. 2006. *The Statistical Analysis of Interval-Censored Failure Time Data*. Springer, New York.

Sun, J. and Kalbfleisch, J.D. 1993. The analysis of current status data on point processes. *Journal of the American Statistical Association*, 88:1449–1454.

Sun, J. and Li, J. 2013. Interval censoring. In Klein, J., van Houwelingen, H., Ibrahim, J., and Scheike, T. (Eds.) *Handbook of Survival Analysis*, Chapter 18. Chapman & Hall/CRC, Boca Raton, FL.

Sun, J. and Shen, J. 2009. Efficient estimation for the proportional hazards model with competing risks and current status data. *Canadian Journal of Statistics*, 37:592–606.

Sun, J., Sun, L., and Zhu, C. 2007. Testing the proportional odds model for interval-censored data. *Lifetime Data Analysis*, 13:37–50.

Sun, J., Zhao, Q., and Zhao, X. 2005. Generalized log rank tests for interval-censored failure time data. *Scandinavian Journal of Statistics*, 32:49–57.

Sun, L., Wang, L., and Sun, J. 2006. Estimation of the association for bivariate interval-censored failure time data. *Scandinavian Journal of Statistics*, 33:637–649.

Titman, A.C. 2014. A pool-adjacent-violators type algorithm for non-parametric estimation of current status data with dependent censoring. *Lifetime Data Analysis*, 20:444–458.

Tong, X., Chen, M.H., and Sun, J. 2008. Regression analysis of multivariate interval-censored failure time data with application to tumorigenicity experiments. *Biometrical Journal*, 50:364–374.

Turnbull, B.W. 1976. The empirical distribution with arbitrarily grouped censored and truncated data. *Journal of the Royal Statistical Society, Series B*, 38:290–295.

Wang, C., Sun, J., Sun, L., Zhou, J., and Wang, D. 2012. Nonparametric estimation of current status data with dependent censoring. *Lifetime Data Analysis*, 18:434–445.

Wang, L. and Dunson, D. 2011. Semiparametric Bayes proportional odds models for current status data with under-reporting. *Biometrics*, 67:1111–1118.

Wang, L., Lin, X., and Cai, B. 2012a. Bayesian semiparametric regression analysis of interval-censored data with monotone splines. In Chen, D., Sun, J., and Peace, K. (Eds.) *Interval Censored Time-to-Event Data: Methods and Applications* (pp. 149–165). Chapman & Hall/CRC, Boca Raton, FL.

Wang, L., Sun, J., and Tong, X. 2008. Efficient estimation for the proportional hazards model with bivariate current status data. *Lifetime Data Analysis*, 14:134–153.

Wang, L., Sun, J., and Tong, X. 2010. Regression analysis of case II interval-censored failure time data with the additive hazards model. *Statistica Sinica*, 20:1709–1723.

Wang, W. and Ding, A.A. 2000. On assessing the association for bivariate current status data. *Biometrika*, 87:879–893.

Wang, X., Chen, M.H., and Yan, J. 2013. Bayesian dynamic regression models for interval censored survival data with application to children dental health. *Lifetime Data Analysis*, 19:297–316.

Wang, X., Sinka, A., Yan, J., and Chen, M.-H. 2012b. Bayesian inference of interval-censored survival data. In Chen, D., Sun, J., and Peace, K. (Eds.) *Interval Censored Time-to-Event Data: Methods and Applications* (pp. 167–196). Chapman & Hall/CRC, Boca Raton, FL.

Wellner, J.A and Zhan, Y. 1997. A hybrid algorithm for computation of the nonparametric maximum likelihood estimator from censored data. *Journal of the American Statistical Association*, 92:945–959.

Wen, C.C. 2012. Cox regression for mixed case interval-censored data with covariate errors. *Lifetime Data Analysis*, 18:321–338.

Wen, C.C. and Chen, Y.H. 2012. Conditional score approach to errors-in-variable current status data under the proportional odds model. *Scandinavian Journal of Statistics*, 39:635–644.

Wen, C.C. and Chen, Y.H. 2013. Assessing age-at-onset risk factors with incomplete covariate current status data under proportional odds models. *Statistics in Medicine*, 32:2001–2012.

Wen, C.C., Huang, S.Y.H., and Chen, Y.H. 2011. Cox regression for current status data with mismeasured covariates. *Canadian Journal of Statistics*, 39:73–88.

Wen, C.C. and Lin, C.T. 2011. Analysis of current status data with missing covariates. *Biometrics*, 67:760–769.

Xiang, L., Ma, X., and Yau, K.K. 2011. Mixture cure model with random effects for clustered interval-censored survival data. *Statistics in Medicine*, 30:995–1006.

Yang, Y. and Nair, V.N. 2011. Parametric inference for time-to-failure in multistate semi-Markov models: A comparison of marginal and process approaches. *Canadian Journal of Statistics*, 39:537–555.

Yavuz, A.Z. and Lambert, P. 2011. Smooth estimation of survival functions and hazard ratios from interval-censored data using Bayesian penalized B-splines. *Statistics in Medicine*, 30:75–90.

Zhang, B. 2012. Regression analysis for current status data. In Chen, D., Sun, J., and Peace, K. (Eds.) *Interval Censored Time-to-Event Data: Methods and Applications* (pp. 91–112). Chapman & Hall/CRC, Boca Raton, FL.

Zhang, B., Tong, X., and Sun, J. 2009. Efficient estimation for the proportional odds model with bivariate current status data. *Far East Journal of Theoretical Statistics*, 27:113–132.

Zhang, M. and Davidian, M. 2008. "Smooth" semiparametric regression analysis for arbitrarily censored time-to-event data. *Biometrics*, 64:567–576.

Zhang, Y., Hua, L., and Huang, J. 2010. A spline-based semiparametric maximum likelihood estimation method for the cox model with interval-censored data. *Scandinavian Journal of Statistics*, 37:338–354.

Zhang, Z. 2009. Linear transformation models for interval-censored data prediction of survival probability and model checking. *Statistical Modelling*, 9:321–343.

Zhang, Z. and Sun, J. 2010. Interval censoring. *Statistical Methods in Medical Research*, 19:53–70.

Zhang, Z., Sun, L., Sun, J., and Finkelstein, D.M. 2007. Regression analysis of failure time data with informative interval censoring. *Statistics in Medicine*, 26:2533–2546.

Zhang, Z. and Zhao, Y. 2013. Empirical likelihood for linear transformation models with interval-censored failure time data. *Journal of Multivariate Analysis*, 116:398–409.

Zhao, Q. and Sun, J. 2004. Generalized log-rank test for mixed interval-censored failure time data. *Statistics in Medicine*, 23:1621–1629.

Zhao, X., Zhao, Q., Sun, J., and Kim, J.S. 2008. Generalized log-rank tests for partly interval-censored failure time data. *Biometrical Journal*, 50:375–385.

Zhu, L., Tong, X., and Sun, J. 2008. A transformation approach for the analysis of interval-censored failure time data. *Lifetime Data Analysis*, 14:167–178.

Section IV

Multiple Comparisons in Clinical Trials

13

Introduction to Multiple Test Problems, with Applications to Adaptive Designs

Jeff Maca, Frank Bretz, and Willi Maurer

CONTENTS

13.1 Introduction

Multiple hypothesis testing is becoming an ever more prominent issue in clinical development. Much of this stems from the idea that sponsors want to learn as much as possible from every clinical trial. This leads to clinical trials with multiple objectives to be studied. These objectives could include comparing multiple doses, multiple endpoints, and multiple analyses such as interim analyses or futility analyses. The complexities of designs with multiple objectives have caused regulatory concerns as well (ICH 1998; CHMP 2002). One other trial design that will often lead to multiplicity issues is an adaptive trial design. An adaptive design will, by design, have an interim analysis in which decisions can be made and, depending on the type of adaptation, there could be a selection between which doses to continue, what is the primary hypothesis, or primary population of interest among others. In order to ensure the validity of the trial, these issues have to be addressed in the statistical design.

In this chapter, we will focus on the general concepts behind multiplicity and the consequences of not properly addressing them. Some of the basic techniques for correcting for multiplicity will be introduced, and the relative merits of each type of technique will be described. These concepts will be further discussed in the adaptive trial design setting. The causes of multiplicity in adaptive designs as well as techniques to ensure the statistical validity is maintained will be discussed. A case study will be used to illustrate how these methods could be used in practice.

13.2 General Multiplicity Concepts

Multiplicity issues arise when there are multiple hypotheses that must be tested simultaneously. There are many sources of multiplicity, which can be seen in clinical development programs. For example, a confirmatory study might include more than one dose to be compared to a common control. There could also be different subpopulations in which the treatment effect would be studied. Many confirmatory studies also include different endpoints, which allow the sponsor to best display the overall beneficial effects of the treatment. In all these cases, it would be desired to ensure that the overall type I error rate is maintained across all comparisons. That would imply that the maximum probability of rejecting any true null hypothesis is bounded by α. If no adjustments are made to the testing strategy, the type I error rate will be inflated when more than one test is performed. As the number of tests increases, the inflation will be more pronounced (see Figure 13.1).

Along with the type I error rate inflation, there is also a bias associated with multiplicity. For example, if we look at the maximum effect size of

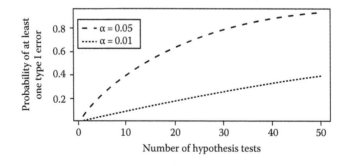

FIGURE 13.1
Effect of number of tests on type I error rate, under independence.

multiple treatment groups, this estimate will be positively biased. For example, suppose that a study had four treatment groups, and all four of these treatment groups had no effect (i.e., a mean effect of zero). However, once the study is completed, the treatment group with the highest positive estimate would have a 50% chance of being over one standard deviation larger than zero (Bretz et al. 2010). Therefore, we must understand the bias, and be able to account for it when making decisions.

The first avenue for reducing multiplicity concerns is to simply try to reduce the number of hypotheses that will be tested. This can be accomplished by focusing on the hypotheses that are truly important, and answer the most critical questions for the study. However, as stated previously, it might not be possible to reduce the test problem to a single hypothesis as sponsors want to be able to make more valid scientific claims for each study, and thus, multiplicity techniques must be implemented.

13.2.1 Error Rate Definitions

Multiplicity methodologies aim at controlling the type I error rates for the entire set of hypotheses that is being tested. However, it should be clear what that statement means. If we assume that there are m different hypotheses to be tested, then the family-wise error rate (FWER) can be defined as: FWER = P(reject at least one true null hypothesis) (Hochberg and Tamhane 1987; Dmitrienko et al. 2009). With this definition for the FWER, there are two ways for which it could be considered to be controlled. The first is the control of the FWER in the weak sense, in which the FWER $\leq \alpha$ under the global null, or in other words, if all m null hypotheses are true. Since this is only one part of the overall null space for the entire family of hypotheses, it is not considered an adequate prerequisite for a multiplicity methodology. Instead, a better one is the control of the FWER in the strong sense. This occurs when FWER $\leq \alpha$ under any configuration of true/false nulls. The family-wise error is also referred to as the overall error rate.

As an example, if a study had m different doses, which would be compared to placebo, then a procedure controlling the type I error rate in the weak sense would only protect the type I error rate if the drug had no effect on any of the doses. However, it would not protect the level if one or more doses had an effect. If the level was protected in the strong sense, then the probability of incorrectly concluding a dose had an effect would be less than α, no matter how many doses did or did not have an effect. Strong control of the type I error rate is considered mandatory in confirmatory clinical trials.

13.2.2 Single Step and Stepwise Methods

Multiple comparison techniques can generally be broken up into two categories, which is *single step* versus *stepwise* procedures. A procedure is classified as a single step procedure if the conclusion about any hypothesis is not affected by the conclusion on any of the others. Many techniques such as the Bonferroni, Dunnett among others fall into this category. If the conclusion on any hypothesis can be affected by the outcome of others, then it would be referred to as a stepwise method. Procedures such as Holm and Hochberg would fall into this category.

13.2.3 General Multiple Comparison Techniques

The simplest method for type I error rate control would be the *Bonferroni method*, whereby the significance level is divided (usually equally) among all the hypothesis tests. For example, if there are four tests to be performed at an overall level $\alpha = 0.05$, then each of the four tests would be tested using $\alpha = 0.05/4 = 0.0125$. Although it is the simplest adjustment to make, it is also the most conservative, since it is applicable irrespective of the correlation between the test statistics. If the correlation between the test statistics is known, for example, in the case of many to one comparisons between several treatments and a control, the adjustments of the single step Dunnett test can be used. The respective adjusted significance levels are slightly larger than those of the Bonferroni test. Another method that would be more powerful than the Bonferroni method is the Simes test procedure for a global hypothesis (Simes 1986). For this test, the m individual p-values are first ordered from smallest to largest: $p_{(1)} \leq p_{(2)} \leq \cdots \leq p_{(m)}$, and then the global null hypothesis would be rejected if $p_{(j)} \leq j\alpha/m$ for at least one j. Under this procedure, the type I error rate is controlled when there is a positive dependency among the test statistics.

A further extension and more powerful method than single step Bonferroni method is the stepwise extension to the *Holm* procedure (Holm 1979), which allows the nonrejected hypotheses to be tested at a higher significance level when other hypotheses have already been rejected. Thus, once a hypothesis has been rejected, it is removed from the set of hypotheses to be tested, and a Bonferroni test is performed on the remaining tests. As in the given example, if one hypothesis, say H_1, was rejected at the Bonferroni adjusted level of 0.0125, then the remaining three hypotheses, H_2, H_3, and H_4 could now be tested at $\alpha = 0.05/3 = 0.0167$. In general, for hypothesis $H_{(1)}, H_{(2)}, \ldots, H_{(m)}$ associated with the ordered p-values, the Holm procedure would be implemented as follows:

- If $p_{(1)} \leq \alpha/m$, then reject $H_{(1)}$, and continue, otherwise stop
- If $p_{(2)} \leq \alpha/(m-1)$, then reject $H_{(2)}$, and continue, otherwise stop
- $\cdots$
- If $p_{(m)} \leq \alpha$, then reject $H_{(m)}$

A stepwise extension of the Simes procedure is known as *Hochberg*'s method (CHMP 2002). As with the Holm procedure, p-values are ordered from

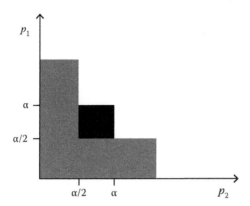

FIGURE 13.2
Comparison of the rejection regions for the Bonferroni and Simes procedures.

smallest to largest: $p_{(1)} \leq p_{(2)} \leq \cdots \leq p_{(m)}$. To implement the Hochberg procedure, the following sequential procedure can be used:

- If $p_{(m)} \leq \alpha$, then reject $H_{(1)}, H_{(2)}, \ldots, H_{(m)}$, and stop, otherwise,
- If $p_{(m-1)} \leq \alpha/2$, then reject $H_{(1)}, H_{(2)}, \ldots, H_{(m-1)}$, and stop, otherwise,
- $\cdots$
- If $p_{(1)} \leq \alpha/m$, then reject $H_{(1)}$.

The rejection regions for the Bonferroni and Simes tests in the case where there are two hypotheses can be seen in Figure 13.2. The darkest shaded region shows the area in which Simes test would be able to reject, whereas the Bonferroni methods could not. Note that the significance levels of the Hochberg procedure are slightly less than those of the Simes test for testing the global null hypothesis, but that the Hochberg procedure does provide test decisions for the m individual null hypotheses, as opposed to the Simes test. Also, the Hochberg procedure is uniformly more powerful than the Holm procedure, which in turn is uniformly more powerful than the single step Bonferroni approach. However, although Hochberg is more powerful than Holm, in order for the type I error rate to be maintained, the test statistics must be positively dependent or independent.

13.2.4 Closed Testing Principle

A key principle that is used in multiple testing is the closure principle (Marcus et al. 1976). This principle allows one to reject a single null hypothesis by looking at every possible intersection of that hypothesis with any other available hypotheses. If all intersection hypothesis can be rejected at level α, then the original hypothesis can also be rejected at the same level α, with the

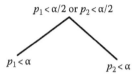

$p_1 < \alpha/2$ or $p_2 < \alpha/2$

$p_1 < \alpha$

$p_2 < \alpha$

FIGURE 13.3
Schematic diagram of closed testing procedure.

type I error rate being maintained. For example, if there are two hypotheses to be tested, H_1 and H_2, then H_1 can be rejected if first the test of $H_1 \cap H_2$ can be rejected at level α, followed by the test for H_1 at level α. The same procedure would hold also for H_2. Therefore, this procedure can be used to simultaneously test H_1 and H_2, protecting the overall type I error rate. Combining this concept with the Bonferroni approach, the Holm procedure can be seen in a formulation of closed testing, by the following schematic diagram (Figure 13.3).

In this diagram, the intersection hypothesis is first tested using the Bonferroni method whereby each hypothesis is tested at the $\alpha/2$ level, and if successful, the individual hypotheses can be tested at the full level α.

13.2.5 Union-Intersections Tests and Intersection-Union Tests

In some settings where there are multiple hypotheses in a clinical trial, the primary objective of the trial would be to have at least one of the tests to be significant. For example, if the study has more than one dose that is being tested, an objective of the study would be to show that at least one dose can be considered effective. These types of situations would lead to using the *union-intersection* testing strategy (Roy 1953). In these situations, the single individual hypotheses might be of less concern, as compared to rejecting the overall null of no effect. An example of a test of this form would be a Dunnett test comparing treatments to a common control.

A methodology that takes the opposite approach is the *intersection-union* test (Berger 1982). This test would assume that the objective would not be met if any of the component hypotheses are not considered significant. Therefore, all components must be considered significant to meet the overall objective. Examples of this type could be if there are multiple endpoints, which must be shown to be significant, such as coprimary endpoints, in order to demonstrate efficacy.

13.3 Introduction to Adaptive Designs

Adaptive designs can be defined as those designs that use accruing data to make changes in an ongoing trial according to a prespecified plan.

These designs can potentially offer great efficiencies for clinical development by allowing for corrections in trial assumptions, or combining studies and/or phases into a single trial. However, once the trial design is changed based on interim data, potential biases can arise, which must be accounted for in the final analysis (CHMP 2007; FDA 2010). In some situations, such as a design using an interim sample size reassessment, the bias would result in an inflation of the overall type I error rate. This can be addressed by modifying the test statistic appropriately to account for the bias. Two commonly used methods are Fisher's combination test (Bauer and Köhne 1994), and the inverse normal methodology (Lehmacher and Wassmer 1999). To use these methods, the patients are separated into two groups of patients enrolled prior to and after the interim analysis. The test statistic would be calculated for both groups separately. For the inverse normal method, the final test statistic

$$Z_f = w_1 \Phi^{-1}(1 - p_1) + w_2 \Phi^{-1}(1 - p_2)$$

is standard normally distributed, where p_1 and p_2 are the p-values from the hypothesis tests from the two groups of patients, Φ^{-1} is the inverse of the standard normal function, and $w_1^2 + w_2^2 = 1$. The weights w_1 and w_2 generally reflect the information fraction for each of the two groups. Properly using these methods would correct and maintain the desired type I error rate. In cases where the interim analysis involves a selection of a treatment, population, etc., bias would result from the fact that only the most promising hypothesis is being pursued. In these situations, the bias can be addressed using multiple testing techniques, as well as adaptive combination tests techniques.

One example of multiple hypotheses that occur in adaptive designs would be that of a treatment selection study (Maca et al. 2006). It should be noted that although selecting a dose is the most common adaptation being made in these types of designs, it could also be desired to choose between treatment regimens or modes of application among others. For simplicity, we will assume that it is a dose, which is being selected. With these studies, the efficacy of the doses would be evaluated at a prespecified time point during the study, and the dose(s) that is considered the most promising would be continued in the study. Further enrollment would be restricted only to the chosen dose(s). This would be in contrast to a traditional design, where the treatment selection and confirmation would be accomplished in two separate studies. A pictorial representation of this design can be seen in Figure 13.4.

Although in this example, there would only be one hypothesis that would be tested at the conclusion of the study, that is, the selected treatment versus placebo, there would have originally been three hypotheses. If the final testing strategy ignores the fact that the final treatment had been selected out of three possible candidate treatments, then the test would inflate the overall type I error rate.

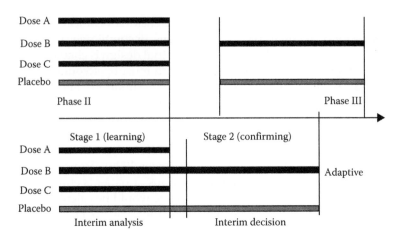

FIGURE 13.4
Pictorial representation of two-stage adaptive designs.

Another adaptive design that can make a type of selection at interim would be what is referred to as an enrichment design (Simon 2004; Temple 2005; FDA 2012). In these types of designs, the decision at interim would not involve the treatment dose but rather to select the primary population for which a final claim will be made. A subpopulation would usually be defined a priori, often defined by a biomarker or other demographic trait. The decision that would be made at interim would then be to either pursue the overall population or focus only on the subpopulation. As in the previous design, even though only one hypothesis would be of interest at the conclusion of the study, the final analysis must account that this hypothesis was selected as the most promising from two hypotheses.

13.4 Multiple Testing Methodologies in Adaptive Designs

In the examples given in the previous section, adjustments must be made in the final testing strategy to account for the multiple hypotheses, which were implicit in the design. The simplest and most straightforward adjustment that could be made would be to use a single step method such as the Bonferroni adjustment specified in Section 13.2.2. This adjustment would simply divide the significance level among all of the hypotheses, including those that are not tested at the final analysis. There is also an added benefit to this type of adjustment, in that the type I error rate is controlled whether only one or more than one hypothesis is tested at the final stage. This flexibility allows one to decide at the interim analysis the number of doses, to be continued in the second stage, without affecting the testing strategy. Although this method could be considered the easiest to implement, it also has the disadvantage that no other adaptations can be made during the interim analysis, such as sample size reestimation.

13.4.1 Closed Testing Procedures in Adaptive Designs

An alternative approach would be to use a closed testing procedure (see Section 13.2.4) that would allow more powerful multiplicity methods, such as Simes' or Dunnett's methods, to be used. In the treatment selection example, the closed testing procedure would imply that in order to reject a hypothesis for a single treatment, every hypothesis that contains that treatment must also be rejected. For example, if there were two treatments and a control, one could define H_1 to be the hypothesis for the pair-wise comparison of Treatment 1 to control, H_2 for the pairwise comparison of Treatment 2 to control, and finally H_{12} to be the global test of either Treatment 1 or Treatment 2 being better than control. Therefore in this case, in order to claim significance for Treatment 1, we must reject the hypothesis H_1. According to the closure principle, this would be accomplished by first testing H_{12} at level α. If significant, then H_1 would also be tested at level α. However, in order to maintain the validity of the final test, the test statistic must account for the selection bias arising from the fact that the most promising treatment can be selected at interim. This can be accomplished by using combination functions that will combine the data obtained prior to the adaption, and the data obtained after the adaption. The strategy is to keep the data from the two stages of the trial separate and then combine them at the conclusion of the trial. As seen in Section 13.3, this can be accomplished by combining the stagewise p-values to create a single test statistics for the hypothesis. The idea of combining p-values from independent data to produce a single test was originally considered by Fisher, generally referred to as Fisher's combination test, and was first described for use in the adaptive designs world by Bauer and Köhne (1994). This methodology was further improved on using the inverse normal method of combining p-values, where p-values from independent sets of data can be combined to a single p-value:

$$C(p_1, p_2) = 1 - \Phi\left(\sqrt{t_0} \times \Phi^{-1}(1 - p_1) + \sqrt{1 - t_0} \times \Phi^{-1}(1 - p_2)\right)$$

Here, p_1 and p_2 can be the multiplicity adjusted p-values obtained from, for example, the Dunnett or Simes method. The weight, t_0, is prespecified and is typically the fraction of data included in the first stage, which is

$$t_0 = \frac{n_1}{n_1 + n_2}$$

Using a combination test, the hypotheses that form the closed testing procedure can be tested using these combined p-values. It should be noted that for each intersection hypothesis, an adjusted p-value must be computed and used in the combination function. If a treatment has been dropped from the study at interim, the second stage p-value can be assumed to be 1; see Bretz et al. (2006) for more details.

13.4.2 Case Study of Multiplicity in an Adaptive Design

As an example of a treatment selection design, consider a study in which three doses (low, medium, and high) are compared to placebo. It is planned to have 75 patients in each treatment group available at interim, which would have completed the primary endpoint. Assume that at interim, the best treatment is selected and continued into the second stage, together with placebo. Another 75 patients will be enrolled in the one selected treatment group as well as placebo. At the interim analysis, the unadjusted p-values for the three pairwise comparisons against placebo are assumed to be $p_L = 0.23$, $p_M = 0.018$, and $p_H = 0.08$. Assume that it is decided to pursue only the high dose, and enrollment will only continue in the high dose and placebo. For this example, the Simes procedure will be used to test the intersection hypotheses using the Simes' adjusted p-values

$$q = \min \frac{S}{i} * p_{(i)}$$

where

    S is the number of doses in the intersection hypothesis
    $p_{(i)}$ are the ordered p-values

In the example, the adjusted p-value for the global test, $H_{L,M,H}$ for the first stage only would be

$$q_{1,LMH} = \min \left(\frac{3}{1} \times 0.08, \frac{3}{2} \times 0.18, \frac{3}{3} \times 0.23 \right) = 0.23$$

Since only the high dose is present in the second stage, the adjusted p-value for this hypothesis would simply be the p-value seen for this hypothesis in the second stage, which we assume to be $q_{2,LMH} = 0.01$.

The use of the inverse normal combination function allows us to now calculate the p-value for the global test for the entire study using the values of $q_{1,LMH}$ and $q_{2,LMH}$, that is, $C(q_{1,LMH}, q_{2,LMH}) = 0.015$. Thus, at the one-sided significance level of $\alpha = 0.025$, the global test H_{LMH} can be rejected. Similarly, one would test the two pairwise intersection hypotheses H_{LH} and H_{MH}, resulting in the combination functions of $C(q_{1,LH}, q_{2,LH}) = C(q_{1,MH}, q_{2,MH}) = 0.0094$ for both tests. Since both of these could be rejected at the $\alpha = 0.025$ level, the test of interest, namely H_H, can be performed, which results in $C(q_{1,H}, q_{2,H}) = 0.0042$. Therefore, the test for the high dose versus the placebo would be rejected at the one sided $\alpha = 0.025$ level.

13.5 Conclusions

The issue of multiple comparisons in the design and analysis of clinical trials has been seen to be of great importance in clinical development. Not accounting and adjusting for the biases, which can arise with multiple comparisons could cause misleading results or inaccurate conclusions.

In this chapter, several methods have been briefly introduced, which could be used to address these concerns. The simplest methods of single step methods, which although more conservative, are the most straightforward to implement. More powerful stepwise methods are further extensions to single step methods and include methods such as Holm and Hochberg. More based on graphical approaches that are tailored to structured clinical trial objectives are reviewed in Chapter 14. Multiplicity concerns can also arise in the adaptive designs framework, where biases stemming from adaptations made during the trial must be addressed using multiple comparison techniques. Using these techniques along with typical adaptive design techniques, such as combining data across multiple stages using combination functions, will ensure the validity of the trial is maintained and the objectives of the trial can be met.

References

Bauer, P. and Köhne, K. (1994). Evaluation of experiments with adaptive interim analysis. *Biometrics*, 50:1029–1041.

Berger, R. L. (1982). Multi-parameter hypothesis testing and acceptance sampling. *Technometrics*, 24:295–300.

Bretz, F., Hothorn, T., and Westfall, P. (2010). *Multiple Comparisons Using R*. CRC Press, Boca Raton, FL.

Bretz, F., Schmidli, H., Koenig, F., Racine, A., and Maurer, W. (2006). Confirmatory seamless phase II/III clinical trials with hypotheses selection at interim: General concepts (with discussion). *Biometrical Journal*, 48(4):623–634.

Committee for Medical Product for Human Use (CHMP). (2002). Points to consider on "Multiplicity issues in clinical trials."

Committee for Medical Product for Human Use (CHMP). (2007). On methodological issues in confirmatory clinical trials planned with an adaptive design.

Dmitrienko, A., Tamhane, A. C., and Bretz, F. (2009). *Multiple Testing Problems in Pharmaceutical Statistics.* Taylor & Francis, Boca Raton, FL.

FDA. (2010). *Draft Guidance for Industry—Adaptive Design Clinical Trials for Drug and Biologics.* US FDA, Rockville, MD.

FDA. (2012). *Enrichment Strategies for Clinical Trials to Support Approval of Human Drugs and Biological Products.* Draft Guidance. US FDA, Rockville, MD.

Hochberg, Y. and Tamhane, A. C. (1987). *Multiple Comparison Procedures.* Wiley, New York.

Holm, S. (1979). A simple sequentially rejective multiple test procedure. *Scandinavian Journal of Statistics,* 6:65–70.

International Conference on Harmonization (ICH). (1998). Topic E9: Statistical Principles for Clinical Trials.

Lehmacher, W. and Wassmer, G. (1999). Adaptive sample size calculations in group sequential trials. *Biometrics* 55:1286–1290.

Maca, J., Battacharya, S., Dragalin, V., Gallo, P., and Krams, M. (2006). Adaptive seamless Phase II/III designs—Background, operational aspects, and examples. *Drug Information Journal* 40:463–473.

Marcus, R., Peritz, E., and Gabrial, K. R. (1976). On closed testing procedure with special reference to ordered analysis of variance. *Biometrika* 63:655–660.

Roy, S. N. (1953). On a heuristic method for test construction and its use in multivariate analysis. *The Annals of Statistics* 24:220–238.

Simes, R. J. (1986). An improved Bonferroni procedure for multiple tests of significance. *Biometrika* 73:751–754.

Simon, R. J. (2004). An agenda for clinical trials in the genomic era. *Clinical Trials* 1:468–470.

Temple, R. (2005). Enrichment designs: Efficiency in development of cancer treatments. *Journal of Clinical Oncology* 23:4838–4839.

14

Graphical Approaches to Multiple Testing

Frank Bretz, Willi Maurer, and Jeff Maca

CONTENTS

14.1 Introduction

Regulatory guidelines for drug development suggest a strong control of the familywise error rate (FWER), when multiple hypotheses are simultaneously tested in confirmatory clinical trials (ICH, 1998; CHMP, 2002). That is, the probability to erroneously reject at least one true null hypothesis is controlled at a prespecified significance level $\alpha \in (0, 1)$ under any configuration of true and false null hypotheses. A variety of multiple test procedures exist that control the FWER at the designated level α and the underlying theory is well developed (Dmitrienko et al., 2009; Bretz et al., 2010; Westfall et al., 2011). However, confirmatory studies are becoming increasingly more complex and often involve multiple statistical hypotheses that reflect structured clinical study objectives. Typical examples include the simultaneous investigation of multiple doses or regimens of a new treatment, two or more clinical endpoints, several populations, noninferiority and superiority, or any combination thereof. Clinical teams are then faced with the difficult task of structuring these hypotheses to best reflect the clinical study's objectives. This task comprises, but is not restricted to, the identification of the study's primary objective(s), its secondary objective(s), a decision about whether only a single hypothesis is of paramount importance or several of them are equally relevant, the degree of controlling incorrect decisions, etc. In addition, pairs of primary and secondary objectives might be coupled and should thus be investigated hierarchically. For example, in a diabetes trial, a reduction in the patients' body weight may only be of interest if a reduction in the glycated hemoglobin (HbA1c) level is achieved, but two different doses of a treatment are equally relevant contenders for a dosage recommendation of a specific drug. In this case, the hypothesis involving body weight reduction in the low dose is a *descendant* of its *parent'* primary hypothesis (HbA1c level reduction in the low dose).

A variety of standard multiple test procedures are available, such as those by Bonferroni, Holm, Hochberg, and Dunnett. However, these procedures are often not suitable for the advanced structured hypotheses test problems mentioned earlier, because they treat all hypotheses equally and do not address the underlying structure of the test problem. At the same time, great care is advised when using ad hoc extensions of standard test procedures, as they may not control the FWER at the designated level α.

For example, consider a clinical trial comparing two doses of a new compound with placebo for a primary and a secondary endpoint, resulting in four null hypotheses of interest, $H_i, i = 1, \ldots, 4$. Let H_1, H_2 denote the two primary hypotheses (both dose-placebo comparisons for the primary endpoint) and H_3, H_4 the two secondary hypotheses (both dose-placebo comparisons for the secondary endpoint). Consider the following intuitive extension of the Holm procedure: Test H_1 and H_2 with the Holm procedure at level α; if at least one hypothesis is rejected, test the *descendant* secondary hypothesis at level $\alpha/2$. This procedure (and many variants thereof) does not control the FWER at level α. The actual FWER can be up to $3\alpha/2$ in this particular example.

In view of the increasing complexity of confirmatory study designs and objectives, new classes of multiple test procedures have been developed in the past years, such as fixed sequence, fallback, and gatekeeping procedures; see Alosh et al. (2014) for a recent review of advanced multiple test procedures applied to clinical trials. Such procedures reflect the difference in importance as well as the relationship between the various study objectives; see Hommel et al. (2007); Guilbaud (2007); Dmitrienko et al. (2008); Li and Mehrotra (2008); Alosh and Huque (2010); Dmitrienko and Tamhane (2011); Dmitrienko et al. (2011); Kim et al. (2011); Luo et al. (2013), among many others. In this paper, we focus on the graphical approaches proposed by Bretz et al. (2009) and Burman et al. (2009) to construct, visualize, and perform multiple test procedures that are tailored to the structured families of hypotheses of interest. Using graphical approaches, vertices with associated weights denote the individual null hypotheses and their local significance levels. Directed edges between the vertices specify how the local significance levels are propagated in case of significant results. The resulting procedures control the FWER in the strong sense at the designated level α across all hypotheses. Many standard multiple test procedures, including some of the recently developed gatekeeping procedures, can be visualized and performed intuitively using graphical approaches.

In the meantime, graphical methods have been applied to different test problems, such as combined noninferiority and superiority testing (Hung and Wang, 2010; Guilbaud, 2011; Lawrence, 2011), testing of composite endpoints and their components (Huque et al., 2011; Rauch and Beyersmann, 2013), and subgroup analyses (Bretz et al., 2011a). The description of these methods has mostly focused on Bonferroni-based test procedures, although extensions have been proposed that include weighted Simes' or parametric tests (Bretz et al., 2011b; Maurer et al., 2011; Millen and Dmitrienko, 2011). Further methodological extensions have been described for group sequential trials (Maurer and Bretz, 2013b), adaptive designs (Sugitani et al., 2013, 2014; Klinglmueller et al., 2014), families of hypotheses (Kordzakhia and Dmitrienko, 2013; Maurer and Bretz, 2014), and entangled graphical test procedures (Maurer and Bretz, 2013a). Power and sample size considerations to optimize a graphical multiple test procedure for given study objectives were given in Bretz et al. (2011a). Software solutions in

SAS and R were described in Bretz et al. (2011a,b). In the following sections, we describe in detail the graphical approach and illustrate it with the visualization of several common gatekeeping strategies. We also present several case studies to illustrate how the approach can be used in practice. We illustrate the methods using the graphical user interface (GUI) from the gMCP package in R (Rohmeyer and Klinglmueller, 2014), which is freely available on the Comprehensive R Archive Network (CRAN).

14.2 Case Studies

In this section, we introduce three case studies that will be revisited and extended later to illustrate some of the methods described in the sequel. These case studies motivate the need for advanced methods to address multiplicity issues in clinical trials.

14.2.1 Comparing Two Doses with a Control for Two Hierarchical Endpoints

Consider a diabetes trial comparing two doses (low and high) against placebo for two hierarchically ordered endpoints (HbA1c level and body weight), resulting in two levels of multiplicity and four null hypotheses H_1, H_2, H_3, and H_4. In addition to the given family of null hypotheses, clinical considerations often lead to a structured hypotheses test problem subject to certain logical constraints. Assume for our example that HbA1c is more important than body weight. Thus, the four hypotheses are grouped into two primary hypotheses H_1, H_2 (both dose-placebo comparisons for HbA1c) and two secondary hypotheses H_3, H_4 (both dose-placebo comparisons for body weight). Both doses are considered equally important, which rules out a full hierarchy of testing first the high dose and, conditional on its significance, then the low dose. In addition, it is required that a secondary hypothesis is not tested without having rejected the associated primary hypothesis (successiveness property; see Maurer et al. 2011; O'Neill 1997). That is, we consider $\{H_1, H_3\}$ and $\{H_2, H_4\}$ as pairs of parent–descendant hypotheses to reflect the hierarchy among the two endpoints within a same dose. The objective is to test all four hypotheses under strong FWER control while reflecting the clinical considerations mentioned above and without leading to illogical decisions (Hung and Wang, 2009, 2010). Standard multiple comparison procedures, such as those by Bonferroni, Holm, or Dunnett, are not suitable here, because they treat all four hypotheses equally and do not address the underlying structure of the test problem. Instead, one needs to construct test strategies that reflect the complex clinical requirements on the structured hypotheses.

14.2.2 Test Strategies for Composite Endpoints and Their Components

Composite endpoints are defined as a collection of (usually) low-incidence outcomes of interest and often analyzed as time-to-first occurrence of any individual component. They have to be distinguished from multicomponent endpoints, where the individual components are not of interest per se and rather used to assign a score or a responder status to each subject, such as sum scores (e.g., the positive and negative syndrome scale [PANSS] for schizophrenia) or responder definitions (e.g., the American College of Rheumatology scores [ACR-N] to measure change in rheumatoid arthritis symptoms).

One common area of application are cardiovascular (CV) trials, where the composite endpoints consist of events of different types such as CV mortality (*CV death*), stroke, myocardial infarction, and hospitalization for urgent revascularization procedures. The use of such composite endpoints may lead to a reduction of trial size and duration, avoiding at first glance the multiplicity issue arising from testing the individual outcomes. However, interpretation of study findings based on a composite endpoint can be problematic. This is in particular true if a statistically significant treatment effect for the composite endpoint was driven mainly by a *soft* endpoint, such as hospitalization, and treatment effects for the *hard* endpoints are small or even in the opposite direction than that for the composite endpoint. Consequently, for an appropriate interpretation of study results, it is important to analyze the treatment effects for the individual components.

The analysis of individual components is merely descriptive if it is based on reporting mean responses, nominal confidence intervals and p-values or forest plots, together with the findings for the composite endpoint. In this case, no formal claim is intended for the component endpoints. In contrast, it is sometimes of interest to establish an efficacy claim for a key individual component, such as CV death, by testing its corresponding hypothesis, say H_2, after the composite endpoint hypothesis, say H_1, is either rejected or has missed slightly the significance level α. In such cases, multiplicity needs to be formally taken into account. Standard multiple test procedures may not be appropriate again, as they treat both hypotheses H_1 and H_2 as equally important and test them individually regardless of the findings of the other hypothesis. We revisit this case study later and discuss alternative test strategies that can be employed, depending on the underlying trial objectives.

14.2.3 Testing Noninferiority and Superiority for Multiple Endpoints in a Combination Trial

The aim of the Aliskiren Trial of Minimizing OutcomeS for Patients with HEart failuRE (ATMOSPHERE) study (Krum et al., 2011) is to evaluate the effect of both aliskiren and enalapril monotherapy and aliskiren/enalapril combination therapy on CV death and heart failure hospitalization in patients with chronic systolic heart failure, New York Heart Association (NYHA) functional class II–IV symptoms, and elevated plasma levels of B-type natriuretic peptide (BNP).

Patients tolerant to at least 10 mg or equivalent of enalapril will undergo an open-label run-in period where they receive enalapril then aliskiren. Approximately 7000 patients tolerating this run-in period will then be randomized 1:1:1 to aliskiren monotherapy, enalapril monotherapy, or the combination. The primary objectives of ATMOSPHERE are to investigate whether (1) the aliskiren/enalapril combination is superior to enalapril monotherapy in delaying time-to-first occurrence of CV death or heart failure hospitalization and (2) aliskiren monotherapy is superior or at least noninferior to enalapril monotherapy on this endpoint. The secondary objectives are to evaluate whether aliskiren monotherapy and/or the combination of aliskiren/enalapril is superior to enalapril monotherapy in (1) reducing the BNP level from baseline to 4 months and (2) improving the clinical summary score as assessed by the Kansas city cardiomyopathy questionnaire from baseline to 12 months. Other efficacy objectives are analyzed in an exploratory manner. Safety data will be collected during the ongoing trial and be part of the overall assessment.

This is a trial with complex and highly structured objectives. For the primary comparison between the combination of aliskiren and enalapril with enalapril monotherapy, a superiority test will be performed. Further comparisons between aliskiren and enalapril monotherapy include both superiority and noninferiority assessments and will be formally investigated, together with the additional secondary hypotheses. We will illustrate how to use the graphical approach to visualize a suitable test strategy when revisiting this case study.

14.3 Main Approach

In this section, we introduce the core graphical approach, which can be used to construct new and extend existing multiple test procedures. More specifically, we describe the graphical approach to Bonferroni-based sequentially rejective multiple test procedures from Bretz et al. (2009).

14.3.1 Bonferroni-Based Graphical Test Procedures

14.3.1.1 Heuristics

Assume that we are interested in testing m elementary null hypotheses $H_1, \ldots, H_m$, which may include primary, secondary, or any other hypotheses of interest. Let $0 \le \alpha_i \le \alpha, i \in I = \{1, \ldots, m\}$, denote the local significance levels.

That is, the overall significance level α is split across the m hypotheses such that $\sum_{i=1}^{m} \alpha_i \leq \alpha$. Finally, let p_i denote the unadjusted p-value for hypothesis $H_i, i \in I$.

Any multiple test procedure for $H_1, \dots, H_m$ should guarantee strong FWER control at level α, be tailored to the structured trial objectives (as reflected through the elementary hypotheses and their relationships), and have good power. To this end, consider the following heuristic approach. Test the m null hypotheses, each at its local significance level $\alpha_i, i \in I$. If a hypothesis H_i can be rejected at level α_i (i.e., $p_i \leq \alpha_i$), propagate its local level α_i to the remaining, not yet rejected hypotheses according to a prespecified rule. Continue testing the remaining hypotheses with the updated local significance levels, thus possibly leading to further rejections with subsequent further propagation of the local levels. This procedure is repeated until no further hypothesis can be rejected. In Section 14.3.1.2, we show that, after a suitable formalization, the resulting sequentially rejective multiple test procedures indeed control the FWER strongly at level α.

We can visualize the resulting multiple test procedures using the following conventions; see Figure 14.1 that also includes an example for $m = 2$ hypotheses. The m null hypotheses are represented as m weighted nodes, where the weights are given by the local significance levels $\alpha_i, i \in I$. The *α-propagation* rules are determined through weighted, directed edges: The weight associated with a directed edge between any two nodes indicates the fraction of the local significance level at the initial node (tail) that is added to the significance level at the terminal node (head) if the hypothesis at the tail is rejected. We will illustrate the resulting graphical test procedures by revisiting in Section 14.3.2 the case studies from Section 14.2 and by visualizing a variety of common multiple test procedures in Section 14.3.4.

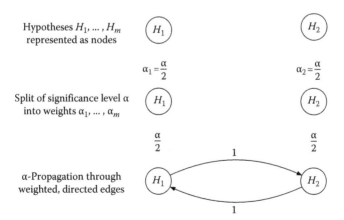

FIGURE 14.1

Conventions for the graphical approach. Right column: Example with $m = 2$ null hypotheses.

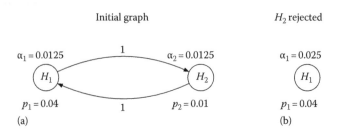

FIGURE 14.2
Numerical example of the graphical Holm procedure with $m=2$ hypotheses and $\alpha=0.025$. (a) Initial graph. (b) Updated graph after rejecting H_2.

As a matter of fact, Figure 14.1 already displays two common multiple test procedures for $m=2$ hypotheses. The middle graph visualizes the Bonferroni procedure: Each of the two hypotheses is tested at level $\alpha/2$; if one of them is rejected (i.e., $p_i \leq \alpha_i = \alpha/2$ for at least one $i \in \{1,2\}$), the other one continues being tested at level $\alpha/2$, since there are no edges connecting the two nodes and therefore, no α-propagation is foreseen. If $\alpha_1 \neq \alpha_2$, we obtain the weighted Bonferroni test for $m=2$. In the bottom graph, however, both nodes are connected and if H_2 (say) is rejected at level $\alpha/2$, its local level is propagated to H_1 along the outgoing edge with weight 1. Consequently, H_1 is tested at updated level $\alpha/2+\alpha/2=\alpha$. This is exactly the Holm procedure, which for $m=2$ rejects the null hypothesis with the smaller p-value if it is less than $\alpha/2$ and continues testing the other hypothesis at level α. Figure 14.2 provides a numerical example of the graphical Holm procedure with $m=2$ hypotheses, $\alpha=0.025$ and unadjusted p-values $p_1=0.04$, $p_2=0.01$ for H_1, H_2, respectively. A numerical example of the graphical Holm procedure with $m=3$ hypotheses, together with the visualization of the updated graphs after each rejection, is given in Bretz et al. (2009).

14.3.1.2 Graphical Approach to Multiple Testing

We now formalize the heuristic approach from Section 14.3.1.1 and state the main result. Let $\alpha=(\alpha_1,\ldots,\alpha_m)$ denote the vector of local significance levels, such that $\sum_{i=1}^{m} \alpha_i \leq \alpha$. Let $\mathbf{G}=(g_{ij})$ denote an $m \times m$ transition matrix with freely chosen entries g_{ij}. The transition weight g_{ij} determines the fraction of the local level α_i that is allocated to H_j in case H_i was rejected. The transition matrix $\mathbf{G}$ thus fully determines the directed edges in a graph. We require the transition weights to satisfy the regularity conditions

$$0 \leq g_{ij} \leq 1, \quad g_{ii}=0 \quad \text{and} \quad \sum_{k=1}^{m} g_{ik} \leq 1 \quad \text{for all } i, \quad j=1,\ldots,m. \tag{14.1}$$

That is, the transition weights should be nonnegative, the sum of the transition weights with tail on a same node is bounded by 1 and there are no elementary loops (edges where head and tail coincide). Based on the

---

Algorithm 14.1 (Weighted Bonferroni tests)

---

0. Set $I = \{1, 2, \ldots, m\}$.

1. Select a $j \in I$ such that $p_j \leq \alpha_j$ and reject H_j; otherwise stop.

2. Update the graph:

$$I \rightarrow I \setminus \{j\}$$

$$\alpha_\ell \rightarrow \begin{cases} \alpha_\ell + \alpha_j g_{j\ell}, & \text{for } \ell \in I, \\ 0, & \text{otherwise}, \end{cases}$$

$$g_{\ell k} \rightarrow \begin{cases} \dfrac{g_{\ell k} + g_{\ell j} g_{jk}}{1 - g_{\ell j} g_{j\ell}}, & \text{for } \ell, k \in I, \ell \neq k, g_{\ell j} g_{j\ell} < 1, \\ 0, & \text{otherwise}. \end{cases}$$

3. If $|I| \geq 1$, go to Step 1; otherwise stop.

---

observed unadjusted p-values p_i, $i \in I = \{1, \ldots, m\}$, we then define a sequentially rejective test procedure through the following algorithm.

The initial levels α, the transition matrix $\mathbf{G}$, and *Algorithm 14.1* define a unique sequentially rejective test procedure that controls the FWER strongly at level α (Bretz et al., 2009). The proof of this statement uses the fact that the graph $(\alpha, \mathbf{G})$ and *Algorithm 14.1* define a closed test procedure with weighted Bonferroni tests for each intersection hypothesis. Moreover, the updated significance levels generated by *Algorithm 14.1* fulfill a mild monotonicity condition that enables the construction of shortcuts for the resulting consonant closed test procedures. For the interested reader, we provide some technical background and relevant references in Section 14.3.5.

Note that sometimes several hypotheses H_i with $p_i \leq \alpha_i$ could be rejected at the same iteration. Step 1 of *Algorithm 14.1* does not specify how to select j in such cases. As a matter of fact, the resulting final set of rejected hypotheses is independent of how the index j is chosen. Thus, for all practical purposes, setting $j = \mathrm{argmin}_{i \in I} p_i / \alpha_i$ in Step 1 of *Algorithm 14.1* is a convenient solution but can be replaced by any other selection rule.

To illustrate the connection between *Algorithm 14.1* and the proposed iterative graphs, consider Figure 14.3 for an example with $m = 3$ hypotheses. For the sake of concreteness, we assume that H_1, H_2 are two primary hypotheses (such as comparing two doses with a control for a primary endpoint) and H_3 a single secondary hypothesis (such as comparing the pooled data from both doses with a control for a secondary endpoint). For the left graph in Figure 14.3, we have $\alpha = \left(\frac{\alpha}{2}, \frac{\alpha}{2}, 0\right)$ and

$$\mathbf{G} = \begin{pmatrix} 0 & \frac{1}{2} & \frac{1}{2} \\ \frac{1}{2} & 0 & \frac{1}{2} \\ 0 & 0 & 0 \end{pmatrix}.$$

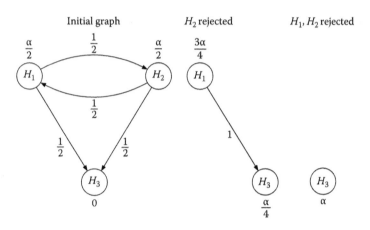

FIGURE 14.3
Example of a graphical multiple test procedure to illustrate *Algorithm 14.1* with $\alpha = 0.025$ and unadjusted p-values $p_1 = 0.015, p_2 = 0.001$, and $p_3 = 0.1$.

Assume $\alpha = 0.025$ and that the unadjusted p-values $p_1 = 0.015, p_2 = 0.001$ and $p_3 = 0.1$ have been observed. Then, $p_2 = 0.001 < 0.0125 = \frac{\alpha}{2} = \alpha_2$, so that $j = 2$ and we can reject H_2 according to Step 1 of *Algorithm 14.1*. Applying the graph iteratively, node H_2 is deleted and the associated significance level α_2 is propagated along the edges with tail on node H_2. In our example, the associated transition weights are $g_{2\ell} = \frac{1}{2}$ for $\ell = 1, 3$ and the updated vector of significance levels becomes $\alpha = \left(\frac{\alpha}{2} + \frac{\alpha}{4}, 0, \frac{\alpha}{4}\right)$ for $\alpha = 0.025$. At the same time, *loose* edges are reconnected and their weights renormalized to satisfy the regularity conditions (14.1), ultimately leading to the middle graph in Figure 14.3. These updates in the graph are essentially reflected and formalized in Step 2 of *Algorithm 14.1*. That is, the transition weight for the edge connecting the two remaining hypotheses H_1 and H_3 becomes

$$g_{13} = \frac{\frac{1}{2} + \frac{1}{2}\frac{1}{2}}{1 - \frac{1}{2}\frac{1}{2}} = 1.$$

Now, we return to Step 1 and test H_1 and H_3 at the updated local significance levels. Since $p_1 = 0.015 < 0.01875 = \frac{3\alpha}{4} = \alpha_1$, we have $j = 1$ and can reject H_1. After updating the graph again, we obtain the right graph in Figure 14.3 and H_3 is tested at level α. Since $p_3 = 0.1 > 0.025 = \alpha = \alpha_3$, we retain H_3 and the procedure stops as no further rejection is possible.

We conclude this example with two remarks. First, assume that both H_1 and H_2 could be rejected in the first iteration, for example, $p_1 = p_2 = 0.001 < \frac{\alpha}{2}$. As mentioned before, Step 1 of *Algorithm 14.1* does not specify how to select j. The test decisions remain the same, whether one first rejects H_1 and proceeds with updating the graph, or starts in the reversed order by first rejecting H_2. In Figure 14.3, if one had decided to first remove the node for H_1, the updated graph would look different to the middle graph. However, the proof in Bretz et al. (2009) ensures that the final test decisions on all three hypotheses remain the same, regardless of the rejection sequence. Second, assume the unadjusted p-values $p_1 = 0.02, p_2 = 0.001$, and $p_3 = 0.005$. Then we could reject both H_2 and H_3 (in this sequence). The last remaining hypothesis H_1 would then be tested at level $\alpha_1 = \frac{3\alpha}{4}$, instead of $\alpha_1 = \alpha$. The reason that the level is not exhausted is that there are no directed edges connecting H_3 with either H_1 or H_2. That is, the third row of $\mathbf{G}$ has only 0 elements and once H_3 is rejected, its level is not further propagated. Figure 14.3 is thus an example of a graph that can be improved immediately (in this case, by inserting edges with tail on H_3). While there are only few results available on optimally selecting the weights for multiple test procedures (Westfall and Krishen, 2001), a sufficient condition for a graph being complete (in the sense that it cannot be improved by adding additional edges) is that the weights of outgoing edges sum to 1 at each node and every node is accessible from any of the other nodes. If $\alpha_i > 0, i = 1, \ldots, k$, this is also a necessary condition for completeness.

14.3.1.3 Adjusted p-Values

Adjusted p-values are often used to describe the results of a multiple test procedure. Adjusted p-values inherently incorporate the structure of the underlying multiple test procedure, and once they are computed, they can be compared directly with the overall significance level α. More formally, an adjusted p-value is the smallest significance level at which a given hypothesis is significant as part of the multiple test procedure (Westfall and Young, 1993).

In the following, we show that a slight modification of *Algorithm 14.1* allows the calculation of m adjusted p-values $p_1^{\text{adj}}, \ldots, p_m^{\text{adj}}$ (Bretz et al., 2009). To this end, we assume for each $J \subseteq I = \{1, \ldots, m\}$ a collection of weights $w_j(J)$ such that $0 \leq w_j(J) \leq 1$ and $\sum_{j \in J} w_j(J) \leq 1$. Setting $\mathbf{w}(I) = (w_1(I), \ldots, w_m(I)) = \left(\frac{\alpha_1}{\alpha}, \ldots, \frac{\alpha_m}{\alpha}\right)$, we obtain the same initial local significance levels when multiplying the weights with α, as used in *Algorithm 14.1*. The remaining weights $w_j(J), J \subsetneq I$, are obtained by updating the graph iteratively in the same way as before.

To illustrate *Algorithm 14.2*, we revisit the numerical example from Figure 14.3. Here, $\mathbf{w}(I) = \left(\frac{1}{2}, \frac{1}{2}, 0\right)$. At the first iteration, $j = 2$ and $p_2^{\text{adj}} = \max\left\{\frac{0.001}{0.5}, 0\right\} = 0.002$. After updating the graph, we obtain at the second iteration $j = 1$ with $p_1^{\text{adj}} = \max\left\{\frac{0.015}{0.75}, 0.002\right\} = 0.02$. Finally, $p_3^{\text{adj}} = \max\{0.1, 0.02\} = 0.1$. Thus, we can reject H_1 for any significance level $\alpha \geq 0.02$, reject H_2 for any $\alpha \geq 0.002$, and reject H_3 for any $\alpha \geq 0.1$. It should be noted

---

Algorithm 14.2 (Adjusted p-values)

0. Set $I = \{1, 2, \ldots, m\}$ and $p_{max} = 0$.

1. Let $j = \arg\min_{i \in I} \frac{p_i}{w_i(I)}$, calculate $p_j^{adj} = \max\left\{\frac{p_j}{w_j(I)}, p_{max}\right\}$, and set $p_{max} = p_j^{adj}$.

2. Update the graph:

$$I \to I \setminus \{j\}$$

$$w_\ell(I) \to \begin{cases} w_\ell(I) + w_j(I)g_{j\ell}, & \text{for } \ell \in I, \\ 0, & \text{otherwise}, \end{cases}$$

$$g_{\ell k} \to \begin{cases} \frac{g_{\ell k} + g_{\ell j}g_{jk}}{1 - g_{\ell j}g_{j\ell}}, & \text{for } \ell, k \in I, \ell \neq k, g_{\ell j}g_{j\ell} < 1, \\ 0, & \text{otherwise}. \end{cases}$$

3. If $|I| \geq 1$, go to Step 1; otherwise stop.

4. Reject all hypotheses H_j with $p_j^{adj} \leq \alpha$.

---

that the test decisions obtained from *Algorithm 14.2* are exactly the same as those from *Algorithm 14.1* for a fixed α.

14.3.1.4 Simultaneous Confidence Intervals

Algorithm 14.1 can also be used to construct compatible simultaneous confidence intervals (Guilbaud, 2008; Strassburger and Bretz, 2008). Consider the one-sided null hypotheses $H_i : \theta_i \leq \delta_i$, $i \in I = \{1, \ldots, m\}$, where θ_i are the parameters of interest (e.g., treatment means or contrasts thereof) and δ_i are prespecified constants (e.g., noninferiority margins). Let $\alpha_j(J) = \alpha w_j(J)$ denote local significance levels with $j \in J \subseteq I$. Further, let $L_i(\gamma)$ denote local (i.e., marginal) lower confidence bounds for θ_i at level $1 - \gamma$ for $i \in I$. Finally, let R denote the index set of hypotheses rejected by a multiple test procedure specified through a graph $(\alpha, \mathbf{G})$.

Following Strassburger and Bretz (2008), lower one-sided confidence bounds for $\theta_1, \ldots, \theta_m$ with simultaneous coverage probability of at least $1 - \alpha$ are given by

$$\bar{L}_i = \begin{cases} \delta_i, & \text{for } i \in R \text{ and } R \neq I, \\ L_i(\bar{\alpha}_i), & \text{for } i \notin R, \\ \max(\delta_i, L_i(\bar{\alpha}_i)), & \text{for } R = I, \end{cases}$$

where $\bar{\alpha}_i = \alpha_i(I \setminus R)$ for $i \notin R \neq I$ denotes the local significance level for H_i in the final graph when applying *Algorithm 14.1*. If all hypotheses can be rejected (i.e., $R = I$), the choice of the local levels $\bar{\alpha}_i = \alpha_i(\emptyset)$ is free. Thus, in order to compute the simultaneous confidence bounds, one only needs to know the set R of rejected hypotheses and the corresponding local levels $\bar{\alpha}_i$ for all indices i of retained hypotheses. Note that if not all hypotheses are rejected, the confidence bounds associated with the rejected hypotheses reflect the test decision $\theta_i > \delta_i$ and the confidence limits associated with the retained hypotheses are the marginal confidence limits at level $\alpha_i(I \setminus R)$. In other words, unless $R = I$, the simultaneous confidence intervals for the rejected hypotheses do not provide any further information beyond the test decision, which limits their use in practice.

To illustrate the calculation of the simultaneous confidence intervals, we revisit the numerical example from Figure 14.3, assuming $\delta_i = 0$ for all i. Recall from Section 14.3.1.2 that both H_1 and H_2 can be rejected at level $\alpha = 0.025$. Thus, $R = \{1, 2\} \subsetneq I$ and $\bar{L}_1 = \bar{L}_2 = 0$. As seen from Figure 14.3, $\bar{\alpha}_3 = \alpha$ and $\bar{L}_3$ reduces to the marginal lower confidence bound at level $1 - \alpha$.

14.3.1.5 Power and Sample Size Calculation

Determining the sample size is an integral part of designing clinical studies. If a single null hypothesis is tested, for example, when comparing a new treatment against a control for a single primary endpoint in a two-armed trial, sample size is usually based on achieving a prespecified power for a specific parameter configuration under the alternative hypothesis. But assume in this example that in addition to the primary endpoint, there is interest in assessing the benefit of the new treatment for a single secondary endpoint. Should then the sample size be determined on achieving a prespecified power to declare both the primary and secondary endpoints significant or just the primary endpoint? The traditional power concept can be generalized in various ways when moving from single to multiple hypotheses test problems. Several authors have introduced a variety of power concepts related to different *win criteria*; see Maurer and Mellein (1988); Xiong et al. (2005); Senn and Bretz (2007); Sozu et al. (2010); Chen et al. (2011); Julious and McIntyre (2012), among many others. It is not always clear which of these criteria is best suited in practice. In this section, we provide some considerations for power and sample size calculation in clinical trials with multiple objectives that are divided into primary and secondary objectives.

Having multiple primary and secondary objectives in a single trial, it becomes important to distinguish between the probability for a successful trial, as driven by the primary objectives, and the power to reject the individual null hypotheses. To reflect these two objectives, we propose to (1) first select a general test strategy addressing the study objectives specified in the protocol, and (2) subsequently fine-tune it based on the importance relationships among the primary and secondary hypotheses, as induced by the study objectives, and the prior assumptions about the effect sizes for all primary and secondary variables (Bretz et al., 2011a). Using a graphical approach, the

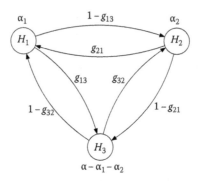

FIGURE 14.4
Graphical multiple test procedure from Figure 14.3 revisited.

initial significance levels α_i and transition weights g_{ij} define the multiple test procedure and thus the sample size under a fixed parameter configuration. Clinical considerations will guide the discussions about the general test strategy and possibly support the weight specifications. Clinical trial simulations are then necessary to further fine-tune the weights, understand the operating characteristics of the resulting multiple test procedure, including its robustness properties against deviations of the initial assumptions, and based on this determine an appropriate sample size.

To illustrate these concepts, consider again the example in Figure 14.3. Once the clinical team has agreed that H_1, H_2 are the two primary and H_3 is the single secondary hypothesis, this leaves five weights to be specified: α_1, α_2 (which determine $\alpha_3 = \alpha - \alpha_1 - \alpha_2$) and g_{13}, g_{21}, g_{32} (which determine $g_{12} = 1 - g_{13}$, $g_{23} = 1 - g_{21}$, $g_{31} = 1 - g_{32}$, respectively); see Figure 14.4. It seems natural to declare this trial successful, if at least one of the two primary hypotheses H_1 and H_2 is rejected. Further significant results are nice to have but not essential for claiming trial success. Consequently, the relevant primary power measure should be the probability for a successful trial, that is, the probability of rejecting either H_1 or H_2 at their local significance levels α_1 and α_2, if they are in fact not true, which in turn implies setting $\alpha_3 = 0$ to maximize that probability.

However, there may be situations where success in H_1 and H_2 is not sufficient. For example, if H_3 is critical to achieve an important label claim, a dream outcome of the trial would be to reject $\{H_1$ and $H_3\}$ or $\{H_2$ and $H_3\}$. That is, the trial is considered truly successful if at least one of the two primary hypotheses H_1 and H_2 is rejected, followed by a rejection of H_3. The sample size needed to achieve this dream outcome is obviously larger than achieving a success in one of the primary hypotheses alone. Thus, it is critical to understand from clinical team discussions what a successful trial truly means and formalize accordingly the objective function for the sample size calculation.

The gMCP package described in Section 14.3.3.2 offers a convenient inter-face to perform power calculations for any graphical test procedures. The user can enter any tailored power function using available buttons from the icon panel. For example, the expression

```
(x[1] || x[2]) && x[3]
```

calculates the probability that the first or (not exclusive) second hypotheses is rejected, together with the third one, where x[i] specifies the proposition that hypothesis H_i is rejected. In addition, any valid R command can be used, such as any(x) to see whether any hypothesis is rejected or all(x[1:3]) to see whether all of the first three hypotheses are rejected.

14.3.2 Case Studies Revisited

In this section, we revisit the case studies from Section 14.2 to illustrate the graphical approach described previously.

14.3.2.1 Comparing Two Doses with a Control for Two Hierarchical Endpoints

We follow the general outline proposed in Section 14.3.1.5 to first identify a suitable general test strategy addressing the study objectives and subse-quently fine-tune it. Such an approach reduces the problem of specifying a suitable multiple test procedure to the determination of a graph $(\alpha, \mathbf{G})$, accounting for the importance relationships among the primary and sec-ondary hypotheses. To start with, we consider the initial levels α_i. In order to reflect the hierarchy between the two endpoints within a given dose, we assign weights 0 to the secondary hypotheses and split the significance level α equally across both doses, since both doses are considered equally important in this case study. Therefore, $\alpha_3 = \alpha_4 = 0$ and $\alpha_1 = \alpha_2 = \frac{\alpha}{2}$.

Next, we determine the initial transition weights g_{ij}. There are in total 12 possible edges to connect any two nodes in order to specify how the signif-icance levels are propagated after a hypothesis has been rejected. However, this number of edges can be reduced substantially by taking clinical consider-ations into account. In our example, the edges $H_3 \rightarrow H_1$ and $H_4 \rightarrow H_2$ receive weight 0 because of the hierarchy among the two endpoints within a given dose. In addition, if successiveness is required (Section 14.2.1), there are no edges $H_1 \rightarrow H_4$, $H_2 \rightarrow H_3$, $H_3 \rightarrow H_4$, and $H_4 \rightarrow H_3$, as otherwise one can always construct examples where for a given dose, the secondary hypothe-sis is rejected, but the associated primary hypothesis is not. This leaves us with the six edges displayed in the left graph of Figure 14.5. As the sum of the weights over all outgoing edges for a given node should not be greater than 1, this gives $g_{41} = g_{32} = 1$ and we are left with two remaining weights g_{12}, g_{21} to be determined. Their choice can be based on different considerations.

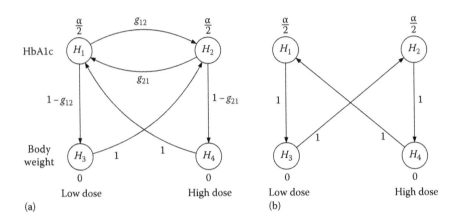

FIGURE 14.5
Graphical visualization of a viable multiple test procedure for the diabetes case study. (a) Graphical visualization of a viable multiple test procedure for the diabetes case study. (b) Resulting graph for the case $g_{12} = g_{21} = 0$.

If, for example, safety is a major concern, one might prefer testing the primary endpoint for the low dose, in case high dose is significant but not safe (leading to a large value of g_{21}). Otherwise, if safety is not of major concern, one might prefer giving the secondary endpoints more weight instead of propagating a large fraction of the significance level to the other primary hypothesis, thus leading to small values for g_{12} and g_{21}. The right graph in Figure 14.5 displays the resulting graph for the extreme case $g_{12} = g_{21} = 0$. If both the primary and secondary endpoints for a given dose are rejected at $\frac{\alpha}{2}$, then the other dose can be tested at a level α. This can be interpreted as a Holm type procedure applied to the families of hypotheses per dose, $\{H_1, H_3\}$ and $\{H_2, H_4\}$. Finally, if no preference for the choice of g_{12} and g_{21} is at hand, based on the available clinical considerations, numerical optimization can be used to determine their values in order to maximize the power of the multiple test procedure (Bretz et al., 2011a), although in many cases, the power does not depend dramatically on the selected weights (Wiens et al., 2013).

14.3.2.2 Test Strategies for Composite Endpoints and Their Components

A simple way of decomposing a composite endpoint is to prioritize its components and employ a hierarchical test procedure. As in Section 14.2.2, let H_1 denote the composite endpoint hypothesis and H_2 the key individual component hypothesis (e.g., for CV death). Accordingly, we test H_1 at level α and only if this is rejected, we proceed with testing H_2, also at level α; see Figure 14.6a. However, a hierarchical test procedure might not always be appropriate, as failure to reject the hypothesis of the composite endpoint H_1 prohibits testing the individual endpoint hypothesis H_2, even if the

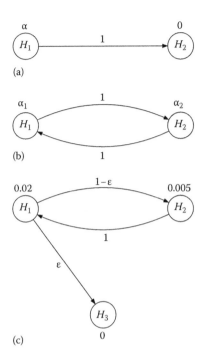

FIGURE 14.6
Test strategies for composite endpoints and their components (a–c details are given in the text).

associated p-value p_2 is very small. One might instead prefer $\alpha_2 > 0$ so that H_2 can be tested even if H_1 is not rejected. Such approach in turn comes only at the cost of reduced power for the composite endpoint, since H_1 will have to be tested at the smaller level $\alpha_1 = \alpha - \alpha_2$, although the overall power to reject at least one of the two hypotheses increases if $\alpha_2 > 0$, in particular if the correlation between the respective test statistics is small. Figure 14.6b displays the graphical visualization of the resulting test procedure, which turns out to be a weighted version of the Holm procedure from Figure 14.2. Note that applying such strategy may lead to difficulties in interpretation if the critical component trends in the wrong direction. In such cases, one may consider introducing a prespecified consistency criterion to ensure clinically meaningful results (Huque et al., 2011). Similarly, Alosh et al. (2014) considered multiple testing strategies that allow testing H_2 as long as the result of testing H_1 establishes a prespecified minimum level of efficacy. Furthermore, the significance level for testing the mortality hypothesis H_2 can be adapted to the findings of testing the composite endpoint hypothesis H_1. Extensions to more complex testing strategies involving multiple components are also possible (Rauch and Beyersmann, 2013).

One question that remains for future discussion is on the rationale of using composite endpoints. Instead of using an aggregate measure that needs to be decomposed in many cases anyway, one may wonder whether the individual components can be treated as multiple primary endpoints (and thus replace the composite endpoint), where study success is defined as at least one of the multiple components being significant. In such cases, the classical multiplicity problem is at hand, since there are multiple chances of winning on at least one endpoint. But having clearly separated endpoints and decision criteria (i.e., win scenarios) might outweigh the disadvantage of having to adjust the endpoints for multiplicity.

More recently, recurrent event data analyses have attracted the interest as an alternative to traditional time-to-first event analyses. For the sake of concreteness, assume two components, one being a recurrent event process (e.g., hospitalization) and the other one being a terminal event process (e.g., CV death). Testing these two components as a composite endpoint does not require multiplicity adjustment and a possible label claim could be *drug X reduces the rate of primary composite endpoint events consisting of CV death and hospitalization*. However, additional testing of the components is highly desirable, but requires multiplicity adjustment (in particular if a claim is sought despite a negative composite endpoint outcome). To this end, let H_1 denote the composite endpoint hypothesis. Further, let H_2 and H_3 denote the two component hypotheses for hospitalization and CV death, respectively, which could be tested using estimates from a joint frailty model (Liu et al., 2004; Cowling et al., 2006). Using the graphical approach, suitable multiple test strategies can be considered and tailored to given clinical study objectives. Figure 14.6c displays a possible sequentially rejective graphical test procedure, which splits the initial significance level of $\alpha = 0.025$ (say) unequally across H_1 and H_2 and tests them using essentially a weighted Holm test. Moreover, CV death is tested only if both the composite and hospitalization endpoints are rejected, which is a consequence of the infinitesimal weight $g_{13} = \epsilon$ chosen for the edge $H_1 \to H_3$ (see Section 14.3.4 for a more formal introduction of such weights). Alternatively, one could choose a truly positive weight g_{13} in order to avoid that a strong effect in CV death is diluted by a lower effect in hospitalization, in which case there would be an increased chance of a nonsignificant composite effect.

14.3.2.3 Testing Noninferiority and Superiority for Multiple Endpoints in a Combination Trial

Recall from Section 14.2.3 that we are interested in comparing the aliskiren/enalapril combination therapy C and aliskiren monotherapy A with the enalapril monotherapy E, resulting in several superiority and noninferiority assessments for the single primary and multiple secondary endpoints. More specifically, we have three single null hypotheses and two subfamilies of null hypotheses for which an appropriate multiple test procedure has to be constructed:

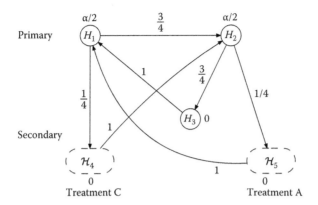

FIGURE 14.7
Graphical illustration of the test strategy for noninferiority and superiority in a combination trial with multiple endpoints.

H_1: superiority of C versus E
H_2: noninferiority of A versus E
H_3: superiority of A versus E
$\mathcal{H}_4$: multiple secondary variables for C versus E
$\mathcal{H}_5$: multiple secondary variables for A versus E

Figure 14.7 visualizes the multiple test procedure resulting from a series of interactive discussion with the clinical team on how to structure the clinically relevant hierarchies reflecting the study objectives. Note that $\mathcal{H}_4$ and $\mathcal{H}_5$ are subfamilies of multiple secondary hypotheses. Appropriate multiple test procedures have to be constructed for each of these two subfamilies but are not discussed here for the sake of brevity.

Some of the considerations leading to Figure 14.7 are as follows. None of the secondary hypotheses in $\mathcal{H}_4$ and $\mathcal{H}_5$ are initially assigned a positive significance level. Any secondary hypothesis can only be tested if at least the associated primary objective has been achieved before. The graph in Figure 14.7 reflects the natural parent–descendant hierarchy within each treatment C and A: $\mathcal{H}_4$ can only be tested if H_1 was significant, and similarly, H_3 and $\mathcal{H}_5$ can only be tested if H_2 was significant. The actual significance levels depend on the test results for the entire family of hypotheses. In particular, the superiority hypothesis H_3 can only be tested if the associated noninferiority hypothesis H_2 was rejected before (any other decision strategy would probably be illogical). Finally, note that if all individual null hypotheses in $\mathcal{H}_4$ or $\mathcal{H}_5$ are rejected, the local significance level is propagated to the hypotheses sequence for the other treatment. If needed, this propagation rule could be modified by displaying the individual secondary hypotheses and specifying propagation rules for each of the nodes. Alternative approaches to propagate significance levels between families of hypotheses are discussed in Section 14.4.4.

14.3.3 Software Implementations

In this section, we review SAS and R implementations of the graphical approach. We revisit some of the previous examples and provide relevant example calls.

14.3.3.1 *SAS*

SAS/IML functions are available for both *Algorithms 14.1* and *14.2*; see Bretz et al. (2011a) and Alosh et al. (2014), respectively. Both functions are straightforward to use and can be applied to any graphical test procedure introduced in Section 14.3.1. To illustrate their functionality, we revisit the diabetes case study from Section 14.2.1. More specifically, we consider the graph in Figure 14.5 with $g_{12} = g_{21} = \frac{1}{2}$. We let $\alpha = 0.025$ and assume the unadjusted p-values $p_1 = 0.1, p_2 = 0.001, p_3 = 0.0001$, and $p_4 = 0.005$, which could be obtained from standard statistical procedures like PROC GLM or PROC MIXED.

In order to execute the mcp function from Bretz et al. (2011a) for *Algorithm 14.1*, we specify

```
h = {0 0 0 0};
a = {0.0125 0.0125 0 0};
w = {0    0.5 0.5 0  ,
     0.5 0   0   0.5,
     0   1   0   0  ,
     1   0   0   0  };
p = {0.1 0.001 0.0001 0.005};
```

where h is a $1 \times m$ vector indicating whether a hypothesis is rejected ($=1$) or not ($=0$), a $1 \times m$ vector α with the initial significance level allocation, w is the $m \times m$ matrix $\mathbf{G}$ with the transition weights, and p is a $1 \times m$ vector with the unadjusted p-values. Calling

```
run mcp(h, a, w, p);
```

we then conclude from the output

	h		
0	1	0	1

that we can reject H_2 and H_4; see Figure 14.8 for the iterated graphs. Note that H_3 cannot be rejected despite its very small p-value, which is consistent with the successiveness requirement stated in Section 14.2.1.

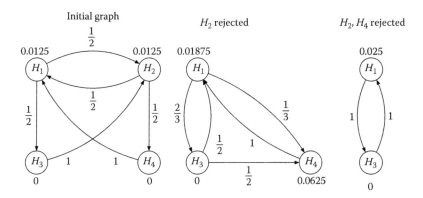

FIGURE 14.8
Numerical example for the diabetes case study with unadjusted p-values $p_1 = 0.1$, $p_2 = 0.001, p_3 = 0.0001$, and $p_4 = 0.005$.

The output for the mcp function also includes the updated significance levels after each iteration as well as the transition matrix after the last iteration, but we omit it here for brevity. Finally, we note that the modified mcp function from Alosh et al. (2014) for *Algorithm 14.2* gives the adjusted p-values $p_1^{\text{adj}} = 0.1, p_2^{\text{adj}} = 0.002, p_3^{\text{adj}} = 0.1$, and $p_4^{\text{adj}} = 0.02$, leading to the same test decisions for $\alpha = 0.025$.

14.3.3.2 R

The gMCP package (Rohmeyer and Klinglmueller, 2014) in R offers a GUI to conveniently construct and perform graphical multiple comparison procedures. The latest version of gMCP is available at CRAN and can be downloaded from http://cran.r-project.org/package=gMCP/; see also the installation instructions at http://cran.r-project.org/web/packages/gMCP/INSTALL.

One way of starting a session is to invoke in R the gMCP package and subsequently call the GUI with

```
> library(gMCP)
> graphGUI()
```

Different buttons are available in the icon panel of the GUI to create a new graph. The main functionality includes the possibility of adding new nodes as well as new edges connecting any two selected nodes. In many cases, the edges will have to be dragged manually in order to improve the readability of the graphs. The associated labels, weights, and significant levels can be edited directly in the graph. Alternatively, the numerical information can be

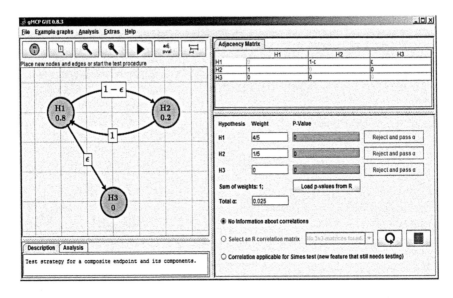

FIGURE 14.9
Screenshot of the GUI from the gMCP package. Left: Display of the graphical Bonferroni-based test procedure from Figure 14.6c. Right: Transition matrix, initial weights, and unadjusted *p*-values.

entered into the transition matrix and other fields on the right-hand side of the GUI. Figure 14.9 provides a screenshot of the GUI from the gMCP package, displaying the graphical Bonferroni-based test procedure from Figure 14.6c.

The gMCP package offers, among other features, the following functionality:

- Create graphs with drag *n* drop or directly in R
- Perform graphical multiple test procedures based on Bonferroni, Simes, and parametric tests
- Compute adjusted *p*-values and simultaneous confidence intervals for Bonferroni-based graphical test procedures
- Perform power calculations based on user-defined objective functions
- Produce S4 objects for the graphs and the corresponding tests
- Export single graphs or produce full reports in LaTeX and PDF/PNG
- Browse through a large collection of example graphs from the literature

We refer to the accompanying vignettes for a complete description of the functionality. A brief illustration of the gMCP package with a cardiovascular clinical trial example is given in Bretz et al. (2011b).

14.3.4 Graphical Visualization of Common Multiple Test Procedures

In the previous sections we already visualized several common multiple test procedures, such as the Bonferroni test in Figure 14.1, the ordinary Holm procedure in Figures 14.1 and 14.2, the weighted Holm procedure in Figure 14.6b, and the hierarchical test procedure in Figure 14.6a and c. In this section, we visualize further Bonferroni-based multiple test procedures from the literature. Other procedures, such as the truncated Holm procedure and k-out-of-n gatekeeping, will be visualized in Section 14.4 after a suitable extension of the graphical approach described so far. While it becomes transparent that many common multiple test procedures can be displayed using the graphical approach, its main advantage is the flexibility to construct and visualize tailored test strategies to address advanced multiplicity issues in clinical trials. This degree of flexibility is demonstrated with the case studies in Sections 14.2 where novel test procedures had to be derived to meet the complex clinical trial objectives.

Figure 14.10a displays the fixed sequence test procedure (Westfall and Krishen, 2001) for $m=3$ hypotheses with $\alpha_1 = \alpha$ and $\alpha_2 = \alpha_3 = 0$. The first hypothesis H_1 is tested at level α. If rejected, its level is propagated to the second hypothesis H_2, and so on. The fixed sequence test procedure controls the FWER in the strong sense and is often used in practice because of its simplicity. However, once a hypothesis is not rejected, no further testing is permitted and care has to be taken when specifying the testing sequence prior to a study.

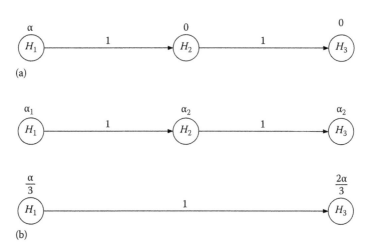

FIGURE 14.10
Visualization of the (a) hierarchical and (b) fallback procedures for $m=3$ hypotheses. Upper graph in (b): Fallback procedure with local levels α_i, $i=1,2,3$. Lower graph in (b): Updated graph after rejecting H_2, with $\alpha_i = \frac{\alpha}{3}$.

The fallback procedure alleviates these concerns. It reserves some fraction of the significance level α for the later hypotheses in the sequence and thus allows one to test those even if the initial hypotheses in the sequence are not rejected; see the upper graph of Figure 14.10b for a visualization, where $\alpha_1 + \alpha_2 + \alpha_3 \leq \alpha$. For illustration, assume equal local significance levels $\alpha_1 = \alpha_2 = \alpha_3 = \frac{\alpha}{3}$ and the three unadjusted p-values $p_1 = 0.015, p_2 = 0.001$, and $p_3 = 0.02$. If we set $\alpha = 0.025$, then $p_2 = 0.001 < 0.0083 = \alpha_2$. Accordingly, we can reject H_2 and propagate its significance level to H_3; see the lower graph in Figure 14.10b. The procedure stops at this stage, since no further hypothesis can be rejected. Alternatively, one can compute the adjusted p-values $p_1^{adj} = 0.045, p_2^{adj} = 0.003$, and $p_3^{adj} = 0.03$ using *Algorithm 14.2*, which lead to the same test decisions.

Note that in Figure 14.10b, the local level α_1 remains unchanged, even if we would have rejected both H_2 and H_3. This is because after rejecting H_3, its local level is not further propagated. Similar to the test procedure from Figure 14.3, the original fallback procedure is not complete and can be improved by adding one or more edges with tail on the last hypothesis in the sequence. Two such improvements are displayed in Figure 14.11. In the upper graph (a), the local significance level α_3 is propagated along the two edges pointing back to H_1 and H_2, where $\gamma = \alpha_2/(\alpha_1 + \alpha_2)$. The resulting test procedure is equivalent to the α-exhaustive extension of the fallback procedure introduced in Wiens and Dmitrienko (2005). Revisiting the numerical example from Figure 14.10b, the adjusted p-values become $p_1^{adj} = 0.03, p_2^{adj} = 0.003$, and $p_3^{adj} = 0.03$ and one observes that p_1^{adj} is now smaller than for the original fallback procedure.

Figure 14.11b displays a second extension by propagating the significance level to the first hypothesis in the hierarchy that has not been rejected so far

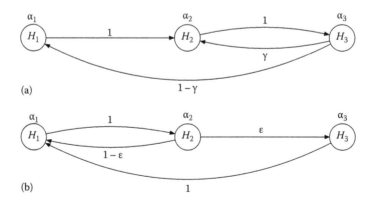

(a)

(b)

FIGURE 14.11
Two extensions of the original fallback procedure for $m = 3$ hypotheses (a,b: details are given in the text).

(Hommel and Bretz, 2008). Here, ϵ denotes an infinitesimally small weight, indicating that the significance level is propagated from H_2 to H_3 only if both H_1 and H_2 are rejected. The motivation for this extension is that H_1 is deemed more important than H_3 as it has been placed earlier in the test sequence. Thus, once H_2 is rejected, its associated significance level is propagated first to H_1 before H_3 gets a second chance to be tested. More formally, when updating the transition weights g_{ij} in the graph according to *Algorithms 14.1* or *14.2*, ϵ is treated as a variable representing some fixed positive real number. For the computation of the updated significance levels α_i (*Algorithm 14.1*) or weights w_i (*Algorithm 14.2*), we let $\epsilon \to 0$. For all real numbers x, we further set $x + \epsilon = x$, $x\epsilon = 0$, $\epsilon^0 = 1$, and for all nonnegative integers k, l,

$$\frac{\epsilon^k}{\epsilon^l} = \begin{cases} 0, & \text{if } k > l, \\ 1, & \text{if } k = l, \\ \infty, & \text{if } k < l; \end{cases}$$

see Bretz et al. (2009) for an introduction of ϵ-edges for the purpose of propagating significance levels between families of hypotheses. With these formalities, we can continue using *Algorithms 14.1* and *14.2* without any changes. In particular, we obtain the adjusted p-values $p_1^{\text{adj}} = 0.0225$, $p_2^{\text{adj}} = 0.003$, $p_3^{\text{adj}} = 0.0225$ for the previous numerical example and can reject all three null hypotheses at level $\alpha = 0.025$.

In a similar way, several gatekeeper procedures can be constructed and visualized using the graphical approach. Applying a serial gatekeeper procedure, all null hypotheses of a family of hypotheses must be rejected before proceeding in the test sequence (Maurer et al., 1995; Bauer et al., 1998; Westfall and Krishen, 2001). Figure 14.6c visualizes an example of a serial gatekeeper procedure with two families $\mathcal{F}_1 = \{H_1, H_2\} \succ \mathcal{F}_2 = \{H_3\}$, where the symbol $\succ$ indicates that all hypotheses of $\mathcal{F}_1$ must be rejected before proceeding with $\mathcal{F}_2$. Figure 14.10a visualizes another example of a serial gatekeeper procedure with $\mathcal{F}_1 = \{H_1\} \succ \mathcal{F}_2 = \{H_2\} \succ \mathcal{F}_3 = \{H_3\}$. In contrast, applying a parallel gatekeeper procedure, at least one null hypothesis of a family must be rejected in order to proceed to the next family Dmitrienko et al. (2003). Consider as an example two families of hypotheses $\mathcal{F}_1 = \{H_1, H_2\} \succ \mathcal{F}_2 = \{H_3, H_4\}$ such that the hypotheses in $\mathcal{F}_2$ are tested only if at least one of the hypotheses in $\mathcal{F}_1$ is rejected. Figure 14.12 displays the parallel gatekeeper procedure from Dmitrienko et al. (2003), which assigns equal levels $\frac{\alpha}{2}$ to the two primary hypotheses H_1, H_2 and levels 0 to the secondary hypotheses H_3, H_4. If H_1 or H_2 is rejected, the corresponding local level $\frac{\alpha}{2}$ is split into half and propagated to H_3 and H_4 as indicated by the directed edges with weights $\frac{1}{2}$. If H_3 (H_4) is rejected in the sequel at its local significance level (either $\frac{\alpha}{2}$ or $\frac{\alpha}{4}$), this level is propagated to H_4 (H_3) as indicated by the directed edges with weights 1. Note that the procedure in Figure 14.12 is neither complete nor successive. That is, it can be improved uniformly by adding

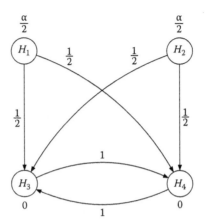

FIGURE 14.12
Graphical visualization of the parallel gatekeeper procedure from Dmitrienko et al. (2003) with two families $\mathcal{F}_1 = \{H_1, H_2\} \succ \mathcal{F}_2 = \{H_3, H_4\}$.

directed edges from $\mathcal{F}_2$ back to $\mathcal{F}_1$ (Bretz et al., 2009), and it does not preserve potential parent–descendant relationships (i.e., the secondary hypotheses can be tested regardless of which primary hypothesis is rejected).

14.3.5 Technical Background

We now provide some technical background for the main results in Section 14.3.1. For the interested reader, we provide references to the literature for further details.

Closed testing offers a general framework to construct powerful multiple test procedures for any finite set of null hypotheses $H_i, i \in I = \{1, \ldots, m\}$ (Marcus and Gabriel, 1976). Using closed testing, we consider the family

$$\mathcal{H} = \left\{ H_J = \cap_{i \in J} H_i \, : \, J \subseteq I, H_J \neq \emptyset \right\},$$

of all nonempty intersection hypotheses H_J. We further prespecify for each $H \in \mathcal{H}$ an α-level test. The resulting closed test procedure rejects $H \in \mathcal{H}$ if all nonempty intersection hypotheses $H' \subseteq H$ are rejected by their corresponding α-level tests. By construction, closed test procedures control the FWER in the strong sense at level α. In what follows, we assume the elementary hypotheses to satisfy the free combination condition, that is, for any subset $J \subseteq I$ the simultaneous truth of $H_i, i \in J$, and falsehood of the remaining hypotheses is possible (Holm, 1979). For related results under restricted combinations, where the previous condition does not hold, we refer to Brannath and Bretz (2010); Maurer and Klinglmüller (2013).

In the following, we assume for each intersection hypothesis H_J a collection of weights $w_i(J)$ such that $0 \leq w_i(J) \leq 1$ and $\sum_{i \in J} w_j(J) \leq 1$ for $J \subseteq I$. These weights quantify the relative importance of the hypotheses H_i included in the intersection H_J. Moreover, we test H_J using a weighted Bonferroni test. That is, an intersection hypothesis H_J is rejected if $p_i \le w_i(J)\alpha = \alpha_i$, for at least one $i \in J \subseteq I$. This defines the class $\mathcal{B}$ of all closed test procedures that use weighted Bonferroni tests for each intersection hypothesis.

A closed test procedures is said to be consonant if the following condition is satisfied: If an intersection hypothesis H_J is rejected, there is an index $i \in J$, such that the elementary hypothesis H_i can be rejected as well (Gabriel, 1969). Consonance is a desirable property as it ensures the rejection of an elementary null after rejecting the global null hypothesis H_I. In particular, consonance enables the construction of sequentially rejective shortcut procedures such that the elementary hypotheses $H_1, \dots, H_m$ are tested in m steps instead that all $2^m - 1$ intersection hypotheses are tested as usually required by closed testing.

Consider now the subclass $\mathcal{S} \subset \mathcal{B}$ of closed weighted Bonferroni tests satisfying the monotonicity condition

$$w_j(J) \le w_j(J') \quad \text{for all } J' \subseteq J \subseteq I \quad \text{and} \quad j \in J'. \tag{14.2}$$

It can be shown that this condition ensures consonance and admits a shortcut procedure (Hommel et al., 2007). That is, any procedure in $\mathcal{S}$ can be performed using the following sequentially rejective algorithm (*Algorithm 14.3*).

What remains is to define a suitable collection of weights $w_i(J), J \subseteq I$, that satisfies the monotonicity condition (14.2) and is tailored to given study objectives. To this end, it can be shown that any initial graph $(\boldsymbol{\alpha}, \mathbf{G})$ applied to a given set of hypotheses (nodes) together with the updating rules in Step 2 of *Algorithm 14.1* generates a unique set of local significance levels. These local significance levels define weighted Bonferroni tests for the corresponding intersection hypotheses satisfying the monotonicity condition (14.2). Applying the shortcut procedure from *Algorithm 14.3* to these local significance levels and the updating rules is then equivalent to *Algorithm 14.1*. In other words, the graphical approach defines a class $\mathcal{G} \subset S$ of sequentially rejective Bonferroni-based closed test procedures, where the vector $\boldsymbol{\alpha}$ specifies a weighted Bonferroni test for the global intersection hypothesis H_I and the transition matrix $\mathbf{G}$ the weighted Bonferroni tests for the $(m - 1)$-way intersection hypotheses $H_{I \setminus \{j\}} = \bigcap_{i \in I \setminus \{j\}} H_i, j = 1, \dots, m$. Note that the graphical

---

Algorithm 14.3 (Shortcut procedures in $\mathcal{S}$)

0. Set $I = \{1, \dots, m\}$.
1. If $\arg\min_{i \in I} \frac{p_i}{w_i(I)} \le \alpha$, reject H_i; otherwise stop.
2. $I \to I \setminus \{i\}$
3. If $|I| \ge 1$, go to Step 1; otherwise stop.

---

approach leads to a specification of m^2 weights. Since closed testing involves $2^m - 1$ intersection hypotheses, consonant Bonferroni-based closed test procedures can be constructed for $m \geq 4$, which are not covered by the graphical approach proposed so far. In Section 14.4, we will extend the graphical approach accordingly to include further test procedures based on weighted Bonferroni and other types of intersection tests.

14.4 Extensions

In this section, we extend the core graphical approach from Section 14.3.1. More specifically, we describe graphical approaches for multiple test procedures using weighted Simes or parametric tests, provide extensions to group sequential trials, and describe entangled graphs that have properties not shared by the approaches considered so far.

14.4.1 Parametric Graphical Test Procedures

The description of the graphical approaches has so far focused on Bonferroni-based test procedures. Following Bretz et al. (2011b); Millen and Dmitrienko (2011), we now discuss how a separation between the weighting strategy and the test procedure facilitates the application of a graphical approach beyond Bonferroni tests.

Graphical weighting strategies are conceptually similar to the graphs proposed in Section 14.3.1. They essentially summarize the complete set of weights for the underlying closed test procedure. Weighted multiple tests can then be applied to the intersection hypotheses $H_J, J \subseteq I = \{1, \ldots, m\}$, such as weighted Bonferroni tests (leading to the graphical test procedures in Section 14.3.1.2), weighted min-p tests accounting for the correlation between the test statistics (this section), or weighted Simes tests (Section 14.4.2). Weighting strategies are formally defined through the weights $w_i(I), i \in I$, for the global null hypothesis H_I and the transition matrix $\mathbf{G} = (g_{ij})$, where the transition weights g_{ij} satisfy the regularity conditions (14.1). We additionally need to determine how the graph is updated once a node is removed. This can be achieved by tailoring *Algorithm 14.1* to the graphical weighting strategies as follows. For a given index set $J \subsetneq I$, let $J^c = I \setminus J$ denote the set of indices that are not contained in J. Then the following algorithm determines the weights $w_j(J), j \in J$. This algorithm has to be repeated for each $J \subseteq I$ to generate the complete set of weights for the underlying closed test procedure.

Similar to what has been stated in Section 14.3.5, the weights $w_j(J), j \in J$, are uniquely determined and do not depend on the sequence in which hypotheses $H_j, j \in J^c$, are removed in Step 1 of *Algorithm 14.4*. We refer to Example 1 in Bretz et al. (2011b) for an illustration of *Algorithm 14.4*.

Algorithm 14.4 (Weighting strategy)

1. Select $j \in J^c$ and remove H_j.

2. Update the graph:

$$I \rightarrow I \setminus \{j\}, J^c \rightarrow J^c \setminus \{j\}$$

$$w_\ell(I) \rightarrow \begin{cases} w_\ell(I) + w_j(I)g_{j\ell}, & \text{for } \ell \in I, \\ 0, & \text{otherwise,} \end{cases}$$

$$g_{\ell k} \rightarrow \begin{cases} \dfrac{g_{\ell k} + g_{\ell j}g_{jk}}{1 - g_{\ell j}g_{j\ell}}, & \text{for } \ell, k \in I, \ell \neq k, g_{\ell j}g_{j\ell} < 1, \\ 0, & \text{otherwise.} \end{cases}$$

3. If $|J^c| \geq 1$, go to Step 1; otherwise $w_\ell(J) = w_\ell(I), \ell \in J$, and stop.

Once a collection of weights has been determined using *Algorithm 14.4*, we can apply any suitable weighted multiple test to the intersection hypotheses $H_J, J \subseteq I$. For example, if for H_J the joint distribution of the p-values $p_j, j \in J$, is known, a weighted min-p test can be defined (Westfall and Young, 1993; Westfall et al., 1998). This test rejects H_J if there exists a $j \in J$ such that $p_j \leq c_J w_j(J)\alpha$, where c_J is the largest constant satisfying

$$P_{H_J}\left(\bigcup_{j \in J}\{p_j \leq c_J w_j(J)\alpha\}\right) \leq \alpha. \tag{14.3}$$

If the p-values are continuously distributed, there is a c_J such that the rejection probability is exactly α. Determination of c_J requires knowledge of the joint null distribution of the p-values and computation of the corresponding multivariate cumulative distribution functions. If the test statistics are multivariate normal or t distributed under the null hypotheses, these probabilities can be calculated using, for example, the mvtnorm package in R (Genz and Bretz, 2009). Alternatively, resampling-based methods may be used to approximate the joint null distribution (Westfall and Young, 1993). If not all, but some of the multivariate distributions of the p-values are known, it is still possible to derive conservative upper bounds of the rejection probability (Bretz et al., 2011b).

It follows immediately from the monotonicity condition (14.2) that the weighted parametric approaches considered here are consonant if

$$c_J w_j(J) \leq c_{J'} w_j(J') \quad \text{for all } J' \subseteq J \subseteq I \quad \text{and} \quad j \in J'. \tag{14.4}$$

If this new monotonicity condition (14.4) is satisfied, a sequentially rejective test procedure similar to the Bonferroni-based graphical tests from Section 14.3.1.2 can be defined.

Algorithm 14.5 (Weighted parametric tests)

---

0. Set $I = \{1, 2, \ldots, m\}$.

1. Choose the maximal constant c_I satisfying (14.3). Select a $j \in I$ such that $p_j \leq c_I w_j(I)\alpha$ and reject H_j; otherwise stop.

2. Update the graph:

$$I \rightarrow I \setminus \{j\}$$

$$w_\ell(I) \rightarrow \begin{cases} w_\ell(I) + w_j(I)g_{j\ell}, & \text{for } \ell \in I, \\ 0, & \text{otherwise,} \end{cases}$$

$$g_{\ell k} \rightarrow \begin{cases} \frac{g_{\ell k} + g_{\ell j}g_{jk}}{1 - g_{\ell j}g_{j\ell}}, & \text{for } \ell, k \in I, \ell \neq k, g_{\ell j}g_{j\ell} < 1, \\ 0, & \text{otherwise.} \end{cases}$$

3. If $|I| \geq 1$, go to Step 1; otherwise stop.

Note that the monotonicity condition (14.4) is often violated in practice when using weighted parametric tests. In such cases, *Algorithm 14.5* no longer applies and one has to go through the entire closed test procedure. That is, the weighting strategies from *Algorithm 14.4* remain applicable, but the connection to a corresponding sequentially rejective test procedure is lost. For a given weighting strategy, however, applying parametric tests exploiting the correlations between the test statistics is uniformly more powerful than the associated Bonferroni-based test procedures from *Algorithm 14.1*.

We conclude this section with an example to illustrate *Algorithm 14.5*. To this end, we consider a simplified version of the ATMOSPHERE study from Section 14.2.3. More specifically, we consider testing noninferiority and superiority for two doses. Assume that H_1, H_2 denote the two noninferiority hypotheses (say, for low and high dose against control) and H_3, H_4 the two superiority hypotheses (for the same two dose-control comparisons).

The left graph in Figure 14.13 visualizes one possible weighting strategy for this example. It is motivated by a strict hierarchy within dose: Superiority will only be assessed if noninferiority was shown previously for a same dose. If for one of the two doses efficacy can be shown for both noninferiority and superiority, the associated weight is propagated to the other dose. A related Bonferroni-based graphical test procedure was used in the diabetes case study in Section 14.3.2.1; see the right graph in Figure 14.5. Other Bonferroni-based graphical approaches for combined noninferiority and superiority testing were investigated by Hung and Wang (2010); Lawrence (2011); Guilbaud (2011).

In the following, we exploit the fact that the correlations between the four test statistics are known. Applying standard analysis-of-variance assumptions with a known common variance, the complete joint distribution is

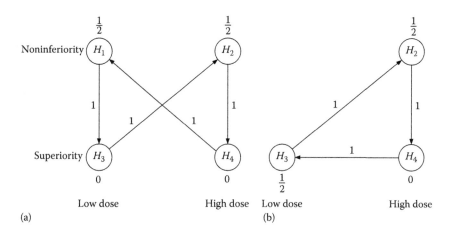

FIGURE 14.13
(a) Weighting strategy to test noninferiority and superiority for two doses. (b) Updated weighting strategy after rejecting H_1.

known and we can apply (14.3), where $\alpha = 0.025$. Note that if $w_j(J) = 0$ for some $j \in J$, the joint distribution degenerates. In our example, it suffices to calculate bivariate or univariate probabilities, where the correlation is determined only by the relative group sample sizes. For simplicity, assume that the group sample sizes are equal. Then the correlation between the noninferiority and superiority tests within a same dose is 1; all other correlations are 0.5. Therefore, $c_J = 1.0783$ for $J = \{1,2\}, \{1,4\}, \{2,3\}$, and $\{3,4\}$; otherwise, $c_J = 1$. Since at most two hypotheses in any intersection can have weight 0.5, condition (14.4) is satisfied and we can apply *Algorithm 14.5*. This leads to a sequentially rejective multiple test procedures, where at each step either bivariate Dunnett z tests or individual z tests are used (Bretz et al., 2011b). This conclusion remains true if the common variance is unknown and Dunnett t tests or individual t tests are used. Note that similar multiple test procedures are immediately applicable to testing for a treatment effect at two different dose levels in an overall population and, if at least one dose is significant, continue testing in a prespecified subpopulation. This could apply to testing, for example, in the global study population and a regional subpopulation or in the enrolled full population and a targeted genetic subpopulation (Bretz et al., 2011b).

To illustrate the procedure, assume the unadjusted p-values $p_1 = 0.01$, $p_2 = 0.02$, $p_3 = 0.005$, and $p_4 = 0.5$. Following *Algorithm 14.5*, we have $p_1 \le c_J w_1(I)\alpha = 0.0135$ and can reject H_1. The update step then leads to the right graph in Figure 14.13. Next, $p_3 \le 0.0135$ and we can reject H_3. This leaves us with H_2, H_4 and the weights $w_2(\{2,4\}) = 1$, $w_4(\{2,4\}) = 0$. Therefore, H_2 is now tested at full level α. Because $p_2 \le \alpha$, we reject H_2 and the procedure stops since H_4 cannot be rejected. These calculations can be reproduced with the gMCP package described in Section 14.3.3.2.

14.4.2 Simes-Based Graphical Test Procedures

Generalizations of the Bonferroni-based graphical test procedures from Section 14.3.1 also apply when the correlations between the test statistics are not exactly known, but certain restriction on them are assumed. In this case, the Simes test is a popular choice. Here, we consider the weighted Simes test introduced by Benjamini and Hochberg (1997), which rejects H_I if for some $j \in I$ $p_j \leq \sum_{i \in I_j} \alpha_i = \alpha \sum_{i \in I_j} w_i$, where $I_j = \{k \in I; p_k \leq p_j\}$. This weighted Simes test reduces to the original (unweighted) Simes test (Simes, 1986) for $w_i = 1/m, i \in I$. The weighted Simes test is conservative if, for example, the test statistics follow a multivariate normal distribution with nonnegative correlations and the tests are one sided (Benjamini and Heller, 2007).

Applying the closure principle, the resulting multiple test procedure rejects $H_i, i \in I$, at level α if for each $J \subseteq I$ with $i \in J$, there exists an index $j \in J$ such that

$$p_j \leq \alpha \sum_{k \in J_j} w_k(J), \tag{14.5}$$

where $J_j = \{k \in J; p_k \leq p_j\}$ (Bretz et al., 2011b). If all weights are equal, this reduces to the Hommel procedure (Hommel, 1988). Although full consonance is generally not available for Simes-based closed test procedures, we can derive a partially sequentially rejective test procedure, which leads to the same test decision as the closed test procedure defined previously. In the following, we assume that the weights are exhaustive, that is, $\sum_{k \in J} w_k(J) = 1$ for all subsets $J \subseteq I$.

Algorithm 14.6 first considers those outcomes that are easy to verify (Steps 1 and 2) or where sequential rejection of the hypotheses is possible (Step 3). Only then one needs to compute for all remaining hypotheses and their

---

Algorithm 14.6 (Weighted Simes tests)

1. If $p_i > \alpha$ for all $i \in I$, stop and retain all m hypotheses.

2. If $p_i \leq \alpha$ for all $i \in I$, stop and reject all m hypotheses.

3. Perform the Bonferroni-based graphical test procedure from Section 14.3.1. Let I_r denote the index set of rejected hypotheses and I_r^c its complement in I. If $|I_r^c| < 3$, stop and retain the remaining hypotheses.

4. If $|I_r^c| \geq 3$ consider the weights $w_i(I_r^c), i \in I_r^c$, and the transition matrix G defined on I_r^c as the new initial graph for the remaining hypotheses. Compute the weights $w_k(J)$ for all $J \subseteq I_r^c$ with *Algorithm 14.4*.

5. Reject $H_i, i \in I_r^c$, if for each $J \subseteq I_r^c$ with $i \in J$ there exists an index $j \in J$ such that $p_j \leq \alpha \sum_{k \in J_j} w_k(J)$.

---

subsets the weights and apply the closed weighted Simes procedure. It can happen though that no hypotheses can be rejected in Steps 2 and 3 and that one has to perform Step 4 with the full set of all m hypotheses. Note that for a given weighting strategy, the Simes-based graphical test procedure is uniformly more powerful than an associated Bonferroni-based procedure from Section 14.3.1. A numerical example applying *Algorithm 14.6* is given in Bretz et al. (2011b).

14.4.3 Graphical Approaches for Group Sequential Designs

We now consider the general situation of testing multiple hypotheses repeatedly in time. More specifically, we extend the scope of the graphical approach to group sequential designs with one or more interim and one final analysis. Under mild monotonicity conditions on the error spending functions, this allows the use of graphical test procedures in group sequential trials in a similar way as described so far.

To this end, we first consider testing a single null hypothesis $H: \theta \leq 0$ at $h - 1$ interim and one final analysis. Following the standard approaches for group sequential designs (Whitehead, 1997; Jennison and Turnbull, 2000; Proschan et al., 2006; Emerson, 2007), we assume (asymptotically) multivariate normal statistics Z_t, $t = 1, \ldots, h$ with $E(Z_t) = \theta\sqrt{I_t}$ and $\text{Cov}(Z_t, Z_{t'}) = \sqrt{I_t/I_{t'}}..., t \leq t'$. Here, I_t denotes the information available at time point t, which is often proportional to the number of patients available up to t and inversely proportional to the standard deviation of the underlying measure of effect. We consider spending functions $a(\gamma, y)$ with information fraction y and significance level $0 < \gamma < 1$ such that $a(\gamma, 0) = 0, a(\gamma, 1) = \gamma$, and $a(\gamma, y) \leq a(\gamma, y')$ for $0 \leq y < y' \leq 1$. For a given time point t, $y_t = I_t/I_{\max}$ and using nominal p-values p_t, we calculate the spent levels as

$$\alpha_t(\gamma) = a\left(\gamma, y_t\right) - a\left(\gamma, y_{t-1}\right) = P\left(\{p_t \leq \alpha_t^*(\gamma)\} \cap \bigcap_{s=1}^{t-1} \{p_s > \alpha_s^*(\gamma)\}\right),$$

where the nominal levels $\alpha_t^*(\gamma)$ serve as the interim decision boundaries. As indicated in Maurer and Bretz (2013b), for many spending functions (including O'Brien-Fleming- and Pocock-type boundaries), it holds for $\gamma' > \gamma$ that for all $t = 1, \ldots, h$

$$\alpha_t(\gamma') \geq \alpha_t(\gamma) \Rightarrow \alpha_t^*(\gamma') \geq \alpha_t^*(\gamma). \tag{14.6}$$

This property is used in the following to ensure the validity of the graphical testing procedure.

We now consider testing m one-sided null hypotheses $H_i, i \in I = \{1, \ldots, m\}$, in a group sequential trial at time points $t = 1, \ldots, h$. For each H_i, define its

---

Algorithm 14.7 (Weighted Bonferroni tests, $h - 1$ interim analyses)

---

0. Set $t = 1$ and $I = \{1, 2, \ldots, m\}$.

1. At interim analysis t compute unadjusted p-values $p_{i,t}$ and nominal significance levels $\alpha_{i,t} = \alpha^*_{i,t}(\alpha w_i(I))$ for $i \in I$.

2. Select a $j \in I$ such that $p_{j,t} \leq \alpha_{j,t}$, reject H_j and go to Step 3. If no such j exists and $t < h$, the trial can be continued with $t \to t + 1$; go to Step 1 in this case, otherwise stop.

3. Update the graph:

$$I \to I \setminus \{j\}$$

$$w_\ell(J) \to \begin{cases} w_\ell(I) + w_j(I)g_{j\ell}, & \text{for } \ell \in I, \\ 0, & \text{otherwise}, \end{cases}$$

$$g_{\ell k} \to \begin{cases} \frac{g_{\ell k} + g_{\ell j}g_{jk}}{1 - g_{\ell j}g_{j\ell}}, & \text{for } \ell, k \in I, \ell \neq k, g_{\ell j}g_{j\ell} < 1, \\ 0, & \text{otherwise}. \end{cases}$$

4. If $|J| \geq 1$, go to Step 1; otherwise stop.

---

spending function $a_i(\gamma, y)$ with spent levels $\alpha_{i,t}(\gamma) = a_i(\gamma, y_t) - a_i(\gamma, y_{t-1})$ and nominal levels $\alpha^*_{i,t}(\gamma)$. As in Section 14.4.1, we separate the weighting strategy from the actual test procedure being employed. Within the framework of closed testing (Section 14.3.5), we reject an intersection hypothesis $H_J, J \subseteq I$, at time point t if $p_{i,t} \leq \alpha^*_{i,t}(\alpha w_i(J))$ for at least one $i \in J$. It then can be shown that the monotonicity conditions on the weights (14.2) and on the spending functions (14.6) ensure sequentially rejective closed group sequential test procedures (Maurer and Bretz, 2013b). In particular, graphical test procedures can be derived by a slight modification of *Algorithm 14.1*.

A numerical example illustrating *Algorithm 14.7* is given in Maurer and Bretz (2013b). The approach mentioned earlier can be generalized to allow more flexibility in the choice of the group sequential boundaries (Xi and Tamhane, 2014a). Extensions of the graphical approach to adaptive group sequential trials with treatment selection and others adaptations at interim are also available; see Sugitani et al. (2013, 2014); Klinglmueller et al. (2014) for details.

14.4.4 Graphical Approaches for Families of Hypotheses

Sometimes the structure for a family of hypotheses is best described by introducing subfamilies of distinct hypotheses. An example of such a situation is given by the ATMOSPHERE case study from Section 14.2.3 and visualized in Figure 14.7. In this example, the hypotheses associated with the secondary objectives are subsumed in two families $\mathcal{H}_4$ and $\mathcal{H}_5$. If in addition, one would consider the remaining individual null hypotheses H_1, H_2, and H_3 also as distinct subfamilies $\mathcal{H}_i = \{H_i\}, i = 1, 2, 3$, then every single node in Figure 14.7 would represent a subfamily. An extended multiple test procedure would then propagate local significance levels between these subfamilies, instead of operating on the individual null hypotheses. More specifically, a local significance level is propagated only if all individual null hypotheses within a subfamily are rejected, followed by an update of the graph according to *Algorithm 14.1* for single null hypotheses.

Such an approach provides an extension of serial gatekeeper procedures to situations with nonhierarchical structures between the subfamilies; see Bauer et al. (1998) for the special case of using a Holm procedures across the subfamilies of hypotheses and any multiple test procedure within each subfamily In Section 14.3.4, we introduced edges with weight ϵ to propagate local significance levels, conditional on rejecting all individual hypotheses in a subfamily. If the multiple test procedures within the subfamilies can also be represented as graphs, an overall multiple test procedure on a staged structure of subfamilies of hypotheses can therefore be visualized and performed as a graphical test procedure as well. In the following, we provide a more general algorithm that is also valid if the test procedures within a subfamily are not necessarily graphical.

Let $\mathcal{H}_i, i \in I = \{1, \ldots, m\}$, denote m families of $k_i \geq 1$ individual null hypotheses $H_{ij}, j = 1, \ldots, k_i$. Further, let $I_i = \{(i, 1), \ldots, (i, k_i)\}$ denote the set of index pairs (i, j) of hypotheses $H_{ij} \in \mathcal{H}_i, i \in I$. Finally, let φ_i denote an α-consistent multiple test procedure defined on $\mathcal{H}_i$, which ensures that any hypothesis in $\mathcal{H}_i$ rejected by φ_i at level α is also rejected at any level $\alpha', \alpha' > \alpha$; see, for example, Hommel and Bretz (2008). This property allows the computation of locally adjusted p-values p_{ij}^*. That is, p_{ij}^* is the smallest significance level at which H_{ij} can be rejected with φ_i. For example, if φ_i denotes the Hochberg procedure for $\mathcal{H}_i$, p_{ij} the unadjusted p-value for H_{ij} and $p_{i(k)}$ the kth ordered p-value within $\mathcal{H}_i$, then $p_{i(k)}^* = \min\left\{1, \min\left[(k_i - k + 1) p_{i(k)}, p_{i(k+1)}^*\right]\right\}$. If locally adjusted p-values p_{ij}^* are available for each hypothesis H_{ij}, then we can define a sequentially rejective test procedure through *Algorithm 14.8*; see also Maurer and Bretz (2014).

A further generalization of the procedure mentioned earlier is described in Kordzakhia and Dmitrienko (2013). They consider separable multiple test procedures φ_i that propagate a certain amount of the level α_i to other subfamilies already after rejecting an individual null hypothesis in $\mathcal{H}_i$. The main difference to *Algorithm 14.8* is that after each rejection of an individual null hypothesis, a certain error fraction is propagated to the other subfamilies according to the transition weights of the graph. The transition weights between the families, however, are updated only if all hypotheses, in a subfamily are rejected, as in *Algorithm 14.8*.

Algorithm 14.8 (Weighted Bonferroni tests on families of hypotheses)

0. Set $I = \{1, 2, \ldots, m\}$ and $I_i = \{(i,j) : j = 1, \ldots, k_i\}, i \in I$.

1. Select a $(i,j) \in I_i, i \in I$, such that $p_{ij}^* \leq \alpha_i$, reject H_{ij} and set $I_i \to I_i \setminus (i,j)$; otherwise stop.

2. If $|I_i| \geq 1$ go to Step 1; otherwise update the graph:

$$I \to I \setminus \{i\}$$

$$\alpha_\ell \to \begin{cases} \alpha_\ell + \alpha_i g_{i\ell}, & \text{for } \ell \in I, \\ 0, & \text{otherwise}, \end{cases}$$

$$g_{\ell k} \to \begin{cases} \frac{g_{\ell k} + g_{\ell i} g_{ijk}}{1 - g_{\ell j} g_{j\ell}}, & \text{for } \ell, k \in I, \ell \neq k, g_{\ell i} g_{i\ell} < 1, \\ 0, & \text{otherwise}. \end{cases}$$

3. If $|I| \geq 1$, go to Step 1; otherwise stop.

14.4.5 Entangled Graphs

The graphical procedures considered so far have no memory in the sense that the origin of the propagated significance level is ignored in subsequent iterations. However, there are clinical trial applications where this property is desirable to reflect the underlying dependence structure of the study objectives. In such cases, it would be desirable that the further propagation of significance levels depends on their origin and thus reflects the grouped parent–descendant structures of the hypotheses.

In the following, we extend the case study from Section 14.2.1 to motivate the need for test procedures with memory. In that case study, assume that both diabetes endpoints HbA1c and body weight are measured in an initial trial period (period 1) and that the trial is continued to a second period (period 2) to investigate potential CV complications by comparing the pooled data from both doses against placebo. Let H_5 denote the additional null hypothesis. With the notation from Section 14.2.1, we have the following requirements:

  i. The primary hypotheses H_1 and H_2 of period 1 are considered to be more important than the period 2 hypothesis H_5, which in turn was considered to be more important than the secondary hypotheses H_3 and H_4.
 ii. Both doses are considered equally important.
iii. A secondary hypothesis should only be rejected if the associated primary hypothesis for the same dose had been rejected.

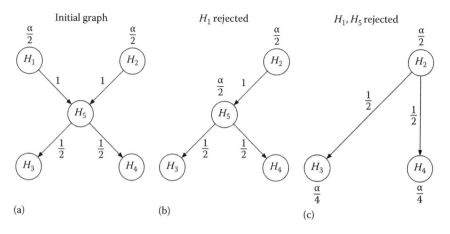

FIGURE 14.14
Graphical test procedure without memory. (a) Initial graph. (b) Updated graph after rejecting H_1. (c) Updated graph after rejecting H_1 and H_5.

We now illustrate that conditions (i) and (iii) cannot be satisfied simultaneously with the graphical test procedures considered so far. Consider the initial graph in Figure 14.14, where for simplicity weights 0 at the nodes are not displayed. Assume that for a given significance level α, we have $p_1 < \alpha/2$ and $p_2 > \alpha$. That is, H_1 can be rejected and its level $\alpha/2$ is propagated to H_5, leading to the middle graph in Figure 14.14. If furthermore $p_5 < \alpha/2$, then H_5 is rejected as well. Its level $\alpha/2$ is halved and propagated to both H_3 and H_4, each of which can now be tested at level $\alpha/4$. This, however, violates condition (iii) mentioned earlier which requires that H_4 should not be tested as long as H_2 is not rejected.

Requirement (iii) can be achieved by defining individual graphs for each parent–descendant relationship and combine them afterward. We can split the initial graph from Figure 14.14 in two separate graphs $(\mathcal{G}_1, \mathcal{G}_2)$, each being defined on the hypotheses H_1 through H_5; see Figure 14.15a. The basic idea is to test each hypothesis according to sum of the significance levels from both individual graphs $\mathcal{G}_1$ and $\mathcal{G}_2$. For example, in Figure 14.15a, we test H_1 at level $\alpha/2 + 0$ and H_2 at level $0 + \alpha/2$. The first step is therefore the same as for the initial graph in Figure 14.14. However, if we assume that H_1 can be rejected, we now update each individual graph using *Algorithm 14.1*. For ease of illustration, Figure 14.15b displays the resulting entangled graph by overlaying the two individually updated graphs $\mathcal{G}_1$ and $\mathcal{G}_2$. The local significance levels are displayed as vectors $(\alpha_{1i}, \alpha_{2i})$ for each hypothesis H_i, unless $\alpha_{1i} = \alpha_{2i} = 0$ (in which case they are omitted for better readability). In the updated graph from Figure 14.15b, H_2 and H_5 can each be tested at level $\alpha/2$. If now again H_5 is rejected, its level $\alpha/2$ is propagated according to the rules for each individual graph, resulting in Figure 14.15c. Note that now H_4 is tested at level $\alpha/2$

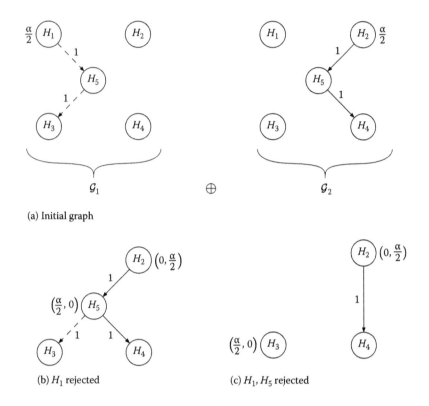

(a) Initial graph

(b) H_1 rejected   (c) H_1, H_5 rejected

FIGURE 14.15

Entangled graphs $(\mathcal{G}_1, \mathcal{G}_2)$ with memory. (a) Initial individual graphs. (b) Updated entangled graph after rejecting H_1. (c) Updated entangled graph after rejecting H_1 and H_5.

originating from H_1, as opposed to the level $\alpha/4$ in Figure 14.14. The example in Figure 14.15 obviously can be improved by propagating further the levels of H_3 and H_4; see Figure 2 in Maurer and Bretz (2013a) for an example.

More formally, entangled graphs and the associated sequentially rejective test procedure can be described as follows. Assume that the structure for m null hypotheses $H_1, \ldots, H_m$ is given by n different parent–descendant relationships. For each of these n relationships, we have tailored structural dependencies between the hypotheses, leading to n individual graphs $\mathcal{G}_1, \ldots, \mathcal{G}_n$. Let $\mathcal{G}_h = (\alpha_h, \mathbf{G}_h), h = 1, \ldots, n$, denote the individual graphs with local significance levels $\alpha_h = (\alpha_{h1}, \ldots, \alpha_{hm})$, such that $\sum_{h=1}^{n} \sum_{i=1}^{m} \alpha_{hi} \leq \alpha$, and $m \times m$ transition matrices $\mathbf{G}_h$. The entries $g_{hij} \in \mathbf{G}_h$ are freely chosen subject to the regularity conditions $0 \leq g_{hij} \leq 1, g_{hii} = 0$, and $\sum_{\ell=1}^{m} g_{hi\ell} \leq 1$ for all $i, j = 1, \ldots, m, h = 1, \ldots, n$.

Each hypothesis $H_i, i = 1, \ldots, m$, is tested at local level $\alpha_i = \sum_{h=1}^{n} \alpha_{hi}$. If for any $j = 1, \ldots, n$, the null hypothesis H_j can be rejected (i.e., $p_j \leq \alpha_j$) then each graph $\mathcal{G}_h$ is updated separately. That is, the node j (i.e., hypothesis H_j)

Algorithm 14.9 (Entangled Bonferroni-based graphs)

0. Set $I = \{1, \ldots, m\}$.
1. Let $\alpha_i = \sum_{h=1}^{n} \alpha_{hi}, i \in I$. Select a $j \in I$ such that $p_j \leq \alpha_j$ and reject H_j; otherwise stop.
2. Update the graph:

$$I \rightarrow I \setminus \{j\}$$

$$\alpha_{h\ell} \rightarrow \begin{cases} \alpha_{h\ell} + \alpha_{hj} g_{hj\ell}, & \text{for } \ell \in I, h = 1, \ldots, n \\ 0, & \text{otherwise,} \end{cases}$$

$$g_{h\ell k} \rightarrow \begin{cases} \frac{g_{h\ell k} + g_{h\ell j} g_{hjk}}{1 - g_{h\ell j} g_{hj\ell}}, & \text{for } \ell, k \in I, \ell \neq k, g_{h\ell j} g_{hj\ell} < 1, h = 1, \ldots, n, \\ 0, & \text{otherwise.} \end{cases}$$

3. If $|I| \geq 1$, go to Step 1; otherwise stop.

---

is removed and the local levels α_h as well as the transition matrices $\mathbf{G}_h$ for the remaining $m - 1$ hypotheses are updated using a modified version of *Algorithm 14.1*; see *Algorithm 14.9*. Once this is achieved, the $m - 1$ remaining hypotheses are tested at the updated significance levels and the previous steps are repeated until no further hypotheses can be rejected.

A formal proof for the validity of this sequentially rejective graphical test procedure is given by Maurer and Bretz (2013a). They also investigated further properties and alternative representations of entangled graphs. In particular, they showed the equivalence between the class of entangled graphs proposed in this section and the default graphs proposed by Burman et al. (2009). The entangled graphs can also be used to visualize gatekeeping procedures using the truncated Holm procedure (Dmitrienko et al., 2008; Strassburger and Bretz, 2008; Maurer and Bretz, 2013a). In the next section, we then show how to use entangled graphs to address the problem of rejecting at least k out of m hypotheses in the context of gatekeeping.

14.4.6 Graphical Approaches for *k*-out-of-*m* Gatekeeper Problems

Multiple test procedures defined by entangled graphs can have properties that a single graph cannot provide. In Section 14.4.5, we have seen how entanglement can create memory. Another property not shared by single graphs in general is the requirement that at least k out of m hypotheses of a primary family of hypotheses should be rejected before secondary hypothesis become testable for $1 \leq k \leq m$. Such a requirement is implicitly given, for example, in the FDA Guidance for Industry on rheumatoid arthritis (FDA, 1999). It states that "... trial results were considered to support a conclusion of effectiveness when statistical evidence of efficacy was shown for at least three of the four measures ..." According to this guideline, the primary objective of a rheumatoid arthritis trial is to demonstrate beneficial effect for at least $k = 3$ out of $m = 4$ primary endpoints.

To discuss the problem formally, let $\mathcal{F}_1$ denote a family of m primary hypotheses and $\mathcal{F}_2$ a family of n secondary hypotheses. We require that $\mathcal{F}_2$ is tested only if at least k of the m primary hypotheses in $\mathcal{F}_1$ have been rejected. For the special case $k = 1$, we can construct multiple test procedures visualized by single graphs as long as there are edges with positive weights connecting each primary hypothesis with at least one secondary hypothesis. For the special case $k = m$, all primary hypotheses have to be rejected before a secondary hypothesis is tested (Maurer et al., 1995). Any FWER controlling multiple test procedure can be used to test $\mathcal{F}_1$ before testing the secondary hypotheses. Such procedures can be constructed with a single graph by using the ϵ-edges introduced in Section 14.3.4. A simple example is the graph displayed in Figure 14.6, where H_3 is only tested after having rejected both H_1 and H_2.

For $2 \leq k \leq m - 1$, it does not seem to be possible to construct a multiple test procedure with a single graph that has the desired k-out-of-m gatekeeping property. It is possible, however, to construct a graph where a particular subset of k primary hypothesis is rejected before a secondary hypothesis can be rejected. In order to allow testing a secondary hypothesis if any subset of primary hypotheses of size k is rejected, we define serial gatekeeping graphs for all $\binom{m}{k}$ subsets of k primary hypotheses and entangle them (Maurer and Bretz, 2013a).

For the sake of concreteness, we revisit the rheumatoid arthritis example discussed earlier and consider the primary hypotheses family $\mathcal{F}_1 = \{H_1, H_2, H_3, H_4\}$ with $m = 4$. Let $k = 3$ and choose a Holm procedure with three hypotheses for each of the four subsets $I_\ell, \ell = 1, \ldots, \binom{4}{3} = 4$. Each of the four subfamilies $\mathcal{F}_{1\ell} = \{H_j; j \in I_\ell\}$ is assigned an initial significance level $\alpha/4$. That level is propagated to $\mathcal{F}_2$ if all three hypotheses are rejected in one of the subfamilies $\mathcal{F}_{1\ell}$. Figure 14.16 visualizes one of the four component graphs with a single secondary hypothesis, that is, $\mathcal{F}_2 = \{H_5\}$. Note that the subgraph on $\{H_1, H_2, H_3\}$ in Figure 14.16 visualizes the Holm procedure at level $\alpha/4$. The other three component graphs are obtained by permuting the indices of the primary hypotheses in $\mathcal{F}_1$. The entangled graph consisting of the four component graphs then has the desired property that at least 3 out of the 4 primary hypotheses have to be rejected before the secondary hypothesis H_5 can be tested. More precisely, the resulting procedure is equivalent to the following test procedure: The ordinary Holm procedure is performed on $\mathcal{F}_1$ at level α until any three of the primary hypotheses are rejected. Then the remaining primary hypothesis can be tested at level $\frac{3}{4}\alpha$. If it cannot be rejected, the secondary hypotheses in $\mathcal{F}_2$ can be tested at level $\frac{1}{4}\alpha$ with any valid multiple test procedure and otherwise at level α.

For large values of m, the number of components graphs becomes difficult to handle. However, if the component graphs are the same up to permutation,

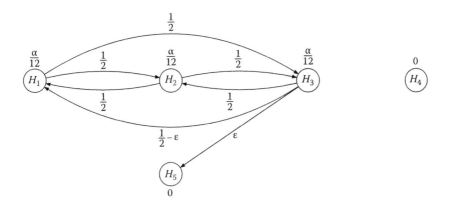

FIGURE 14.16

Graphical visualization of a k-out-of-m gatekeeper procedure with $m=4$ primary hypotheses, $k=3$, and one secondary hypothesis H_5.

it is sufficient to display only one of them, as done in Figure 14.16. In addition, if the component graphs are of a simple structure, the construction method allows one to derive an equivalent sequentially rejective test procedure with the desired properties that can be described in a few sentences or a simple decision tree for any values of k and m.

Another class of k-out-of-m gatekeeper procedures has been proposed by Xi and Tamhane (2014b). This procedures allow one to use Holm, Hochberg, or Hommel tests for the primary hypotheses before proceeding to the secondary hypotheses. The basic idea for their Bonferroni-based k-out-of-m gatekeeper procedure can be described as follows. Let I_1 denote the index set of the m primary hypotheses, I_2 that of the n secondary hypotheses and $I = I_1 \cup I_2$. Consider an index set $J = J_1 \cup J_2 \subseteq I$ where $J_1 \subseteq I_1$ and $J_2 \subseteq I_2$. Let $\alpha_j^*(J_1)$ denote the local significance levels of (any) consonant weighed Bonferroni test on I_1. The local levels $\alpha_j(J)$ of the hypotheses H_j for a weighted Bonferroni test of the intersection hypothesis H_J have the following property: As long as the cardinality $|J_1|$ of J_1 is greater than or equal to k, one sets $\alpha_j(J) = \alpha_j^*(J_1)$ for $j \in J_1$ and $\alpha_j(J) = 0$ for $j \in J_2$. For $|J_1| < k$, one chooses $\alpha_j(J) \leq \alpha_j^*(J_1)$ for $\in J_1$ with inequality for at least one j and $\alpha_j(J) \geq 0$ for $j \in J_2$ with inequality for at least one j. This is possible since the local significance levels satisfy the monotonicity condition (14.2). The resulting closed test procedure then has the k-out-of-m property. The entangled graphs described earlier can represent some but not all of the procedures proposed by Xi and Tamhane (2014b) but remain more flexible to address partially ordered primary and secondary hypotheses.

14.5 Conclusions

In this chapter, we provided an extensive overview of graphical approaches to multiple testing problems that are frequently encountered in clinical studies. The proposed graphical approaches offer the possibility to tailor advanced multiple test procedures to structured families of hypotheses and visualize complex decision strategies in an efficient and easily communicable way while controlling the FWER strongly at a designated significance level α. Many common multiple test procedures can be displayed using the graphical approach, including fixed sequence, fallback, and gatekeeping procedures. The main advantage, however, is the degree of flexibility offered by this approach to meet the given clinical study objectives, as demonstrated by the various case studies in this chapter. The graphical approach covers a broad range applications and extensions, such as the calculation adjusted p-values and simultaneous confidence intervals, the use of weighted Bonferroni, Simes, and parametric tests, the application to group sequential trials, and the creation of entangled graphs for advanced clinical trial applications.

The proposed graphical approach is tailored to confirmatory trials that could potentially serve later as basis for regulatory decision making. The need to control strongly the FWER in this context is clear and mandated by regulatory guidelines (ICH, 1998; CHMP, 2002). Beyond this, multiplicity has a much broader impact that raises challenging problems, which affect almost every decision throughout drug development. Good decision making and reproducibility need to account for multiplicity and might need different solutions at different drug development stages. The details of how to address multiplicity are often not clear-cut and depend on the situation. Thus, on a broader scale, there is a need for statisticians to engage strategically in the related clinical team discussions. There should be a transparent discussion with the medical/commercial colleagues with respect to which endpoints are critically important to approval and label, and which are not as important and should be considered more exploratory. Clinical trials with a rather large number of hypotheses and simple hierarchically ordered test procedures should be avoided. In the end, any methodology is only as good as the business decisions that the teams are making with respect to their identification of important/critical endpoints.

References

Alosh, M., Bretz, F., and Huque, M. (2014). Advanced multiplicity adjustment methods in clinical trials. *Statistics in Medicine*, 33:693–713.

Alosh, M. and Huque, M. (2010). A consistency-adjusted alpha-adaptive strategy for sequential testing. *Statistics in Medicine*, 29:1559–1571.

Bauer, P., Röhmel, J., Maurer, W., and Hothorn, L. (1998). Testing strategies in multi-dose experiments including active control. *Statistics in Medicine*, 17:2133–2146.

Benjamini, Y. and Heller, R. (2007). False discovery rates for spatial signals. *Journal of the American Statistical Association*, 102:1272–1281.

Benjamini, Y. and Hochberg, Y. (1997). Multiple hypothesis testing with weights. *Scandinavian Journal of Statistics*, 24:407–418.

Brannath, W. and Bretz, F. (2010). Shortcuts for locally consonant closed test procedures. *Journal of the American Statistical Association*, 105:660–669.

Bretz, F., Hothorn, T., and Westfall, P. (2010). *Multiple Comparisons Using R*. Taylor & Francis, Boca Raton, FL.

Bretz, F., Maurer, W., Brannath, W., and Posch, M. (2009). A graphical approach to sequentially rejective multiple test procedures. *Statistics in Medicine*, 28:586–604.

Bretz, F., Maurer, W., and Hommel, G. (2011a). Test and power considerations for multiple endpoint analyses using sequentially rejective graphical procedures. *Statistics in Medicine*, 30:1489–1501.

Bretz, F., Posch, M., Glimm, E., Klinglmueller, F., Maurer, W., and Rohmeyer, K. (2011b). Graphical approaches for multiple comparison procedures using weighted Bonferroni, Simes or parametric tests. *Biometrical Journal*, 53:894–913.

Burman, C., Sonesson, C., and Guilbaud, O. (2009). A recycling framework for the construction of Bonferroni-based multiple tests. *Statistics in Medicine*, 28: 739–761.

Chen, J., Luo, J., Liu, K., and Mehrotra, D. (2011). On power and sample size computation for multiple testing procedures. *Computational Statistics & Data Analysis*, 55:110–122.

CHMP (2002). Committee for Medical Product for Human Use (CHMP). Points to consider on "Multiplicity issues in clinical trials". www.ema.europa.eu, accessed July 11, 2014.

Cowling, B., Hutton, J., and Shaw, J. (2006). Joint modeling of event counts and survival times. *Applied Statistics*, 55:31–39.

Dmitrienko, A., Kordzakhia, G., and Tamhane, A. (2011). Multistage and mixture parallel gatekeeping procedures in clinical trials. *Journal of Biopharmaceutical Statistics*, 21:726–747.

Dmitrienko, A., Offen, W., and Westfall, P. (2003). Gatekeeping strategies for clinical trials that do not require all primary effects to be significant. *Statistics in Medicine*, 22:2387–2400.

Dmitrienko, A. and Tamhane, A. (2011). Mixtures of multiple testing procedures for gatekeeping applications in clinical trial applications. *Statistics in Medicine*, 30:1473–1488.

Dmitrienko, A., Tamhane, A., and Bretz, F. (2009). *Multiple Testing Problems in Pharmaceutical Statistics*. Taylor & Francis, Boca Raton, FL.

Dmitrienko, A., Tamhane, A., and Wiens, B. (2008). General multi-stage gatekeeping procedures. *Biometrical Journal*, 50:667–677.

Emerson, S. (2007). Frequentist evaluation of group sequential clinical trial designs. *Statistics in Medicine*, 26:5047–5080.

FDA (1999). U.S. Food and Drug Administration (FDA). Guidance for industry—Clinical development programs for drugs, devices, and biological products for the treatment of rheumatoid arthritis (RA). www.fda.gov, accessed July 11, 2014.

Gabriel, K. (1969). Simultaneous test procedures—Some theory of multiple comparisons. *Annals of Mathematical Statistics*, 40:224–250.

Genz, A. and Bretz, F. (2009). *Computation of Multivariate Normal and t Probabilities.* Springer, Heidelberg, Germany.

Guilbaud, O. (2007). Bonferroni parallel gatekeeping—Transparent generalization, adjusted p-values and short direct proofs. *Biometrical Journal,* 49:917–927.

Guilbaud, O. (2008). Simultaneous confidence regions corresponding to Holm's stepdown procedure and other closed testing procedures. *Biometrical Journal,* 50:678–692.

Guilbaud, O. (2011). Note on simultaneous inferences about non-inferiority and superiority for a primary and a secondary endpoint. *Biometrical Journal,* 53: 927–937.

Holm, S. (1979). A simple sequentially rejective multiple test procedure. *Scandinavian Journal of Statistics,* 6:65–70.

Hommel, G. (1988). A stagewise rejective multiple test procedure based on a modified bonferroni test. *Biometrika,* 75:383–386.

Hommel, G. and Bretz, F. (2008). Aesthetics and power considerations in multiple testing—A contradiction? *Biometrical Journal,* 50:657–666.

Hommel, G., Bretz, F., and Maurer, W. (2007). Powerful short-cuts for multiple testing procedures with special reference to gatekeeping strategies. *Statistics in Medicine,* 26:4063–4073.

Hung, H. and Wang, S. (2009). Some controversial multiple testing problems in regulatory applications. *Journal of Biopharmaceutical Statistics,* 19:1–11.

Hung, H. and Wang, S. (2010). Challenges to multiple testing in clinical trials. *Biometrical Journal,* 52:747–756.

Huque, M., Alosh, M., and Bhore, R. (2011). Addressing multiplicity issues of a composite endpoint and its components in clinical trials. *Journal of Biopharmaceutical Statistics,* 21:610–634.

ICH (1998). *International Conference on Harmonization. Topic E9: Statistical Principles for Clinical Trials.* www.ich.org, accessed July 11, 2014.

Jennison, C. and Turnbull, B. (2000). *Group Sequential Methods with Applications to Clinical Trials.* Chapman and Hall/CRC, Boca Raton, FL.

Julious, S. and McIntyre, N. (2012). Sample sizes for trials involving multiple correlated must-win comparisons. *Pharmaceutical Statistics,* 11:177–185.

Kim, H., Entsuah, R., and Shults, J. (2011). The union closure method for testing a fixed sequence of families of hypotheses. *Biometrika,* 98:391–401.

Klinglmueller, F., Posch, M., and Koenig, F. (2014). Adaptive graph-based multiple testing procedures. *(submitted).*

Kordzakhia, G. and Dmitrienko, A. (2013). Superchain procedures in clinical trials with multiple objectives. *Statistics in Medicine,* 32:486–508.

Krum, H., Massie, B., Abraham, W., and et al. (2011). Direct renin inhibition in addition to or as an alternative to angiotensin converting enzyme inhibition in patients with chronic systolic heart failure: rationale and design of the aliskiren trial to minimize outcomes in patients with heart failure (ATMOSPHERE) study. *European Journal of Heart Failure,* 13:107–114.

Lawrence, J. (2011). Testing non-inferiority and superiority for two endpoints for several treatments with a control. *Pharmaceutical Statistics,* 10:318–324.

Li, J. and Mehrotra, D. (2008). An efficient method for accommodating potentially underpowered primary endpoints. *Statistics in Medicine,* 27:5377–5391.

Liu, L., Wolfe, R., and Huang, X. (2004). Shared frailty models for recurrent events and terminal event. *Biometrics,* 60:747–756.

Luo, X., Chen, G., Ouyang, S., and Turnbull, B. (2013). A multiple comparison procedure for hypotheses with gatekeeping structure. *Biometrika*, 100:301–317.

Marcus, R., Peritz, E., and Gabriel, K. (1976). On closed testing procedure with special reference to ordered analysis of variance. *Biometrika*, 63:655–660.

Maurer, W. and Bretz, F. (2013a). Memory and other properties of multiple test procedures generated by entangled graphs. *Statistics in Medicine*, 32:1739–1753.

Maurer, W. and Bretz, F. (2013b). Multiple testing in group sequential trials using graphical approaches. *Statistics in Biopharmaceutical Research*, 5:4:311–320.

Maurer, W. and Bretz, F. (2014). A note on testing families of hypotheses using graphical procedures. *Statistics in Medicine* (to appear).

Maurer, W., Glimm, E., and Bretz, F. (2011). Multiple and repeated testing of primary, co-primary and secondary hypotheses. *Statistics in Biopharmaceutical Research*, 3:336–352.

Maurer, W., Hothorn, L., and Lehmacher, W. (1995). Multiple comparisons in drug clinical trials and preclinical assays: A-priori ordered hypotheses. In Vollmar, J., ed., *Biometrie in der chemisch-pharmazeutischen Industrie*. Fischer Verlag, Stuttgart, Germany.

Maurer, W. and Klinglmüller, F. (2013). Sequentially rejective test procedures for partially ordered and algebraically dependent systems of hypotheses. Talk given at the *International Conference on Simultaneous Inference*, Hannover, Germany.

Maurer, W. and Mellein, B. (1988). On new multiple tests based on independent p-values and the assessment of their power. In Bauer, P., Hommel, G., and Sonnemann, E., eds., *Multiple Hypothesenprüfung*. Springer Verlag, Berlin, Germany.

Millen, B. and Dmitrienko, A. (2011). A class of flexible closed testing procedures with clinical trial applications. *Journal of Biopharmaceutical Statistics*, 3:14–30.

O'Neill, R. (1997). Secondary endpoints cannot be validly analyzed if the primary endpoint does not demonstrate clear statistical significance. *Controlled Clinical Trials*, 18:550–556.

Proschan, M., Lan, K., and Wittes, J. (2006). *Statistical Monitoring of Clinical Trials: A Unified Approach*. Springer, New York.

Rauch, G. and Beyersmann, J. (2013). Planning and evaluating clinical trials with composite time-to-first-event endpoints in a competing risk framework. *Statistics in Medicine*, 32:3595–3608.

Rohmeyer, K. and Klinglmueller, F. (2014). *gMCP: Graph Based Multiple Test Procedures*. R package version 0.8-6. http://cran.r-project.org/web/packages/gMCP/, accessed July 11, 2014.

Senn, S. and Bretz, F. (2007). Power and sample size when multiple endpoints are considered. *Pharmaceutical Statistics*, 6:161–170.

Simes, R. (1986). An improved Bonferroni procedure for multiple tests of significance. *Biometrika*, 73:751–754.

Sozu, T., Sugimoto, T., and Hamasaki, T. (2010). Sample size determination in clinical trials with multiple co-primary binary endpoints. *Statistics in Medicine*, 29:2169–2179.

Strassburger, K. and Bretz, F. (2008). Compatible simultaneous lower confidence bounds for the holm procedure and other bonferroni based closed tests. *Statistics in Medicine*, 27:4914–4927.

Sugitani, T., Bretz, F., and Maurer (2014). A simple and flexible graphical approach for adaptive group-sequential clinical trials. *(submitted for publication)*.

Sugitani, T., Hamasaki, T., and Hamada, C. (2013). Partition testing in confirmatory adaptive designs with structured objectives. *Biometrical Journal*, 55:341–359.

Westfall, P. and Krishen, A. (2001). Optimally weighted, fixed sequence, and gate-keeping multiple testing procedures. *Journal of Statistical Planning and Inference*, 99:25–40.

Westfall, P., Krishen, A., and Young, S. (1998). Using prior information to allocate significance levels for multiple endpoints. *Statistics in Medicine*, 17:2107–2119.

Westfall, P., Tobias, R., and Wolfinger, R. (2011). *Multiple Comparisons and Multiple Tests Using SAS*. SAS, Cary, NC.

Westfall, P. and Young, S. (1993). *Resampling-Based Multiple Testing: Examples and Methods for p-Value Adjustment.* Wiley, New York.

Whitehead, J. (1997). *The Design and Analysis of Sequential Clinical Trials.* Wiley, Chichester.

Wiens, B. and Dmitrienko, A. (2005). The fallback procedure for evaluating a single family of hypotheses. *Journal of Biopharmaceutical Statistics*, 15:929–942.

Wiens, B., Dmitrienko, A., and Marchenko, O. (2013). Selection of hypothesis weights and ordering when testing multiple hypotheses in clinical trials. *Journal of Biopharmaceutical Statistics*, 23(6):1403–1419.

Xi, D. and Tamhane, A. (2014a). Allocating recycled significance levels in group sequential procedures for multiple endpoints. *(submitted)*.

Xi, D. and Tamhane, A. (2014b). A general multistage procedure for k-out-of-n gatekeeping. *Statistics in Medicine*, 33(8):1321–1335.

Xiong, C., Yu, K., Gao, F., Yan, Y., and Zhang, Z. (2005). Power and sample for clinical trials when efficacy is required size in multiple endpoints: application to an alzheimer's treatment trial. *Clinical Trials*, 2:387–393.

15

Pairwise Comparisons with Binary Responses: Multiplicity-Adjusted P-Values and Simultaneous Confidence Intervals

Bernhard Klingenberg and Faraz Rahman

CONTENTS

15.1 Simultaneous Inference

In many clinical trials, a binary response is measured in several groups, sometimes including a control group. An important question centers around estimating the significance and size of potential group differences, measured by some suitable effect measure, when the groups are compared to each other. In this chapter, we focus on constructing multiplicity adjusted P-values and, more importantly, simultaneous confidence intervals for pairwise comparisons between the groups (such as all pairwise comparisons or all comparisons to control), using the difference of proportion as the effect measure. Simultaneous here refers to the fact that the set (or family) of confidence intervals controls the familywise error rate, that is, the probability that at least one of the confidence intervals fails to cover the true parameter. This is in contrast to ignoring the multiplicities, which results in error rates that are largely *unknown* (a conservative upper bound on the familywise error rate [FWER] can always be provided through Bonferroni's inequality) and that can be quite large. Therefore, it is better to control the FWER at some known level α so that precise (asymptotic) error statements can be given. The goal of this chapter is then to introduce, develop, and demonstrate, through various simulations and real examples, the statistical methods to achieve this. Throughout, we will assume that we have independent binomial observations in K groups.

For the single comparison case, that is, comparing the success probabilities in two groups through their difference, many studies have shown (Newcombe, 1998; Newcombe and Nurminen, 2011) that the score statistic has excellent asymptotic behavior and inverting it leads to confidence intervals with coverage very close to the nominal level, for almost all values of the parameter space. Hence, in this chapter, we will focus on methods that use the *maximum score statistic* to derive multiplicity adjusted P-values and invert it to obtain simultaneous confidence intervals. Simulations show that the FWER when using the maximum score statistic is very well controlled even for relatively small sample sizes or probabilities close to the boundary. Klingenberg (2012) and Agresti et al. (2008) also used the score statistic, while Klingenberg (2012), Schaarschmidt et al. (2009), and Piegorsch (1991) use (adjusted) Wald statistics and others. A disadvantage of inverting the score statistic is that it does not provide closed-form solutions for the confidence bounds and iterative methods are needed to find them. However, all methods aforementioned require the computation of a multivariate normal quantile, so computer implementation is a necessity in any case. For this purpose, in Section 15.5 we present and illustrate publicly available and general R code to implement and replicate all procedures mentioned in this chapter, for both two- and one-sided intervals.

A straightforward procedure to control the FWER is to use Bonferroni- (or Sidak-) adjusted quantiles in constructing simultaneous intervals. However, since these adjustments do not take account of the correlation information in the score statistics (the Sidak adjustment assumes independence while the Bonferroni adjustment is based on a general probability inequality), we might lose information that could yield smaller quantiles, which lead to tighter confidence bounds (and smaller multiplicity adjusted P-values). This is explored in Section 15.2 together with general notation and a discussion of the maximum score statistic. In Section 15.3, we discuss hypothesis testing of multiple pairwise hypotheses using the closure principle and show how to compute multiplicity adjusted P-values for the special case of all comparisons to control and all pairwise comparison. In Section 15.4, we present methods to construct simultaneous confidence intervals for these cases and discuss an example in great detail.

15.2 Score Statistic for Pairwise Comparisons

Let $Y_i, i = 1, \ldots, K$ denote independent binomial(n_i, π_i) observations in K groups, where n_i denotes the number of Bernoulli trials with success probability π_i in group i. Let $\boldsymbol{\pi} = (\pi_1, \ldots, \pi_K)^t$ be the vector of success probabilities. Although the following theory holds for testing any sets of linear combinations of the π_i's, here we only consider pairwise comparisons. To this end, we define pairwise comparisons of interest through a contrast matrix C and are interested in simultaneous inference on the elements of the parameter vector $\boldsymbol{\delta} = C\boldsymbol{\pi}$. The matrix C has rows with exactly one 1, one -1, and the remaining terms 0. Two important special cases are all pairwise comparisons where we are interested in the $\binom{K}{2}$ differences $\delta_{ij} = \pi_i - \pi_j, j = 1, \ldots, (K-1), i = (j+1), \ldots, K$ and all comparisons to control where we are interested in the $K-1$ differences $\delta_{i1} = \pi_i - \pi_1, i = 2, \ldots, K$ (without loss of generality, we let the first group be the control group). But the rows of C can also just contain some preselected pairwise comparisons.

For given pairwise comparisons $\{\delta_{ij}\}$, we want to test null vs alternative hypotheses of the form

$$H_{ij}^0 : \delta_{ij} = \delta_{ij}^0 \quad \text{vs} \quad H_{ij}^a : \delta_{ij} \neq \delta_{ij}^0.$$

In many practical applications, all δ_{ij}^0 are equal to zero, but we need the more general case here in order to incorporate clinically relevant effects (see the example in Section 15.3.2) or superiority/inferiority margins but also to invert the family of hypothesis tests and derive simultaneous confidence intervals for the δ_{ij}'s.

For comparisons to a control or in other settings, one-sided inferences may be more useful (e.g., in noninferiority or superiority trials) and we will consider one-sided hypotheses of the form

$$H_{ij}^0 : \delta_{ij} \leq \delta_{ij}^0 \quad \text{vs} \quad H_{ij}^a : \delta_{ij} > \delta_{ij}^0,$$

which, when inverted, yield lower confidence bounds on the δ_{ij}'s. The methods are easily adjusted to provide upper confidence bounds.

For a particular pairwise comparison (including a comparison to control where $j = 1$), let $d_{ij} = p_i - p_j$ be the difference in the sample proportions between the two pairs, with $p_i = y_i/n_i, i = 1, \ldots, K$ and consider

$$T_{ij} = \frac{d_{ij} - \delta_{ij}^0}{s_{ij}^{1/2}} \tag{15.1}$$

as a test statistic for H_{ij}^0, where s_{ij} is a consistent estimator of $\text{Var}\left[d_{ij}\right] = \pi_i(1 - \pi_i)/n_i + \pi_j(1 - \pi_j)/n_j$. We will consider the score statistic that uses

$$s_{ij} = \frac{\tilde{\pi}_i (1 - \tilde{\pi}_i)}{n_i} + \frac{\tilde{\pi}_j (1 - \tilde{\pi}_j)}{n_j},$$

where $\tilde{\pi}_i$ is the restricted (under $H_{ij}^0: \delta_{ij} = \delta_{ij}^0$) maximum likelihood estimate of π_i, available in closed form (Nurminen, 1986). Other possibilities for s_{ij} exist, such as estimating π_i by adding one pseudosuccess and one pseudo failure in the ith group (Agresti and Caffo, 2000) and then using the resulting adjusted sample proportion instead of $\tilde{\pi}_i$ in the formula for s_{ij}. Although inverting the score statistic does not result in a closed-form solution for the confidence bounds, the asymptotic performance of it is very good, as we will see in Figures 15.1 and 15.2, even for relatively small sample sizes. The resulting FWER control or coverage probability is usually more precise than with other competitors, such as Wald statistics, in particular over the relevant part of the parameter space (see, e.g., Klingenberg, 2012).

In (15.1), we use the form of the score statistic as given in Mee (1984) and do not use the bias correction factor (when $\delta_{ij} = 0$) of $\sqrt{(n_i + n_j)/(n_i + n_j - 1)}$ as given in Miettinen and Nurminen (1985). In the single comparison case, for small sample sizes, this correction leads to a better performance in terms of the coverage being closer to the nominal level. However, this correction does not seem to be necessary in the multiple comparison cases we considered here.

15.2.1 Correlation of Test Statistics

Under H_{ij}^0, T_{ij} is asymptotically standard normal due to the asymptotic normality of the sample proportions, but the T_{ij}'s are not all independent of each other. Two test statistics T_{ij} and T_{kl} that share a common index are correlated with correlation in the product form

$$\text{Cor}[T_{ij}, T_{kl}] = \begin{cases} \lambda_{ij}\lambda_{kl} & j = l, i \neq k \\ \lambda_{ji}\lambda_{lk} & j \neq l, i = k \\ -\lambda_{ij}\lambda_{lk} & j = k, i \neq l \\ -\lambda_{ji}\lambda_{kl} & j \neq k, i = l \\ 0 & j \neq k, i \neq l \end{cases} \tag{15.2}$$

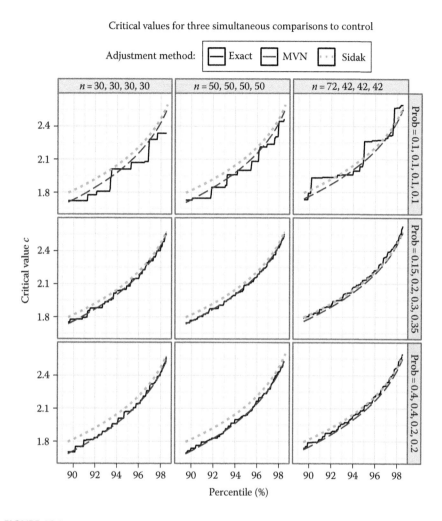

FIGURE 15.1

One-sided critical values (upper quantiles) of $\max_i T_{i1}$ using the exact distribution based on the binomial probabilities (black, full line), the approach based on the asymptotic multivariate normality (MVN) of (T_{21}, T_{31}, T_{41}) (gray, dashed) and the approach based on the Sidak inequality (light-gray, dotted) when comparing three groups to a control, for various sample sizes $n_1, \ldots, n_4$ (denoted n) and true binomial probability vector π (denoted Prob). The first group is taken to be the control.

with

$$\lambda_{ij} = \left[1 + \frac{n_j \, \pi_i (1 - \pi_i)}{n_i \, \pi_j \left(1 - \pi_j\right)} \right]^{-1/2}. \tag{15.3}$$

For all comparisons to control ($j = l = 1$), only the first case is needed in Formula 15.2. Note that in general, the asymptotic covariance matrix of the test

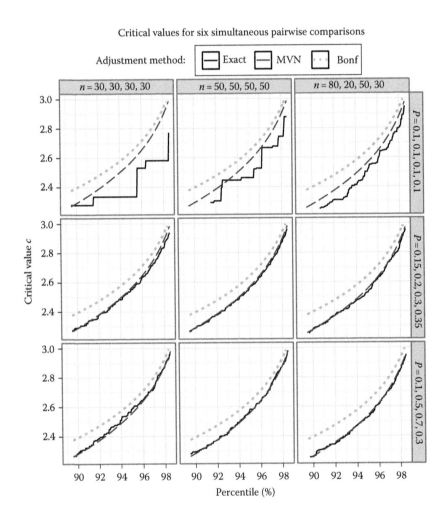

FIGURE 15.2

Two-sided critical values c of $\max_{(i,j)} |T_{ij}|$ using the exact distribution based on the binomial probabilities (black, full line), the approach based on the asymptotic multivariate normality (MVN) of $(T_{21}, T_{31}, \ldots, T_{43})$ (gray, dashed), and the approach based on the regular Bonferroni (Bonf) correction (light-gray, dotted) for all six pairwise comparisons among $K = 4$ groups. The critical values are shown for various sample sizes $n_1, \ldots, n_4$ (denoted n) and true binomial probability vector π (denoted P). Critical values from the $1/\sqrt{2}$-multiple of Tukey's studentized range distribution are not shown. These are exactly equal to the multivariate normal quantiles for the first row and differ only marginally from the multivariate normal quantiles for the other two rows, see Section 15.2.4.

statistics may be written in matrix notation as $C\Sigma C^t$, where Σ is a diagonal matrix with elements $\pi_i(1 - \pi_i)/n_i$ on the diagonal.

15.2.2 Maximum Score Statistic

To construct simultaneous tests and confidence intervals, we consider the distribution of

$$\max_{(i,j)\in\mathcal{I}} |T_{ij}|,$$

where $\mathcal{I}$ is the set of all index pairs (i, j) that appear in the family of null hypotheses $\{H^0_{ij}\}$. For example, for all comparisons to control $\mathcal{I} = \{(2, 1), (3, 1), \dots, (K, 1)\}$ and for all pairwise comparison, $\mathcal{I} = \{(2, 1), (3, 1), \dots, (K, 1), (3, 2), \dots, (K, 2), (4, 3), \dots, (K, K - 1)\}$. To control the FWER for the family of hypotheses $\{H^0_{ij}\}_{(i,j)\in\mathcal{I}}$ at level α, we need to find a critical value c such that

$$P\left(\max_{(i,j)\in\mathcal{I}} |T_{ij}| < c\right) = 1 - \alpha. \tag{15.4}$$

Then, rejecting an individual (also called elementary) hypothesis H^0_{ij} when $|T_{ij}| \geq c$ controls the FWER at level α since the probability of at least one false rejection equals α, that is, $P\left(\bigcup_{(i,j)\in\mathcal{I}} \{|T_{ij}| \geq c\}\right) = \alpha$. Note that for all procedures considered here, the FWER control is in the asymptotic sense as we are going to use the asymptotic joint multivariate normal distribution of the T_{ij}'s as $\min_i n_i \to \infty$ and the π_i's stay away from the boundary values of zero or one.

By taking advantage of the correlation in the T_{ij}'s, we obtain a smaller critical value c using the joint multivariate normal distribution (with correlation matrix determined through (15.2)) than with the straightforward (two-sided) Bonferroni adjustment $c_{\text{Bonf}} = z_{1-\alpha/2r}$, where r is the number of pairwise comparisons. For the case of all comparisons to control, the correlation components λ_{i1} are all positive and the slightly less conservative (one-sided) Sidak adjustment $c_{\text{Sid}} = z_{(1-\alpha)^{1/(K-1)}}$ (i.e., assuming independence) can be used instead of the Bonferroni adjustment.

15.2.3 Exact vs Asymptotic Distribution

In Figures 15.1 and 15.2, we compare the upper tail quantiles (i.e., critical values) of the exact distribution of $\max T_{i1}$ (for the case of all comparisons to control) and $\max |T_{ij}|$ (for the case of all pairwise comparisons) with the ones obtained by approximating the distribution of the maximum via the joint multivariate normal distribution of the T_{ij}'s. This is done for several settings of the sample size (balanced and unbalanced) and true probability vector π. The exact quantiles are computed from the product binomial distribution, $\prod_i \text{Bin}(n_i, \pi_i)$, via complete enumeration of all possible outcomes, which is computationally feasible if the n_i's are not too large. Quantiles resulting from straightforward Sidak (for the case of all comparisons to control) or Bonferroni (for the case of all pairwise comparisons) adjustments are also shown.

The figures show that the distribution of $\max_i T_{i1}$ or $\max_{(i,j)} |T_{ij}|$ approximated via the asymptotic multivariate normality of the T_{ij}'s is remarkably close to the exact distribution of the maximum, even for relatively small sample sizes and somewhat extreme settings for the success probabilities. Further, the conservative nature of the Sidak or Bonferroni adjustments is apparent from these plots. For instance, the true 95th percentile of the distribution of $\max |T_{ij}|$ for all $\binom{4}{2} = 6$ pairwise comparisons when all groups have sample size $n_i = 50$ and $\pi_1 = 0.15, \pi_2 = 0.2, \pi_3 = 0.3, \pi_4 = 0.35$ equals $c_{\text{exact}} = 2.564$, compared to $c_{\text{MVN}} = 2.566$ and $c_{\text{Bonf}} = 2.638$ ($c_{\text{Tukey}} = 2.569$, see the following). For the 90th percentile, $c_{\text{exact}} = 2.288$, compared to $c_{\text{MVN}} = 2.288$ and $c_{\text{Bonf}} = 2.394$ ($c_{\text{Tukey}} = 2.291$, see the following). For the 97.5th percentile, $c_{\text{exact}} = 2.790$, compared to $c_{\text{MVN}} = 2.814$ and $c_{\text{Bonf}} = 2.865$ ($c_{\text{Tukey}} = 2.817$, see the following).

15.2.4 Special Case: Testing Homogeneity with Balanced Data

For the special case of balanced sample sizes $n_1 = \cdots = n_K \equiv n$ and $\delta_{ij}^0 = 0$ for all (i, j) in the all pairwise comparison case (i.e., testing homogeneity $\pi_i = \pi_j \equiv \pi$ for all possible pairs (i, j)), the score statistic becomes

$$T_{ij} = \frac{p_i - p_j}{[p(1-p)(2/n)]^{1/2}},$$

where P is the pooled proportion $p = (y_i + y_j)/2n$. Then $|T_{ij}|$ can be written as

$$|T_{ij}| = \frac{1}{\sqrt{2}} \left[\frac{\pi(1-\pi)}{p(1-p)} \right]^{1/2} \left| \frac{p_1 - \pi}{\sqrt{\pi(1-\pi)/n}} - \frac{p_2 - \pi}{\sqrt{\pi(1-\pi)/n}} \right|.$$

Under the null and as $n \to \infty$, $|T_{ij}|$ is distributed as $\frac{1}{\sqrt{2}}|Z_i - Z_j|$, where the Z_i's are K iid. standard normal random variables.

The maximum of $|Z_i - Z_j|$ over all possible pairs (i, j) equals the range (i.e., the largest minus the smallest) of K standard normal random variables. Hence, the maximum of $|T_{ij}|$ is distributed like the $1/\sqrt{2}$ multiple of the studentized range distribution (with parameter K and infinite degrees of freedom). The critical value c is then given by $c = Q_{K,\infty}^{\alpha}/\sqrt{2}$, where $Q_{K,\infty}^{\alpha}$ denotes the upper α quantile of the studentized range distribution with parameters K and degrees of freedom equal to infinity (implemented, e.g., in R as qtukey). Agresti et al. (2008) used this fact to construct simultaneous confidence intervals for all pairwise comparisons.

Although this result only holds for balanced sample sizes and $\pi_i = \pi_j$ for all pairs (i, j), as long as $\pi_i(1 - \pi_i)/n_i \approx \pi_j(1 - \pi_j)/n_j$, the approximation via the studentized range distribution is going to hold asymptotically. In Figure 15.2, we originally included the plot of the critical values derived via Tukey's studentized range distribution, even for cases where $\pi_i \neq \pi_j$ and unbalanced sample sizes, settings where strictly speaking it does not apply. However, we noticed that the critical values were almost identical to or only slightly more conservative than those derived via the multivariate normal distribution and hence decided not to plot them. For greatly unbalanced cases (in both the true success probabilities and the sample sizes), there is a more pronounced difference between the two. In any case, if a reliable procedure to compute multivariate normal quantiles is available, the quantile c_{mvn} based on the multivariate normal distribution should be used as it applies more generally.

Similarly, for the case of all comparisons to control under the special scenario of balanced sample sizes and $\delta_{i1} = 0$ for all i (i.e., $\pi_i = \pi_1$ for $i = 2, \ldots, K$), the maximum of $\frac{1}{\sqrt{2}}|Z_i - Z_1|$ follows the distribution tabulated by Dunnett (1955) with infinite degrees of freedom, which also is a special case of the multivariate normal distribution approach presented here.

15.2.5 Estimating the Correlation in the General Case

Under the special case considered earlier (balanced sample size and $\delta_{ij}^0 = 0$ for all (i, j)), $\lambda_{ij} = 1/\sqrt{2}$ and the correlation matrix (under the null) of the T_{ij}'s has off-diagonal elements that are either ± 0.5 or zero. In particular, the correlation matrix is fully specified under the null and does not depend on any nuisance parameters. This is not the case in the more general case when $\delta_{ij}^0 \neq 0$, which for instance is needed with inferiority testing or when inverting the family of tests to obtain simultaneous confidence intervals.

To compute the critical value c from the multivariate normal distribution in general, we need to estimate λ_{ij} for all pairs (i, j) in $\mathcal{I}$. One option is to compute the correlation under the null as in the special case discussed earlier (Tukey's studentized range distribution and Dunnett's distribution), but with general δ_{ij}^0's. Using the null correlation may improve the asymptotic behavior of the null distribution of $\max_{(i,j) \in \mathcal{I}} |T_{ij}|$. However, the procedure gets complicated as the null distribution of the maximum is no longer a pivotal statistic, that is, it's null distribution depends on a nuisance parameter. Klingenberg (2010) discusses this for the case of all comparison to control.

In simulation studies we carried out, we did not see any noticeable improvement when using the null correlation. Therefore and because of its serious computational complexity, we abandon estimating the correlation of the T_{ij}'s under the null and use the sample proportions y_i/n_i as plug-in estimators for the unknown probabilities π_i appearing in λ_{ij}. This still yields a consistently estimated correlation matrix, which we use to compute multiplicity adjusted P-values in Section 15.3 or the multivariate normal (equidistant) $1 - \alpha$ quantile c in (15.4) for forming simultaneous confidence intervals in Section 15.4. For cases where $y_i = 0$ or n_i, we use the adjusted sample proportions $(y_i + 1)/(n_i + 2)$.

15.3 Multiplicity Adjusted P-Values

In this section, we explain and demonstrate how we use the maximum score statistic and its asymptotic distribution discussed earlier to derive multiplicity adjusted P-values for the elementary hypothesis H_{ij}^0. For a single hypothesis, the P-value represents the smallest (type I) error rate for which the hypothesis can be rejected. Any smaller type I error rate would not result in a rejection. Similarly, a multiplicity adjusted P-value when testing several hypotheses simultaneously is the smallest (familywise) error rate for which an elementary hypothesis H_{ij} can be rejected. The classification *multiplicity adjusted* means that this P-value already incorporates the multiplicities in the error rates arising from testing multiple hypotheses and that it can directly be compared to a desired FWER.

For instance, a Bonferroni multiplicity adjusted P-value for an elementary hypothesis is simply obtained by multiplying the *raw* (unadjusted) P-value for that hypothesis by the number of hypotheses tested (or setting it equal to one if this results in a value larger than one). Alternatively, when using the maximum score statistic approach as discussed in Section 15.2.2, the multiplicity adjustment is incorporated through computing the P-value based on the distribution of the maximum. For hypothesis H_{ij}^0, a multiplicity adjusted P-value using the maximum approach is given by

$$P\left(\max_{(i,j)\in\mathcal{I}} |T_{ij}| \geq \left|t_{ij}^{\mathrm{obs}}\right|\right) = 1 - P\left(\max_{(i,j)\in\mathcal{I}} |T_{ij}| < \left|t_{ij}^{\mathrm{obs}}\right|\right) = 1 - P\left(\left\{|T_{ij}| < \left|t_{ij}^{\mathrm{obs}}\right|\right\}\right),$$

(15.5)

where the last probability is computed using the estimated joint multivariate normal distribution of the $\{T_{ij}\}$ and where t_{ij}^{obs} is the observed test statistic for hypothesis H_{ij}^0. This provides uniformly smaller P-values than using the Bonferroni approach. However, we will be using an even more powerful method known as the step-down approach to find these multiplicity adjusted P-values, see Section 15.3.1.

No matter which multiplicity adjustment method is chosen, an elementary hypothesis is rejected when its multiplicity adjusted P-value is smaller than α, the FWER. This guarantees control of the FWER at level α. A typical value for the FWER in the two-sided case is $\alpha = 5\%$ or 10%.

15.3.1 Closed Testing and Stepwise Procedures

To derive the multiplicity adjusted P-values based on the maximum score statistic, we make use of a stepwise procedure. This means stepping through an ordered set of elementary null hypotheses H^0_{ij} and calculating the multiplicity adjusted P-value for each elementary hypothesis along the way. The ordering of the null hypotheses is obtained from the magnitude of the score statistics. That is, for the all pairwise comparison case, the elementary hypothesis with the largest observed value of $|T_{ij}|$ is tested first, followed by the one with the second largest observed value of $|T_{ij}|$, etc. For all comparisons to control, the order is based on the largest observed value of T_{i1}, followed by the second largest observed T_{i1} and so on. This stepwise (or, more precisely, step-down) approach is in contrast to (and improves on) a single-step procedure, which computes all adjusted P-values in one step. The Bonferroni adjustment and the one based on the global maximum of $|T_{ij}|$ mentioned in Section 15.3 are prominent examples of a single-step procedure.

The stepwise procedure is based on the methodology of closed testing by Marcus and Gabriel (1976), a general methodology that leads to very efficient multiplicity adjusted inference about the elementary hypotheses. The closure principle and the derivation of multiplicity adjusted P-values for comparing binary proportions can be described as follows:

1. Let $\mathcal{I}$ be the index set containing all pairs (i, j) of groups (treatments) that we are interested in comparing, that is, all index pairs (i, j) that appear in the family of null hypotheses $\left\{ H^0_{ij} \right\}$. For example, for all comparisons to control $\mathcal{I} = \{(2, 1), (3, 1), \ldots, (K, 1)\}$ and for all pairwise comparison, $\mathcal{I} = \{(2, 1), (3, 1), \ldots, (K, 1), (3, 2), \ldots, (K, 2), (4, 3), \ldots, (K, K - 1)\}$.

2. Consider the *intersection hypothesis*

$$H_{\mathcal{J}} = \bigcap_{(i,j) \in \mathcal{J}} H^0_{ij},$$

where $\mathcal{J} \subseteq \mathcal{I}$ and $\mathcal{J} \neq \emptyset$. We test each intersection hypothesis $H_{\mathcal{J}}$ using the maximum score statistic, where the maximum is taken over all test statistics T_{ij} with index pair $(i, j) \in \mathcal{J}$. When $\mathcal{J} \equiv \mathcal{I}, H_{\mathcal{J}}$ is called the global (intersection) hypothesis and the maximum is taken over the entire set of indices (i, j) appearing in $\mathcal{I}$, just as in Section 15.2.2. For lower-dimensional $\mathcal{J} \subset \mathcal{I}$, the maximum is taken over the smaller set of test statistics T_{ij} with index pair $(i, j) \in \mathcal{J}$. Using the multivariate normality of the T_{ij}'s (with dimension determined by the cardinality of $\mathcal{J}$, $|\mathcal{J}|$), a P-value $p_{\mathcal{J}}$ can be computed for each intersection hypothesis (see Point 5).

3. The closure principle says that an elementary hypotheses H_{ij}^0 is rejected if and only if all intersection hypotheses $H_{\mathcal{J}}$ containing H_{ij}^0 are rejected at level α, the FWER. This leads to the construction of a multiplicity adjusted P-value p_{ij}^{adj} for each elementary H_{ij}^0 as the maximum over all p_J with $(i, j) \in \mathcal{J}$:

$$p_{ij}^{adj} = \max_{\mathcal{J} \subseteq \mathcal{I}:(i,j)\in\mathcal{J}} \{p_{\mathcal{J}}\} \qquad (15.6)$$

In other words, the multiplicity adjusted P-value for an elementary hypothesis H_{ij}^0 is the maximum over all those intersection P-values $p_{\mathcal{J}}$ for which (i, j) is an element of $\mathcal{J}$.

4. When using a maximum statistic such as the maximum score statistic for the intersection hypotheses, there are a couple of so-called shortcuts, which greatly reduce the set of intersection hypotheses one has to consider at a particular step. This does depend on the particular application (i.e., all pairwise comparison, all comparisons to control, other pairwise comparisons), but a set of general rules are as follows: In any given step of the stepwise procedure, one can ignore any intersections $\mathcal{J}$ that include the index (i, j) of a previously tested elementary hypothesis. Also, it is not necessary to consider intersection hypotheses that are subsets of larger ones. Their P-values are necessarily smaller and since we are taking a maximum, they do not matter. Finally, because of monotonicity considerations, if the multiplicity adjusted P-value from a previous step is larger than the one computed for the current step, we set the multiplicity adjusted P-value for the current step equal to the one from the previous step. This is called truncation. All rules are illustrated with the two examples given in the following

5. How do we compute the intersection P-values $p_{\mathcal{J}}$ for a given $\mathcal{J} \subseteq \mathcal{I}$? Let t_{ij}^{obs} be the observed value of the test statistic T_{ij} for elementary hypothesis H_{ij}^0. Then the two-sided intersection P-value $p_{\mathcal{J}}$ appearing in (15.6) is given by

$$p_{\mathcal{J}} = P\left(\max_{(i,j)\in\mathcal{J}} |T_{ij}| \geq |t_{ij}^{obs}|\right).$$

This can be computed using the $|\mathcal{J}|$-dimensional multivariate normal distribution of the $\{T_{ij}\}_{(i,j)\in J}$, with correlation matrix estimated as explained in Section 15.2.5. One-sided intersection P-values are computed accordingly, see the following example.

Comparing the resulting p_{ij}^{adj} directly to the nominal FWER α, that is, rejecting H_{ij}^0 if $p_{ij}^{adj} < \alpha$ controls the FWER at level α. A more detailed account on multiplicity adjusted P-values and using a maximum statistic under closed testing can be found in Bretz et al. (2011) and the references therein.

15.3.2 Example: All Comparisons to Control

Klingenberg (2012) mentions the following scenario from a phase II clinical trial measuring whether a new drug provides relief from abdominal pain over a 3-week period for patients suffering from irritable bowl syndrome (IBS): A total of 493 patients were randomized to placebo (0 mg) or four dose groups of a new drug, with dose levels of 1, 4, 12, and 24 mg of the active ingredient. The primary concern was to show superiority of at least one of the dose groups over placebo, where superiority was defined as at least a 10 percentage point increase in the probability of relief from abdominal pain.

In order to identify which dose group, if any, is significantly superior, we compare each active dose to control, leading to four simultaneous hypotheses of the form

$$H_{i1}^0 : \delta_{i1} \leq 0.1 \quad \text{vs} \quad H_{i1}^a : \delta_{i1} > 0.1, \quad i = 2, \ldots, 5,$$

where δ_{i1} is the difference in the probability of relief from abdominal pain between the ith dose group ($i = 2, 3, 4$, and 5 for the four active dose groups) and the control group ($i = 1$). Here, $\delta_{i1}^0 = 10\%$, the margin after which the effect becomes clinically relevant. In this section, we show how to compute multiplicity adjusted P-values for the four hypotheses, while in Section 15.4, we compute simultaneous lower confidence bounds for the $\delta_{i1}, i = 2, \ldots, 5$.

The allocation of patients to the five dose groups (including placebo) was nearly balanced. Table 15.1 shows the sample sizes and corresponding number of patients reporting relief of abdominal pain across the dose groups. The table also reports the observed score statistic t_{i1}^{obs} (15.1) for each elementary hypothesis.

To find the multiplicity adjusted P-value for each elementary hypothesis, we order them according to the value of their score test statistic. The largest score statistic (2.886) occurs for H_{31}^0, so it is tested first. H_{31}^0 is part of the intersection hypotheses corresponding to $\mathcal{J} = \{(2, 1), (3, 1), (4, 1), (5, 1)\}$, $\mathcal{J} = \{(2, 1), (3, 1), (4, 1)\}$, $\mathcal{J} = \{(2, 1), (3, 1), (5, 1)\}$, $\mathcal{J} = \{(3, 1), (4, 1), (5, 1)\}$, $\mathcal{J} = \{(2, 1), (3, 1)\}$, $\mathcal{J} = \{(3, 1), (4, 1)\}$, $\mathcal{J} = \{(3, 1), (5, 1)\}$, and $\mathcal{J} = \{(3, 1)\}$. However, we only need to consider the global intersection hypothesis with $\mathcal{J} = \mathcal{I} = \{(2, 1), (3, 1), (4, 1), (5, 1)\}$, since all others are subsets of it.

To find this $p_{\mathcal{J}}$, we estimate the correlation matrix of the four score statistics $T_{21}, T_{31}, T_{41}, T_{51}$ using formulas (15.2) and (15.3) with sample proportions

TABLE 15.1

Data for the IBS Trial and the Resulting Observed Differences in the Proportions of Relief from Abdominal Pain between Dose Groups $i = 2, \ldots, 5$ and Placebo

Dose Group i	1 (0 mg)	2 (1 mg)	3 (4 mg)	4 (12 mg)	5 (24 mg)
n_i	100	102	98	99	94
y_i	38	52	67	59	58
Sample prop. p_i	0.380	0.510	0.684	0.596	0.617
Diff. to placebo		0.130	0.304	0.216	0.237
Elem. hypoth. H^0_{i1}		$H^0_{21}: \delta_{21} \leq 0.1$	$H^0_{31}: \delta_{31} \leq 0.1$	$H^0_{41}: \delta_{41} \leq 0.1$	$H^0_{51}: \delta_{51} \leq 0.1$
Obs. score statistic t^{obs}_{i1}		0.428	2.886	1.645	1.917
Unadjusted P-value		0.334	0.002	0.050	0.028
Sidak adj. P-value		0.803	0.008	0.186	0.106
Mult. adj. P-value p^{adj}_{ij}		0.334	0.007	0.088	0.069

Notes: The last row of the table shows the multiplicity adjusted P-values for the four simultaneous tests of no difference to placebo. Unadjusted and Sidak-adjusted P-values are also shown.

replacing the unknown π_i's, see Section 15.2.5. The estimated correlation matrix equals

```
        2 - 1  3 - 1  4 - 1  5 - 1
2 - 1   1.00   0.5    0.49   0.49
3 - 1   0.50   1.0    0.50   0.50
4 - 1   0.49   0.5    1.00   0.49
5 - 1   0.49   0.5    0.49   1.00,
```

which is optional output for our R routines explained in Section 15.5. The multiplicity adjusted P-value for H^0_{31} is given by

$$p^{adj}_{31} = p_{\mathcal{J}} = P\left(\max_{(i,j) \in \mathcal{J}} T_{ij} \geq 2.886 \right) = 0.0071.$$

This probability is computed via the mvtnorm package in R and part of the standard output of our R routines.

Next, we find the second largest score statistic, which is T_{51} (with an observed value of 1.917), and hence test hypothesis H^0_{51}. The intersection hypotheses that contain H^0_{51} but not H^0_{31} (the hypothesis from the previous step) are given by $\mathcal{J} = \{(2, 1), (4, 1), (5, 1)\}$, $\mathcal{J} = \{(2, 1), (5, 1)\}$ and $\mathcal{J} = \{(5, 1)\}$; however, we only need to consider $J = \{(2, 1), (4, 1), (5, 1)\}$ as all others are subsets. Estimating the correlation between T_{21}, T_{41}, and T_{51} using (15.2) and (15.3) with sample proportions, we get the multiplicity adjusted P-value for H^0_{51} as

$$p_{51}^{\text{adj}} = p_{\mathcal{J}} = P\left(\max_{(i,j)\in\mathcal{J}} T_{ij} \geq 1.917\right) = 0.0692.$$

Continuing in a similar fashion, H_{41}^0 is tested next (with observed score statistic of 1.645), for which we only have to consider the intersection hypothesis corresponding to $J = \{(2,1),(4,1)\}$. This results in the multiplicity adjusted P-value

$$p_{41}^{\text{adj}} = p_{\mathcal{J}} = P\left(\max_{(i,j)\in\mathcal{J}} T_{ij} \geq 1.645\right) = 0.0881.$$

Finally, for H_{21}^0, $\mathcal{J} = \{(2,1)\}$ and

$$p_{21}^{\text{adj}} = p_{\mathcal{J}} = P\left(\max_{i\in\mathcal{J}} T_{ij} > 1.856\right) = P(T_{21} \geq 0.0428) = 0.3342.$$

All multiplicity adjusted P-values are displayed in Table 15.1 and we see that, with an FWER of 5% (one-sided), we can reject hypothesis H_{31}^0 but not the others. These multiplicity adjusted P-values are smaller than the ones obtained by applying the straightforward but more conservative Sidak adjustment, which is based on an independence assumption of the test statistics. For instance, when the FWER is selected as 10%, we can reject H_{31}, H_{41}, and H_{51} using the closed testing step-down procedure as outlined earlier but would only be able to reject H_{31} with the Sidak approach.

15.3.3 Example: All Pairwise Comparisons

We use the IBS data to illustrate testing all pairwise comparisons, that is, comparing the probability of relief from abdominal pain for every possible combination of two dose groups out of the five considered (including placebo). We still take into account the clinically relevant margin of 10% for those comparisons that involve the control, although we will conduct two-sided tests, using a (two-sided) FWER of $\alpha = 5\%$. The resulting $\binom{5}{2} = 10$ simultaneous null hypotheses to be tested are shown in Table 15.2, together with the corresponding score test statistics.

The largest test statistic occurs for hypothesis H_{31}^0, with $|t_{31}^{\text{obs}}| = 2.886$. To obtain the multiplicity adjusted P-value for this hypothesis, the only relevant intersection hypothesis to test is the global one: $\mathcal{J} = \mathcal{I} = \{(2,1), (3,1),\dots,(5,3),(5,4)\}$. Using the estimated correlation matrix of the 10 score statistics (the 10 T_{ij}'s shown in the next page results in in a P-value of

$$p_{31}^{\text{adj}} = p_{\mathcal{I}} = P\left(\max_{(i,j)\in\mathcal{I}} |T_{ij}| \geq 2.886\right) = 0.0318.$$

TABLE 15.2

Multiplicity Adjusted P-values p_{ij}^{adj} Corresponding to the Family of Null Hypotheses $\{H_{ij}^0\}$ for All 10 Pairwise Comparisons between the 5 Dose Groups of the IBS Trial

Hypothesis	d_{ij}	$|t_{ij}^{obs}|$	Raw	Bonferroni	p_{ij}^{adj}
$H_{21}^0: \delta_{21} = 0.1$	0.130	0.428	0.668	1.000	0.890
$H_{31}^0: \delta_{31} = 0.1$	0.304	2.886	0.004	0.039	0.032
$H_{41}^0: \delta_{41} = 0.1$	0.216	1.645	0.100	1.000	0.304
$H_{51}^0: \delta_{51} = 0.1$	0.237	1.917	0.056	0.552	0.221
$H_{32}^0: \delta_{32} = 0$	0.174	2.504	0.012	0.123	0.059
$H_{42}^0: \delta_{42} = 0$	0.086	1.228	0.219	1.000	0.524
$H_{52}^0: \delta_{52} = 0$	0.107	1.511	0.131	1.000	0.304
$H_{43}^0: \delta_{43} = 0$	−0.088	1.282	0.200	1.000	0.524
$H_{53}^0: \delta_{53} = 0$	−0.067	0.969	0.333	1.000	0.555
$H_{54}^0: \delta_{54} = 0$	0.021	0.299	0.765	1.000	0.890

Notes: Bonferroni adjusted and unadjusted (raw) P-values are also shown. $|t_{ij}^{obs}|$ is the absolute value of the observed score statistic (15.1) and d_{ij} is the difference in the sample proportions.

	2 – 1	3 – 1	4 – 1	5 – 1	3 – 2	4 – 2	5 – 2	4 – 3	5 – 3	5 – 4
2 – 1	1.00	0.50	0.49	0.49	−0.52	−0.51	−0.50	0.00	0.00	0.00
3 – 1	0.50	1.00	0.50	0.50	0.48	0.00	0.00	−0.48	−0.48	0.00
4 – 1	0.49	0.50	1.00	0.49	0.00	0.50	0.00	0.52	0.00	−0.50
5 – 1	0.49	0.50	0.49	1.00	0.00	0.00	0.51	0.00	0.52	0.51
3 – 2	−0.52	0.48	0.00	0.00	1.00	0.51	0.51	−0.47	−0.47	0.00
4 – 2	−0.51	0.00	0.50	0.00	0.51	1.00	0.50	0.51	0.00	−0.49
5 – 2	−0.50	0.00	0.00	0.51	0.51	0.50	1.00	0.00	0.52	0.51
4 – 3	0.00	−0.48	0.52	0.00	−0.47	0.51	0.00	1.00	0.47	−0.51
5 – 3	0.00	−0.48	0.00	0.52	−0.47	0.00	0.52	0.47	1.00	0.52
5 – 4	0.00	0.00	−0.50	0.51	0.00	−0.49	0.51	−0.51	0.52	1.00

In the second step, we test H_{32}^0, which has the second largest test statistic, with $|t_{32}| = 2.504$. The relevant subsets to test are $\mathcal{J}_1 = \{(3, 2), (5, 2), (4, 3), (4, 2), (5, 3), (5, 4)\}$, $\mathcal{J}_2 = \{(3, 2), (5, 1), (4, 1), (5, 4)\}$, $\mathcal{J}_3 = \{(3, 2), (4, 1), (5, 2), (5, 3)\}$, and $\mathcal{J}_4 = \{(3, 2), (5, 1), (4, 3), (4, 2)\}$. Here, we already have excluded those $\mathcal{J}$'s that are subsets of the ones listed previously (and the ones that contain the previously tested pair $(3, 1)$). Also, note that if an intersection hypothesis contains, for example, the two elements $(2, 1)$ and $(3, 2)$, it is not listed earlier and does not need to be tested. This is because $H_{21} \cap H_{32}$ implies H_{31}, which was tested at the previous stage.

The intersection P-values $p_{\mathcal{J}} = P\left(\max_{(i,j)\in\mathcal{J}} |T_{ij}| \geq 2.504\right)$ for each of these relevant intersections are $p_{\mathcal{J}_1} = 0.059$, $p_{\mathcal{J}_2} = 0.045$, $p_{\mathcal{J}_3} = 0.045$, and $p_{\mathcal{J}_4} = 0.045$, leading to the multiplicity adjusted P-value

$$p_{32}^{adj} = \max\left\{p_{\mathcal{J}_1}, p_{\mathcal{J}_2}, p_{\mathcal{J}_3}, p_{\mathcal{J}_4}\right\} = \max\{0.059, 0.045, 0.045, 0.045\} = 0.059.$$

Since $p_{32}^{adj} > p_{31}^{adj}$, no truncation occurs.

In the third step, we test H_{51}^0, which has the third largest test statistic, with $\left|t_{51}^{obs}\right| = 1.917$. The relevant subsets to test are $\mathcal{J}_1 = \{(5, 1), (4, 1), (5, 2), (4, 2), (2, 1), (5, 4)\}$ and $\mathcal{J}_2 = \{(5, 1), (5, 2), (4, 3), (2, 1)\}$. The intersection P-values $p_{\mathcal{J}} = P\left(\max_{(i,j) \in \mathcal{J}} |T_{ij}| \geq 1.917\right)$ for each of these relevant intersections are $p_{\mathcal{J}_1} = 0.221$ and $p_{\mathcal{J}_2} = 0.182$, leading to the multiplicity adjusted P-value (no truncation occurs)

$$p_{51}^{adj} = \max\left\{p_{\mathcal{J}_1}, p_{\mathcal{J}_2}\right\} = \max\{0.221, 0.182\} = 0.221.$$

In the fourth step, we test H_{41}^0, which has the fourth largest statistic, with $\left|t_{41}^{obs}\right| = 1.645$. The relevant subsets to test are $\mathcal{J}_1 = \{(4, 1), (5, 2)\}$ and $\mathcal{J}_2 = \{(4, 1), (4, 2), (5, 3), (2, 1)\}$. The intersection P-values $p_{\mathcal{J}} = P\left(\max_{(i,j) \in \mathcal{J}}|T_{ij}| \geq 1.645\right)$ are given by $p_{\mathcal{J}_1} = 0.190$ and $p_{\mathcal{J}_2} = 0.304$, leading to the multiplicity adjusted P-value (no truncation occurs)

$$p_{41}^{adj} = \max\left\{p_{\mathcal{J}_1}, p_{\mathcal{J}_2}\right\} = \max\{0.190, 0.304\} = 0.304.$$

In the fifth step, we test H_{52}^0. The relevant subsets to test are $\mathcal{J}_1 = \{(5, 2), (43)\}$ and $\mathcal{J}_2 = \{(5, 2), (4, 2), (5, 4)\}$ with intersection P-values $p_{\mathcal{J}_1} = 0.244$ and $p_{\mathcal{J}_2} = 0.286$, leading to the multiplicity adjusted P-value

$$p_{52}^{adj} = \max\left\{p_{\mathcal{J}_1}, p_{\mathcal{J}_2}\right\} = \max\{0.244, 0.286\} = 0.286.$$

Here, since $p_{52}^{adj} < p_{41}^{adj}$, we need to truncate and set $p_{52}^{adj} = 0.304$, the adjusted P-value from the previous step.

In the sixth step, we test H_{43}^0. The only relevant subset to test is $\mathcal{J}_1 = \{(4, 3), (5, 3), (2, 1), (5, 4)\}$ with intersection P-value $p_{\mathcal{J}_1} = 0.524$ and multiplicity adjusted P-value

$$p_{43}^{adj} = p_{\mathcal{J}_1} = 0.524.$$

Since $p_{43}^{adj} > p_{52}^{adj}$, no truncation occurs.

In the seventh step, we test H_{42}^0 using $|t_{42}^{obs}| = 1.228$. The only relevant subset to test is $\mathcal{J}_1 = \{(4, 2), (5, 3)\}$. The intersection P-value $p_{\mathcal{J}} = P\left(\max_{(i,j) \in \mathcal{J}} |T_{ij}| \geq 1.228\right)$ for this intersection equals 0.391, and the multiplicity adjusted P-value is equal to

$$p_{42}^{adj} = p_{\mathcal{J}_1} = 0.391.$$

Here, since $p_{42}^{adj} < p_{43}^{adj}$, we set $p_{42}^{adj} = 0.524$, the adjusted P-value from the previous step.

In the eighth step, we test H_{53}^0 using the only relevant subset $\mathcal{J}_1 = \{(5,3),(2,1)\}$ with intersection P-value equal to 0.554. The multiplicity adjusted P-value (no truncation occurs) is

$$p_{53}^{adj} = p_{\mathcal{J}_1} = 0.554.$$

In the ninth step, we test H_{21}^0 using the only relevant subset $\mathcal{J}_1 = \{(2,1),(5,4)\}$. The intersection P-value is 0.890, and the multiplicity adjusted P-value (no truncation) equals

$$p_{21}^{adj} = p_{\mathcal{J}_1} = 0.890.$$

In the final stage, we test the last remaining hypothesis, H_{54}^0, with $|t_{54}^{obs}| = 0.299$. This can be tested using the univariate normal distribution, yielding

$$p_{54}^{adj} = P\left(|T_{54}| \geq 0.299\right) = 0.765.$$

Since $p_{54}^{adj} < p_{21}^{adj}$, truncation occurs at this last stage and we set $p_{54}^{adj} = 0.890$, the adjusted P-value from the previous stage. Overall, we see that the multiplicity adjusted P-values in the last column of Table 15.2 are quite a bit smaller than the ones resulting from using the Bonferroni correction.

15.4 Simultaneous Confidence Intervals

We can invert a multiple testing procedure for the family of hypotheses $H_{ij}^0 : \delta_{ij} = \delta_{ij}^0$ vs $H_{ij}^a : \delta_{ij} \neq \delta_{ij}^0$ to derive simultaneous confidence intervals for the $\{\delta_{ij}\}$'s. For simultaneous lower bounds, we consider the one-sided alternatives $H_{ij}^a : \delta_{ij} > \delta_{ij}^0$, see the example in Section 15.4.1. We again use the maximum score statistic approach as the multiple testing procedure for the pairwise comparison of binary proportions.

In general, any combination of null values $\{\delta_{ij}^0\}_{(i,j)\in\mathcal{I}}$ that is *not rejected* by the simultaneous test procedure is a member of the ($|\mathcal{I}|$-dimensional) simultaneous confidence set for the $\{\delta_{ij}\}_{(i,j)\in\mathcal{I}}$. As before, $\mathcal{I}$ is any set of pairwise comparisons of interest among K groups, such as all comparisons to control with $\mathcal{I} = \{(2,1),(3,1),\ldots,(K,1)\}$ or all pairwise comparisons with $\mathcal{I} = \{(2,1),(3,1),\ldots,(K,1),(3,2),\ldots,(K,K-1)\}$.

In order to not reject any of a given combination of null values $\{\delta_{ij}^0\}_{(i,j)\in\mathcal{I}}$ at FWER α, all multiplicity adjusted P-values p_{ij}^{adj} for the H_{ij}^0 need to be larger than α. Here, $1-\alpha$ is the desired simultaneous coverage probability of the intervals for the $\{\delta_{ij}\}_{(i,j)\in\mathcal{I}}$. The simultaneous coverage probability is the joint probability that each interval (constructed with the method discussed in the following) contains its respective true parameter δ_{ij}. When all p_{ij}^{adj} need to be larger than α, necessarily the smallest needs to be larger than α. The smallest, however, is the one that corresponds to the global intersection hypothesis, testing the elementary hypothesis H_{ij}^0 with the largest observed test statistic. This P-value is given by $P\left(\max_{(i,j)\in\mathcal{I}}|T_{ij}| \geq t_{\max}\right)$, where $t_{\max} = \max_{(i,j)\in\mathcal{I}}|t_{ij}^{\text{obs}}(\delta_{ij}^0)|$ is the largest observed test statistic (in the two-sided case) for any H_{ij}^0. (With this notation, we purposely indicated that the observed test statistics t_{ij}^{obs} depend on the null values δ_{ij}^0.) This smallest P-value needs to be larger than α, or, using the complement,

$$P\left(\max_{(i,j)\in\mathcal{I}}|T_{ij}| < t_{\max}\right) \leq 1-\alpha. \tag{15.7}$$

In Section 15.2.2, we let c be the (equidistant) multivariate normal two-sided $1-\alpha$ quantile (with correlation matrix estimated as in Section 15.2.5) such that $P\left(\max_{(i,j)\in\mathcal{I}}|T_{ij}| < c\right) = \alpha$. Hence, in order for Equation 15.7 to hold (and using the monotonicity of the score statistic), we need

$$t_{\max} = \max_{(i,j)\in\mathcal{I}}|t_{ij}^{\text{obs}}(\delta_{ij}^0)| \leq c$$

or, equivalently,

$$|t_{ij}^{\text{obs}}(\delta_{ij}^0)| \leq c \quad \text{for all } (i,j) \in \mathcal{I}.$$

This means that the smallest multiplicity adjusted P-value will be larger than α if and only if the observed score statistics $|t_{ij}^{\text{obs}}(\delta_{ij}^0)| \leq c$ for all pairwise comparisons of interest. From this relation, we can determine all combination of null values $\{\delta_{ij}^0\}$ that will not be rejected: They are simply those for which

$$|t_{ij}^{\text{obs}}(\delta_{ij}^0)| = \left|\frac{d_{ij} - \delta_{ij}^0}{\left[s_{ij}(\delta_{ij}^0)\right]^{1/2}}\right| \leq c \tag{15.8}$$

holds for all $(i,j) \in \mathcal{I}$, where $s_{ij}(\delta_{ij}^0)$ is the standard error using restricted MLEs defined in the text following equation (15.1).

Summing up, we create the $|\mathcal{I}|$-dimensional acceptance region as $\{$all δ_{ij}^0 with$(i,j) \in \mathcal{I}: \max_{(i,j) \in \mathcal{I}} |t_{ij}^{obs}(\delta_{ij}^0)| < c\}$. This implies that the simultaneous interval for each δ_{ij} is (implicitly) given by all $\delta_{ij}^0 \in [-1,1]$ (the permissible range) for which $|t_{ij}^{obs}(\delta_{ij}^0)| \leq c$, where c is the equidistant, two-sided quantile of the multivariate normal distribution of the T_{ij}'s, with correlation matrix estimated as discussed at the end of Section 15.2.5.

Inequality (15.8) cannot be solved explicitly in terms of the δ_{ij}^0. However, straightforward numerical routines such as interval halving can be used to find all δ_{ij}^0 for which $|t_{ij}^{obs}(\delta_{ij}^0)| \leq c$. Alternatively, one can simply compute each interval with existing routines as the regular univariate score interval for the difference of two binomial proportions (with y_i and y_j successes out of n_i and n_j observations, respectively) using an adjusted (for multiplicity) confidence coefficient of $1 - 2(1 - \Phi(c))$ for each interval, where $\Phi()$ is the standard normal cdf.

The adjusted confidence coefficient corresponding to a Bonferroni adjustment is given by $1 - \alpha/r$, where r is the number of pairwise comparisons. For the all pairwise (with $r = 10$) example given in Section 15.4.2, $c = 2.728$ and the adjusted confidence coefficient (for nominal 95% simultaneous coverage) for each interval equals $1 - 2(1 - \Phi(2.728)) = 99.36\%$, while the Bonferroni adjusted confidence coefficient is given by $1 - 0.05/10 = 99.50\%$, leading to wider intervals than necessary.

Figure 15.3 shows an example of a simultaneous confidence region for both the case of all comparisons to control (top plot) and for all pairwise comparisons (bottom plot), when the number of groups $K = 3$. To construct these, we used the data for the first three dose groups of the IBS example. For the case of two comparisons to control, the plot shows the simultaneous region (rectangle) defined by all $(\delta_{21}^0, \delta_{31}^0)$: $t_{21}(\delta_{21}^0) \leq c, t_{31}(\delta_{31}^0) \leq c$, where $c = 1.916$. The resulting simultaneous lower bounds for δ_{21} and δ_{31} are indicated. For the case of all three pairwise comparisons, we projected the simultaneous confidence region $(\delta_{21}^0, \delta_{31}^0, \delta_{32}^0)$: $|t_{21}(\delta_{21}^0)| \leq c, |t_{31}(\delta_{31}^0)| \leq c, |t_{32}(\delta_{32}^0)| \leq c$, where $c = 2.344$ into the $(\delta_{21}^0, \delta_{31}^0)$ plane. Again, the resulting simultaneous lower and upper confidence bounds for δ_{21} and δ_{31} are indicated. A similar picture can be constructed by projecting the simultaneous region into, for example, the $(\delta_{31}^0, \delta_{32}^0)$ plane.

15.4.1 Example: All Comparisons to Control

Refer to the IBS dataset introduced in Section 15.3.2. Since we are interested in lower bounds on the differences δ_{i1} between the probability of abdominal pain in dose group i vs control, we invert the family of hypotheses $H_{ij}^0: \delta_{ij} = \delta_{ij}^0$ vs $H_{ij}^a: \delta_{ij} > \delta_{ij}^0$. Larger values of the score test statistics T_{ij} are in favor of the alternative hypothesis. The multivariate normal quantile c is computed from the equation $P\left(\max_{(i,1) \in \mathcal{I}} T_{i1} < c\right) = 1 - \alpha$. Using the estimated

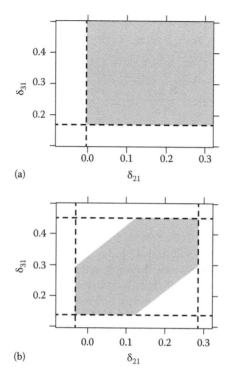

(a)

(b)

FIGURE 15.3
Simultaneous confidence region for the case of two comparisons to control (a) and three pairwise comparisons (b, projected into the two-dimensional space), using the first three dose groups of the IBS trial as data. The bounds for the intervals for the two parameters δ_{21} and δ_{31} are given by where the dashed line meets the x- and y-axes.

correlation matrix of the four score test statistics displayed in Section 15.3.2 and setting $\alpha = 5\%$, $c = 2.161$. For each $i = 2, \ldots, 5$, the largest $\delta_{i1}^0 \in [-1, 1]$ for which $t_{i1}^{obs}(\delta_{i1}^0) \leq 2.161$ is shown in Table 15.3 (last line). These are the 95% simultaneous lower confidence bounds on the δ_{i1}'s.

The 95% simultaneous confidence coefficient means that the joint probability of the true value of δ_{i1} being larger than the corresponding lower bound for each of the four comparison is at least 95%. We see that the lower bound for δ_{31} is larger than the required clinical relevant effect of 10%, but the other lower bounds are smaller. Of the four comparisons, only the one with dose group 3 yields a significant and clinically relevant improvement over control. The improvement is such that with dose group 3, the probability of relief from abdominal pain is at least 15.2 percentage points higher when compared to control. There is no improvement with dose group 2 (lower bound is below zero), at least a 6.3 percentage point improvement with dose group 4 and at least a 8.2 percentage point improvement with dose

TABLE 15.3

Simultaneous 95% Lower Bounds for the Difference in the Probability of Relief from Abdominal Pain Comparing Dose Group $i = 2, \ldots, 5$ to Placebo ($i = 1$) for the IBS Data

Dose Group i	2 (1 mg)	3 (4 mg)	4 (12 mg)	5 (24 mg)
Obs. diff. to control	0.130	0.304	0.216	0.237
Unadj. lower bounds	0.015	0.189	0.100	0.120
Sidak adj. lower bounds	−0.026	0.147	0.058	0.077
Simult. lower bounds	−0.021	0.152	0.063	0.082

Note: The last row shows the simultaneous bounds derived by inverting the maximum score statistic. Unadjusted and Sidak-adjusted lower score bounds are also shown.

group 5. The probability that at least one of these statements is wrong is controlled at 5%. The ability to not only judge the significance but also estimate (a lower bound for) the size of the improvement makes simultaneous confidence intervals more useful than the corresponding multiplicity adjusted P-values from Section 15.3.2. However, since we only used the global intersection hypothesis in the construction of these intervals, that is, the single-step procedure P-values (15.5), they are less powerful compared to the multiplicity adjusted P-values constructed with the step-down procedure of the previous section. That is, there is a one-to-one correspondence between the simultaneous intervals and the single-step multiplicity adjusted P-values based on the maximum distribution but not with the step-down multiplicity adjusted P-values in Table 15.1.

Table 15.3 also shows the simultaneous lower bounds one obtains with the Sidak adjustment, where one uses an adjusted confidence coefficient of $(1 - \alpha)^{1/r} = (1 - 0.05)^{1/4} = 98.73\%$ for each of the four individual score confidence intervals. For comparison, the maximum score approach implies an adjusted confidence coefficient of $1 - (1 - \Phi(c)) = 98.46\%$. The Sidak method (and similarly the Bonferroni method) results in bounds that are lower than necessary.

15.4.2 Example: All Pairwise Comparisons

As before, let δ_{ij} be the difference in the probability of relief from abdominal pain between dose group i and j. To find simultaneous confidence intervals for all 10 possible dose group comparisons (including placebo), we invert the family of hypotheses $H_{ij}^0: \delta_{ij} = \delta_{ij}^0$ vs $H_{ij}^a: \delta_{ij} \neq \delta_{ij}^0$, using the maximum score test. With the estimated correlation matrix as shown in Section 15.3.3, the c in (15.4) equals $c = 2.728$. The simultaneous confidence intervals for all pairwise comparison are then given by all values $\delta_{ij}^0 \in [-1, 1]$ such that $|t_{ij}^{obs}(\delta_{ij}^0)| \leq 2.728$, for all $(i, j) \in \mathcal{I}$. Equivalently, for each pairwise comparison, we can compute the regular univariate score confidence interval

TABLE 15.4

Simultaneous 95% Confidence Intervals for All 10 Pairwise Comparisons among the 5 Dose Groups for the IBS Data

i vs j	Diff.	Unadj. LB	Unadj. UB	Bonf. LB	Bonf. UB	Simul. LB	Simul. UB
2 − 1	13.0	−0.7	26.2	−6.6	31.6	−6.1	31.1
3 − 1	30.4	16.7	42.9	10.6	47.8	11.1	47.4
4 − 1	21.6	7.7	34.6	1.7	39.8	2.3	39.4
5 − 1	23.7	9.7	36.8	3.6	42.0	4.1	41.5
3 − 2	17.4	3.8	30.3	−2.1	35.6	−1.6	35.1
4 − 2	8.6	−5.1	22.0	−11.0	27.5	−10.4	27.0
5 − 2	10.7	−3.2	24.2	−9.1	29.7	−8.6	29.2
4 − 3	−8.8	−21.9	4.6	−27.3	10.4	−26.8	9.9
5 − 3	−6.7	−20.0	6.8	−25.5	12.5	−25.0	12.0
5 − 4	2.1	−11.6	15.7	−17.4	21.4	−16.9	20.9

Note: The last two columns show the simultaneous interval derived from inverting the maximum score statistic. Bonferroni-adjusted and Bonferroni-unadjusted score intervals are also shown.

for the difference of proportion with adjusted confidence coefficient $1 - 2(1 - \Phi(2.728)) = 99.36\%$. The resulting simultaneous intervals for all pairwise δ_{ij} are shown in Table 15.4.

We see that dose groups 3, 4, and 5 (but not dose group 2) are significantly better (in terms of relief of abdominal pain) than placebo (dose group 1). However, only for dose group 3, the improvement is sufficient to meet clinical relevance, as the probability of relief from abdominal pain is at least 11.1 percentage points higher than in the placebo group. The other two groups show improvements of at least 2.3 and 4.1 percentage points over placebo, respectively. Further, there was no evidence of a significant difference between the active dose groups, although the interval for the comparison of groups 3 and 2 indicates that the probability may be as much as 35 percentage points larger in group 3. The probability that at least one of the 10 intervals displayed in Table 15.4 does not capture the true difference is bounded by 5%. Similar to the all comparison to control case, the Bonferroni-adjusted intervals are unnecessarily wide.

15.5 R Implementations

Here, we illustrate how to use our costume R functions to obtain and reproduce all results in this chapter. For the code to run, the `multcomp` package needs to be installed. After scouring the necessary functions (first line), we use the function `pair.test()` to obtain multiplicity adjusted *P*-values for all comparisons to control (`contrasts = "Dunnett"`) or all pairwise comparisons (`contrasts = "Tukey"`), for the number of successes (y) and sample sizes (n) as given in Table 15.1. The null values δ_{ij}^0 can be specified via the `delta0 = ...` option. For both type of comparisons, they default to all zeros.

The function `pair.ci()` can be used to obtain simultaneous confidence intervals for the δ_{ij}, again for both types of pairwise comparisons, as selected by the `contrasts = ...` option. The simultaneous confidence level can be specified via the `conflev = ...` option, which defaults to 0.95. Incorporating other types of pairwise comparisons into the R code is under construction.

```
> source("http://sites.williams.edu/bklingen/files/2013/08/
           pairwise.r")
> # Example using IBS data:
> # Multiplicity adjusted P-values:
> # All comparisons to control (=first group):
> pair.test(
    delta0 = c(0.1,0.1,0.1,0.1),
    y = c(38,52,67,59,58),
    n = c(100,102,98,99,94),
    contrasts = "Dunnett",
    alternative = "greater"
    )
$p.adj
       2 - 1        3 - 1        4 - 1        5 - 1
0.334177115 0.007091914 0.088130962 0.069199269
>
> # All pairwise comparisons:
> pair.test(
+ delta0 = c(0.1,0.1,0.1,0.1,0,0,0,0,0,0),
+ y = c(38,52,67,59,58),
+ n = c(100,102,98,99,94),
+ contrasts = "Tukey",
+ alternative = "two.sided"
+ )
$p.adj
       2 - 1        3 - 1        4 - 1        5 - 1        3 - 2
0.89001108 0.03186191 0.30433737 0.22070294 0.05925578
       4 - 2        5 - 2        4 - 3        5 - 3        5 - 4
0.52407229 0.30433737 0.52407229 0.55476285 0.89001108
>
> # Simultaneous confidence intervals:
```

```
> # All comparisons to control (=first group):
> # Simultaneous lower bounds:
> pair.ci(
+ y = c(38,52,67,59,58),
+ n = c(100,102,98,99,94),
+ contrasts = "Dunnett",
+ conflev = 0.95,
+ type = "lower"
+ )
$CI.adj
            [,1]       [,2]        [,3]        [,4]
[1,] -0.02138243 0.1521732 0.06306853 0.08227054
>
> # All pairwise comparisons:
> pair.ci(
+ y = c(38,52,67,59,58),
+ n = c(100,102,98,99,94),
+ contrasts = "Tukey",
+ conflev = 0.95,
+ type = "two.sided"
+ )
$CI.adj
              [,1]        [,2]
2 - 1 -0.06069078 0.31113163
3 - 1  0.11145717 0.47398147
4 - 1  0.02272053 0.39366390
5 - 1  0.04127673 0.41524286
3 - 2 -0.01556079 0.35116990
4 - 2 -0.10428158 0.27035608
5 - 2 -0.08582727 0.29194055
4 - 3 -0.26831383 0.09855288
5 - 3 -0.25009780 0.12012208
5 - 4 -0.16887197 0.20881911
```

References

Agresti, A., Bini, M., Bertaccini, B., and Ryu, E. (2008). Simultaneous confidence intervals for comparing binomial parameters. *Biometrics*, 64:1270–1275.

Agresti, A. and Caffo, B. (2000). Simple and effective confidence intervals for proportions and differences of proportions result from adding two successes and two failures. *American Statistician*, 54:280–288.

Bretz, F., Hothorn, T., and Westfall, P. (2011). *Multiple Comparison Using R*. Chapman & Hall/CRC Press, Boca Raton, FL.

Dunnett, C. (1955). A multiple comparison procedure for comparing several treatments with a control. *Journal of the American Statistical Association*, 50:1096–1121.

Klingenberg, B. (2010). Simultaneous confidence bounds for relative risks in multiple comparisons to control. *Statistics in Medicine*, 29:3232–3244.

Klingenberg, B. (2012). Simultaneous score confidence bounds for risk differences in multiple comparisons to a control. *Computational Statistics and Data Analysis*, 56:1079–1089.

Marcus, R., Peritz, E., and Gabriel, K. (1976). On closed testing procedures with special reference to ordered analysis of variance. *Biometrika*, 63:655–660.

Mee, R. (1984). Confidence bounds for the difference between two probabilities (letter). *Biometrics*, 40:1175–1176.

Miettinen, O. and Nurminen, M. (1985). Comparative analysis of two rates. *Statistics in Medicine*, 4:213–226.

Newcombe, R. (1998). Interval estimation for the difference between independent proportions: Comparison of eleven methods. *Statistics in Medicine*, 17:873–890.

Newcombe, R. and Nurminen, M. (2011). In defence of score intervals for proportions and their differences. *Communication in Statistics: Theory and Methods*, 40:1271–1282.

Nurminen, M. (1986). Confidence intervals for the ratio and difference of two binomial proportions. *Biometrics*, 42:675–676.

Piegorsch, W. (1991). Multiple comparisons for analyzing dichotomous response. *Biometrics*, 47:45–52.

Schaarschmidt, F., Biesheuvel, E., and Hothorn, L. (2009). Asymptotic simultaneous confidence intervals for many-to-one comparisons of binary proportions in randomized clinical trials. *Journal of Biopharmaceutical Statistics*, 19:292–310.

16

Comparative Study of Five Weighted Parametric Multiple Testing Methods for Correlated Multiple Endpoints in Clinical Trials

Changchun Xie, Xuewen Lu, and Ding-Geng (Din) Chen

CONTENTS

16.1 Introduction

In order to obtain better overall knowledge of a treatment effect in clinical trials, the clinical trialists often collect many medically related or correlated endpoints and test the treatment effect for each. However, the problem of multiplicity arises when multiple hypotheses are tested. Ignoring this problem can cause false positive results. A lot of statistical methods have been proposed to handle multiplicity issues in clinical trials. However, many commonly used multiple testing correction methods proposed to control family-wise type I errors (FWER) disregard the correlation among the test statistics, for example, the Bonferroni correction and Holm procedure.

When some hypotheses are more important than others, weighted multiple testing correction methods are required. Many weighted multiple testing correction methods have been proposed in the literature, such as weighted Holm-Bonferroni method (Holm, 1979), the prospective alpha allocation scheme (PAAS) (Moyé, 1998, 2000), Bonferroni fixed sequence procedure (BFS) (Wiens, 2003), and alpha-exhaustive fallback (AEF) (Wiens and Dmitrienko, 2005, 2010). However, all these methods do not take into account correlation among the test statistics either. Westfall and Young (1993) have shown that substantial improvements can be obtained by using permutation method instead of Bonferroni correction since the dependence structures of the multiple endpoints are automatically incorporated into the analysis. However, this permutation method is computationally intensive and is rarely used to set the alpha level for sample size calculation at the time of protocol development because the data required by permutation have not been collected yet (Xie, 2014), and hence, this method will not be discussed further in this chapter.

Recently, weighted parametric multiple testing methods have been proposed to take into account correlations among the test statistics. Huque and Alosh (2008) suggested a flexible fixed-sequence (FFS) testing method to improve the BFS procedure by considering correlation among endpoints, which can be viewed as a parametric version of the fallback procedure. Li and Mehrotra (2008) developed an adaptive alpha allocation approach (4A), where the first endpoint is tested at a prespecified level, the second endpoint at an adaptive level based on the p-value for the first endpoint and the correlation between the first endpoint and the second endpoint. Xie (2012) proposed a weighted multiple testing correction for correlated tests (WMTCc), assuming that test statistics are asymptotically distributed as multivariate normal with known correlation matrix or the correlation matrix, which can be correctly estimated from the data. This WMTCc method can be viewed as a parametric version of the weighted Holm procedure (Holm, 1979). Although this method was proposed for correlated continuous endpoints, it has been evaluated when correlated binary endpoints are used (Xie et al., 2013a). Simulations show it still has higher power for testing each hypothesis than the weighted Holm method, especially when the correlation between endpoints is high (Xie et al., 2013a). The other two methods are Bretz et al. (2011) graphical approaches and Millen and Dmitrienko (2011) chain procedures. The graphical approach has been incorporated into the *gMCP* package (Rohmeyer and Klinglmueller, 2011) in R software (R Development Core Team, 2010) for one-sided tests. So far, it cannot control FWER at a prespecified level if two-sided tests are used (Xie and Chen, 2013). The chain procedures give weights on test statistics, while the FFs, 4A, and WMTCc methods give weights on p-values. Xie (2013) proposed a retested FFS method and a retested 4A method and compared them with the WMTCc and the original FFS and 4A methods. All these weighted parametric multiple testing methods assume the correlation among endpoints are known or can be correctly estimated from the data.

However, the exact correlations among endpoints are usually unknown. If the correlations are misspecified, the type I error rate can be out of control. We usually think the type I error rate will be inflated if the correlations are overestimated (Wiens and Dmitrienko, 2010). From our simulations, this is not true for 4A method (Xie et al., 2013b). When the inflation occurs and how large the inflation can be might depend on the method and the magnitude of the misspecification of the correlations among the endpoints.

In this chapter, a comparative study is conducted to compare five weighted multiple testing methods when the correlations are correctly specified as well as when the correlations are misspecified. The graphical approach and chain procedures are not included in the comparisons since for the time being the implement of graphical approach (gMCP) is based on one-sided test only and the chain procedure gives weights on test statistics instead of p-values, which change the interpretation due to a nonlinear relationship between p-values and test statistics. The outline of this chapter is as follows. The reviews of the original FFS, the retested FFS, the original 4A, the retested 4A, and the WMTCc are presented in Section 16.2. Comparisons of these five weighted parametric multiple testing methods are provided in Section 16.3. Examples are given in Section 16.4 to illustrate these parametric multiple testing methods. Finally, some discussions and tentative guidelines to help choosing an appropriate method are provided in Section 16.5.

16.2 Weighted Parametric Multiple Testing Methods

16.2.1 FFS Method

Suppose that T_1 and T_2 are test statistics for hypothesis $H_0^{(1)}$ and $H_0^{(2)}$, respectively. The FFS procedure is as follows:

1. Test $H_0^{(1)}$ at the significance level $\alpha_1 (<\alpha)$ and reject $H_0^{(1)}$ if $T_1 \geq C_{1;\alpha_1}$, where $C_{1;\alpha_1}$ satisfies $P\left(T_1 \geq C_{1;\alpha_1} \mid H_0^{(1)}\right) = \alpha_1$.

2. If $H_0^{(1)}$ is rejected, test $H_0^{(2)}$ at the significance level α. If $H_0^{(1)}$ is not rejected, test $H_0^{(2)}$ at the significance level α_2 and reject $H_0^{(2)}$ if $T_2 \geq C_{2;\alpha_2}$, where $C_{2;\alpha_2}$ satisfies $P\left(T_1 < C_{1;\alpha_1}, T_2 \geq C_{2;\alpha_2} \mid H_0^{(1)}, H_0^{(2)}\right) = \alpha - \alpha_1$.

16.2.2 4A Method

Assume p_1 and p_2 are the p-values for the first endpoint and the second endpoint respectively. Let ρ be the correlation between the two endpoints. The 4A method tests the null hypothesis for the first endpoint at the prespecified level $\alpha_1 (< \alpha)$ and tests the null hypothesis for the second endpoint at the adaptive level $\alpha_2 = \alpha$ if $p_1 \leq \alpha_1$ and $\alpha_2 = \min(\alpha_1, \lambda(\alpha_1, \alpha, \rho)\alpha_t/p_1^2)$ if $p_1 > \alpha_1$, where $\lambda(\alpha_1, \alpha, \rho)$ is the largest constant such that $P\left(p_1 > \alpha_1, p_2 \leq \alpha_2\left(p_1, \alpha_1, \alpha, \rho\right)\right) \leq \alpha - \alpha_1$ and $\alpha_t = \alpha_1(\alpha - \alpha_1)/(1 - \alpha_1)$ if $\alpha_1 + \alpha_1^2 - \alpha_1^3 > \alpha$; otherwise,

$$\alpha_t = \alpha_1 \left(1 - \sqrt{\left(2\alpha_1 - \alpha - \alpha_1^2\right)/\alpha_1}\right)^2.$$

16.2.3 WMTCc Method

Let $p_1, \ldots, p_m$ be the observed p-values for null hypotheses $H_0^{(1)}, \ldots, H_0^{(m)}$, respectively, and $w_i > 0$, $i = 1, \ldots, m$ be the weight for null hypothesis $H_0^{(i)}$. Note that we do not assume that $\sum_{i=1}^m w_i = 1$. Freedom from this constraint is taken from (16.1), where we will see that the adjusted p-values only depend on the ratios of the weights. For each $i = 1, \ldots, m$, let $q_i = p_i/w_i$. (Note p_i and q_i are realizations of random variables P_i and Q_i, respectively.) Then the adjusted p-value for the null hypothesis $H_0^{(i)}$ is

$$
\begin{aligned}
P_{adj\_i}^m &= P\left(\min_j \; Q_j \leq q_i\right) \\
&= 1 - P\left(\text{all} \;\; Q_j > q_i\right) \\
&= 1 - P\left(\text{all} \;\; P_j/w_j > p_i/w_i\right) \\
&= 1 - P\left(\text{all} \;\; P_j > p_i w_j/w_i\right) \\
&= 1 - P\left(\bigcap_{j=1}^m a_j \leq X_j \leq b_j\right),
\end{aligned}
\tag{16.1}
$$

where $X_j, j = 1, \ldots, m$ are standardized multivariate normal with correlation matrix Σ and

$$a_j = \Phi^{-1}\left(\frac{p_i w_j}{2w_i}\right), \quad b_j = \Phi^{-1}\left(1 - \frac{p_i w_j}{2w_i}\right) \tag{16.2}$$

for the two-sided case and

$$a_j = -\infty, \quad b_j = \Phi^{-1}\left(1 - \frac{p_i w_j}{w_i}\right) \tag{16.3}$$

for the one-sided case.

If $P^m_{adj\_i} \leq \alpha$, we will reject the corresponding null hypothesis $H_0^{(i)}$. Suppose k_1 null hypotheses have been rejected, we then adjust the remaining $m - k_1$ observed p-values for multiple testing after removing the rejected k_1 null hypotheses, using the corresponding correlation matrix and weights. For example, if we begin with $m = 3$, the correlation matrix $\begin{pmatrix} 1 & \rho_{12} & \rho_{13} \\ \rho_{12} & 1 & \rho_{23} \\ \rho_{13} & \rho_{23} & 1 \end{pmatrix}$, and the following weights (w_1, w_2, w_3), but discover that only $P^3_{adj\_3} \leq \alpha$. After removing $H_0^{(3)}$, the remaining correlation matrix is $\begin{pmatrix} 1 & \rho_{12} \\ \rho_{12} & 1 \end{pmatrix}$ and the corresponding weights are now (w_1, w_2). We continue these procedures until there are no remaining null hypotheses, which can be rejected. The package *mvtnorm* in R software (Genz et al., 2010; R Development Core Team, 2010) can be used to calculate the adjusted p-values in (16.1).

16.2.4 Retested FFS and 4A Methods

The retested FFS procedure is as follows: (1) and (2) are the same as those in the FFS procedure. (3) If $H_0^{(1)}$ is not rejected and $H_0^{(2)}$ is rejected, test $H_0^{(1)}$ at the significance level α.

The retested 4A procedure is as follows:

1. Perform the 4A procedure.
2. If $p_1 > \alpha_1$ and the null hypothesis for the second endpoint is rejected, test the null hypothesis for the first endpoint at the significance level α. Note that the FWER of the retested FFS procedure is the same as the FWER of the FFS procedure and the FWER of the retest 4A procedure is the same as the FWER of the 4A procedure.

16.3 Comparison of Weighted Parametric Multiple Testing Methods

16.3.1 Correlations are Correctly Specified

The original FFS and 4A methods have the same power for testing the first hypothesis in the testing sequence and do not change when correlation increases. The retested FFS procedure has more power to reject the null hypothesis for the first and usually more important endpoint than the FFS procedure due to the retest. Also, the retested 4A procedure has more power to reject the null hypothesis for the first endpoint than the 4A procedure. However, when correlation is high ($\rho \geq 0.7$), the WMTCc method has higher power for testing the first hypothesis than the retested FFS and retested 4A method (Xie, 2013).

In the FFS and 4A methods, when the first and second endpoints have the same effect size, the power for testing the first hypothesis can be lower than the power for the second hypothesis although the α allocation given to the first hypothesis is higher (Xie, 2014). Given that hypotheses are ranked in order of importance, this potential reversal of desired power for the two hypotheses is inappropriate. The relative importance of the two hypotheses is clear in the retested FFS, retested 4A, and WMTCc methods, whereby the hypothesis with larger weight always has higher power when the endpoints have the same effect size (Xie, 2013).

The retested FFS and the retested 4A methods have the same power for the second endpoint compared to the original FFS and 4A methods, respectively. When the second endpoint is adequately powered and the first endpoint is underpowered, the retested 4A method performs badly compared to the retested FFS and WMTCc methods, especially when the correlation between the two endpoints is high. The retested FFS method has higher power for the second endpoint and higher power to reject at least one null hypothesis compared to the WMTCc method (Xie, 2013). When the first endpoint is adequately powered and the second endpoint is potentially underpowered, the retested 4A method shows higher power for the second endpoint compared to the retested FFS and WMTCc method, when the correlation between the two endpoints is low. However, when the correlation between the two endpoints is high, all the methods show similar power for the second endpoint although the WMTCc method has higher power to reject at least one null hypothesis compared to the retested FFS and retested 4A method (Xie, 2013). When the first and the second endpoints have the same effect size, the retested 4A method has higher power for the second endpoint and higher power to reject at least one null hypothesis compared to the retested FFS and WMTCc methods.

The original FFS and 4A methods have the property that the rejection of earlier hypothesis (usually more important hypothesis) does not depend on the rejection of the later hypothesis (less important hypothesis). The retested FFS, re-tested 4A, and WMTCc do not have this property. Like the original FFS and 4A methods, the retested FFS and retested 4A methods require the prespecified testing sequence and face a computational difficulty if there are more than three endpoints, while the WMTCc method does not need the prespecified testing sequence and can handle up to a thousand endpoints, using the package *mvtnorm* in R (Genz et al., 2010; R Development Core Team, 2010).

16.3.2 Correlations are Misspecified

Simulations were conducted to compare the family-wise type I error rate of the parametric multiple testing methods when the correlations are misspecified (Xie et al., 2013a). Since the original FFS and the retested FFS methods have the same family-wise type I error rate and the original 4A and the retested 4A methods have the same family-wise type I error rate, we will not distinguish the original FFS/4A from the retested FFS/4A and only use FFS and 4A in this section.

We considered trials with two endpoints. Each trial has 240 individuals. Each individual had probability 0.5 to receive the active treatment and probability 0.5 to receive placebo. The endpoints were generated from a bivariate normal distribution with the correlation between the two endpoints, ρ chosen as 0.0, 0.1, 0.2, 0.3, 0.4, 0.5, 0.6, 0.7, 0.8, and 0.9. The treatment effect size (per unit standard deviation) was assumed as 0.0. The weights for the two endpoints were (4, 1), which correspond to alpha allocations (0.04, 0.01). The observed p-values were calculated using two-sided t-tests for the coefficient of the treatment, $\beta = 0$, in linear regressions. For each correlation given in data generation, the specified correlations, 0.0, 0.1, 0.2, 0.3, 0.4, 0.5, 0.6, 0.7, 0.8, and 0.9 were used in multiple testing adjustments. The simulation results are shown in Table 16.1.

From these simulations, we can conclude the following:

1. All the FFS, 4A, and WMTCc methods have the simulated family-wise type I error rate at 5.0% if the correlation is correctly specified.

2. Both WMTCc and FFS have inflated family-wise type I error rate if the correlation among the endpoints is overspecified and the family-wise type I error rate increases with the magnitude of the misspecification of the correlations. Family-wise type I error rate of FFS increases faster than that of WMTCc. Both WMTCc and FFS have deflated family-wise type I error rate if the correlation among the endpoints is underspecified and the family-wise type I error rate decreases with the magnitude of the misspecification of the correlations. Family-wise type I error rate of FFS decreases a little bit faster than that of WMTCc.

3. Unlike WMTCc and FES, 4A has deflated family-wise type I error rate if the correlation among the endpoints is overspecified and the family-wise type I error rate decreases with the magnitude of the misspecification of the correlations. 4A has inflated family-wise type I error rate if the correlation among the endpoints is underspecified and the family-wise type I error rate increases with the magnitude of the misspecification of the correlations.

4. When the true correlation among the endpoints >0.5, the effect of the misspecified correlation in both WMTCc and FFS methods can be larger than the effect of the misspecified correlation in the 4A method.

TABLE 16.1

Two Endpoints: Simulated Family-Wise Type I Error Rate (%) Based on 1,000,000 Runs for WMTC, FFS, and 4A When Different Correlations Are Specified in Multiple Testing Adjustments

True Correlations	Methods	Specified Correlations in Multiple Testing Adjustments									
		0.0	0.1	0.2	0.3	0.4	0.5	0.6	0.7	0.8	0.9
0.0	WMTCc	5.0	5.0	5.0	5.1	5.1	5.2	5.3	5.5	5.7	6.0
	FFS	5.0	5.0	5.0	5.1	5.1	5.2	5.4	5.6	6.0	6.8
	4A	5.0	5.0	4.9	4.8	4.7	4.6	4.5	4.4	4.4	4.4
0.1	WMTCc	5.0	5.0	5.0	5.0	5.1	5.2	5.3	5.5	5.7	6.0
	FFS	5.0	5.0	5.0	5.0	5.1	5.2	5.4	5.6	6.0	6.8
	4A	5.0	5.0	4.9	4.8	4.7	4.6	4.5	4.4	4.4	4.4
0.2	WMTCc	5.0	5.0	5.0	5.0	5.1	5.2	5.3	5.5	5.7	6.0
	FFS	5.0	5.0	5.0	5.0	5.1	5.2	5.4	5.6	6.0	6.7
	4A	5.1	5.1	5.0	4.9	4.8	4.7	4.6	4.5	4.5	4.5
0.3	WMTCc	4.9	4.9	5.0	5.0	5.1	5.1	5.3	5.4	5.6	5.9
	FFS	4.9	4.9	5.0	5.0	5.1	5.1	5.3	5.5	5.9	6.7
	4A	5.2	5.1	5.1	5.0	4.9	4.8	4.6	4.6	4.5	4.5
0.4	WMTCc	4.9	4.9	4.9	5.0	5.0	5.1	5.2	5.4	5.6	5.8
	FFS	4.9	4.9	4.9	4.9	5.0	5.1	5.2	5.5	5.8	6.5
	4A	5.3	5.3	5.2	5.1	5.0	4.9	4.8	4.7	4.6	4.6
0.5	WMTCc	4.8	4.8	4.8	4.9	4.9	5.0	5.1	5.3	5.5	5.7
	FFS	4.8	4.8	4.8	4.9	4.9	5.0	5.1	5.4	5.7	6.4
	4A	5.5	5.4	5.4	5.3	5.1	5.0	4.9	4.8	4.7	4.8
0.6	WMTCc	4.7	4.7	4.7	4.8	4.8	4.9	5.0	5.2	5.4	5.6
	FFS	4.7	4.7	4.7	4.7	4.8	4.9	5.0	5.2	5.5	6.2
	4A	5.6	5.6	5.5	5.4	5.3	5.1	5.0	4.9	4.8	4.8
0.7	WMTCc	4.6	4.6	4.6	4.6	4.7	4.8	4.9	5.0	5.2	5.5
	FFS	4.6	4.6	4.6	4.6	4.7	4.7	4.8	5.0	5.3	5.9
	4A	5.7	5.7	5.6	5.5	5.4	5.3	5.1	5.0	4.9	4.9
0.8	WMTCc	4.4	4.4	4.4	4.5	4.5	4.6	4.7	4.8	5.0	5.3
	FFS	4.4	4.4	4.4	4.4	4.5	4.5	4.6	4.8	5.0	5.6
	4A	5.7	5.7	5.7	5.6	5.5	5.3	5.2	5.1	5.0	5.0
0.9	WMTCc	4.2	4.2	4.2	4.2	4.3	4.4	4.5	4.6	4.8	5.0
	FFS	4.2	4.2	4.2	4.2	4.2	4.2	4.3	4.4	4.6	5.0
	4A	5.5	5.5	5.5	5.4	5.4	5.3	5.2	5.1	5.0	5.0

Note: The total sample size is 240. α allocation is (0.04, 0.01) or weight is (4, 1).

16.4 Examples

In this section, examples have been provided to illustrate the performances of the weighted parametric multiple testing methods for different clinical trial scenarios.

16.4.1 Example 1

Assume a two-arm trial with two-sided $\alpha = 0.05$, two endpoints, the correlation between the two endpoints, $\rho = 0.7$, a corresponding weights (4, 1), and the observed p-values: 0.043, 0.016.

Both the original FFS method and 4A method reject the second null hypothesis of no treatment difference for the second endpoint, but they cannot reject the first null hypothesis of no treatment difference for this first and most important endpoint since 0.043 > 0.04. In fact, both the FFS and 4A methods cannot reject the first null hypothesis, even as ρ increases to 0.99. However, the retested FFS and the re-tested 4A can also reject the first hypothesis of no treatment difference since 0.043 < 0.05. In this example, the WMTCc method can reject both the first and the second null hypotheses of no treatment difference.

16.4.2 Example 2

This example is the same as Example 1 except that the correlation between the two endpoints, ρ, is 0.3.

The FFS method and the retested FFS method cannot reject the first and the second null hypotheses of no treatment difference since 0.043 > 0.04 and 0.016 > 0.0111977. The 4A method cannot reject the first null hypothesis but can reject the second null hypothesis since 0.016 < 0.04. The retested 4A method can reject both the first and the second null hypotheses. The WMTCc method using $\rho = 0.3$ and the weight (4,1) gives the working adjusted p-values: 0.0525 and 0.0777, respectively. Since all the adjusted p-values > 0.05, the WMTCc cannot reject both the first and the second null hypotheses.

16.4.3 Example 3

This example is the same as Example 1 except that the observed p-values for the first endpoint is 0.5.

As in the Example 1, the FFS and the retested FFS methods cannot reject the first null hypothesis but can reject the second null hypothesis. The 4A method cannot reject either the first or the second null hypothesis since 0.5 > 0.04 and 0.016 > 0.0009. The WMTCc method using $\rho = 0.7$ and the weight (4,1) gives the *working* adjusted p-values: 0.522 and 0.0715, respectively. Since all the adjusted p-values >0.05, the WMTCc cannot reject either the first or the second null hypothesis.

16.4.4 Example 4

This example is the same as Example 1 except that the observed p-values for the first and the second endpoints are 0.042 and 0.043, respectively.

Like the FFS method, the retested FFS method cannot reject either the first or the second null hypothesis of no treatment difference, since $0.042 > 0.04$ and $0.043 > 0.0171479$. Like the 4A method, the retested 4A method cannot reject either the first or the second null hypothesis since $0.042 > 0.04$ and $0.043 > 0.04$. The WMTCc method, using $\rho = 0.7$ and the weight (4, 1), gives the *working* adjusted p-values: 0.0474 and 0.187, respectively. The first null hypothesis can be rejected. After removing the rejected null hypothesis, only one test (for the second endpoint) remains. Since $0.043 > 0.05$, the second null hypothesis is rejected. The WMTCc can reject both null hypotheses, which cannot be rejected by the FFS, the retested FFS, the 4A, and the retested 4A methods.

16.5 Discussion

In this chapter, we have compared five weighted parametric multiple testing procedures. We conclude that there is no universally superior procedure.

As a tentative guideline, if the investigator needs the property that the rejection of earlier hypothesis (usually more important hypothesis) does not depend on the rejection of the later hypothesis (less important hypothesis), the original FFS method or the original 4A method might be used. Otherwise, the retested FFS or the retested 4A methods should be chosen since they have more power than the original FFS and 4A methods respectively. The retested FFS method may be preferred if the investigator's primary interest is to reject any of the null hypotheses when the earlier endpoints are underpowered and the later endpoints are adequately powered. The retested 4A method may be preferred if the investigator's primary interest is to reject any of the null hypotheses when all the endpoints have the same effect size. The WMTCc method may be preferred when one of the following conditions is satisfied: (1) There are more than three endpoints. (2) The testing sequence is not easy to be specified. (3) The endpoint with the more weight potentially has larger effect size and the correlation between endpoints is high. When correlation is unknown, to be conservative, we may use the lower limit of the correlation for the WMTCc or FFS method and the upper limit for the 4A method.

Acknowledgment

This project was partially supported by the National Center for Research Resources and the National Center for Advancing Translational Sciences, National Institutes of Health, through Grant 8 UL1 TR000077-04. The content is solely the responsibility of the authors and does not necessarily represent the official views of the NIH.

References

Bretz, F., Posch, M., Glimm, E., Klinglmueller, F., Maurer, W., and Rohmeyer, K. (2011). Graphical approaches for multiple comparison procedures using weighted Bonferroni, Simes, or parametric tests. *Biometrical Journal* 53:894–913.

Genz, A., Bretz, F., and Hothorn, T. (2010). mvtnorm: multivariate normal and t distribution. R package version 2.12.0. http://cran.r project.org/web/packages/mvtnorm/index.html, accessed June 9, 2010.

Huque, M. F. and Alosh, M. (2008). A flexible fixed-sequence testing method for hierarchically ordered correlated multiple endpoints in clinical trials. *Journal of Statistical Planning and Inference* 138:321–335.

Holm, S. (1979). A simple sequentially rejective multiple test procedure. *Scandinavian Journal of Statistics* 6:65–70.

Li, J. and Mehrotra, D. V. (2008). An efficient method for accommodating potentially underpowered primary endpoints. *Statistics in Medicine* 27:5377–5391.

Millen, B. A. and Dmitrienko, A. (2011). A class of flexible closed testing procedures with clinical trial applications. *Statistics in Biopharmaceutical Research* 3:14–30.

Moyé, L. A. (1998). P-value interpretation and alpha allocation in clinical trials. *Annals of Epidemiology* 8:351–357.

Moyé, L. A. (2000). Alpha calculus in clinical trials: Considerations and commentary for the new millennium. *Statistics in Medicine* 19:767–779.

R Development Core Team. (2010). R: A language and environment for statistical computing. R Foundation for Statistical Computing. http://www.r-project.org/, accessed June 9, 2010.

Rohmeyer, K. and Klinglmueller, F. (2011). gMCP: A graphical approach to sequentially rejective multiple test procedures. R package version 0.6-5. http://cran.r-project.org/package=gMCP, accessed December 18, 2011.

Westfall, P. H. and Young, S. S. (1993). *Resampling-Based Multiple Testing: Examples and Methods for P-Value Adjustment*. Wiley: New York.

Wiens, B. L. (2003). A fixed sequence Bonferroni procedure for testing multiple endpoints. *Pharmaceutical Statistics* 2:211–215.

Wiens, B. L. and Dmitrienko, A. (2005). The fallback procedure for evaluating a single family of hypotheses. *Journal of Biopharmaceutical Statistics* 15:929–942.

Wiens, B. L. and Dmitrienko, A. (2010). On selecting a multiple comparison procedure for analysis of a clinical trial: Fallback, fixed sequence, and related procedures. *Statistics in Biopharmaceutical Research* 2:22–32.

Xie, C. (2012). Weighted multiple testing correction for correlated tests. *Statistics in Medicine* 31:341–352.

Xie, C. (2013). Re-tested parametric multiple testing methods for correlated tests. *Quantitative Bio-Science* 32(1):1–5.

Xie, C. (2014). Relations among three parametric multiple testing methods for correlated tests. *Journal of Statistical Computation and Simulation* 84(4):812–818.

Xie, C. and Chen, D. (2013). Letter to the editor: On weighted parametric tests with "gMCP" R package. *Biometrical Journal* 55(2):264–265. doi:10.1002/bimj.201200116.

Xie, C., Lu, X., Pogue, J., and Chen, D. (2013a). Weighted multiple testing corrections for correlated binary endpoints. *Communications in Statistics-Simulation and Computation* 42(8):1693–1702. doi:10.1080/03610918.2012.674599.

Xie, C., Lu, X., Singh, R., and Chen, D. (2013b). Effect of misspecified correlations in parametric multiple testing. In *JSM Proceedings, Biopharmaceutical Section, American Statistical Association* 1753–1764.

Section V

Clinical Trials in a Genomic Era

17

Statistical Analysis of Biomarkers from -Omics Technologies

Herbert Pang and Hongyu Zhao

CONTENTS

17.1 Introduction

17.1.1 Technological Development

Rapid developments and applications of various genomics and proteomics platforms in recent years have revolutionized biomedical research. Instead of focusing on one or a few candidate genes, researchers now routinely collect the expression information of all transcripts in the genome, study their coexpression patterns, and correlate expression profiles with disease phenotypes and clinical outcomes. One of the most cited papers is Golub et al. (1999), where the authors demonstrated that gene expression profiles can be used to reliably define leukemia subtypes (Golub et al. 1999). This approach has been applied to essentially all cancer types where gene expression levels from a small number of genes can be used to characterize tumor heterogeneity. Earlier platforms were based on gene expression microarrays, whereas next-generation sequencing is more commonly used now to measure gene expressions as well as other molecular phenotypes, for example, methylation to discover and define cancer subtypes. In these cases, the genes selected that represent cancer subtypes can be regarded signature biomarkers.

In addition to using expression levels of a subset of genes to define disease subtypes, researchers have also been successful to correlate genomic features with treatment responses. For example, for breast cancer, MammaPrint (which is based on the expression levels of 70 genes) and Oncotype DX (which is based on the expression levels of 21 genes) are two genomics-based tests for treatment choices. These 70 genes and 21 genes are biomarkers for these two genomics tests. It is likely with more samples collected and analyzed, and with advances in statistical and computational methods, many successes will be achieved within the next several years where genomics features can be used to identify more effective treatment targets and to assign the most effective treatments to a patient.

17.1.2 Popular Data Repository

Even in the early phase of gathering and analyzing genomics data, the genomics research community realized the importance of open data access and sharing. Common data formats were adopted, and various databases were established. More general purpose microarray databases include Gene Expression Omnibus (http://www.ncbi.nlm.nih.gov/geo/)

(Edgar et al. 2002) and Array Express (http://www.ebi.ac.uk/arrayexpress/) (Parkinson et al. 2007). There are many databases focusing on disease areas, for example, Oncomine (https://www.oncomine.org/) (Rhodes et al. 2004) was developed to collect and curate data sets related to cancer. Various community-based genomics projects have also created and maintain public databases, for example, The Cancer Genome Atlas Project (http://cancergenome.nih.gov), (Chin et al. 2011) Genotype-Tissue Expression Portal (http://www.broadinstitute.org/gtex/) (GTEx Consortium 2013), and Connectivity Map (http://www.broadinstitute.org/cmap/) (Lamb et al. 2006), that offer rich information as well as some informatics tools for mining their data.

17.2 Classification Methods

In this section, we will introduce several commonly used classification methods, which are popular alternatives for binary outcomes data in place of model-based approaches like logistic regression. In Section 17.2.1, we introduce discriminant analysis and its variations. In Section 17.2.2, we introduce two well-known nonparametric approaches to classification problems: k-nearest neighbor and random forests. In Section 17.2.3, we briefly review multiclass problems. In Section 17.2.4, we describe commonly used performance measures in a binary classification setting.

17.2.1 Discriminant Analysis

The main goal of discriminant analysis is to assign an unknown subject to one of G classes on the basis of observed subjects from each class. Let $X_{g,1}, \ldots, X_{g,n_g}$ be independent and identically distributed random vectors from a p-dimensional multivariate normal distribution with mean vector μ_g and covariance matrix Σ_g for class $g = 1, \ldots, G$. Let $n = n_1 + \cdots + n_G$ be the total number of observations. Note that the sample covariance matrices are singular when p is larger than n. Therefore, traditional methods such as linear discriminant analysis (LDA) and quadratic discriminant analysis (QDA) are not applicable to high-dimensional data classification directly.

To overcome the singularity problem, Dudoit et al. (2002) introduced two simplified discriminant rules by assuming independence between covariates. For each class g, let $\bar{X}_g = (\bar{X}_{1g}, \ldots, \bar{X}_{pg})^T = \sum_{j=1}^{n_g} X_{g,j}/n_g$ be the sample mean, and $\hat{\Sigma}_g = \text{diag}(\hat{\sigma}_{1g}^2, \ldots, \hat{\sigma}_{pg}^2)$ be the sample covariance matrix where the off-diagonal elements are all set to be zero. Also, let $\hat{\pi}_g = n_g/n$ be the estimated prior probability of observing a class g subject. The first rule developed in Dudoit et al. (2002) is called diagonal quadratic discriminant analysis

(DQDA). It classifies a new subject $X = (X_1, \ldots, X_p)^T$ to class g that minimizes the discriminant score $\hat{d}_g^Q(X) = \sum_{i=1}^{p} (X_i - \bar{X}_{ig})^2 / \hat{\sigma}_{ig}^2 + \sum_{i=1}^{p} \log \hat{\sigma}_{ig}^2 - 2 \log \hat{\pi}_g$. The second rule is called diagonal linear discriminant analysis (DLDA) that classifies the new subject analogously according to the discriminant score $\hat{d}_g^L(X) = \sum_{i=1}^{p} (X_i - \bar{X}_{ig})^2 / \hat{\sigma}_i^2 - 2 \log \hat{\pi}_g$, where $\hat{\sigma}_i^2$ are the pooled variances across the G classes. DQDA and DLDA are sometimes called *naive Bayes* classifiers because they can arise in a Bayesian setting. Due to the small sample size, DLDA and DQDA, which ignore correlations among genes, perform well in many instances compared to some more sophisticated classifiers in terms of both accuracy and stability. In addition, DQDA and DLDA are easy to implement and have been adopted to analyze *-omics* data.

Despite the popularity of the independence assumption used in DQDA and DLDA in the literature, it is unlikely to be true in practice. Recently, Hu et al. (2011) and Pang et al. (2013) have proposed gene module–based linear discriminant analysis strategy by utilizing the correlation among genes in discriminant analysis.

17.2.1.1 Shrinkage and Regularization

Due to the small sample sizes, another direction to improve the diagonal discriminant analysis is by shrinkage (Pang et al. 2009). For instance, Pang et al. (2009) applied the shrinkage estimates of variances in Tong and Wang (2007) into the diagonal discriminant scores and formed two shrinkage-based rules called shrinkage-based DQDA (SDQDA) and shrinkage-based DLDA (SDLDA). Pang et al. (2009) also developed RSDDA that incorporates regularization as in Friedman (1989) to further improve the performance of SDQDA and SDLDA. Combining shrinkage-based variances in diagonal discriminant analysis and regularization in a new classification scheme may result in better performance over other standard methods.

17.2.2 Nonparametric Methods

In some *-omics* studies, it may not be appropriate to make strong parametric assumptions. The following are two popular nonparametric classification methods.

17.2.2.1 k-Nearest Neighbor

The k-nearest neighbor classifier finds the k-nearest samples in the data set and takes a majority vote among the classes of these k samples (Ripley 1996). In the case of two predictors of interest x_1 and x_2, it is essentially calculating $E(Y|(x_1, x_2)) = \sum_{i \in \text{class } g \cap \text{ksamples}} i/k$ the proportion of the class g among the k samples. Distance among the samples is usually Euclidean-based. For k-nearest neighbor, researchers commonly would use 1, 3, or 5 neighbors.

17.2.2.2 Random Forests

Random forests (Breiman 2001) is an improved classification and regression trees (CART) method (Breiman et al. 1984). It grows many classification trees or regression trees and thus the name *forests*. Every tree is built using a deterministic algorithm and the trees are different owing to two factors. First, at each node, a best split is chosen from a random subset of the predictors rather than all of them. Second, every tree is built using a bootstrap sample of the observations. The out-of-bag (OOB) data, approximately one-third of the observations, are then used to estimate the prediction accuracy. Unlike other tree algorithms, no pruning, trimming of the fully grown tree, is involved. Each observation is assigned to a leaf, the terminal node of a tree, according to the order and values of the predictor variables. For a particular tree, the predictions for observations are given only for the OOB data. The overall prediction is then calculated by averaging over all the trees built. A classification tree is built as follows:

Step (I): Draw bootstrap samples from the original data *ntree* times. For each bootstrap sample, this leaves approximately one-third of the samples OOB.

Step (II): A classification tree is grown for each bootstrap sample.

Step (IIa): At each node of the tree, select predictors ($\sqrt{p}$ for classification) at random for splitting.

Step (IIb): Using the gini impurity criterion described later, a node is split using the single predictor from step (IIa). Gini impurity criterion for a binary classification problem, is $1 - p_1^2 - p_2^2$ with p_1 and p_2 the proportions of individuals in class 1 and 2, respectively.

Step (IIc): Repeat steps (IIa and b) until each terminal node contains samples in the same class or only one sample.

Step (III): Calculate OOB error rate for each classification tree built. Aggregate the *ntree* trees to obtain the ensemble OOB error rate.

17.2.3 Multiclass Problems

So far, we have discussed problems involving a binary outcome or two classes. There are instances where we would be interested in classifying more than two cancer subtypes, multiple cancer tissues, or comparing genes of multiclasses of a drug. The methods described earlier, discriminant analysis, *k*-nearest neighbor, and random forests, have the ability to analyze multiclasses outcomes.

17.2.4 Classification Prediction Performance Measures

To evaluate the accuracy of classification, we would consider the use of a confusion matrix. For example, Table 17.1 illustrates the case for $n = 100$ with a misclassification rate of $10\% = 4\% + 6\%$. Several other measures

TABLE 17.1

Confusion Matrix Example

	Predicted	Class
True class	Positive	Negative
Positive	51	6
Negative	4	39

TABLE 17.2

Confusion Matrix Definitions

	Predicted	Class
True class	Positive	Negative
Positive	TP	FN
Negative	FP	TN

are also important for consideration. Based on the values from Table 17.2, these measures can be defined as (1) positive predictive value (PPV) = $TP/(TP+FP)$, (2) negative predictive value (NPV) = $TN/(FN+TN)$, (3) sensitivity = $TP/(TP+FN)$, and (4) Specificity $(TN)/(FP+TN)$. The area under the receiver operating characteristics (ROC) curve (AUC) is also commonly used. A value of 0.5 represents a random guess while a value of 1 represents a perfect prediction.

17.3 Validation Methods

To properly build a classifier, we need to use proper internal validation. From these internal validation procedures, we can also assess the performance of the resulting classifier. Internal validation is by no means a replacement for external (independent) validation, which is the gold standard for validation. In Section 17.3.1, training and testing framework will be described. In Section 17.3.2, common ways of feature selection will be presented.

17.3.1 Training and Testing

Assessing the accuracy of a fitted prediction model based on the same data set that was used to develop the model can result in an overly optimistic performance assessment of the model for future validation. Doing so can result in overfitting, which refers to when too many parameters compared to the number of data points are used in the training set. A complicated classifier that fits the training set well does not imply that it will perform well in external validation using an independent data set. To prevent overfitting the data, validation methods such as cross-validation can be employed. We later describe several resampling techniques that are commonly used to perform proper internal validation.

17.3.1.1 Hold-Out or Split Sample Method

The classifier is first built or trained using an initial set of data. This is usually called the *training* data. Once the classifier is *trained*, then it is applied to an *independent* set of data. This is usually called the *test* (or validation) set. The hold-out method or split sample method is the simplest of all the resampling methods. It involves a single random split or partition of the data into a training set with proportion P and a test set with proportion $1 - P$. For example, we can use 50% for the training set and the rest as the test set. Choices for the split can be of other proportions such as 60% training and 40% testing, or 70% training and 30% testing. Because we do not reuse the training set for testing, overfitting is not an issue for the hold-out method.

17.3.1.2 k-Fold Cross-Validation (CV)

The k-fold CV method consists of splitting the data set randomly into k partitions that are close to equal in size. At the kth iteration, $k - 1$ partitions will be used as the training set and the left out partition will be used as the test set. Five and ten fold CVs are commonly used.

17.3.1.3 Leave-One-Out Cross Validation (LOOCV)

LOOCV is a special case of k-fold CV in which $k = n$, the sample size of the data set. At each of the n iterations, the whole data set is used as the training except one sample that is left out as test set. Because it requires the largest number of iterations, it is the most computationally expensive resampling method introduced here. In addition, there is strong dependency among the training sets as any two training sets share $n - 2$ samples in common. This is recommended when n is relatively small.

17.3.1.4 Permutation

Cross-validation can be used with permutation. Among the aforementioned resampling methods, only the hold-out method is free of an overfitting potential because the resulting training and validation sets are mutually exclusive. All other methods reuse part of the data for training. In general, the idea is to generate the permuted data sets by shuffling the outcome and randomly matching them with the -omics data. The observed test statistic is then compared with those from the permuted data sets. The proportion of the permuted data sets that are more extreme than the observed will serve as a measure of significance.

17.3.2 Feature Selection

Feature selection, sometimes called feature extraction, is a critical component of building a classifier. We cannot expect to build a good classifier if poor features are selected. Classifiers that are parsimonious are easier to interpret. Complicated ones with too many features can degrade the performance of the classifier.

For instance, one can compute the two-sample t-test for all m features based on the training set and then identify the top 20 features by ranking the p-values. A classifier is then built on these top 20 features for the training set. One can also use the nonparametric test statistic like the Wilcoxon rank-sum test. Other commonly used feature selection includes the ratio of between-group to within-group sums of squares (BSS-WSS) and is defined as follows:

$$\frac{\text{BSS}(j)}{\text{WSS}(j)} = \frac{\sum_i \sum_g I\left(y_i = g\right)\left(\bar{x}_{gj} - \bar{x}_{*j}\right)^2}{\sum_i \sum_g I\left(y_i = g\right)\left(x_{ij} - \bar{x}_{gj}\right)^2}, \tag{17.1}$$

where
$\bar{x}_{gj}$ denotes the average level of covariate j (e.g., genes) across samples being in class g
$\bar{x}_{*j}$ denotes the average level of covariate j (e.g., genes) across all the samples

17.3.2.1 Tuning

Building a classifier using default values provided by a program may not work. For example, how many top features are desired, what is the k for k-NN, and what is the number of trees for random forests. In some instances, these tuning parameters can be identified using another layer of cross-validation in the training set. This is sometimes called *nested cross-validation*.

17.4 Survival Prediction Methods

In this section, we present two machine learning tools based on random forests for building survival models. We note that there are other extensions to survival outcomes using the following approaches: gradient boosting, recursive partitioning trees, neural networks, and support vector machine that were utilized in Pang et al. (2010). In Section 17.4.1, we present random conditional inference forest. In Section 17.4.2, we present random survival forests.

17.4.1 Random Conditional Inference Forests

The conditional inference forests (CIFs) *cforest* implemented in R differs from original random forests in two ways. The base learners used are conditional inference trees (Hothorn et al. 2006b). Also, the aggregation scheme works by averaging observation weights extracted from each of the trees built instead of averaging predictions as in the original version.

CIFs consists of the construction of many conditional inference trees, which are built using a regression framework by binary recursive partitioning algorithm. This algorithm has three basic steps:

Step (1): Variable selection by testing the global null hypothesis of independence between any of the predictors and the survival outcome.
Step (1a): If the hypothesis cannot be rejected, then terminate the algorithm.
Step (1b): Otherwise, select the predictor with the strongest association to the survival outcome.
Step (2): Implement a binary split based on split criteria using the selected predictors from step (1b).
Step (3): Recursively repeat steps 1 and 2.

For the case of censored data, the association is measured by a *p*-value corresponding to a linear test statistic defined later for a single predictor and the survival outcome. Let X_{Ji} be the *J*th covariate for the *i*th subject. a_i is the log-rank score (LRS) (Hothorn and Lausen 2003) of subject *i* defined as follows.

Let $X_{(1)} \leq X_{(2)} \leq \cdots \leq X_{(n)}$ be the set of ordered predictors, Y_i be the response for individual *i*, and for each survival time of observation *i*,

$$a_i = \mathbb{1}_i - \sum_l^{\gamma_l} \frac{\mathbb{1}_l}{N - \gamma_l + 1}, \qquad (17.2)$$

where $\mathbb{1}_i = 1$ if an event is observed for individual *i* and 0 otherwise, $\gamma_l = \sum_i^n \#t : S_t \leq S_l$ is the number of observed events or censored, occurring at S_l or before.

The linear statistic $T_J^R = \sum_i^n \mathbf{1}(\text{in node}) \mathbf{1}(X_{Ji} \in R) a_i$ measures the difference between the samples (I) $X_{Ji} \in R$ for $i = 1, \ldots, n$ and $\mathbf{1}(\text{in node}) Y_i$ and (II) $X_{Ji} \notin R$ for $i = 1, \ldots, n$ and $\mathbf{1}(\text{in node}) Y_i$; where $\mathbf{1}(\text{in node}) = 1$ if the individual is in node, and 0 otherwise; $\mathbf{1}(X_{Ji} \in R) = 1$ if $X_{Ji} \in R$, and 0 otherwise; and a_i is the log-rank scores of subject *i* as defined earlier. A split is implemented once the predictor has been selected from step (1) described earlier and that it meets its criterion. Let X_J be the predictor *J* that was chosen, and let R_k be one of all possible subsets of the sample space of the predictor X_J. The best split B_k is chosen from B such that $B_k = \arg\max \left(t_J^K - \mu_J^K \right) \Sigma_J^K \left(t_J^K - \mu_J^K \right)^T$ for all possible subsets B_K. A split is established when the sum of the weights in both child nodes is larger than a prespecified minisplit. This helps avoid splits that are borderline and reduces the number of subsets to be evaluated. One advantage is that the approach taken ensures that the right-sized tree is grown without the need of pruning. Since the distribution of T_j is unknown in most situations, permutation tests are employed.

The survival ensemble of the conditional inference tree are formed as described in the following. Let $\mathbf{w}$ be the weight vector where $w_i = 0$ iff it corresponds to a censored observation. The random conditional inference forest (Hothorn et al. 2006a) is constructed as follows:

Step (A): Draw a bootstrap sample of $c = (c_1, \ldots, c_n)$ from the multinomial distribution $Mult(n, p_1, \ldots, p_n)$, where $p_i = \left(\sum_{i=1}^n w_i \right)^{-1} \mathbf{w}$.

Step (B): The base learner is constructed using the conditional inference trees algorithm described in the aforementioned section. The training sample comes from step (A).

Step (C): Repeat steps (A) and (B) until it reaches the desired number of trees (ntree).

The base learners are aggregated by averaging observation weights extracted from each of the *ntree* trees. This is different from the original random forests, which averages the predictions directly. Specifically, for a survival tree of bootstrap sample b, $Tree^b$ from step (A), we can determine the observations that are elements of the same leaf of a survival tree as an observation x_{new} that was left out of the training sample. The aggregated sample is then $Tree_A(X_{new}) = \cup_{b=1}^{ntree} Tree^b(X_{new})$ for $b = 1, \ldots, ntree$. Then an aggregated estimator of the true conditional survival function of $F(.|x_{new})$ is the Kaplan–Meier curve of $\hat{F}_{Tree_A(X_{new})}(.)$.

For prediction, it is the weighted average of the observed (log)-survival times under the quadratic loss function:

$$\hat{Y} = \left(\sum_i^n pw_i(x) \right)^{-1} \sum_i^n pw_i(x) Y_i, \tag{17.3}$$

$$pw_i(x) = \sum_{b=1}^{ntree} c_{(tree,i)} \sum_m^{\#partitions} \mathbb{1}(*),$$

and $\mathbb{1}(*) = 1$ when X_i and x_{new} are both in the partition for $i = 1, \ldots, n$ and 0 otherwise.

17.4.2 Random Survival Forests

The random survival forest (RSF) algorithm (Ishwaran et al. 2008) implemented in R more closely resembles the original survival random forests by Breiman. Like random forests, RSF builds many binary trees. However, the aggregation scheme is now based on a cumulative hazard function (CHF). The CHF estimate for a terminal node L is the Nelson-Aalen estimator $\hat{\Lambda}_L(t) = \sum_{t_{i,L} \leq t} d_{t_i,L}/R_{t_i,L}$.

Step (I): Draw bootstrap samples from the original data ntree times. For each bootstrap sample, this leaves approximately one-third of the samples OOB.

Step (II): A survival tree is grown for each of the bootstrap sample.

Step (IIa): At each node of the tree, select $\sqrt{p}$ predictors at random for splitting.

Step (IIb): Using one of the splitting criteria described in the following, a node is split using the single predictor from step (IIa) that maximizes the survival differences between daughter nodes.

Step (IIc): Repeat steps (IIa and b) until each terminal node contains no more than 0.632 times the number of events.

Step (III): Calculate a CHF for each survival tree built. Aggregate the ntree trees to obtain the ensemble cumulative hazard estimate.

An extra step for calculating the OOB error rate for the ensemble is available in the software, but we will not make use of this when we compare the random conditional inference trees with RSF as the former does not have an analogous way for calculating the prediction error.

17.4.2.1 Split Criteria

The RSF algorithm has four split criteria available. The most commonly used two are log-rank (LR) (Segal 1988) and standardized LRS (Hothorn and Lausen 2003). We will describe LR and LRS later. For a proposed node split, let c be the cutoff used, which is of the form $x \leq c$ and $x > c$. For an individual i, let $n = n_1 + n_2$, where $n_1 = \sum_{i=1}^{n} \mathbb{1}_{(X_i \leq c)}$.

The LR test for splitting (Segal 1988) is defined as follows:

$$LR(X, c) = \frac{\sum_{i=1}^{E} d_{t_i,child_1} - R_{t_i,child_1} \frac{d_{t_i}}{R_{t_i}}}{\sqrt{\sum_{i=1}^{E} \frac{d_{t_i}(R_{t_i}-d_{t_i})}{R_{t_i}-1} \frac{R_{t_i,child_1}}{R_{t_i}} \left[1 - \frac{R_{t_i,child_1}}{R_{t_i}}\right]}}, \qquad (17.4)$$

where

E is the number of distinct event times $T_{(1)} \leq T_{(2)} \leq \cdots \leq T_{(E)}$ in the parent node

$d_{t_i,child_j}$ is the number of events at time t_i in the child nodes $j = 1, 2$

$R_{t_i,child_j}$ is the number of individuals at risk at time t_i in the child nodes $j = 1, 2$, that is then number of individuals who are alive or died at time t_i

$R_{t_i} = R_{t_i,child_1} + R_{t_i,child_2}$ and $d_{t_i} = d_{t_i,child_1} + d_{t_i,child_2}$

The absolute value of $LR(X, c)$ measures the node separation. The best split is chosen such that it maximizes the absolute value of Equation 17.4.

The LRS test based on the log-rank score defined earlier in (17.2) is defined as

$$LRS(X, c) = \frac{\sum_{X_i \leq c} a_i - n_1 \mu_a}{\sqrt{n_1 \left(1 - \frac{n_1}{n}\right) s_a^2}},$$ (17.5)

where μ_a and s_a^2 are the sample mean and sample variance of a_i respectively. $LRS(X, c)$ measures the node separation. The best split is chosen such that it maximizes the absolute value of Equation 17.5. All the aforementioned procedures use univariate variable selection procedure.

17.4.2.2 Bivariate Split Criteria

In a pathway-based setting like our real data example near the end of this chapter, a bivariate splitting criterion is feasible. The strategy is to split on the best pair of covariates at every node split by changing

Step (II): Portion of the aforementioned algorithm.

Step (IIa): At each node of the tree, select $\sqrt{p}$ predictors at random for splitting.

Step (IIb): Using LR splitting criterion, a node is split using the predictor pair from step (IIa) that maximizes the survival differences between daughter nodes by finding best split of the form $x_i + x_j \leq c$ for $i \neq j$ (Pang et al. 2010).

17.4.3 Feature Selection for Survival Outcomes

There are some implicit feature selection techniques such as boosting for survival outcomes (Binder and Schumacher 2008). Others developed Bayesian approach using model averaging (Lee and Mallick 2004). Current approaches are mostly univariate approaches, and the number of genes is usually chosen with an arbitrary cutoff. Pang et al. (2012) introduced an iterative feature elimination algorithm for gene selection using RSF in the survival setting. This approach provides an alternative to univariate gene selection (such as Dunkler et al. 2010).

17.4.4 Survival Prediction Performance Measures

One way to assess survival prediction performance is to compare the predicted survival of a high- and low-risk group using a log-rank test. This can be coupled with permutation when appropriate. To evaluate the accuracy of survival prediction without dichotomizing, we can employ the area under the ROC curve (AUC) approach for censored data (Heagerty et al. 2000). Time-dependent ROC analysis is an extension of the concept of ROC curves for time-dependent binary disease variables in censored data. In particular, we want to see how well the covariates predict the survival of individuals. For expected survival times or expected number of events, E, derived from a set of predictive markers, sensitivity and specificity are defined as a function of time t as follows: $sensitivity(c, t) = P(E > c|D(t) = 1)$ and $specificity(c, t) = P(E \leq c|D(t) = 0)$, where the disease variable $D_i(t) = 1$, if patient i has died before time t and $D_i(t) = 0$ otherwise. An ROC(t) is a function of t at different cutoffs c. Time-dependent ROC curve is a plot of $sensitivity(c, t)$ versus $1 - specificity(c, t)$ and AUC is an accuracy measure of the ROC curve. A higher prediction accuracy is supported by a larger AUC value. An alternative would be to use the concordance index (C-index) (Newson 2006). A measure on how well the prediction ranks the survival of any pair of individuals C takes value from [0,1]. A C-index of 0.5 corresponds to a random guess and 1 means perfect concordance.

17.5 Real Data Examples

17.5.1 GDSC Data Set

In this section, we demonstrate some of the classification methods in two scenarios: a binary outcome and a multiclass outcome. Genomics of Drug Sensitivity in Cancer (GDSC) Release 4 (March 2013) from Yang et al. (2013) was used. The database is a resource for information on drug sensitivity in cancer cells and molecular markers of drug response.

17.5.1.1 Binary Outcome

The question of interest for the binary outcome is similar to Yang et al. (2013), which is to investigate the effect of gene expression measure features on sensitivity to the BRAF-inhibitor PLX4720 IC50. Top 20 and bottom 20 based on the BRAF-inhibitor PLX4720 IC50 measure will serve as an outcome. A total of 13,321 genes are measured in this study. Using BSS-WSS with top 10 genes selected at each iteration with LOOCV, DLDA produced a misclassification rate of 22.5% while k-NN had 29.8% and 26.6%, for $k = 3$ and $k = 5$, respectively. The top two genes proteolipid protein 1 (PLP1) and serinethreonine kinase 10 (STK10) have been identified by Masica and Karchin (2013) to be overexpressed for PLX4720.RAF with *The Cancer Cell Line Encyclopedia* database.

TABLE 17.3

Confusion Matrix of Results

	GI Tract	Kidney	Pancreas	Upper Aerodigestive
GI tract	52	0	0	1
Kidney	2	18	0	0
Pancreas	14	0	0	1
Upper aerodigestive	4	0	0	37

17.5.1.2 Multiclass Outcome

The question of interest for the multiclass outcome is to investigate the effect of gene expression measure features on four different cancer types: GI tract, kidney, pancreas, and upper aerodigestive. The same number of genes 13,321 is available. In the multiclass classification, we used random forests classification with OOB error rate as the accuracy measure. Random forests importance measure was used to identify important genes.

Confusion matrix results is available in Table 17.3. The OOB error rate is $(1 + 2 + 14 + 1 + 4)/129 = 0.17\%$.

The top gene Plakophilin-1 (PKP1) was discovered as an oncogene by Knauf et al. (2011).

17.5.2 TCGA Data Set

In this section, we demonstrate pathway-based prediction with survival outcomes. The Cancer Genome Atlas (TCGA) from TCGA Research Network (2011) was used. The data used consist of messenger RNA expression of 578 patients with ovarian cancer. Affymetrix Hgu-133a gene chip with 22,215 probes are available for investigation. The median survival was 3.7 years with 47.7% of the data being censored.

17.5.2.1 Survival Outcome: Pathway Analysis

With the simultaneous measurement of thousands of genes, microarrays allow us to ask formal hypotheses that go beyond a single gene because real biology is not about individual genes (Pang et al. 2010). Researchers have started looking at groupings of genes. A total of 283 Biocarta pathways are investigated in our example. Pathway-based random survival forests analysis with 10-fold CV was performed. Table 17.4 shows a list of pathways with q-values of less than 0.005. For more about q-values, please refer to Storey and Tibshirani (2003).

TABLE 17.4

Top Pathways with p-Values ≤ 0.0002 and q-Values ≤ 0.005

Pathways	No. of Genes	p-Values	q-Values
Granzyme A–mediated apoptosis	20	0.000008	0.0014
Toll-like receptor	54	0.000007	0.0029
TNFR1 signaling	57	0.000007	0.0029
Phospholipids as signaling intermediaries	58	0.000008	0.0029
p38 MAPK signaling	85	0.000009	0.0029
MAPKinase signaling	168	0.00014	0.0039
Proepithelin conversion to epithelin and wound repair control	11	0.00019	0.0046

These methodologies can be extended to investigate genome-wide association study (GWAS) or single-nucleotide polymorphism (SNP) studies (Pang et al. 2011).

17.6 Conclusion and Discussion

In this chapter, we have given an overview of classification, validation, and survival prediction methodologies, and real data examples for analyzing -*omics* data. Classification method is helpful in identifying genes or markers that are important in classifying a categorical outcome. We have mainly focused on discriminant analysis and a couple of nonparametric methods, k-nearest neighbor, and random forests. It is hard to show that one method is superior to the other. Depending on the setting, other classification strategies may have better performance. However, the methods described are commonly used and have shown great results in many case studies. In the development of classifiers or survival predictors, sound internal validation and feature selection approaches are essential. We have introduced several ways for this. Without doing these steps properly, an independent external validation on the findings will become very difficult. Finally, we have illustrated the methods introduced to several real data examples including a binary and multiclass classification on expression data related to drug sensitivity, and pathway-based survival analysis and prediction for microarray data. New methods for sample size calculation in the survival setting have recently been developed, such as those considered in the article on validation-prediction (Pang and Jung 2013). As new data types evolve, there is ample opportunity for statisticians and bioinformaticians to develop novel methodologies for solving these big data challenges.

References

Binder, H. and Schumacher, M. (2008). Allowing for mandatory covariates in boosting estimation of sparse high-dimensional survival models. *BMC Bioinformatics*, 10:9–14.

Breiman, L. (2001). Random forests. *Machine Learning*, 45:5–32.

Breiman, L., Friedman, J. H., Olshen, R., and Stone, C. J. (1984). *Classification and Regression Trees*. Chapman and Hall, New York.

Cancer Genome Atlas Research Network (2011). Integrated genomic analyses of ovarian carcinoma. *Nature*, 474:609–615.

Chin, L., Andersen, J. N., and Futreal, P. A. (2011). Cancer genomics: From discovery science to personalized medicine. *Nature Medicine*, 17:297–303.

Dudoit, S., Fridlyand, J., and Speed, T. P. (2002). Comparison of discrimination methods for the classification of tumors using gene expression data. *JASA*, 97:77–87.

Dunkler, D., Schemper, M., and Heinze, G. (2010). Gene selection in micro-array survival studies under possibly non-proportional hazards. *Bioinformatics*, 26:784–790.

Edgar, R., Domrachev, M., and Lash, A. E. (2002). Gene expression omnibus: NCBI gene expression and hybridization array data repository. *Nucleic Acids Research*, 30:207–210.

Friedman, J. H. (1989). Regularized discriminant analysis. *JASA*, 84:165–175.

Golub, T., Slonim, D., Tamayo, P., Huard, C., Gaasenbeek, M., Mesirov, J., Coller, H., Loh, M., Downing, J., Caligiuri, M., Bloomfield, C., and Lander, E. (1999). Molecular classification of cancer: Class discovery and class prediction by gene expression monitoring. *Science*, 286:531–537.

GTEx Consortium. (2013). The genotype-tissue expression (GTEx) project. *Nature Genetics*, 45:580–585.

Heagerty, P., Lumley, T., and Pepe, M. (2000). Time-dependent ROC curves for censored survival data and a diagnostic marker. *Biometrics*, 56:337–344.

Hothorn, T., Bühlmann, P., Dudoit, S., Molinaro, A., and van der Laan, M. (2006a). Survival ensembles. *Biostatistics*, 7:355–373.

Hothorn, T., Hornik, K., and Zeileis, A. (2006b). Unbiased recursive partitioning: A conditional inference framework. *Journal of Computational and Graphical Statistics*, 15:651–674.

Hothorn, T. and Lausen, B. (2003). On the exact distribution of maximally selected rank statistics. *Computational Statistics & Data Analysis*, 43:121–137.

Hu, P., Bull, S., and Jiang, H. (2011). Gene network modules-based liner discriminant analysis of microarray gene expression data. *Lecture Notes in Computer Science*, 6674:286–296.

Ishwaran, H., Kogalur, U., Blackstone, E., and Lauer, M. (2008). Random survival forests. *Annals of Applied Statistics*, 43:841–860.

Knauf, J., Sartor, M., Medvedovic, M., Lundsmith, E., Ryder, M., Salzano, M., Nikiforov, Y., Giordano, T., Ghossein, T., and Fagin, J. (2011). Progression of BRAF-induced thyroid cancer is associated with epithelial-mesenchymal transition requiring concomitant MAP kinase and TGF? signaling. *Oncogene*, 30:3153–3162.

Lamb, J., Crawford, E. D., Peck, D., Modell, J. W., Blat, I. C., Wrobel, M. J., Lerner, J., Brunet, J. P., Subramanian, A., Ross, K. N., Reich, M., Hieronymus, H., Wei, G., Armstrong, S. A., Haggarty, S. J., Clemons, P. A., Wei, R., Carr, S. A., Lander, E. S., and Golub, T. R. (2006). The connectivity map: Using gene-expression signatures to connect small molecules, genes, and disease. *Science*, 313:1929–1935.

Lee, K. and Mallick, B. (2004). Bayesian methods for variable selection in survival models with application to DNA microarray data. *Sankhya*, 66:756–778.

Masica, D. and Karchin, R. (2013). Collections of simultaneously altered genes as biomarkers of cancer cell drug response. *Cancer Research*, 73:1699–1708.

Newson, R. (2006). Confidence intervals for rank statistics: Somers' D and extensions. *Stata Journal*, 6:309–334.

Pang, H., Datta, D., and Zhao, H. (2010). Pathway analysis using random forests with bivariate node-split for survival outcomes. *Bioinformatics*, 26:250–258.

Pang, H., Hauser, M., and Minvielle, S. (2011). Pathway-based identification of SNPs predictive of survival. *European Journal of Human Genetics*, 19:704–709.

Pang, H., George, S. L., Hui, K., Tong, T. (2012). Gene selection using iterative feature elimination random forests for survival outcomes. *IEEE Transactions on Computational Biology and Bioinformatics*, 9(5):1422–1431.

Pang, H. and Jung, S. (2013). Sample size considerations of prediction-validation methods in high-dimensional data for survival outcomes. *Genetic Epidemiology*, 37:276–282.

Pang, H., Tong, T., and Ng, M. (2013). Block-diagonal discriminant analysis and its bias-corrected rules. *Statistical Applications in Genetics and Molecular Biology*, 12:347–359.

Pang, H., Tong, T., and Zhao, H. (2009). Shrinkage-based diagonal discriminant analysis and its applications in high-dimensional data. *Biometrics*, 65:1021–1029.

Parkinson, H., Kapushesky, M., Shojatalab, M., Abeygunawardena, N., Coulson, R., Farne, A., Holloway, E., Kolesnykov, N., Lilja, P., Lukk, M., Mani, R., Rayner, T., Sharma, A., William, E., Sarkans, U., and Brazma, A. (2007). ArrayExpress–a public database of microarray experiments and gene expression profiles. *Nucleic Acids Research*, 35(Database issue):D747–D750.

Rhodes, D. R., Yu, J., Shanker, K., Deshpande, N., Varambally, R., Ghosh, D., Barrette, T., Pandey, A., and Chinnaiyan, A. M. (2004). ONCOMINE: A cancer microarray database and integrated data-mining platform. *Neoplasia*, 6:1–6.

Ripley, B. (1996). *Pattern Recognition and Neural Networks*. Cambridge University Press, Cambridge, U.K.

Segal, M. (1988). Regression trees for censored data. *Biometrics*, 44:35–47.

Storey, J. and Tibshirani, R. (2003). Statistical significance for genome-wide studies. *PNAS*, 100:9440–9445.

Tong, T. and Wang, Y. (2007). Optimal shrinkage estimation of variances with applications to microarray data analysis. *JASA*, 102:113–122.

Yang, W., Soares, J., Greninger, P., Edelman, E., Lightfoot, H., Forbes, S., Bindal, N., Beare, D., Smith, J., Thompson, I., Ramaswamy, S., Futreal, P., Haber, D., Stratton, M., Benes, C., McDermott, U., and Garnett, M. (2013). Genomics of drug sensitivity in cancer (GDSC): A resource for therapeutic biomarker discovery in cancer cells. *Nucleic Acids Research*, 41:D955–961.

18

Understanding Therapeutic Pathways via Biomarkers and Other Uses of Biomarkers in Clinical Studies

Michael D. Hale and Scott D. Patterson

CONTENTS

18.1 Introduction

Biomarkers have been an important part of clinical trials for decades. The recent surge of interest in biomarkers is due to advances in knowledge of molecular pathways and mechanisms that have provided new opportunities for their use and enabled new ways of designing trials.

The topic of biomarkers encompasses many dimensions and is much more than simply adding another measurement to a clinical trial. It begins with the development of biological insight of how an investigational molecule may interact with its intended target and pathway to clinical effect and includes many considerations: statistical (study design, analysis, interpretation), technical (sample acquisition, handling, assay performance), practical (consent, availability of tissue, timing of results), and regulatory and commercial (these include most of the other issues).

Biomarkers may be used for different purposes, and the intended use is an effective way of classifying biomarkers; what does one hope to achieve with a biomarker?

In our experience, there are four primary purposes for which biomarkers may be usefully employed in clinical drug development, and three of those are the subject of this chapter. Those four uses roughly correspond to time frames from a subject's screening and initial therapeutic intervention through to clinical outcome.

The first purpose is patient selection, before a therapeutic intervention occurs. The *selection biomarker* is used to identify subjects most likely to have a favorable benefit/risk ratio when a specific investigational therapy is administered, relative to a specific control and for a specific therapeutic purpose. Once well established, this marker could be used to guide therapy. It is important to note that the "selection" properties of this biomarker only apply in the context of the specific proposed investigational and control therapies; with a different investigational agent, or control, this biomarker might offer no aid in a treatment decision. Among biomarker uses, this use is currently receiving the most attention for several compelling reasons: it could dramatically reduce the size of phase 3 clinical trials, it embodies *personalized medicine*, and its use in routine clinical practice is favorable from a public health point of view. (Note: It also addresses a legal requirement, US 21CFR201.57 (2010), in appropriate labeling for selected subgroups expected to benefit.) Essentially it informs the best treatment choice for a specific patient for the relevant therapies. This use is the basis for a companion diagnostic and will be discussed later in this chapter at some length.

The second use occurs very soon after treatment administration, when a biomarker might indicate whether a target pathway is successfully being modulated. The biomarker could potentially be any marker between the point of intervention and a clinical outcome, though it would ideally be near the point of pathway intervention. This *pathway biomarker* may be very useful in early trials to identify whether a molecule is reaching and modulating its intended target, to provide objective evidence of pathway intervention. If this does not occur in essentially all eligible subjects in a trial, then the value of the intervention may be very limited. If the biomarker is directly on the targeted pathway, it may be thought of as necessary, but not sufficient, for a favorable clinical outcome. If the marker is too distal, its modulation may be impacted by signal cross-talk, thus making it potentially less reliable for evidence of pathway intervention; however, this reason might also make such a marker useful as a response biomarker (see next paragraph). An early trial that reliably shows little or no pathway intervention may allow for a drug program to be terminated early due to unlikely chance of success. Further, showing that the desired pathway is modulated in the manner desired in early clinical studies is critically important for the case where subsequent studies in patients reveal no impact to the disease symptoms. If this is the case, one can distinguish between a drug that has failed to impact its target from a pathway that does not impact the course of the disease in humans.

The third use also occurs early after treatment administration, where a *response biomarker* is used to identify whether a subject is expected to experience a favorable outcome. In general, the pathway biomarker is near the drug target, while the response biomarker is nearer the clinical response, from a chain of causality point of view. A pathway biomarker might also be a response biomarker. The difference between the two might be very subtle or perhaps nonexistent, with the pathway biomarker being more about evidence that the investigational drug reaches and modulates the target, while the response biomarker is generally considered to suggest desirable activity further along on the causal pathway and nearer to the clinical response. Generally speaking, one would expect a pathway biomarker signal in all suitably chosen subjects in a trial, while the response biomarker signal would be expected to be associated with subjects who respond to the intervention (a subset of the study population). A response biomarker may or may not be a good candidate to be a surrogate. An example of a response biomarker that is not a pathway biomarker and that is also not currently a surrogate marker (though it has often been suggested, and may one day be) is the appearance of rash in oncology patients receiving EGFR therapy. This example is very interesting because the association of rash with clinical effectiveness was found by observation, rather than being suggested as a marker because of known or presumed biology, and there is currently still no general agreement regarding mechanism for rash (Liu et al., 2013) or for this association. The primary use of a response biomarker would be to determine whether a subject ought to continue therapy or not.

The fourth biomarker use is when a biomarker is used as a substitute for a clinical endpoint, also known as a "surrogate" (e.g., cholesterol, blood pressure). *Surrogate biomarkers* will not be covered in this chapter as they have been thoroughly covered elsewhere (Prentice, 1989; Fleming and Powers, 2012, Chapter 19 of this book; US FDA DRAFT, 2010).

This chapter is organized to consider pathway intervention and response biomarkers first, followed by consideration of selection biomarkers. Each section will be divided into practical considerations we believe are important for awareness by the statistician followed by some statistical considerations. The practical considerations sections are not meant to be stand-alone detailed and comprehensive descriptions for a bench scientist to use as a checklist, but to present a range of issues faced in the design of experiments to employ biomarkers in the clinical development setting. The aim is to set these in context for the subsequent statistical discussion that will describe in greater detail how to address and account for these challenges. For the experienced bench scientist, the statistical sections will help them understand additional issues they may need to consider in the design of their experiments. Overall it was our intention and hope in presenting our thoughts in this manner that it would provoke and enable effective communication between the statistician and the bench scientist because these interactions are critical to the successful implementation of biomarker strategies in clinical development.

Key Summary Points:

- Pathway biomarkers may measure the more proximal pharmacodynamic (PD) effects of the drug in question and can provide confidence that the target pathway is interdicted in the manner anticipated.
- Pharmacodynamic biomarker assessment is key to ultimately determining whether a drug that has failed to elicit an improvement in disease symptoms is the wrong drug (i.e., does not modulate or reach the target) or targets the wrong pathway; similarly, a drug that apparently elicits the desired clinical response but has not impacted the target pathway is not doing so through the target mechanism of action.
- Development of selection biomarkers requires careful attention to assay development and preanalytic and sampling considerations, and these should ideally be addressed early in clinical development of a new compound.
- Selection biomarkers may help reduce the size of trials and enable limiting use of a given therapeutic to those patients most likely to benefit.

- It is critically important to give careful consideration to both the feasibility of sampling in the clinic and which tissue is to be collected.
- Understanding preanalytical variables for the biomarkers to be interrogated is necessary to be confident in the data generated.
- Careful planning of the biomarker protocol is key to successful biomarker assessment.

18.2 Pathway Intervention and Response Biomarkers

18.2.1 Considerations for Biomarker Utilization in Early- and Late-Phase Clinical Trials

18.2.1.1 Early-Phase Trials: Understanding Whether the Desired Pathway is Interdicted

The conduct of early-phase trials, with the increased sampling that is often undertaken, provides a unique opportunity that should not be missed to explore the biological impact of the therapeutic intervention (Severino et al., 2006). This opportunity enables interrogation of the pathway in humans if appropriate samples can be ethically accessed at least twice, predose and subsequent to dosing. Biochemical measurements of these samples have the potential to reveal whether specific molecules in the pathway, in the appropriate (or acceptable surrogate) tissue, have been altered as anticipated (concentration or post-translational modification) following drug exposure. Ideally many such measures could be undertaken over time to provide a greater understanding of the response kinetics of the drug exposure. The classification of such pharmacodynamic assessments of pathway biomarkers can be further subdivided into target coverage, pathway coverage, mechanism of action biomarkers, etc., or collectively as markers of biochemical coverage. But whether one subdivides these biomarkers to define exactly where they occur in the pathway is less relevant than the purpose for such measurement. To understand this aim, let us consider drug development without pathway biomarkers.

To put it very simply, in the absence of pathway biomarkers, early-phase studies are guided by drug exposure (assessed by pharmacokinetics (PK)), safety, and tolerability). Preclinical data in a model species will provide preliminary models of the relationship among drug exposure levels (usually in blood), what is achieved in the target organ, and the effect that concentration will have on the drug target, that is, the minimum amount required to achieve maximum blockade or stimulation of that target. Thus, achieving a desired drug concentration in blood is associated with expected impact on the drug target.

Outside of oncology, such early studies are usually conducted in healthy volunteers. If the drug appears reasonably safe and tolerable with acceptable PK in healthy volunteers, patients with the target disease will typically be enrolled in phase 1b or phase 2 studies, though increasingly they may participate in later cohorts in a FIH trial. It is at this stage where the first feedback on PD can be assessed, but if there is no impact on the disease symptoms do you know whether the drug has failed to impact its target or does intervention of the pathway not really favorably alter the course of the disease in man? Do you take a back-up compound into the clinic to try again?

Without PD biomarker data, it will be unclear whether you have a drug that does not impact or reach the target or a target in a pathway that does not impact the course of disease, hence why inclusion of PD biomarkers is a valuable addition to early-phase studies.

18.2.1.2 *Biochemical Coverage: All in the Translation*

Of course, it is often far easier to state the desire to employ PD biomarkers in early-phase trials than to achieve that aim. Let us consider some of the challenges listed as follows and discuss them.

1. Is the target of the therapeutic expressed in healthy volunteers?
2. Is the tissue in which the target expressed accessible—repeatedly if necessary?
3. Is modulation of the pathway by the drug in a healthy volunteer expected to cause a measureable response or will the pathway need to be stimulated to evoke a response?
4. What assays exist to measure the desired pathway element(s) and do they have good performance characteristics?
5. Do assays exist for measurement of more proximal or distal biomarkers in the pathway?
6. Are there known off-target pathway elements that could and should be measured?
7. Has the assay only been employed for in vitro studies? Or is there clinical experience?
8. What knowledge exists regarding the preanalytical variables that could impact measurement of the biomarker?
9. How well understood is the timing of the biomarker modulation in man?

Although a rarer case, biomarkers expressed exclusively in the disease state are more challenging, and in such cases PD biomarker assessment in healthy volunteers may be restricted to biomarkers of off-target interactions identified from preclinical studies. However, in the more common situation where the target of the therapeutic is present, PD biomarker assessments are more likely to establish the minimum exposure levels needed to interdict the pathway to the desired degree than to identify one specific desirable exposure level.

A more common challenge arises when the target tissue is not one in which repeated measurements (or even one) are possible. Consider neurology-specific targets, which are an extreme example from a direct biochemical measurement perspective. The development of tracers to measure their displacement by the drug is one way to mitigate this challenge but still requires a considerable investment and parallel development pathway for the investigational molecule.

When the target tissue is poorly accessible, one must ask whether the pathway exists in a more accessible surrogate tissue, such a blood, for example, peripheral blood mononuclear cells (PBMCs)? If so, how relevant is pathway modulation in PBMCs—aside from potential off-target effects—to interpret pathway modulation in the target tissue? Repeated sampling of PBMCs is feasible and helps better define the time course and magnitude of PD effect. Associated measurement of drug concentration in those samples (at judiciously selected times) enables exposure-(biomarker) response modeling.

If the target tissue is accessible but only a single repeated sample/biopsy is possible, then one may employ a strategy where the sampling times are varied among the subjects in the trial to better characterize the time course of exposure–response. The predose measurements among all subjects will provide some insight as to variability in the population before treatment and our ability to reliably detect changes can be evaluated. Preclinical studies may provide a greater degree of confidence in the assessment of PBMCs, or other alternate tissues more easily sampled, to serve as well representing measurement of the pathway in the target tissue, if such is possible. This will require the development of these assays and the conduct of appropriate studies to test this concordance.

After determining which tissue is to be sampled, the ability to measure the pathway elements of interest has to be addressed. Multiple markers, including target occupancy, proximal, and distal, are preferred but are not always possible. Target engagement measured by receptor occupancy assays provides a direct measure of drug–target interaction. However, receptor occupancy assays are not simple to establish, for example, they require ensuring the binding kinetics of the probe and drug are similar. Such an assessment requires only target expression and does not require pathway activity in healthy volunteers for successful measurement.

Measures of pathway impact, for example, change in phosphorylation of a membrane receptor bound by a drug, provide the most proximal downstream PD assessment. The further downstream a pathway element is interrogated, the greater the potential for pathway cross-talk to contaminate the signal due to uncontrolled extraneous inputs. And as such PD, biomarkers

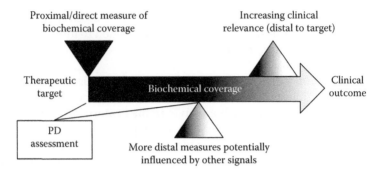

FIGURE 18.1

Biochemical coverage, pharmacodynamic assessment and clinical outcome. Schematic representation of a hypothetical signaling pathway beginning with the site of therapeutic target interdiction and ending with clinical outcome. Pharmacodynamic (PD) assessment at the site of therapeutic intervention represents the highest degree of biochemical coverage. PD assessment further downstream (more distal) can be impacted by unrelated signals. The degree of clinical relevance increases to the ultimate clinical outcome.

can be loosely considered to reflect a greater or lesser degree of biochemical coverage (e.g., biomarker evidence to provide confidence the drug is modulating the target as intended) depending upon whether they are more proximal or distal to the site of therapeutic interdiction (see Figure 18.1). The more distal the measure, the closer it may be to the measured clinical endpoint, but the greater the opportunity for additional signals to impact the measure, making exposure–response measures more difficult to interpret.

There is one aspect of target engagement that bears mentioning in the context of pathway biomarkers and that is whether the measurement truly reflects a PD measure or is a surrogate for systemic drug concentration? For example, consider an antibody directed against a soluble ligand. In the simplest case, measurement of the amount of bound ligand may be considered an exposure measure, although subject to factors such as whether the half-life of the bound ligand is different from the free ligand, or whether there is any upregulation of the ligand in response to its sequestration. It is not always a simple matter to accurately measure total ligand and free or bound ligand. In some cases, the same may be said for the measurement of target engagement of cell surface receptors by antibodies directed to them. In such a case, you may wish to measure the amount of total receptor that is bound by the drug since this total amount of receptor may also change over the course of the drug exposure due to feedback.

Understanding the intrinsic pathway activity in healthy volunteers or in surrogate tissue of patients where the target tissue is not amenable to sampling is critical for design of PD biomarker assays. If the pathway needs to be stimulated to enable a drug-induced modulation to be measured, does that need to be achieved by an ex vivo stimulation or is there a compound that can be safely administered to the subject? Physiological rather than biochemical assays are more likely to have been developed in this manner but those methods may also be applicable. A physiological assessment of this type is the cold-pressor test, a well-known test for pain induction, where a subject's arm is rapidly immersed in ice water to determine if a drug can attenuate the pain response (Wallace and Schulteis, 2008). Another such example would be the impact of a drug on the attenuation of capsaicin-induced changes in dermal blood flow (Sinclair et al., 2010). One could also design experiments where skin biopsies could be taken before and after topical administration of a compound to measure the impact on specific pathway elements prior to and after drug exposure (Hanneman et al., 2006). Some biochemical pathway markers may require use of ligands that can neither be administered systemically nor locally. In these cases, samples (or extracts thereof) may be challenged ex vivo, typically to measure reduction in response to stimulation.

This begins our discussion of the preanalytic issues that one needs to consider to successfully implement PD biomarker assays in the clinic. Assays can be considered to have three parts: preanalytic (everything that happens to the sample from collection until it is rendered incapable of change by its environment, i.e., *fixed*), analytical (the actual test), and postanalytical (interpretation and reporting of the test results). In this section, we will not further consider the analytical or postanalytical portions as these are the same for either clinical or nonclinical measurements. In this section, we will focus on those aspects of assay design and execution that require more attention in clinical drug development.

Preanalytics is rarely a consideration for in vitro experimentation. For testing of clinical samples, it is a well known issue. However, what may not be well-known is the impact that preanalytic factors could have on your specific assay. Any sample that is taken from the body will respond to its perturbed environment until it is rendered incapable of doing so by being *fixed*. Consider one of the simplest and most commonly encountered clinical samples, blood. Once drawn, blood cells are responding to decreasing oxygenation, and different shear forces to name just two factors. However, many analytes that are tested in blood can be successfully measured from samples that have sat at room temperature for hours or even shipped at ambient temperature, and others are equally stable in serum or plasma after the cellular component has been separated by centrifugation.

However, many assays that are developed for PD biomarker assessment are measuring very labile analytes, such as phosphorylation sites or even mRNA transcript levels. Simple microarray experiments of whole blood collected into tubes designed to *fix* RNA, PAXgene tubes, and stored at room temperature for various periods of time up to 48 h have shown that the mRNA levels (transcript profile) for 30% of genes changed more than two-fold (Russell et al., 2008). If the transcript(s) of interest was one of those so affected, effects due to sample storage time and temperature could be confounded with the PD biomarker assessment. It should be noted that although one should always conduct training of the clinical site, mistakes will occur and use of intrinsic quality controls is highly desirable; consideration must be given to any potential protocol deviations that would impact the preanalytics of the analyte to be tested.

One also has to consider the complexity of what is being asked of a clinical site. Take for example the instructions for use of a particular blood collection tube that specify to invert the tube several times and then leave at room temperature for 4 h before freezing. This seems like a straightforward request, but when one considers the schedule for sampling and the multiple tasks such staff have to undertake during a subject visit, it is not a surprise that this 4 h timeframe can be challenging to meet and as such tubes can be left for much longer periods of time (or shorter). Hence, in this case, the solution may be to conduct experiments to determine if a shorter incubation period can be employed to allow immediate freezing after multiple tube inversions and also differences in the manner in which the frozen tubes are thawed. There are many other examples of such preanalytical considerations, and some can be found in Russell et al. (2008).

Finally, to manage the preanalytic aspects of the assay, it is important to provide clear instructions that also require the recording of the time (not elapsed time) particular steps were undertaken so that auditing of the preanalytics can occur to determine whether there were any deviations from the protocol, to inform whether results from that sample can be accepted.

As stated earlier, this section is not meant to represent a comprehensive review of all potential methodologies that are applied to PD biomarker assessment but it does seem worthwhile mentioning some of the other sample types that may be obtained in a clinical program. Blood, as well as being the easiest to collect, also has many components that can be measured separately: PBMCs, defined cell subsets (as well as signaling molecules within these cells), plasma proteins, microvesicles, exosomes, and circulating nucleic acids, and metabolites. Skin punch biopsies from healthy volunteers and/or patients can be obtained in some situations. In the oncology setting, biopsies or fine needle aspirates of tumors may also be possible to sample, but repeated sampling could be challenging. Hair follicles can also be obtained and have proven valuable to evaluate the impact of drugs on cell cycle and other pathways (Camidge et al., 2005; Yap et al., 2011). From these sample types, an array of biomarkers can be interrogated including, but not limited to, the following:

- Cytokines, chemokines, other plasma proteins or signaling molecules in cell lysates using ELISA methodologies, single or multiplexed planar or bead arrays
- Signaling molecules in permeabilized cells by flow cytometry or immunohistochemistry
- Second messengers in cell lysates by a range of methods

- Enzymatic activity in tissue/fluids using biochemical methods
- Cell surface protein analysis and circulating tumor cell enumeration by flow cytometry transcript levels in cell lysates
- Nucleic acid measures (DNA, miRNA) in plasma as leakage markers or tumor burden

Previously we mentioned the case where the mechanism of action of the drug is to attenuate a pathway and that pathway is not active under normal physiological conditions in healthy volunteers. And further, the pathway cannot be stimulated by systemic or local administration of a specific compound and as such a sample has to be taken for ex vivo stimulation. The simplest substrate for such an ex vivo stimulation experiment is whole blood. Blood can be readily collected at multiple time points, but one has to be confident that these cells are an acceptable surrogate for the pathway in the tissue that is the target (unless blood cells are the target). One has to consider that drug exposure in blood could well be different from drug exposure in different tissues of the body. As mentioned previously, experiments in preclinical species may be able to increase one's confidence in the correlation of pathway responsiveness to drug exposure between the target tissue and cells of blood. When such confidence is sufficient to move forward with this sample type, the design of the assay will be undertaken and the following are some of the questions that one should consider:

- What is the stability of the drug in blood ex vivo?
- What is the stability of the analyte/pathway to be measured from blood ex vivo?
- Can the blood sample be transported from the site or will the stimulation have to be performed on site?
- Can the sample be *fixed* after the stimulation for subsequent testing at a central site?
- What controls can be incorporated to ensure all procedures have been undertaken per protocol?
- How simply can the assay be designed for the minimum number of steps to be conducted outside of the central laboratory site?

Obviously, the ideal situation would be one in which the analyte and pathway were stable for shipment and the assay in its entirety could be conducted at a central site. That way, training at the site is limited to ensuring appropriate sample collection, handling, and shipment. In the case where that is not possible, either trained staff will have to be deployed to the site to conduct the assays or site staff will need to be trained to perform the assay, a considerable undertaking. The bench scientist needs to have a detailed understanding of all aspects of potential deviations from the assay process and to anticipate their ultimate impact on the results. This will often require deployment of staff experienced in the assay to the site to train staff, together with detailed protocols and checklists. A training video may also prove helpful to reinforce onsite training and as a reference for those conducting the experimental procedure at the site. Test runs are required to ensure the protocol and sample shipment can be performed successfully. Finally, attention should be paid to ensure that those being trained will be those who are performing the procedures at the site; the trial may progress for an extended period of time and the same staff may not always be available, so a contingency for training of new staff has to be contemplated.

In designing the experimental protocol, one should take account of the expertise of the staff implementing the protocol and their familiarity with the available equipment. Further, will additional equipment have to be supplied, and if so, how difficult is shipping that equipment to the country? Can steps be eliminated from the standard protocol through the supply of materials partially prepared in advance? For example, we have utilized 96-well plates that have reagents already distributed in the wells (either frozen or lyophilized) such that all that is required of the site is to accurately pipette whole blood into the wells and incubate (with adequate mixing) the plate for a given period of time prior to the fixation step. Does the fixation step require the cellular contents to be separated for ultimate analysis of the plasma or is the cellular component (PBMCs) the substrate for testing? How is that to be accomplished? Finally for shipment, inclusion of a temperature tracking device with the sample provides an important level of confidence in sample handling. All aspects of the experimental protocol have to be carefully evaluated and understood to develop an extremely robust assay. This will take time, but if such planning is undertaken, PD biomarker assessments utilizing ex vivo stimulation may be deployed globally.

Finally, an understanding of the timing of the change in the PD biomarker has to be determined such that sampling is appropriate and ideally can be optimized. This is dependent on a number of factors including how proximal the PD biomarker is to the site of drug interdiction. If the measure is receptor occupancy, the timing will be that of the drug exposure. However, as the PD biomarker is further downstream, one may expect the effect to occur later (e.g., is the change in a protein analyte a result of new transcription and/or new translation, or release of stored protein or displaced from some other store such as a stroma?).

Take for example the measurement of systemic biomarkers of angiogenesis inhibition, the so-called angiogenic cytokines. In this example, we are considering angiogenesis inhibitors, drugs that target the vascular endothelial growth factor receptors (VEGFR). These receptors are present on the endothelial cells of the vasculature, which represents a very small fraction of any tissue sample and as such, changes in receptor activity or downstream pathway biomarkers within the endothelial cells are not amenable to direct measurement. However, blockade of these receptors does result in increases of angiogenic cytokines in a positive feedback loop. Although the impact of the drug on the target occurs at the same time, the measurement of the pharmacodynamic change of these biomarkers is variable in timing (mechanism of induction), magnitude (impact of multikinase inhibitor spectrum), and duration (half-life of analyte).

Although some of these analytes have proven of value as PD biomarkers when measured as circulating factors, they are essentially measuring a systemic response rather than a specific tumor response, as has been revealed by preclinical and clinical studies of healthy Vs tumor bearing subjects (Ebos et al., 2007; Lindauer et al., 2010). However, a comparison of these PD changes following administration of various drugs in this class did provide consistent evidence for the value of these PD biomarkers. In some cases, the degree of pharmacodynamic change appeared to correlate with outcome and as such had the potential to become a therapeutic response biomarker (Rosen et al., 2007; Batchelor et al., 2010). Further, in one case, a combination of imaging response one day after dosing combined with plasma biomarker analysis appeared to predict ultimate patient benefit (Batchelor et al., 2010). The key to any response biomarker is the confidence that the measured pharmacodynamic effect is correlated with clinical response and that this can be observed earlier than any standard measure of therapeutic efficacy. If this is achieved, then the response biomarker can be a meaningful guide to patient therapy. An example of this is CD4 or HIV RNA levels. Those are used early after treatment initiation for informing therapeutic decisions, long before the clinical consequences of HIV are observable. In this case, the same marker (HIV RNA levels) that may give an early indication of drug effectiveness is also a validated surrogate when measured at 24 or 48 weeks for predicting efficacy of antiretrovirals (Medecins Sans Frontieres, 2012; US FDA DRAFT, 2013).

In summary, pathway biomarkers have the potential to provide valuable information for a clinical drug development program, particularly in the early phases where more intensive sampling is possible. As described earlier, there are a number of important parameters that one needs to consider with respect to the biochemical response to ensure the data are truly representative of the pharmacodynamic effect and are not an artifact of how the sample was collected or analyzed.

18.2.2 Statistical Considerations for Pathway Intervention and Response Biomarkers

An important goal of early clinical development is to uncover the therapeutic potential of a new drug candidate. To achieve this, one may design and analyze clinical trials to help answer the following key questions:

- Does the drug reach the intended target?
  - o For example, receptor occupancy, such as imaging for neurology applications
- Does the drug modulate the intended target?
  - o Ideally to show intended initial effect at the target

- What is the relationship between drug concentration and the magnitude of (pharmacological) effect?
  - o Demonstration of concentration/effect increases confidence in the biological hypothesis and guides choice of desirable drug exposure (and hence, dose)

This last point takes one into the field of quantitative pharmacology and is the subject of the following section. We provide a greatly abbreviated overview with a view to aid study design and basic analysis, but will not cover the details of methods. It is not the intention to replace the pharmacologist but rather to help the analyst better understand some of the principles and overall approach and methods.

18.2.2.1 Pathway Intervention: Pharmacology and Some Mechanistic Models

For every drug, it is desirable to understand the relationship between dose administered and clinical outcome. There are usually a number of intermediate steps between the dose and clinical outcome, and a series of models describing progression of the steps may guide dose selection and extension to other dosage forms, populations, etc. Even better, one may identify fundamental problems with a compound.

A useful paradigm is

- Drug concentration is a function of dose administered
- Pharmacological effect is a function of drug concentration
- Clinical outcome is a function of pharmacological effect

This is a high-level representation, and each of these three may have a much greater level of detail as needed to be helpful. The first is about bioavailability, and will include time and method of administration, and often may include subject-specific factors such as age, race, body weight, etc. Models describing this first relationship fall within the field of pharmacokinetics.

The third relationship would include functions that are well understood (e.g., surrogate markers) and also include some functions that may be poorly understood. This set of relationships is not explored further in this chapter.

The second group of functions is the focus of this section, with the aim of developing an understanding of the relationship between pharmacological effect and drug concentration.

The pharmacological effect might be measured near the drug target, or it might be the clinical observation (e.g., blood pressure), or it could be anywhere in-between.

Measurements near the target, virtually always a biomarker, may help answer whether we are modulating the target and by how much. This information is useful because it isolates the drug–target interaction; there are potentially multiple points of "failure" in the pathway from target to clinical effect, and the proximal biomarker helps understand if the first part of that pathway intervention is successful or not. A pathway biomarker that fails to meet its goal may inform a go/no go decision. The degree of pathway modulation, measured by pharmacodynamic response, may also aid in selecting a dose.

Definitions: The building blocks for the models we discuss will involve receptors, agonists, and antagonists. Our discussion will involve informal definitions related to cell surface receptors, and the interested reader may find more formal definitions in some of the references (e.g., Neubig et al., 2003). The following definitions were found at www.everythingbio.com (accessed September 23, 2013):

- *Receptor*: A specialized protein on a cell's surface that binds to substances that effect the activities of the cell.
- *Agonist*: A drug that binds to a receptor and activates it, producing a pharmacological response (e.g., contraction, relaxation, secretion, enzyme activation, etc.).
- *Antagonist*: A drug that attenuates the effects of an agonist.
- *Ligand*: A substance that binds specifically and noncovalently; as in a ligand-gated ion channel.

A ligand may refer to either an agonist or an antagonist. There are numerous endogenous agonists that stimulate a biological response (such as hormones), and many drugs are developed to either block or mitigate the effect of those agonists by binding (and blocking) the associated receptor. An example of an endogenous agonist is dopamine, and there are numerous marketed drugs that are antagonists for one or more of the dopamine receptors. Prochlorperazine is one such dopamine receptor antagonist (against D2), used as an antipsychotic or for treatment of migraines, for example.

Please note that the study of receptors is a rich and complex field, and our simple presentation here is only a rudimentary introduction, intended to help provide some background to aid with collaborations for study planning and analysis; whether the approach discussed here will potentially be useful in a particular situation will require engagement of the appropriate subject matter expert.

Rang (2006) described receptor theory as pharmacology's *big idea*, where he provided a good overview of the development of the field. The earliest quantitative work appears to be due to Hill in 1909, with initial focus on describing drug effect as a function of the logarithm of drug concentration. An influential quantitative development of receptor theory was given by A. J. Clark in 1926, with J. H. Gaddum publishing a similar development at about the same time in 1926. Clark is generally credited with originating modern receptor theory (Kenakin, 2009). In 1937, Gaddum published the binding equation for two ligands competing for the same receptor.

The Hill model may often have direct application for a drug that acts as an agonist. Things are more complex for drugs that are antagonists, however, since their activity is a result of attenuating the effects of an agonist. In the following, we will incorporate Gaddum's competitive binding equation into the Hill model to quantify the effect of an antagonist via its impact on the agonist binding.

A. V. Hill introduced the equilibrium concentration–effect curve in 1909, and in 1913 used it to describe oxygen binding to hemoglobin using the equation commonly written as

$$E = M \times \frac{[A]^H}{[A]^H + [A_{50}]^H} \qquad (18.1)$$

where
  E is the (pharmacological) effect
  $[A_{50}]$ is the concentration yielding half of the maximum effect M
  H is the Hill constant

So once the Hill equation is known, the desired effect E can be achieved by choosing the appropriate concentration for A. Gesztelyi et al. (2012) provide a detailed history of the Hill equation and variations.

J. H. Gaddum proposed the "mechanistic model" of two ligands A and B competing for the same receptor, based on theory and scientific principles, as

$$f = \frac{[A] \times K_B}{[A] \times K_B + K_A(K_B + [B])}$$

where
  $[B]$ is the concentration of B
  K_B is the equilibrium dissociation constant for B
  $[A]$ is the concentration of A
  K_A is the dissociation constant for A
  f is the fractional occupancy of A (i.e., the proportion of receptors occupied by A)

(Note: Inspection of this equation when B is not present reveals that the equilibrium dissociation constant K_A is the concentration of A at which half of the receptors are occupied by A.)

We prefer to rearrange terms as in Equation 18.2 as this form has a ready interpretation of the impact of a competitor, that is, this makes it clear that the effect of the competitive antagonist B on the fractional occupancy of A is the same as altering the dissociation constant of A by the factor $1 + [B]/K_B$.

Thus, the presence of B effectively alters the potency of A by reducing the number of receptors occupied by A.

$$f = \frac{[A]}{[A] + K_A \left(1 + \frac{[B]}{K_B}\right)} \tag{18.2}$$

A common drug development usage of Equation 18.2 is as follows. The ligand A is an endogenous agonist and we want to inhibit its activity by administering drug B, an antagonist, where A and B are in competition for the same receptor. The dissociation constant K_B is an intrinsic property of B, and our aim is to estimate K_B since that allows us to evaluate the impact of various concentrations of B, via its practical impact on the fractional occupancy of A.

Waud et al. (1978) combined Equations 18.1 and 18.2 to yield Equation 18.3, which allows us to answer our problem at hand, namely, what concentration of B do we need to have in order to achieve effect E?

$$E = M \times \frac{[A]^H}{[A]^H + \left[A_{50} \left(1 + \frac{[B]}{K_B}\right)\right]^H} \tag{18.3}$$

It should be noted that in early evaluation of a compound one often includes an exponent for the term $[B]/K_B$, which is S, the Schild constant. Schild (1957) introduced the Schild plot of log(dose ratio -1) versus log(antagonist concentration), with slope S. When the antagonist is competitive, the slope should be 1. (Note: The term *dose ratio* was used in the original development, and its usage often persists even when talking about the ratio of concentrations.)

For practical purposes, when the Schild constant is 1, then one may plot the quantity, $\log_{10}(1 + [B]/K_B)$ versus $\log_{10}([B])$. This basically reveals how the effective endogenous agonist A_{50} value (the concentration of A yielding half of maximal effect) changes as a function of the antagonist B concentration. An overlay of expected drug (i.e., antagonist) concentrations on the x-axis and the desired multiplier (i.e., $1 + [B]/K_B$) on the y-axis aids evaluation of feasibility of achieving the desired degree of inhibition, noting that the multiplier is for A_{50}, not directly for the effect.

Figure 18.2 shows this relationship, obtained by dividing concentration by K_B. For values of this "standardized" concentration less than about 0.1, there is little appreciable impact on A_{50}, so that large changes in concentration in this region will usually lead to small changes in effect. The effect change may be evaluated by the Hill equation, which is a function of both the endogenous level of A and the modified A_{50}; if these effect changes are smaller than desired, then it may make further development infeasible due to required high concentrations of the investigational drug. For values of the

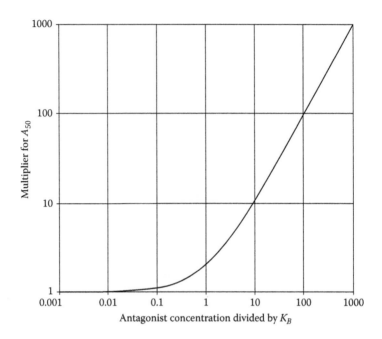

FIGURE 18.2
Graphical representation of the attenuation of an endogenous agonist by an antagonist. A competitive antagonist attenuates the effect of an endogenous agonist by increasing the agonist's A_{50} value (i.e., the concentration of the agonist that produces half of maximal effect) in the manner show in this curve, when the Schild constant is 1. The x-axis has been *standardized* by dividing the concentration of the antagonist B by the equilibrium dissociation constant of B (i.e., K_B is the concentration of B at which half of the receptors are occupied by B).

"standardized" concentration greater than about 10, fold changes in concentration yield approximately the same fold change in the multiplier for A_{50}, so variation in exposure directly translates into variation in the A_{50} multiplier and may guide decisions about further development.

We have used such plots for evaluation of compounds from early human trial data. It should be noted that this approach requires the existence of the drug target pathway in the subjects studied, in addition to drug concentration data and biomarker data from the same samples, with appropriate assays. A similar useful plot uses the proportional impact on pharmacodynamic effect for the y-axis, perhaps with an overlaid horizontal band of desired effect to aid target concentration (and thus dose) selection. Barlow et al. (1997) provide more information on the topic of this section, with various methods of estimating K_B, and also include several examples of useful and more traditional graphs. Kenakin (2009) is another excellent source for far greater depth on this topic, while Bindslev (2007) provides an extensive commentary and comparison of methods.

18.2.2.2 Response Biomarkers

There are clinical situations where an early indicator that a drug is working for a particular patient would be very valuable. For example, if a subject has a potentially fatal illness and a choice of treatments, it would be highly desirable to know whether the chosen therapy is working or not. The choice might be a different intervention or perhaps a different dose of the same drug.

This has similarities to the practice of therapeutic drug monitoring (TDM), which operates on the principle of trying to keep a subject's drug concentration within a certain range by periodic concentration measurement and dose adjustment. TDM is common in immunosuppression in the organ transplant setting, where oversuppression leads to infections and malignancy, and undersuppression leads to organ rejection.

The same principle may also be employed using a pharmacodynamic readout. An example of this is monitoring the effect of warfarin (an anticoagulant commonly used for the prevention of thrombosis and thromboembolism) dosage, using INR (international normalized ratio, the patient prothrombin time divided by the mean of the prothrombin time reference interval) to evaluate warfarin's impact on the extrinsic pathway of coagulation. Warfarin dosage is adjusted as needed to keep INR in an appropriate range.

A major issue for a response biomarker is the strength of association between the biomarker and the clinical response. In the warfarin example, that association is very high; the response biomarker (INR) is a surrogate for the clinical response (bleeding risk) in this specific case (patients receiving warfarin) (Crowther, 2009).

In the ideal response biomarker situation, one has

- A strong association with clinical outcomes
- Rapid biomarker assessment (so a timely change may be made)
- Biomarker assessment undertaken sufficiently ahead of any accepted clinical or surrogate indication of patient outcome
- Clear decision criteria and next steps

A major advantage of a biomarker-informed decision to change therapy is that one has objective evidence of biomarker (or pharmacodynamic) response for a specific patient. The choice to start treatment is informed by statistical considerations, such as a population response rate observed in prospective controlled clinical trials. For any given particular subject, however, the situation changes once therapy has begun and a biomarker response (or nonresponse) is observed. This may be viewed as a practical application of a Bayesian perspective, with potential for more quantitative usage in routine clinical practice.

In the oncology solid tumor case, subjects typically have a scan every 8 weeks to monitor the target lesion(s). While this is a response biomarker, a marker that was observable a day or two after treatment initiation or at least after one cycle of therapy (if that were only a few weeks) would potentially be very valuable. Sorenson et al. (2009) reported that MRI done 1 day after starting cediranib for glioblastoma correlated with survival, and that was improved by a composite measure that incorporated plasma markers (Batchelor et al., 2010). Another example involves a marker for early detection of the development of resistance to tivozanib (AACR Press release, 2010; Hepgur et al., 2013).

At present it appears the primary use of response biomarkers in drug development programs might be to shorten the time required for proof of concept trials where the approvable clinical endpoint takes much longer to observe. One example of this use would be measurement of glucose for a type 2 diabetes development program to inform a go/no go to phase 3.

18.3 Selection Biomarkers

18.3.1 Ways to Find a Selection Biomarker

For a given investigational therapeutic intervention, there are three primary ways that selection biomarkers may be discovered: (1) motivated by known or postulated biology, (2) brute force examination of numerous candidates, and (3) an intermediate approach between (1) and (2). After consideration of these hypothesis-generating approaches to finding selection candidates, we will present some approaches to confirmation.

Discussion of selection biomarkers requires four components:

- A therapeutic indication (or disease state and desired clinical outcome)
- A control (or standard) therapy
- An investigational therapy (or drug)
- One or more candidate biomarkers

The selection potential of a biomarker is only relevant in the context of informing a clinical choice to utilize either the investigational or control therapy for a given therapeutic purpose.

The importance of disease state in biomarker consideration is becoming more prominent in oncology, for example, where the paradigm is shifting (Benes, 2013) from organ-defined (e.g., non-small cell lung cancer) to genetically defined (e.g., EGFR, KRAS, ALK, ROS, etc., tumors Raparia et al., 2013); a genetic-based tumor taxonomy naturally suggests potential selection biomarkers in a way which an organ-based taxonomy cannot. A diagnostic marker for disease may sometimes also serve as a selection biomarker, or it may become a therapeutic target, as in the case of development of bosutinib targeting the Philadelphia chromosome for treatment of chronic myelogenous leukemia (Rassi and Khoury, 2013).

Selection biomarkers may often be viewed as either positive or negative. Positive selection biomarkers may identify those patients who are more likely to experience greater benefit than those receiving standard therapy (or are untreated). Likewise, negative selection biomarkers may identify patients unlikely to benefit compared with standard therapy (or perhaps known unable to benefit). An example of the "unable to benefit" situation would be KRAS and panitumumab or cetuximab monotherapy for colorectal cancer, while an example of harm (negative benefit) would be abacavir for HIV patients, with the HLA-B*5701 allele appearing to confer a high risk of abacavir hypersensitivity (Burns et al., 2010). Although positive selection biomarkers may seem to be ideal, negative selection biomarkers, if they clearly identify those patients who cannot benefit or who might suffer harm, are equally valuable.

It should be noted that benefit is defined relative to a standard therapy, and so benefit related to a biomarker may derive from an association between the biomarker and outcome on either standard therapy or on the investigational therapeutic, or on both. A biomarker that is associated with benefit (positive or negative) from a therapeutic intervention is said to be predictive. A biomarker is prognostic if outcome is associated with the biomarker in the presence of standard therapy, or no therapy. All selection biomarkers will be predictive biomarkers, and some may also be prognostic biomarkers. Consideration of whether a selection biomarker is also prognostic or not may have implications for study design, but otherwise we will not be making the distinction in the following discussion. We briefly discuss characteristics of the three approaches (known biology, brute force, and intermediate), associated methods, and likely next steps.

18.3.1.1 Motivated by Known or Postulated Biology

The most promising candidate selection biomarkers are those motivated by biology. These are either the target of the therapeutic or may be pathway biomarkers; those more proximal to the target of therapeutic interdiction are likely more promising. This means the pool may have very few potential selection candidates, and there will often be an existing body of scientific literature related to these markers. Both in vitro and animal model studies may be employed to further narrow the pool.

Potential problems with this approach include

- The models may be wrong
- In vitro and animal models may not translate to humans
- Sample acquisition may be very challenging
- Assays may lack sufficient sensitivity, there may be interfering substances, etc.

Markers chosen for biological reasons will often be studied at some length before human trials for the targeted agent of interest are started and are often the result of studies aimed at understanding a particular biological process, the therapeutic target being a result of such studies. Once candidate biomarkers are chosen, additional targeted studies are usually undertaken either to characterize a subject population (e.g., distribution of a biomarker by disease state) or to characterize and develop the measurement methodology (e.g., collection of atypical samples, assay, etc.). Such studies are often undertaken by accessing sources of tissues (tissue banks) with associated patient treatment and outcome information (Shaw and Patterson, 2011). This is an efficient approach if a suitable tissue (or bio-) bank can be accessed, otherwise a trial (sometimes referred to as a phase 0 trial where no investigational drug is delivered but samples are obtained) may need to be undertaken to obtain the relevant tissue in the same manner as is anticipated for the clinical trial. It could involve recruitment of subjects covering a wide range of disease states (or severity), examining the distribution of the biomarker for each state, and seeing whether the biomarker correlates with disease severity. Another way to describe the overall goals of such studies is to evaluate the ability to measure a biomarker and to get an early sense of its potential for clinical use.

Ideally such studies would occur in parallel with preclinical phases of drug development and would be further explored during early-phase human trials, phase 0 or 1, with the most promising candidates being studied in later phase trials. However, if the potential selection marker is not the target of the therapeutic, then the evidence will likely come later in the drug development process.

18.3.1.2 Brute Force Methods

Brute force methods typically examine hundreds or thousands of candidates. The studies are usually single nucleotide polymorphism (SNP) or genome-wide association studies (GWAS), of case–control or cross-sectional/cohort design. A more recent trend in the oncology field is to interrogate large numbers of oncogenes using next-generation sequencing technologies. Commonly the odds ratio for cases and controls is used to identify potential associations of marker and disease. For time-to-event endpoints, one might examine hazard ratios corresponding to various potential biomarker classifiers.

Drawbacks with this brute force approach include

- False discovery, leading to wasted money, time, and effort
- Bias, common in observational studies and also bias unique to GWAS (Pearson, 2008)
- Design issues, including selection of control group
- Identified marker may not be relevant to the proposed intervention
- Odds ratio usually too small to be useful

Perhaps the most notable drawback to the odds ratio approach is that the odds ratio needs to be rather large for the marker to be useful as a selection marker (e.g., >16 per unit increase in the marker (Pepe et al., 2004)). In practice one seldom finds an odds ratio greater than 3 in such studies.

Another issue is whether a marker associated with disease is relevant to the proposed intervention; the marker motivated by biology has a large advantage in this regard, typically being on the putative causal pathway.

Sample size and the observational nature (i.e., subjects are not randomized to biomarker) of such trials remain a challenge to this approach. Additionally, the huge number of tests in the typical SNP/GWAS study requires care in analysis and interpretation. Although methods such as false discovery rate (FDR) control have been well known and practiced for years, many reported findings appear to be spurious, and lack of reproducibility remains a concern. Further, the usual cautions about subset analysis apply, as Simon (2010) reminds us (Bonetti and Gelber (2004) also describes the ISIS-1 trial of atenolol for suspected acute myocardial infarction, citing an overall study mortality reduction of 30% but 71% in the astrological sign Leo subgroup, with no apparent benefit in the other astrological sign subgroups!).

Besides the GWAS approach, other frequently used methods include quantile-quantile and Manhattan plots, and logistic regression. Control of the FDR is common, although not universal (see Box 1 in Wellcome Trust Case Control Consortium (2007) for an excellent discussion). Zhong and Prentice (2008) provide advice for bias reduction and confidence interval adjustment, and advice (Zhong and Prentice, 2010) for comparison with replication studies. Interacting loci presents a more difficult problem without completely satisfactory solutions (see Chen et al. (2011) for a comparative evaluation of methods). Pepe and Feng (2011) provide recommendations for design and reporting.

Next steps for candidate markers found by brute force could be additional SNP/GWAS work in another population or other existing data source for confirmation, or (preferred) to prospectively plan experiments using animal or in vitro models, or human trials to reproduce the discovered association between biomarker and clinical effect.

18.3.1.3 Intermediate Approach

The intermediate approach to discovering a selection biomarker would typically involve retrospective analysis of prospective randomized clinical trials, usually phase 3, using a small number of candidates. This may often happen due to evolving knowledge of markers and availability of assays. Since we are discussing "finding" a selection biomarker, we will primarily discuss exploratory and hypothesis-generating efforts before discussing trials for confirmation.

Potential concerns and limitations of the intermediate approach include

- Sample acquisition and storage
- Relevant assay (marker of interest, reference standards, changing technology)
- Ascertainment (availability of marker values for subjects), potentially impacted by subject consent, and the issues given earlier
- Outcome-defined subsets
- Postrandomization measurements
- Power to detect a true association, alpha (false association)
- May be a hunt rather than a prospectively defined marker assessment
- Needs replication

Inference regarding selection is problematic since marker-driven subject selection would typically not be part of the study design for such trials. There are cases where this approach has been very useful, for example, KRAS for panitumumab, and usually regulatory authorities want follow-up phase 3 trials to confirm the clinical utility of the selection marker. However, in cases where there appears to be evidence of lack of benefit, or potential harm to subjects, these may be difficult or unethical to perform.

Marker evaluation methods are generally better developed and understood for an intermediate number of markers. The emphasis is usually on finding well-performing classification rules. Common (and less common) methods include logistic regression, forest plots, artificial neural networks, support vector machines, random forest, cross-validation, bootstrap and approximate randomization tests, and weighted correlation network analysis. Unfortunately many findings are not replicated. Berry (2012) says "the biomarker literature is replete with studies that cannot be reproduced." While not specifically about biomarker discovery, Ioannidis (2005, 2008) provides a very relevant and a worthwhile read on the topic of irreproducibility. The literature comparing these exploratory methods for marker selection is rather sparse, with a few notable exceptions, such as Hsieh et al. (2011), who compared five methods for selecting among 16 potential markers, a mix of phenotype and biomarkers.

Next steps for such markers are highly variable, ranging from discussions with regulators to additional phase 3 trials.

18.3.2 Building Sufficient Evidence to Trigger Testing a Patient Selection Hypothesis Including Biomarker Measurement Considerations

From a practical perspective, the decision to formally test a hypothesis that includes a patient selection biomarker either prospectively or as a prospective-retrospective analysis (Simon et al., 2009a; Patterson et al., 2011; Hayes et al., 2013) requires the development of a sufficient level of evidence to convince all stakeholders (both within the therapeutics company and external regulators) of the value of such a study. Four scenarios are described as follows: (1) selecting patients during Phase 1, (2) selecting patients at enrollment in the pivotal trial, (3) applying patient selection categorization during the pivotal trial, and (4) prospective-retrospective studies. Testing a patient selection biomarker also requires the following with critical stakeholder agreement on these points:

- An effective drug (for a desired outcome, possibly yet to be confirmed)
- A well-defined biological hypothesis that rationalizes why the treatment of interest might offer benefit for subjects with a certain biomarker status (association without biological plausibility is unlikely to reach a level of evidence to result in testing the hypothesis)
- An appropriate clinical trial setting that allows the desired treatment effect to be observed (prospective or if retrospective, sufficient sample ascertainment)
- A validated assay that can robustly measure the biomarker of interest

18.3.2.1 Selecting Patients during Phase 1

If the target of the therapeutic is a measurable biomarker, then it will usually be in the pool of candidate selection biomarkers, and the path is more straightforward. This is a likely scenario if the target displays a restricted expression within the specific disease setting (i.e., identifies a fraction of colorectal cancer tumors). In such cases, patient selection may begin at first dose in humans depending upon the benefit–risk to the patients. This is more likely to be the case for life-threatening illness, such as in oncology.

18.3.2.2 Selecting Patients at Enrollment in the Pivotal Trial

The ideal situation occurs when phase 1 and 2 clinical trials yield a sufficient level of evidence to justify the necessary diagnostic development efforts to enable a robust confirmatory phase 3 drug-diagnostic co-development clinical trial, and when there is sufficient time for this diagnostic development. Allowing sufficient time for the diagnostic development is the single biggest challenge for drug-diagnostic co-development. Once one has sufficient confidence in the biomarker to move to a pivotal trial, there are time pressures due to patient care, economics, and competition that force initiation of that trial at the earliest possible date; the diagnostic development timeline is inevitably squeezed, being on the critical path. Development of diagnostic tests follows what is referred to as design control (US FDA, March 1997).

Design control is a formal process whereby user needs are gathered to design an assay with a clearly defined intended use (how the device is to be used: what analyte is to be measured; whether the test is quantitative, semi-quantitative or qualitative; what specimen type or matrix is to be measured) and indication for use (for what or whom the device is to be used: disease or condition to be evaluated; target population frequency of use). Thus, requirements are developed, the device is designed to meet those requirements, the design is evaluated (a potentially iterative process) and transferred to production for manufacturing. Typically a prototype undergoes verification to determine whether the design requirements have been meet. If so, validation is undertaken to demonstrate that the device meets the predefined acceptance criteria required for the intended use and indications for use. At this point, the design and manufacturing are *locked in* so that the *market ready* diagnostic devices produced are the same as intended for commercial clinical usage. It should also be noted that the cut-off/clinical decision point (discussed later) is one of the key variables that is tested during analytical verification and validation and as such needs to be set before these diagnostic development studies are undertaken. These studies will typically employ clinical samples with outcome data.

All of this work needs to be conducted ahead of the start of the pivotal trial that will use the diagnostic device to test for the biomarker chosen for patient selection. The finished *market-ready* diagnostic device should be employed for the pivotal trial since any changes prior to the regulatory submission of the diagnostic device will likely lead to a requirement for analytical bridging studies to demonstrate that the changes have little impact on patient selection. The analytical bridging should usually be a comparison of the pre- and postchange diagnostic device results derived from the same set of tissue specimens. This becomes challenging due to the requirement for high ascertainment of results from patients in the trial. Baseline samples for all screened subjects should be banked and available for testing. Even if only one biomarker category was enrolled (i.e., biomarker positive), samples from both the screen positive and screen negative (i.e., screening failure) patients should have been banked and available for testing. Besides availability of tissue, one must also ensure comparable preanalytical factors, such as fresh versus archived samples; for oncology, one must be aware of issues related to tissue heterogeneity, as different portions of a specimen may have different biomarker profiles.

Though difficult, one needs to coordinate the timeline to conduct diagnostic device development in parallel with efforts to identify a candidate patient selection biomarker, while also ensuring alignment of all stakeholders. How many times will a sufficient level of evidence be obtained in time to conduct the necessary diagnostics development studies? Careful consideration of these timelines and what risk-based approaches may be undertaken to try to meet them needs to be well understood by the drug development organization. Even in the first scenario described earlier with patient selection by biomarker status beginning during the phase 1 studies, one has to ensure all of the diagnostic development studies have been completed ahead of the pivotal trial.

18.3.2.3 Applying Patient Selection Categorization during the Pivotal Trial

Another scenario that is more common but also challenging is where a specific biomarker has been contemplated for patient selection, but the evidence (or diagnostic development) was insufficient at the time of pivotal trial initiation to include a *market-ready* diagnostic device as part of the patient screening for biomarker selection. It may be possible in such a scenario, if regulators are consulted and agree with the approach, to include in the protocol the specification of hypothesis tests using the patient selection biomarker, using banked samples (collected before randomization) from all subjects in the trial, analyzed for the biomarker using the *market-ready* diagnostic device after it has met the required diagnostic development standards. The evidence would need to be compelling to convince both internal and external stakeholders of such an approach but it is one way to initiate the pivotal trial without waiting for the availability of the market ready diagnostic. An advantage of this approach is the clarity around prespecification of the biomarker and management of alpha in the analysis. Note that regardless of when the biomarker assay is performed, it is important for samples to be collected before randomization, to enable a valid randomized comparison in the biomarker-defined subgroups. In many situations, it might be better to wait for completion of the market-ready diagnostic device if the wait is anticipated to be short, for risk mitigation. It is also important to mention the planned biomarker evaluation in the patient informed consent; more generally one should routinely consider whether specimens should be banked for future testing and if so, allowing for that in the patient informed consent. When such samples are collected, they should be for all subjects (US FDA January 2013a).

A similar but perhaps importantly different situation may occur when there is sufficient confidence to take a drug forward to a pivotal trial without a clearly identified selection marker (though there may be one or more good candidate selection biomarkers). Tissue samples from the clinical trial subjects should be banked in anticipation of future availability of a candidate selection biomarker and associated market-ready diagnostic. If a marker is found (not using data from the current study!) and a market-ready assay becomes available before the study completes, a protocol amendment may be considered, to include a suitable hypothesis involving the biomarker and describe how the study alpha will be managed. There may be little practical difference between this, and the case where the marker is known before study initiation (but without a market-ready assay), since in both cases the trial is of "all comers" design, with analysis of banked samples to occur late in (or after completion of) the trial.

An example of the protocol amendment approach involved two ongoing trials of the anti-EGFR therapy panitumumab in combination with chemotherapy in first and second line metastatic colorectal cancer, where the primary analysis was amended to include only subjects who were classified as wild-type KRAS using banked samples (Douillard et al., 2010; Peeters et al., 2010). This occurred after the hypothesis of KRAS exon 2 as a predictor of nonresponse to anti-EGFR therapy had been independently demonstrated in the third-line monotherapy trial (Amado et al., 2008) (see next section). These trials, which had not selected patients based upon KRAS exon 2 mutational status, had essentially completed enrollment at the time this evidence was developed and so the primary analysis population was changed to include only subjects whose analysis for KRAS exon 2 somatic mutation status was negative (Douillard et al., 2010; Peeters et al., 2010). So although patients had not been enrolled based upon their KRAS exon 2 somatic mutation status, the analysis set defined by KRAS exon 2 negative status was used for the primary analysis; the study was therefore considered prospective for the hypothesis and analysis of panitumumab in this biomarker-defined population.

The point of presenting these possibilities, which arise due to lack of a biomarker with market-ready assay, is not to argue in favor of routinely working this way, but rather to point out options that may be appropriate in the given circumstances. If the evidence from other sources is convincing and all the elements are in place to enable an appropriate protocol amendment, such an approach should be considered. It may be that the primary analysis is not altered, allowing for the drug effect in the unselected population to be tested, perhaps with an additional analysis specified to enable testing of a hypothesis involving the biomarker. This will require careful consideration of alpha management and may require specification of a sequential strategy of testing. We note that both trials described in the previous paragraph specified a sequential testing approach, similar to one of the approaches described by Simon (2008; see the section "Analysis plan: analysis of test negatives contingent on significance in test positives"). Proceeding to a phase 3 trial without the *market-ready* diagnostic device or delaying the study start involves consideration of risks, cost, and other factors, so that each situation will need to be evaluated on its own merits. Ultimately, patients will benefit if the hypothesis can be shown to be correct and yield benefit.

18.3.2.4 Prospective-Retrospective Studies

In cases where a selection biomarker has not been prospectively confirmed, if the evidence for a patient selection hypothesis has been generated in independent studies and is sufficiently compelling for all stakeholders, then a prospective-retrospective analysis (PRA) study may be undertaken for studies already completed. The requirements for such an approach have been outlined (Simon et al., 2009a,b, 2013; Patterson et al., 2011) and are as follows:

- An adequate, well-conducted, well-controlled trial
- A predefined statistical analysis plan containing analysis methods and statistical tests corresponding to the hypothesis, and the associated power outlined in the original approved clinical study protocol
- A description of the validated analytical assay (preferably using a candidate diagnostic test that has completed verification and validation) and the application of relevant quality assurance compliance rules to the biomarker testing and analysis (e.g., lab blinded to patient outcomes)
- The imperative for pre-randomization sample procurement and sufficient ascertainment (including approximate random allocation of factors not used as stratification variables for randomization) such that the available samples are representative of the patients in the trial
- The ability to determine statistical power to assess both drug and biomarker assay/device effectiveness (i.e., biomarker clinical utility)

A strength of the prospective-retrospective approach is that it provides the opportunity to employ emerging biology to add a valuable and timely dimension to a completed trial.

As alluded to earlier, the prime example of this PRA approach was the testing of KRAS exon 2 as a negative selection biomarker for the anti-EGFR therapy panitumumab, which was achieved in the third-line monotherapy setting and has been described in detail elsewhere (Burns et al., 2010; Patterson et al., 2011) but will be briefly presented here. The evidence for KRAS as a negative selection biomarker had been emerging during the conduct of the monotherapy trial and continued after its completion; panitumumab had been granted conditional approval in the United States in 2006 and the biomarker selection hypothesis gained a sufficient level of evidence to convince internal stakeholders to test the hypothesis in a PRA study during 2007. The statistical test for interaction supported a differential treatment versus control effect in the wild-type and mutant subgroups, and the results confirmed the hypothesis that there were no benefit to patients who received panitumumab if their tumors harbored mutations in KRAS exon 2 (Amado et al., 2008). This was also a case where data from another molecule in the class (cetuximab) were generated in a similar manner (Patterson et al., 2011) and collectively these data resulted in a rapid change to the practice of medicine with treatment guidelines incorporating this information being issued by the American Society of Clinical Oncology (ASCO) and the National Comprehensive Cancer Network (NCCN) as well as test advice from the College of American Pathologists (CAP)(NCCN, 2008; Allegra et al., 2009; CAP, 2009). Independent confirmation of findings for a specific biomarker (rather than from a single trial), such as this case with panitumumab and cetuximab, are considered to be important and constitute Level of Evidence (LOE) I according to Simon et al. (2009a). The evidence was considered sufficiently strong that it led the ASCO to issue their first

"Provisional Clinical Opinion (PCO)" in April 2009, encouraging KRAS codon 12/13 testing of tumors from metastatic colorectal cancer patients in a CLIA-accredited laboratory before starting anti-EGFR therapy (Allegra et al., 2009). This opinion was based on five randomized controlled trials and five single-arm studies.

In this age of rapid advancement in the understanding of biology, and with the desire to see these advances translate quickly into patient benefit, opportunities to employ PRA-based approaches may occur with increasing frequency. By adhering to the criteria outlined earlier and through transparent interactions with regulatory authorities, this approach may be used more frequently.

18.3.3 Analysis Issues for Finding Candidate Selection Biomarkers

Analysts need to be alert to the potential for confounding factors while searching for selection biomarkers. Several baseline covariates may be highly correlated, possibly being different ways of measuring an attribute, or biologically driven. This is further complicated if postrandomization measurements are included as potential causal variables; besides confusing interpretation, a common cause factor could drive an association between a postrandomization explanatory variable and a study endpoint. For this reason, a finding for a postrandomization measurement of a biomarker may have little potential for baseline measurement selection and should be handled with caution.

One particularly difficult confounding problem involves prognostic factors. These may be known or unknown, and if unaccounted for they increase variability in the analysis and could mislead. Inclusion as covariates is common, and they may inform subsets, as well. Unfortunately, many reported prognostic factors are not supported by replication, and different researchers often report different sets of prognostic factors. Thus, it is often necessary to evaluate one's own datasets for evidence of prognostic factors, either confirmation or (riskier) discovery.

One of the most significant challenges is how to define the result of the biomarker assay as positive or negative. Every assay used for patient selection, whether quantitative, semi-quantitative or qualitative, requires the definition of the clinical decision point. The clinical decision point, or cut-off, defines whether a sample is declared positive or negative. There will often be no a priori rationale for a specific cut-off value, and the cut-off selection is usually based on empirical clinical observations.

However, such clinical data may be scarce or nonexistent (such as for a drug's first in human trial). In such circumstances, it is advisable to initially set the cut-off to the lowest level supported by preclinical data or perhaps to the limit of detection or limit of quantitation for the assay. As evidence accumulates during a development program, one may gain sufficient conviction to adjust the cut-off to a value thought to better discriminate for potential benefit.

While identifying a cut-point for one identified biomarker can be difficult with limited data, the problem is compounded when several biomarkers are under consideration for their subject selection potential. It will often be necessary to incorporate cut-off determination in the evaluation process so that the definition of a biomarker includes its candidate cut-point; the problem of choosing among a group of markers is greatly increased by the uncertainty of an appropriate cut-off for continuous markers. Since relevant datasets are likely few, it is often difficult to separate the discovery process from the validation process for cut-points. Faraggi and Simon (1996) presented the results of a simulation study to evaluate a cross-validation procedure for cut-point selection, and this may be helpful in evaluating uncertainty, though an independent study to confirm the clinical utility of the cut-point will usually be needed.

More generally one needs to clearly separate the processes of discovery and validation of biomarker. Regulators have made public presentations in which they have stressed that training sets have to be separated from testing sets; some of the issues around using the same dataset for both training and evaluation are discussed by Simon (2013, p.13). Pepe et al. (2008) and Tzankov et al. (2010) also provide strong warnings about separating these processes. It is prudent to clarify in advance whether an analysis is a discovery activity or for validation.

Last, it is helpful to consider the types of biomarker (continuous or classification) and response (continuous or classification) in choosing a method of analysis. A table such as in Figure 18.3 may help organize thinking, though in many cases one has mixes of types.

		Type of response	
		+/− Classification	Continuous
Type of biomarker	+/− Classification	• Contingency table • Classifiers	• Subgroups • ANOVA
	Continuous	• Logistic regression • ROC • Covariate • *Predictiveness* curve	• Regression

Note 1: An *extremely* important topic is how to convert a continuous biomarker measurement into a +/− classification. This involves finding a *threshold* or *cut-point*, covered later.

Note 2: Sometimes a continuous response may be reexamined as a +/− variable, such as time to event (months) versus event (yes/no) in 6 months.

FIGURE 18.3
A few analysis possibilities for various combinations of continuous and binary biomarkers and responses (noncomprehensive).

18.3.3.1 Criteria for Evaluation

As mentioned earlier, there are many factors that may impact assay performance, and characterizing performance is important for identifying and controlling those factors to ensure that assay expectations are consistently met. Woodcock (2010) said, "Drug regulators worldwide have generally been extremely reluctant to accept results from new diagnostics that are intended to be part of a drug's evidentiary base, unless the performance of the assay is very well understood. Generally, regulators have waited (in some cases, decades) until the biomarker's clinical utility is worked out by the medical community." Our consideration of biomarker assay performance will be from the perspective of its potential ability to identify patients likely (or unlikely) to experience favorable clinical outcomes under well-defined conditions.

Figure 18.4 shows the typical approach to summarizing the performance of a screening (used to identify a population at risk, that is, asymptomatic people) or diagnostic assay intended to identify disease state in subjects, by producing a positive/negative marker classification. The true positive fraction (TPF) and false positive fraction (FPF) are useful measures for describing the marker's ability to identify disease. The positive predictive value (PPV) and negative predictive value (NPV) depend on the population prevalence of disease, and so are very relevant when considering deployment for screening.

A biomarker classification rule used for selection may sometimes also be evaluated by a 2×2 table of biomarker status (positive/negative) and clinical outcome (success/failure). This may require conversion of a continuous outcome into a binary outcome, such as survival at 6 months, for example. This table for evaluation has similarities to the diagnostic problem of using a marker to determine disease state, but there may be some important interpretation differences.

The manner of sampling to construct the table is important, for example, and needs to be guided by intended use; methods appropriate for an assay for screening or diagnostic purposes (to detect or confirm disease) may not be suitable for selecting patients mostly likely to benefit from a therapy. For example, Pepe et al. (2008) recommend a *case–control* sampling scheme for "studies of biomarkers intended for use in disease diagnosis, screening, or prognosis," which yield the proportion of subjects who are biomarker positive in the *bad outcomes* (case) group (true positive rate) and in the *good outcomes* (control) group (false positive rate). That would not seem to be a natural way to sample for a selection marker, as our interest is in finding (or demonstrating) a high proportion of good outcomes in the biomarker positive group compared with that seen in the biomarker negative group.

A continuous marker may be evaluated for its classification potential by examination of the receiver operating characteristic (ROC) curve (Hanley, 1989). This is a graph that is constructed from a dataset that contains a

		True disease state	
		Healthy	Disease
Biomarker status	Negative	**TN**	**FN**
	Positive	**FP**	**TP**

True negative (TN), false negative (FN)

False positive (FP), true positive (TP)

Measures of biomarker association with disease	*True positive fraction (**TPF**) = TP/(FN + TP)* *also known as **sensitivity*** *False positive fraction (**FPF**) = FP/(TN + FP)* *also known as 1 – **specificity***
Sometimes reported:	***Odds ratio** = [TPF/(1 – TPF)] * [(1 – FPF)/FPF]*
Other definitions	*Positive predictive value (**PPV**)* $= \rho * TPF / [\rho * TPF + (1 - \rho) * FPF]$ *Negative predictive value (**NPV**)* $= (1 - \rho) * (1 - FPF) / [(1 - \rho) * (1 - FPF) + \rho * (1 - TPF)]$ *where ρ is the population **prevalence** of disease*

FIGURE 18.4
Common performance characteristics of screening and diagnostic assays.

biomarker value and success/failure designation for each subject. The graph is constructed by calculating sensitivity and 1 − specificity for each potential cut-point from the smallest to the largest value of the biomarker, and plotting sensitivity on the y-axis versus 1 − specificity on the x-axis. (See Figure 18.5). One wants this curve to rise sharply from zero to near the upper left-hand corner (staying near the y-axis), then asymptotically approach the $y = 1$ value. Such a curve would be considered a good classifier. A curve near the line of identity would mean the marker has little predictive value. The area under the ROC curve is a common measure for comparison of markers. This method has been used to select variables *best* associated with response on occasion (e.g., Hale and Reeve, 1994) and has close ties to logistic regression. This may be helpful in exploring potential cut-points for a potential selection marker.

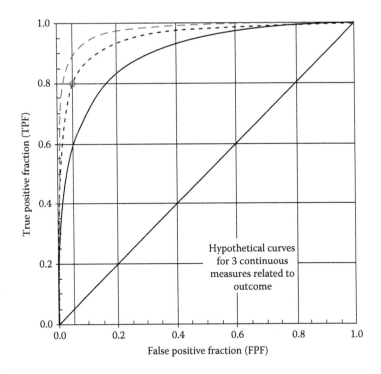

FIGURE 18.5

Illustration of a receiver operating characteristic (ROC) curve. Such curves are monotone non-decreasing from (0, 0) to (1, 1). Curves which are near (0, 1) provide the best classification capability.

A more recent graphical tool is the Treatment X Marker Predictiveness plot, suggested by Janes et al. (2011). It is used to compare curves of treatment effect as a function of a biomarker. This tool is good for facilitating discussion and helps suggest biomarker potential, as well as regions for a potential cut-point. (see Figure 18.6). Ideally the curves would cross, suggesting the expectation that subjects with biomarker values to the right of the crossing would do better on the treatment that dominates on the right (TRT in the example Figure 18.6), with those to the left expected to do better on the alternate treatment (PLA in Figure 18.6). In our experience, the crossing point (or other statistically favorable point) should not be taken as the sole consideration in selecting a cut-point; it is probably worthwhile to think about benefit/risk, cost, and other factors related to therapeutic choice, increasing or decreasing the point as appropriate and practical. If the curves do not cross, one might look for a difference of magnitude sufficient to justify a treatment choice; a plot of difference versus biomarker is recommended. At present, it is unclear how the population distribution of the biomarker figures into this approach,

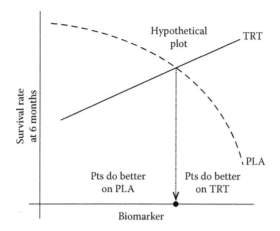

FIGURE 18.6

Hypothetical treatment X marker predictiveness plot. For this plot, subjects to the right of the curve's crossing are expected to have a better outcome when on TRT than on PLA; their expected outcome is worse to the left. See Janes et al. (2011) for several other potential scenarios.

though the curves should not be considered to be *true,* and the distribution may impact the cut-point selection.

At present, there does not seem to be universal agreement regarding the best performance measures for biomarkers; the evaluation method needs to be appropriate to the intended use. We have found the ROC curve and Treatment X Marker Predictiveness plot approaches useful during biomarker discovery and cut-point evaluation efforts. Pepe (2003) provides an excellent exposition of tests used for classification and prediction. We emphasize that prediction is not the same problem as patient selection, recalling that clinical context is important for selection (e.g., what is the default course of action?) In Chapter 16, Steyerberg (2010) addresses this aspect of prediction models, pointing out that discrimination is important for prediction, while calibration is a requirement for making decisions. See also Steyerberg et al. (2010) for additional considerations regarding metrics for marker performance.

18.3.3.2 Demonstrating the Clinical Utility of a Candidate Selection Biomarker

Once a candidate selection biomarker is chosen, clinical studies need to be performed to confirm the clinical utility of using this marker to inform a treatment choice. Choosing the candidate is primarily a discovery (or hypothesis generating) activity. Confirming clinical utility needs to be a clearly differentiated activity from discovery and will normally involve one or more new trials. This is also the situation for a biomarker cut-point, where choice of the cut-point and confirmation of the clinical utility of the biomarker using that cut-point are different activities.

This confirmation is more than simple replication and involves making the biomarker selection a part of the fabric of study design and analysis.

The diagnostic assay and the investigational drug are both evaluated in a trial performed to confirm clinical utility of a selection biomarker for a specific indication, and in many cases an investigational device exemption (IDE) will be required for the assay before the trial can begin for US-based sites. The entire measurement process, including sample acquisition, handling, shipping and storage, evaluation and reporting, etc., needs to be well established before embarking on this trial. This process will need to essentially be "locked down" (as described earlier) prior to embarking upon the confirmatory trial, with any changes only made for compelling reasons as trial interpretation could be compromised.

18.3.3.3 Study Planning

Understanding the prevalence of biomarker positive subjects in the intended population will impact study planning. If the rate is very low, one might have to screen a large number of subjects and it may take a long time to full enrollment. Simon (2008) provides several considerations for sample size for various trial scenarios. The performance of the assay also plays a role, with false positives and false negatives potentially impacting study power, depending on the study design. Logistical considerations can impact site selection, perhaps due to issues with sample collection or time to receive the biomarker assessment. Informed consent will need consideration, as will interaction with regulatory authorities; in the United States, the FDA CDRH will need to be included in discussions with CDER/CBER, and sponsors will need to include their assay experts in many such discussions.

Interestingly, one approach to confirmation of a selection biomarker does not require a new clinical trial, but rather utilizes existing studies. This case of retrospectively examining an existing trial for "confirmatory" purposes, rather than exploratory purposes, involves the "prospective-retrospective" approach described by Simon and colleagues (Simon, 2009a,b, 2013), the elements of which were described under "Prospective-Retrospective Studies." Their approach involves a given, analytically validated, completely specified marker (for classification) where one then (1) designs a study to address the question of interest with detailed protocol and analysis, (2) finds an existing completed study closely matching the newly designed one, and (3) performs the assay on the archived samples from the existing study and analyzes the data as if from the newly designed trial. An important caveat is that the assay must be analytically validated for archived tissue, whereas fresh tissue may often be used if the marker proceeds to routine clinical usage. Simon et al. (2009a) also took some care to point out that "retrospective" might be misleading, saying "The most serious limitation of epidemiological studies is their non-experimental nature, not whether they are retrospective or prospective." A key feature of clinical trials used for a prospective-retrospective study would be randomization to either investigational therapy or control in those original trials. One would also want a reasonably high ascertainment rate and a sufficient balance so there is not a notable association between biomarker status and treatment assignment.

Examples of a prospective-retrospective approach are few at present; one of the earliest and most well known involves the KRAS mutation for colorectal cancer for cetuximab and panitumumab (Amado et al., 2008). Another case was reported by Sargent et al. (2010), involving a meta-analysis of five previous trials where colon cancer patients were randomized to either fluorouracil-based (FU-based) adjuvant therapy or to no postsurgical treatment. They prepared a protocol for the pooled analysis and analyzed tissues (where available) from patients not previously included in the statistical analysis using microsatellite instability (MSI) as a classifier. The authors concluded that mismatch repair (MMR) status (the focus of the study, as determined by MSI) was appropriate for a clinical decision for use of FU therapy alone. Kerr and Midgley (2010), in an editorial accompanying the Sargent et al. (2010) article, pointed out the significance of this effort and cited other markers (p. 53, thymidylate synthase, loss of heterozygosity of 18q) previously proposed for this purpose that had failed, due to "underpowered studies, technical variation in means of measurement, and absence of validating data sets with adequate controls."

The literature contains several trial design proposals for biomarker-focused trials (e.g., see Buyse et al. (2011) for several such designs), though few are appropriate for confirmation of a selection marker. All of those will include classification of subjects as either biomarker positive or biomarker negative using samples collected at baseline; the collection should occur before randomization. Some studies enroll and randomize subjects without consideration of biomarker status, delaying the biomarker test and classification until sometime following full enrollment. In such situations, the ascertainment rate can jeopardize interpretation of a study since subjects with biomarker data may not be representative of the randomized population due to consent or availability of tissue, etc. Prospective biomarker evaluation and stratified randomization (perhaps including enrollment of only biomarker positive subjects, the "enriched design") is considered the best course in most situations.

One of the two simplest approaches is the interaction design, where subjects are classified as biomarker positive or negative and then randomly assigned to treatment or control within the positive and negative subgroups (i.e., stratified by biomarker status, sometimes also called a "biomarker-stratified design"). The statistical test for interaction is then performed to look for a differential treatment versus control difference in the two biomarker subgroups. This will typically have very low power. This design would also be unfavorable if the investigational treatment were thought to have potential to cause harm in biomarker negative subjects.

The other simple approach utilizes the targeted or selection trial and only enrolls biomarker positive subjects, randomizing them to either treatment or control. This can have the advantage of reasonably good power for showing a treatment advantage over control in the biomarker positive subjects, but does not quantify clinical utility, in terms of a strategy where biomarker positive subjects receive the investigational treatment and biomarker negative subjects receive control. Rather, this design can confirm that a treatment offers advantage over control in subjects who measure biomarker positive. This targeted design is typically the preferred approach for a confirmatory trial for a companion diagnostic/drug co-development program. As of this writing, some regulators prefer to see enrollment of some biomarker negative subjects, to confirm lack of treatment benefit in those people; this is controversial, adding additional cost and involves assigning some subjects to treatment who are expected to not benefit (see Buyse et al. (2011) for a discussion of the pros and cons; Simon (2013) includes a chapter on this topic). If studies before that confirmatory trial contain biomarker negative subjects, those data may have bearing on the discussions around this point, possibly reducing or eliminating the need for including biomarker negative subjects in future trials.

An additional consideration for the targeted trial is the possibility of "prescreening." Subjects may have been evaluated by a test other than the market-ready diagnostic device before presenting to one of the clinical trial sites for potential enrollment. There are internet sites that encourage subjects to utilize their services (i.e., marker assessment by a pre-screening assay, but typically not the market-ready diagnostic device) to see whether they might be qualified for a specific trial. Additionally, there may well be other trials involving a test for the same biomarker and the same clinical indication so that subjects may have already been tested for your selection biomarker before consideration for your trial, but not necessarily using your specific test. Prescreening can lead to bias since the population studied in the trial may not be representative of the population for clinical use, and so steps should be taken to discourage prescreening. This is becoming increasingly difficult for somatic mutation biomarkers that are tested as part of next-generation sequencing panels that encompass a number of genes implicated in tumorigenesis.

More recently, there have been some collaborative efforts that involve multiple markers and multiple investigational agents. Subjects are assessed on those markers and then matched to the therapy as indicated for their marker profile (or possibly randomized to control). An example of this is the I-SPY 2 trial, which uses an adaptive Bayesian design to compare each of five therapies against a common control. A more general implementation would have subjects routed through a laboratory portal to the treatment best suited for each specific subject. This approach is currently being proposed in the "Master Protocol Multi-drug Registration Trial Design Proposal" for metastatic non-small cell lung cancer refractory to prior chemotherapy utilizing drug/biomarker pairs for a phase 3 trial (Herbst et al., 2012).

18.4 Conclusion

Biomarkers should play a prominent role in most drug development programs. In some cases, a pathway biomarker may aid a decision for an early termination of a compound or suggest a target dosing range. For some therapeutic situations, such as a potentially terminal illness, a response biomarker that is an early indicator of drug effectiveness could offer substantial benefit in a decision to continue or change therapy. Perhaps the most important use for a biomarker will be the ongoing clinical application of a test to guide choice of therapy, using a selection biomarker. In such cases, not only is the marker an important part of the development program, but it will remain important for each subject tested for that marker in clinical practice. Along with phenotype information, this biomarker use is central to achieving a health-care system based on personalized medicine. For these aims to be successful, the statistician and biomarker scientist need to work closely together.

References

AACR Press Release (September 28, 2010). Novel biomarker may predict response to new VEGF receptor inhibitor. (http://www.aacr.org/home/public–media/aacr-press-releases.aspx?d=2071) [accessed September 22, 2013].

Allegra CJ, Jessup JM, Somerfield, MR, Hamilton SR, Hammond EH, Hayes DF, McAllister PK, Morton RF, and Schilsky RL. (2009). ASCO provisional clinical opinion: Testing for KRAS gene mutations in patients with metastatic colorectal carcinoma to predict response to anti-epidermal growth factor receptor monoclonal antibody therapy. *J Clin Oncol* 27:2091–2096.

Amado RG, Wolf M, Peeters M, Van Cutsem E, Siena S, Freeman DJ, Juan T et al. (2008). Wild-Type KRAS is required for Panitumumab efficacy in patients with metastatic colorectal cancer: Results from a randomized, controlled trial. *J Clin Oncol* 26:1626–1634.

Barlow RB, Bond SM, Bream E, Macfarlane L, and McQueen DS (1997). Antagonist inhibition curves and the measurement of dissociation constants. *Br J Pharmacol* 120: 13–18.

Batchelor TT, Duda DG, di Tomaso E, Ancukiewicz M, Plotkin SR, Gerstner E, Eichler AF et al. (2010). Phase II study of cediranib, an oral pan-vascular endothelial growth factor receptor tyrosine kinase inhibitor, in patients with recurrent glioblastoma. *J Clin Oncol* 28:2817–2823.

Benes CH (2013). Functionalizing genomic data for personalization of medicine. *Clin Pharm Ther* 93:309–311.

Berry D (2012). Multiplicities in cancer research: Ubiquitous and necessary evils. *J Natl Cancer Inst* 104:1124–1132.

Bindslev N (2007). Drug-acceptor interactions. Co-action publishing. (http://co-action.net/books/Bindslev.php)[accessed September 22, 2013].

Bonetti M and Gelber RD (2004). Patterns of treatment effects in subsets of patients in clinical trials. *Biostatistics* 5: 465–481.

Burns DK, Hughes AR, Power A, Wang S-J, and Patterson SD (2010). Designing pharmacogenomic studies to be fit for purpose. *Pharmacogenomics* 11: 1657–1667.

Buyse M (2011). An overview of biomarker-based trial designs. *Oral presentation to PSI Biomarkers SIG* Cambridge, U.K., November 11, 2011.

Buyse M, Michiels S, Sargent DJ, Grothey A, Matheson A, and de Gramont A (2011). Integrating biomarkers in clinical trials. *Expert Rev Mol Diagn* 11: 171–182.

Camidge DR, Randall KR, Foster JR, Sadler CJ, Wright JA, Soames AR, Laud PJ, Smith PD and Hughes AM (2005). Plucked human hair as a tissue in which to assess pharmacodynamic end points during drug development studies. *Br J Cancer* 92: 1837–1841.

Chen L, Yu G, Langefeld CD, Miller DJ, Guy RT, Raghuram J, Yuan X, Herrington DM, and Wang Y (2011). Comparative analysis of methods for detecting interacting loci. *BMC Genomics* 12:344.

College of American Pathologists (September 2009). Perspectives on emerging technology Report: KRAS testing for colorectal cancer. (http://www.cap.org/apps/docs/committees/technology/KRAS.pdf) [accessed September 23, 2013].

Crowther MA (2009). Introduction to surrogates and evidence-based mini-reviews. *ASH Educ Program Book* 2009(1): 15–16.

Douillard JY, Siena S, Cassidy J, Tabernero J, Burkes R, Barugel M, Humblet Y et al. (2010). Randomized, phase III trial of panitumumab with infusional fluorouracil, leucovorin, and oxaliplatin (FOLFOX4) versus FOLFOX4 alone as first-line treatment in patients with previously untreated metastatic colorectal cancer: The PRIME study. *J Clin Onco* 28:4697–4705. (doi: 10.1200/JCO.2009.27.4860)

Ebos JM, Lee CR, Christensen JG, Mutsaers AJ, and Kerbel RS (2007). Multiple circulating proangiogenic factors induced by sunitinib malate are tumor-independent and correlate with antitumor efficacy. *Proc Natl Acad Sci USA* 23:17069–17074.

Faraggi D and Simon R (1996) Simulation study of cross-validation for selecting an optimal cutpoint in univeriate survial analysis. *Stat in Med* 15:2203–2213.

Fleming TR and Powers JH (2012). Biomarkers and surrogate endpoints in clinical trials. *Stat in Med* 31:2973–2984.

Gesztelyi R, Zsuga J, Kemeny-Beke A, Varga B, Juhasz B, and Tosaki A (2012). The Hill equation and the origin of quantitative pharmacology. *Arch Hist Exact Sci* 66:427–438.

Hale MD and Reeve RL (1994). Planning a randomized concentration controlled trial with a binary response. *ASA Proc Biopharm Sect* 3:19–323.

Hanley JA (1989). Receiver operating characteristic (ROC) methodology: The state of the art. *Crit Rev Diagnostic Imaging* 29:307–335.

Hanneman KK, Scull HM, Cooper KD, and Baron ED (2006). Effect of topical vitamin D analogue on in vivo contact sensitization. *Arch Dermatol* 142:1332–1334.

Hayes DF, Allen J, Compton C, Gustavsen G, Leonard DG, McCormack R, Newcomer L et al. (2013). Breaking a vicious cycle. *Sci Transl Med* 5:196cm6. (doi:10.1126/scitranslmed.3005950)

Hepgur M, Sadeghi S, Dorff TB, and Quinn DI (2013). Tivozanib in the treatment of renal cell carcinoma. *Biologics* 7:139–148.

Herbst R, Rubin E, LaVange L, Abrams J, Wholley D, Arscott K, and Malik S (2012). Issue brief: Design of a disease-specific master protocol. *Conference on Clinical Cancer Research*, Washington DC, November 2012. http://www.focr.org/sites/default/files/CCCR12MasterProtocol.pdf.

Hsieh C-H, Lu R-H, Lee N-H, Chiu W-T, Hsu M-H, and Li Y-C (2011). Novel solutions for an old disease: Diagnosis of acute appendicitis with random forest, support vector machines, and artificial neural networks. *Surgery* 149:87–93.

Ioannidis JPA (2005). Why most published research findings are false. *PLoS Med* 2:e124.

Ioannidis JPA (2008). Why most discovered true associations are inflated. *Epidemiology* 19:640–648.

Janes H, Pepe MS, Bossuyt PM, and Barlow WE (2011). Measuring the performance of markers for guiding treatment decisions. *Ann Intern Med* 154:253–259.

Kenakin T (2009). *A Pharmacology Primer: Theory, Applications, and Methods*. 3rd ed. Amsterdam, the Netherlands: *Elsevier Academic Press*.

Kerr DJ and Midgley R (2010). Defective mismatch repair in colon cancer: A prognostic or predictive biomarker? *J Clin Oncol* 28:3210–3212.

Lindauer A, Di Gion P, Kanefendt F, Tomalik-Scharte D, Kinzig M, Rodamer M, Dodos F, SÃűrgel F, Fuhr U, and Jaehde U. (2010) Pharmacokinetic/pharmacodynamic modeling of biomarker response to sunitinib in healthy volunteers. *Clin Pharmacol Ther* 87:601–608.

Liu H-B, Wu Y, Lv T-F, Yao Y-W, Xiao Y-Y, Yuan D-M, and Song Y (2013). Skin rash could predict the response to EGFR tyrosine kinase inhibitor and the prognosis for patients with non-small cell lung cancer: A systematic review and meta-analysis. *PLoS One* 8:e55128.

Medecins Sans Frontieres (July 24, 2012). Report: Viral load monitoring improves HIV treatment in developing countries. (http://www.msf.org/article/viral-load-monitoring-improves-hiv-treatment-developing-countries) [accessed September 22, 2013].

National Comprehensive Cancer Network, (NCCN) (November 3, 2008). Updates guidelines for colorectal cancer. (http://www.nccn.org/about/news/newsinfo.asp?NewsID=194) [accessed August 25, 2013].

Neubig RR, Spedding M, Kenakin T, and Christopoulos A (2003). International union of pharmacology committee on receptor nomenclature and drug classification. XXXVIII. Update on terms and symbols in quantitative pharmacology. *Pharmacol Rev* 55:597–606.

Patterson SD, Cohen N, Karnoub M, Louis Truter S, Emison E, Khambata-Ford S, Spear B et al. (2011). Prospective-retrospective biomarker analysis for regulatory consideration: White paper from the industry pharmacogenomics working group. *Pharmacogenomics* 12:939–951.

Pearson TA (2008). Bias in studies of the human genome. Oral presentation at genetics for epidemiologists: Application of human genomics to population sciences, May 13–14, 2008, Northwestern University Evanston, IL. (http://www.genome.gov/Pages/About/OD/OPG/GeneticsforEpidemiologists/GFELecture6.ppt) [accessed September 22, 2013].

Peeters M, Price TJ, Cervantes A, Sobrero AF, Ducreux M, Hotko Y, André T et al. (2010). Randomized phase III study of panitumumab with fluorouracil, leucovorin, and irinotecan (FOLFIRI) compared with FOLFIRI alone as second-line treatment in patients with metastatic colorectal cancer. *J Clin Oncol* 28:4706–4713.

Pepe MS (2003). *The Statistical Evaluation of Medical Tests for Classification and Prediction.* Oxford, U.K.: Oxford University Press.

Pepe MS and Feng Z (2011). Improving biomarker identification with better designs and reporting. *Clin Chem* 57:1093–1095.

Pepe MS, Feng Z, Janes H, Bossuyt PM, and Potter JD (2008). Pivotal evaluation of the accuracy of a biomarker used for classification or prediction: Standards for study design. *J Natl Cancer Inst* 100:1432–1438.

Pepe MS, Janes H, Longton G, Leisenring W, and Newcomb P (2004). Limitations of the odds ratio in gauging the performance of a diagnostic, prognostic, or screening marker. *Am J Epidemiol* 159:882–890.

Prentice, R. L. (1989). Surrogate endpoints in clinical trials: Definition and operational criteria. *Stat Med* 8:431-440.

Rang HP (2006). The receptor concept: Pharmacology's big idea. *Br J Pharmacol* 147: S9–S16.

Raparia K, Villa C, DeCamp MM, Patel JD, and Mehta MP (2013) Molecular profiling in non-small cell lung cancer: A step toward personalized medicine. *Arch Pathol Lab Med* 137:481–491.

Rassi FE and Khoury HJ (2013). Bosutinib: A SRC-ABL tyrosine kinase inhibitor for treatment of chronic myeloid leukemia. *Pharmgenomics Pers Med* 6:57–62.

Rosen LS, Kurzrock R, Mulay M, Van Vugt A, Purdom M, Ng C, Silverman J et al. (2007). Safety, pharmacokinetics, and efficacy of AMG 706, an oral multikinase inhibitor, in patients with advanced solid tumors. *J Clin Oncol* 25:2369–2376.

Russell CB, Suggs S, Robson KM, Kerkof K, Kivman LD, Notari KH, Rees WA, Leshinsky N, and Patterson SD (2008). Biomarker sample collection and handling in the clinical setting to support early phase drug development. In: *Biomarker Methods in Drug Discovery and Development.* F. Wang (ed.). The Humana Press Inc, Totowa, NJ, pp. 1–26.

Sargent DJ, Marsoni S, Monges G, Thibodeau SN, Labianca R, Hamilton SR, French AJ et al. (2010). Defective mismatch repair as a predictive marker for lack of efficacy of 5-fu-based adjuvant therapy in colon cancer. *J Clin Oncol* 28:3219–3227.

Schild HO (1957). Drug antagonism and pA2. *Pharmacol Rev* 9:242–246.

Severino ME, DuBose RF, and Patterson SD (2006). Pharmacodynamic biomarkers in early clinical drug development. *IDrugs* 9:849–853.

Shaw PM and Patterson SD (2011). The value of banked samples for oncology drug discovery and development. *J Natl Cancer Inst Monogr* 42:46–49.

Simon R (2008). The use of genomics in clinical trial design. *Clin Cancer Res* 14(19): 5984–5993.

Simon RM, Paik S, and Hayes DF (2009a). Use of archived specimens in evaluation of prognostic and predictive biomarkers. *J Natl Cancer Inst* 101:1–7.

Simon RM (2009b). Prospective, retrospective and prospective-retrospective designs for evaluating prognostic and predictive Biomarkers. (http://ctep.cancer.gov/investigatorResources/nabcg/docs/200903/workshop_12.pdf) [accessed September 22, 2013].

Simon RM (2010). Clinical trials for predictive medicine: New challenges and paradigms. *Clin. Trials* 7:516–524.

Simon RM (2013). *Genomic Clinical Trials and Predictive Medicine.* Cambridge, U.K.: Cambridge University Press.

Sinclair SR, Kane SA, Van der Schueren BJ, Xiao A, Willson KJ, Boyle J, de Lepeleire I et al. (2010). Inhibition of capsaicin-induced increase in dermal blood flow by the oral CGRP receptor antagonist, telcagepant (MK-0974). *Br J Pharmacol* 69:15–22.

Sorensen AG, Batchelor TT, Zhang WT, Chen PJ, Yeo P, Wang M, Jennings D et al. (2009). A "vascular normalization index" as potential mechanistic biomarker to predict survival after a single dose of cediranib in recurrent glioblastoma patients. *Cancer Res* 69:5296–5300.

Steyerberg EW (2010). *Clinical Prediction Models: A Practical Approach to Development, Validation, and Updating.* New York: Springer.

Steyerberg, EW, Vickers AJ, Cook NR, Gerds T, Gonen M, Obuchowski N, Pencina MJ, and Kattan MW (2010). Assessing the performance of prediction models: A framework for traditional and novel measures. *Epidemiology* 21:128–138.

Tzankov A, Zlobec I, Went P, Robl H, Hoeller S, and Dirnhofer S (2010). Prognostic immunophenotypic biomarker studies in diffuse large B cell lymphoma with special emphasis on rational determination of cut-off scores. *Leukemia & Lymphoma* 51:199–212.

United States 21CFR201.57 (2010). Sec. 201.57 Specific requirements on content and format of labeling for human prescription drug and biological products described in 201.56(b)(1). (http://www.accessdata.fda.gov/scripts/cdrh/cfdocs/cfcfr/CFRSearch.cfm?fr=201.57) [accessed October 14, 2010].

United States Food and Drug Administration (March 1997) Design control guidance for medical device manufacturers. (http://www.fda.gov/medicaldevices/deviceregulationandguidance/guidancedocuments/ucm070627.htm) [accessed September 9, 2013].

United States Food and Drug Administration DRAFT (October 2010). Guidance for industry: Qualification process for drug development tools. (http://www.fda.gov/downloads/Drugs/GuidanceComplianceRegulatoryInformation/Guidances/UCM230597.pdf) [accessed September 23, 2013].

United States Food and Drug Administration (January 2013a). Guidance for industry: Clinical pharmacogenomics: Premarket evaluation in early-phase clinical studies and recommendations for labeling. (http://www.fda.gov/downloads/Drugs/GuidanceComplianceRegulatoryInformation/Guidances/UCM337169.pdf) [accessed September 23, 2013].

United States Food and Drug Administration DRAFT (June 2013b). Guidance for industry: Human immunodeficiency virus-1 infection: Developing antiretroviral drugs for treatment. Revision 1. (http://www.fda.gov/downloads/Drugs/GuidanceComplianceRegulatoryInformation/Guidances/UCM355128.pdf) [accessed September 23, 2013].

Wallace MS and Schulteis G (2008). Effect of chronic oral gabapentin on capsaicin-induced pain and hyperalgesia: A double-blind, placebo-controlled, crossover study. *Clin J Pain* 24:544–549.

Waud DR, Son LS, and Waud BE (1978). Kinetic and empirical analysis of dose-response curves illustrated with a cardiac example. *Life Sci* 22:1275–1286.

Wellcome Trust Case Control Consortium (2007). Genome-wide association study of 14,000 cases of seven common diseases and 3,000 shared controls. *Nature* 447:661–678.

Woodcock J (2010). Assessing the clinical utility of diagnostics used in drug therapy. *Clin Pharm Ther* 88:765–773.

Yap TA, Yan L, Patnaik A, Fearen I, Olmos D, Papadopoulos K, Baird RD et al. (2011). First-in-man clinical trial of the oral pan-AKT inhibitor MK-2206 in patients with advanced solid tumors. *J Clin Oncol* 29:4688–4795.

Zhong H and Prentice RL (2008). Bias-reduced estimators and confidence intervals for odds ratios in genome-wide association studies. *Biostatistics* 9:621–634.

Zhong H and Prentice RL (2010). Correcting "winner's curse" in odds ratios from genomewide association findings for major complex human diseases. *Genet Epidemiol* 34:78–91.

19

Statistical Evaluation of Surrogate Endpoints in Clinical Studies

Geert Molenberghs, Ariel Alonso Abad, Wim Van der Elst,
Tomasz Burzykowski, and Marc Buyse

CONTENTS

19.1 Introduction

The rising costs of drug development and the challenges of new and reemerging diseases are putting considerable demands on efficiency in the drug candidates selection process. A very important factor influencing duration and complexity of this process is the choice of endpoint used to assess drug efficacy. Often, the most sensitive and relevant clinical endpoint might be difficult to use in a trial. This happens if measurement of the clinical endpoint (1) is costly (e.g., to diagnose *cachexia*, a condition associated with malnutrition and involving loss of muscle and fat tissue, expensive equipment measuring content of nitrogen, potassium, and water in patients' body is required); (2) is difficult (e.g., involving compound measures such as encountered in quality-of-life or pain assessment); (3) requires a long follow-up time (e.g., survival in early stage cancers); or (4) requires a large sample size because of low event incidence (e.g., short-term mortality in patients with suspected acute myocardial infarction). An effective strategy is then proper selection and application of biomarkers for efficacy, replacing the clinical endpoint by a biomarker that is measured more cheaply, more conveniently, more frequently, or earlier. From a regulatory perspective, a biomarker is considered acceptable for efficacy determination only after its establishment as a valid indicator of clinical benefit, that is, after its validation as a surrogate marker (Burzykowski et al. 2005).

These considerations naturally lead to the need of proper definitions. An important step came from the Biomarker Definitions Working Group Group (2001); Ellenberg and Hamilton (1989), their definitions nowadays being widely accepted and adopted. A clinical endpoint is considered the most credible indicator of drug response and defined as a characteristic or variable that reflects how a patient feels, functions, or survives. During clinical trials, endpoints should be used, unless a biomarker is available that has risen to the status of surrogate endpoint. A biomarker is defined as a characteristic that can be objectively measured as an indicator of healthy or pathological biological processes, or pharmacological responses to therapeutic intervention. A surrogate endpoint is a biomarker, intended for substituting a clinical endpoint. A surrogate endpoint is expected to predict clinical benefit, harm, or lack of these.

Surrogate endpoints have been used in medical research for a long time (Investigators 1989; Fleming and DeMets 1996). Owing to unfortunate historical events and in spite of potential advantages, their use has been surrounded by controversy. The best known case is the approval by the Food and Drug Administration (FDA) of three antiarrhythmic drugs: encainide, flecainide, and moricizine. The drugs were approved because of their capacity to effectively suppress arrhythmias. It was believed that, because arrhythmia is associated with an almost fourfold increase in the rate of cardiac-complication-related death, the drugs would reduce the death rate. However, a postmarketing trial showed that the active-treatment death rate was double the placebo rate. A risk was also detected for moricizine (DeGruttola and Tu 1994). Another example came with the surge of the AIDS epidemic. The impressive early therapeutic results obtained with zidovudine, and the pressure for accelerated evaluation of new therapies, led to the use of CD4 blood count as a surrogate endpoint for time to clinical events and overall survival (Lagakos and Hoth 1992), in spite of concern about its limitations as a surrogate marker for clinically relevant endpoints (DeGruttola et al. 1997).

The main reason behind failures was the incorrect perception that surrogacy simply follows from the association between a potential surrogate endpoint and the corresponding clinical endpoint, the mere existence of which is insufficient for surrogacy (Investigators 1989). Even though the existence of an association between the potential surrogate and the clinical endpoint is undoubtedly a desirable property, what is required to replace the clinical endpoint by the surrogate is that the effect of the treatment on the surrogate endpoint reliably predicts the effect on the clinical endpoint. Owing to a large extent the lack of appropriate methodology, this condition was not checked in the early attempts and, consequently, negative opinions about the use of surrogates in the evaluation of treatment efficacy emerged (Investigators 1989; Fleming 1994; Ferentz 2002).

Currently, the steady advance in many medical and biological fields is dramatically increasing the number of biomarkers and hence potential surrogate endpoints. The genetics and 'omics revolutions have largely contributed to this. Indeed, ever more new drugs have well-defined mechanisms of action at molecular level, allowing drug developers to measure the effect of these drugs on the relevant biomarkers (Lesko and Atkinson 2001). There is also increasing public pressure for fast approval of promising drugs, so it is naturally to then base the approval process, at least in part, on biomarkers rather than on long-term, costly clinical endpoints (Dunn and Mann 1999). Obviously, the pressure will be especially high when a rapidly increasing incidence of the targeted disease could become a serious threat to public health or the patient's (quality of) life. Shortening the duration of clinical trials not only can decrease the cost of the evaluation process but also limit potential problems with noncompliance and missing data, which are more likely in longer studies (Verbeke and Molenberghs 2000; Burzykowski et al. 2005).

Surrogate endpoints can play a role in the earlier detection of safety signals that could point to toxic problems with new drugs. The duration and sample size of clinical trials aimed at evaluating the therapeutic efficacy of new drugs are often insufficient to detect rare or late adverse effects (Heise et al. 1997; Jones 2001); using surrogate endpoints in this context might allow one to obtain information about such effects even during the clinical testing phase. Discoveries in medicine and biology are further creating an exciting range of possibilities for the development of potentially effective treatments. This is an achievement, but it also confronts us with the challenge of coping with a large number of new promising treatments that should be rapidly evaluated. This is already clear in oncology, because the increased knowledge about the genetic mechanisms operating in cancer cells led to the proposing of novel cancer therapies, such as the use of a genetically modified virus that selectively attacks p53-deficient cells, sparing normal cells (Prentice 1989). Validated surrogate endpoints can offer an efficient route. The role of surrogate endpoints may depend on the trial's phase. Nowadays, their use is more accepted in early phases of clinical research, such as in phase II or early phase III clinical trials. Using them to substitute for the clinical endpoint in pivotal phase III trials or to replace the clinical endpoint altogether in all clinical research past a certain point is still controversial and the subject of scientific debate. It is difficult to precisely define the future role of surrogate endpoints in the various trial phases. Ultimately, the combination of medical and statistical elements, together with practical and economical considerations, will help answer this question. While the huge potential of surrogate endpoints to accelerate and improve the quality of clinical trials is unquestioned, these considerations indicate that only thoroughly evaluated surrogates should be used.

Evidently, surrogates should be used only when they have been properly evaluated. Sometimes, the term *validation* is used, but this requires careful qualification (Schatzkin and Gail 2002). Like in many clinical decisions, statistical arguments will play a major role, but ought to be considered in conjunction with clinical and biological evidence. At the same time, surrogate endpoints can play different roles in different phases of drug development. While it may be more acceptable to use surrogates in early phases of research, there should be much more restraint in using them as substitutes for the true endpoint in pivotal phase III trials, since the latter might imply replacing the true endpoint by a surrogate for all future studies as well, a far-reaching decision. For a biomarker to be used as a *valid* surrogate, a number of conditions must be fulfilled. The International Conference on Harmonization (ICH) Guidelines on Statistical Principles for Clinical Trials state that "In practice, the strength of the evidence for surrogacy depends upon (i) the biological plausibility of the relationship, (ii) the demonstration in epidemiological studies of the prognostic value of the surrogate for the clinical outcome, and (iii) evidence from clinical trials that treatment effects on the surrogate correspond to effects on the clinical outcome" on harmonization of technical requirements for registration of pharmaceuticals for human use (ICH 1998).

Motivating case studies are introduced in Section 19.2. A perspective on data from a single trial is given in Section 19.3. The meta-analytic evaluation framework is presented in Section 19.4, in the context of normally distributed outcomes. Extensions to a variety of non-Gaussian settings are discussed in Section 19.5. Efforts for unifying the scattered suite of validation measures are reviewed in Section 19.6. Some alternative computational techniques and validation paradigms are presented in Section 19.7. Implications for prediction of the effect in a new trial and for designing studies based on surrogates are the topics of Section 19.8. The developments presented here are based to a large extent on Burzykowski et al. (2005) and Molenberghs et al. (2010).

19.2 Meta-Analysis of Five Clinical Trials in Schizophrenia

The data come from a meta-analysis of five double-blind randomized clinical trials, comparing the effects of risperidone to conventional antipsychotic agents for the treatment of chronic schizophrenia. The treatment indicator for risperidone versus conventional treatment will be denoted by Z. Schizophrenia has long been recognized as a heterogeneous disorder with patients suffering from both *negative* and *positive* symptoms. Negative symptoms are characterized by deficits in cognitive, affective, and social functions, for example, poverty of speech, apathy, and emotional withdrawal. Positive symptoms entail more florid symptoms such as delusions, hallucinations, and disorganized thinking, which are superimposed on mental status (Kay et al. 1987). Several measures can be considered to assess a patient's global condition. Clinician's global impression (CGI) is generally accepted as a clinical measure of change, even though it is somewhat subjective. Here, the change of CGI versus baseline will be considered as the true endpoint T. It is scored on a 7-grade scale used by the treating physician to characterize how well a subject has improved since baseline. Another useful and sufficiently sensitive assessment scales is the positive and negative syndrome scale (PANSS) (Kay et al. 1988). The PANSS consists of 30 items that provide an operationalized, drug-sensitive instrument, which is highly useful for both typological and dimensional assessments of schizophrenia. We will use the change versus baseline in PANSS as our surrogate S. The data contain five trials and in all trials, information is available on the investigators that treated the patients. This information is helpful to define group of patients that will become units of analysis. Figure 19.1 displays the individual profiles (some of them have been highlighted) for each scale by treatment group. It seems that, on average, these profiles follow a linear trend over time and the variability seems to be constant over time.

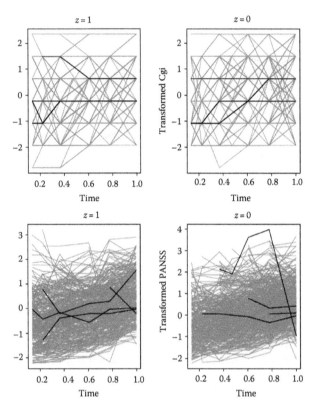

FIGURE 19.1
Psychiatric study. Individual and mean profiles for each scale by treatment group.

19.3 Data from a Single Unit

In this section, we will discuss the single unit setting (e.g., a single trial). The notation and modeling concepts introduced are useful to present and critically discuss the key ingredients of the Prentice–Freedman framework. Therefore, this section should not be seen as setting the scene for the rest of the paper. For that, we refer to the multiunit case (Section 19.4).

Throughout the paper, we will adopt the following notation: T and S are random variables that denote the true and surrogate endpoints, respectively, and Z is an indicator variable for treatment. For ease of exposition, we will assume that S and T are normally distributed. The effect of treatment on S and T can be modeled as follows:

$$S_j = \mu_S + \alpha Z_j + \varepsilon_{Sj}, \tag{19.1}$$

$$T_j = \mu_T + \beta Z_j + \varepsilon_{Tj}, \tag{19.2}$$

where $j = 1, \ldots, n$ indicates patients, and the error terms have a joint zero-mean normal distribution with covariance matrix

$$\Sigma = \begin{pmatrix} \sigma_{SS} & \sigma_{ST} \\ & \sigma_{TT} \end{pmatrix}. \tag{19.3}$$

In addition, the relationship between S and T can be described by a regression of the form

$$T_j = \mu + \gamma S_j + \varepsilon_j. \tag{19.4}$$

Note that this model is introduced because it is a component of the Prentice–Freedman framework. Given that the fourth criterion will involve a dependence on the treatment as well, as in (19.5), it is of legitimate concern to doubt whether (19.4) and (19.5) are simultaneously plausible. Also, the introduction of (19.4) should *not* be seen as an implicit explicit assumption about the absence of treatment effect in the regression relationship, but rather as a model that can be used, when the uncorrected association between both endpoints is of interest.

We will assume later (Section 19.4) that the n patients come from N different experimental units, but for now, the simple situation of a single experiment will suffice to explore some fundamental difficulties with the validation of surrogate endpoints.

19.3.1 Definition and Criteria

Prentice (1989) proposed to define a surrogate endpoint as "a response variable for which a test of the null hypothesis of no relationship to the treatment groups under comparison is also a valid test of the corresponding null hypothesis based on the true endpoint" (Prentice 1989, p. 432). In terms of our simple model (19.1) and (19.2), the definition states that for S to be a valid surrogate for T, parameters α and β must simultaneously be equal to, or different from, zero. This definition is not consistent with the availability of a single experiment only, since it requires a large number of experiments to be available, each with tests of hypothesis on both the surrogate and true endpoints. An important drawback is also that evidence from trials with nonsignificant treatment effects cannot be used, even though such trials may be consistent with a desirable relationship between both endpoints. Prentice derived operational criteria that are equivalent to his definition. These criteria require that

- Treatment has a significant impact on the surrogate endpoint (parameter α differs significantly from zero in (19.1))
- Treatment has a significant impact on the true endpoint (parameter β differs significantly from zero in (19.2))

- The surrogate endpoint has a significant impact on the true endpoint (parameter γ differs significantly from zero in (19.4))
- The full effect of treatment upon the true endpoint is captured by the surrogate

The last criterion is verified through the conditional distribution of the true endpoint, given treatment *and* surrogate endpoint, derived from (19.1) and (19.2)

$$T_j = \tilde{\mu}_T + \beta_S Z_j + \gamma_Z S_j + \tilde{\varepsilon}_{Tj}, \tag{19.5}$$

where the treatment effect (corrected for the surrogate S), β_S, and the surrogate effect (corrected for treatment Z), γ_Z, are

$$\beta_S = \beta - \sigma_{TS}\sigma_{SS}^{-1}\alpha, \tag{19.6}$$

$$\gamma_Z = \sigma_{TS}\sigma_{SS}^{-1}, \tag{19.7}$$

and the variance of $\tilde{\varepsilon}_{Tj}$ is given by

$$\sigma_{TT} - \sigma_{TS}^2\sigma_{SS}^{-1}. \tag{19.8}$$

It is usually stated that the fourth criterion requires that the parameter β_S be equal to zero (we return to this notion in Section 19.3.3). Essentially, this last criterion states that the true endpoint T is completely determined by knowledge of the surrogate endpoint S. Buyse and Molenberghs (1998) showed that the last two criteria are necessary and sufficient for binary responses, but not in general. Several authors, including Prentice, pointed out that the criteria are too stringent to be fulfilled in real situations (Prentice 1989).

In spite of these criticisms, the spirit of the fourth criterion is very appealing. This is especially true if it can be considered in the light of an underlying biological mechanism. For example, it is interesting to explore whether the surrogate is part of the causal chain leading from treatment exposure to the final endpoint. While this issue is beyond the scope of the current paper, the connection between statistical validation (with emphasis on association) and biological relevance (with emphasis on causation) deserves further reflection.

19.3.2 Proportion Explained

Freedman et al. (1992) argued that the last Prentice criterion raises a conceptual difficulty, since it requires the statistical test for treatment effect on the true endpoint to be *nonsignificant* after adjustment for the surrogate. The nonsignificance of this test does not prove that the effect of treatment upon the true endpoint is *fully* captured by the surrogate, and therefore,

Freedman et al. (1992) proposed the proportion of the treatment effect mediated by the surrogate:

$$PE = \frac{\beta - \beta_S}{\beta},$$

with β_S and β obtained, respectively, from (19.5) and (19.2). In this paradigm, a valid surrogate would be one for which the proportion explained (*PE*) is equal to one. In practice, a surrogate would be deemed acceptable if the lower limit of its confidence interval of *PE* was *sufficiently* large.

Some difficulties surrounding the *PE* have been described in the literature (Volberding et al. 1990; Choi et al. 1993; Daniels and Hughes 1997; Lin et al. 1997; Buyse and Molenberghs 1998; Flandre and Saidi 1999). The *PE* will tend to be unstable when β is close to zero, a situation that is likely to occur in practice. As Freedman et al. (1992) themselves acknowledged, the confidence limits of *PE* will tend to be rather wide (and sometimes even unbounded if Fieller confidence intervals are used), unless large sample sizes are available or a very strong effect of treatment on the true endpoint is observed. Note that large sample sizes are typically available in epidemiological studies or in meta-analyses of clinical trials. Another complication arises when (19.5) is not the correct conditional model, and an interaction term between Z_i and S_i needs to be included. In that case, defining the *PE* becomes problematic.

19.3.3 Relative Effect

Buyse and Molenberghs (1998) proposed another quantity for the validation of a surrogate endpoint: the relative effect (*RE*) that is the ratio of the effects of treatment upon the final and the surrogate endpoint. Formally,

$$RE = \frac{\beta}{\alpha}. \tag{19.9}$$

They also suggested the treatment-adjusted association between the surrogate and the true endpoint, ρ_z:

$$\rho_z = \frac{\sigma_{ST}}{\sqrt{\sigma_{SS}\sigma_{TT}}}. \tag{19.10}$$

Now, a simple relationship can be derived between *PE*, *RE*, and ρ_z. Let us define $\lambda^2 = \sigma_{TT}\sigma_{SS}^{-1}$. It follows that $\lambda\rho_z = \sigma_{ST}\sigma_{SS}^{-1}$ and, from (19.6), $\beta_S = \beta - \rho_z\lambda\alpha$. As a result, we obtain

$$PE = \lambda\rho_z\frac{\alpha}{\beta} = \lambda\rho_z\frac{1}{RE}. \tag{19.11}$$

A similar relationship was derived by Buyse and Molenberghs (1998) and by Begg and Leung (2000) for standardized surrogate and true endpoints. Let us now turn to the more promising meta-analytic framework.

19.4 Meta-Analytic Framework for Normally Distributed Outcomes

Several methods have been suggested for the formal evaluation of surrogate markers, some based on a single-trial with others of a meta-analytic nature. The first formal single-trial approach to evaluate markers is laid out in the seminal paper of Prentice (1989), who gave a definition of the concept of a surrogate endpoint, followed by a set of operational criteria. Freedman et al. (1992) augmented Prentice's hypothesis-testing based approach with the estimation paradigm, through the so-called *proportion of treatment effect explained* (PE or PTE). In turn, Buyse and Molenberghs (1998) added two further measures: the RE and the *adjusted association* (AA). The PE and RE are hampered by the fact that they are single-trial based, in which there evidently is replication at the patient level (which is fine for the AA), but not at the level of the trial. There are further issues surrounding the PE, to which we return.

19.4.1 Meta-Analytic Approach

Although the single trial–based methods are relatively easy in terms of implementation, they are surrounded with the difficulties alluded to at the end of the previous section. Therefore, several authors, such as Daniels and Hughes (1997), Buyse et al. (2000), and Gail et al. (2000), have introduced the meta-analytic approach. This section briefly outlines the methodology, followed by simplified modeling approaches as suggested by Tibaldi et al. (2003).

The meta-analytic approach was formulated originally for two continuous, normally distributed outcomes, and extended in the meantime to a large collection of outcome types, ranging from continuous, binary, ordinal, time-to-event, and longitudinally measured outcomes (Burzykowski et al. 2005). First, we focus on the continuous case, where the surrogate and true endpoints are jointly normally distributed.

The method is based on a hierarchical two-level model. Both a fixed-effects and a random-effects view can be taken. Let T_{ij} and S_{ij} be the random variables denoting the true and surrogate endpoints for the jth subject in the ith trial, respectively, and let Z_{ij} be the indicator variable for treatment. First, consider the following fixed-effects models:

$$S_{ij} = \mu_{Si} + \alpha_i Z_{ij} + \varepsilon_{Sij}, \tag{19.12}$$

$$T_{ij} = \mu_{Ti} + \beta_i Z_{ij} + \varepsilon_{Tij}, \tag{19.13}$$

where
  μ_{Si} and μ_{Ti} are trial-specific intercepts
  α_i and β_i are trial-specific effects of treatment Z_{ij} on the endpoints in trial i

ε_{Si} and ε_{Ti} are correlated error terms, assumed to be zero-mean normally distributed with covariance matrix

$$\Sigma = \begin{pmatrix} \sigma_{SS} & \sigma_{ST} \\ & \sigma_{TT} \end{pmatrix}. \tag{19.14}$$

In addition, we can decompose

$$\begin{pmatrix} \mu_{Si} \\ \mu_{Ti} \\ \alpha_i \\ \beta_i \end{pmatrix} = \begin{pmatrix} \mu_S \\ \mu_T \\ \alpha \\ \beta \end{pmatrix} + \begin{pmatrix} m_{Si} \\ m_{Ti} \\ a_i \\ b_i \end{pmatrix}, \tag{19.15}$$

where the second term on the right hand side of (19.15) is assumed to follow a zero-mean normal distribution with covariance matrix

$$D = \begin{pmatrix} d_{SS} & d_{ST} & d_{Sa} & d_{Sb} \\ & d_{TT} & d_{Ta} & d_{Tb} \\ & & d_{aa} & d_{ab} \\ & & & d_{bb} \end{pmatrix}. \tag{19.16}$$

A classical hierarchical, random-effects modeling strategy results from the combination of the earlier two steps into a single one:

$$S_{ij} = \mu_S + m_{Si} + \alpha Z_{ij} + a_i Z_{ij} + \varepsilon_{Sij}, \tag{19.17}$$

$$T_{ij} = \mu_T + m_{Ti} + \beta Z_{ij} + b_i Z_{ij} + \varepsilon_{Tij}. \tag{19.18}$$

Here, μ_S and μ_T are fixed intercepts, α and β are fixed treatment effects, m_{Si} and m_{Ti} are random intercepts, and a_i and b_i are random treatment effects in trial i for the surrogate and true endpoints, respectively. The vector of random effects $(m_{Si}, m_{Ti}, a_i, b_i)$ are assumed to be mean-zero normally distributed with covariance matrix (19.16). The error terms ε_{Sij} and ε_{Tij} follow the same assumptions as in the fixed effects models.

After fitting the earlier models, surrogacy is captured by means of two quantities: trial-level and individual-level coefficients of determination. The former quantifies the association between the treatment effects on the true and surrogate endpoints at the trial level, while the latter measures the association at the level of the individual patient, after adjustment for the treatment effect. The former is given by

$$R^2_{trial} = R^2_{b_i|m_{Si}, a_i} = \frac{\begin{pmatrix} d_{sb} \\ d_{ab} \end{pmatrix}^T \begin{pmatrix} d_{SS} & d_{Sa} \\ d_{Sa} & d_{aa} \end{pmatrix}^{-1} \begin{pmatrix} d_{sb} \\ d_{ab} \end{pmatrix}}{d_{bb}}. \tag{19.19}$$

The mentioned quantity is unitless and, at the condition that the corresponding variance–covariance matrix is positive definite, lies within the unit interval.

Apart from estimating the strength of surrogacy, this model can also be used for prediction purposes. To this end, observe that $(\beta + b_0 | m_{S0}, a_0)$ follows a normal distribution with mean and variance:

$$E(\beta + b_0 | m_{S0}, a_0) = \beta + \begin{pmatrix} d_{sb} \\ d_{ab} \end{pmatrix}^T \begin{pmatrix} d_{ss} & d_{sa} \\ d_{sa} & d_{aa} \end{pmatrix}^{-1} \begin{pmatrix} \mu_{S0} - \mu_s \\ \alpha_0 - \alpha \end{pmatrix}, \tag{19.20}$$

$$\text{Var}(\beta + b_0 | m_{S0}, a_0) = d_{bb} \begin{pmatrix} d_{sb} \\ d_{ab} \end{pmatrix}^T \begin{pmatrix} d_{ss} & d_{sa} \\ d_{sa} & d_{aa} \end{pmatrix}^{-1} \begin{pmatrix} d_{sb} \\ d_{ab} \end{pmatrix}. \tag{19.21}$$

A prediction can be made using (19.20), with prediction variance (19.21). Of course, one has to properly acknowledge the uncertainty resulting from the fact that parameters are not known but merely estimated. We return to this issue in Section 19.8.

Models (19.12) and (19.13) are referred to as the full fixed-effects models. It is sometimes necessary, for computational reasons, to contemplate a simplified version. A reduced version of these models is obtained by replacing the fixed trial-specific intercepts by a common one. Thus, the reduced mixed effect models result from removing the random trial-specific intercepts m_{Si} and m_{Ti} from models (19.17) and (19.18). The R^2 for the reduced models then is

$$R^2_{\text{trial}(r)} = R^2_{b_i | a_i} = \frac{d_{ab}^2}{d_{aa} d_{bb}}.$$

A surrogate could be adopted when R^2_{trial} is sufficiently large. Arguably, rather than using a fixed cutoff above which a surrogate would be adopted, there always will be clinical and other judgment involved in the decision process. The R^2_{indiv} is based on (19.14) and takes the following form:

$$R^2_{\text{indiv}} = R^2_{\varepsilon_{Ti} | \varepsilon_{Si}} = \frac{\sigma_{ST}^2}{\sigma_{SS} \sigma_{TT}}. \tag{19.22}$$

19.4.2 Simplified Modeling Strategies

Though the hierarchical modeling is elegant, it often poses a considerable computational challenge (Burzykowski et al. 2005). To address this problem, Tibaldi et al. (2003) suggested several simplifications, briefly outlined here. These authors considered three possible dimensions along which simplifications can be undertaken.

The first choice is between treating the trial-specific effects as fixed or random. If the trial-specific effects are chosen to be fixed, a two-stage approach is adopted. The first-stage model will take the form (19.12) and (19.13) and at the second stage, the estimated treatment effect on the true endpoint is regressed on the treatment effect on the surrogate and the intercept associated with the surrogate endpoint as

$$\widehat{\beta}_i = \widehat{\lambda}_0 + \widehat{\lambda}_1 \widehat{\mu}_{Si} + \widehat{\lambda}_2 \widehat{\alpha}_i + \varepsilon_i. \tag{19.23}$$

The trial-level $R^2_{\text{trial}(f)}$ then is obtained by regressing $\widehat{\beta}_i$ on $\widehat{\mu}_{Si}$ and $\widehat{\alpha}_i$, whereas $R^2_{\text{trial}(r)}$ is obtained from regressing $\widehat{\beta}_i$ on $\widehat{\alpha}_i$ only. The individual-level value is calculated as in (19.22), using the estimates from (19.14).

The second option is to consider the trial-specific effects as random. Depending on whether the endpoints are considered jointly or separately (see next paragraph), two directions can be followed. The first one involves a two-stage approach with at the first stage univariate models (19.17) and (19.18). The second stage model consists of a normal regression with the random treatment effect on the true endpoint as response and the random intercept and random treatment effect on the surrogate as covariates. The second direction is based on a fully specified random effects model.

Though natural to assume the two endpoints correlated, this can lead to computational difficulties in fitting the models. The need for a bivariate model is associated with R^2_{indiv}, which is in some cases of secondary importance. In addition, there is also a possibility to estimate it by making use of the correlation between the residuals from two separate univariate models. Thus, further simplification can be achieved by fitting separate models for the true and surrogate endpoints, the so-called univariate approach.

If in the trial dimension, the trial-specific effects are considered fixed, models (19.12) and (19.13) are fitted separately. Similarly, if the trial-specific effects are considered random, models (19.17) and (19.18) are fitted separately, that is, the corresponding error terms in the two models are assumed independent.

When the univariate approach and/or the fixed-effects approach are chosen, there is a need to adjust for the heterogeneity in information content between trial-specific contributions. One way of doing so is weighting the contributions according to trial size. This gives rise to a weighted linear regression model (19.23) in the second stage.

In summary, the simplified strategies perform rather well, especially when outcomes are of a continuous nature (Cortinas Abrahantes et al. 2004), and are a valuable addition to the fully specified hierarchical model, for those situations where the latter is infeasible or less reliable.

19.4.3 Some Reflections

A key consideration of the meta-analytic method is the choice of unit of analysis such as, for example, trial, center, or investigator. This choice may depend on practical considerations, such as the information available in the data, experts' considerations about the most suitable unit for a specific problem, the amount of replication at a potential unit's level, and the number of patients per unit. From a technical point of view, the most desirable situation is where the number of units and the number of patients per unit is sufficiently large. This issue has been discussed by (Cortinas Abrahantes et al. 2004). Of course, in cases where one has to resort to simplified strategies, one has to reflect carefully on the status of the results obtained. Arguably, they may not be as reliable as one might hope for, and one should undertake every effort possible to increase the amount of information available. Clearly, even an analysis based on a simplified strategy, especially in the light of good performance, may support efforts to make more data available for analysis.

Most of the work reported in Burzykowski et al. (2005) is for a dichotomous treatment indicator. Two choices need to be made for analysis. First, the treatment variable can be considered continuous or discrete. Second, when a continuous route is chosen, it is relevant to reflect on the actual coding, $0/1$ and $-1/+1$ being the most commonly encountered ones. For models with treatment occurring as a fixed effect only, there is no difference, since all choices lead to an equivalent model fit, with parameters connected by simple linear transformations. Note that this is not the case, of course, for more than three treatment arms. However, of more importance for us here is the impact the choices can have on the hierarchical model. Indeed, while the marginal model resulting from (19.17) and (19.18) is invariant under such choices, this is not true for the hierarchical aspects of the model, such as, the R^2 measures derived at the trial level. Indeed, a $-1/+1$ coding ensures the same components of variability operate in both arms, whereas a $0/1$ coding, for a positive-definite D matrix, forces the variability in the experimental arm to be greater than or equal to the variability in the standard arm. Both situations may be relevant, and it is of importance to elicit views from the study's investigators.

When the full bivariate random effect is used, the R^2_{trial} is computed from the variance–covariance matrix (19.16). It is possible that this matrix be ill-conditioned and/or nonpositive definite. In such cases, the resulting quantities computed based on this matrix might not be trustworthy. One way to assess the ill-conditioning of a matrix is by reporting its condition number, that is, the ratio of the largest over the smallest eigenvalue. A large condition number is an indication of ill-conditioning. The most pathological situation occurs when at least one eigenvalue is equal to zero. This corresponds to a positive semidefinite matrix, which occurs, for example, when a boundary solution is obtained. While it is hard to definitively identify the reason for a zero eigenvalue, insufficient information, either in terms of the number of trials, the sample size within trials, or both, may often be the cause and deserving of careful assessment. Using the simplified methods is certainly an option in this case; apart from providing a solution to the problem, it may give a handle on the problem at hand.

19.4.4 Analysis of the Meta-Analysis of Five Clinical Trials in Schizophernia

Let us analyze the schizophrenia study. Here, trial seems the natural unit of analysis. Unfortunately, the number of trials is not sufficient to apply the full meta-analytic approach. The use of trial as unit of analysis for the simplified methods might also entail problems. The second stage involves a regression model based on only five points, which might give overly optimistic or at least unreliable R^2 values. The other possible unit of analysis for this study is *investigator*. There were 176 investigators, each treating between 2 and 60 patients. The use of investigator as unit of analysis is also surrounded with problems. Although a large number of investigators are convenient to explain the between-investigator variability, because some investigators treated few patients, the resulting within-unit variability might not be estimated correctly.

The basic meta-analytic approach and the corresponding simplified strategies have been applied, with results displayed in Table 19.1. Investigator and trial were both used as units of analysis. However, as there were only five trials, it became difficult to base the analysis on trial as unit of analysis in the case of the full bivariate random-effects approach. The results have shown a remarkable difference in the two cases. Consistently, in all of the different

TABLE 19.1

Schizophrenia Study: Results of the Trial-Level (R^2_{trial}) Surrogacy Analysis

Unit of Analysis	Fixed Effects		Random Effects	
	Unweighted	Weighted	Unweighted	Weighted
Full model				
Univariate approach				
Investigator	0.5887	0.5608	0.5488	0.5447
Trial	0.9641	0.9636	0.9849	0.9909
Bivariate approach				
Investigator	0.5887	0.5608	0.9898[a]	
Trial	0.9641	0.9636	—	
Reduced model				
Univariate approach				
Investigator	0.6707	0.5927	0.5392	0.5354
Trial	0.8910	0.8519	0.7778	0.8487
Bivariate approach				
Investigator	0.6707	0.5927	0.9999[a]	
Trial	0.7418	0.8367	0.9999[a]	

[a] The 7 variance–covariance matrix is ill-conditioned; in particular, at least one eigenvalue is very close to zero. The condition numbers for the three models with ill-condition matrices, from top to bottom, are 3.415E+18, 2.384E+18, and 1.563E+18, respectively.

simplifications, the R^2_{trial} values were found to be higher when trial was used as unit of analysis. The bivariate full random-effects model does not converge when trial is used as the unit of analysis. This might be due to lack of sufficient information to compute all sources of variability, or to the fact that sample sizes tend to vary across trials. The reduced bivariate random effects model converged for both cases, but the resulting variance–covariance matrices were not positive-definite and were ill-conditioned, as can be seen from the very large value of the condition number. Consequently, the results of the bivariate random effects model should be treated with caution. Such issues are the topic of ongoing research. If we concentrate on the results based on investigator as unit of analysis, we observe a low level of surrogacy of PANSS for CGI, with R^2_{trial} ranging roughly between 0.5 and 0.68 for the different simplified models. This result, however, has to be coupled with other findings based on expert opinion to fully guarantee the validation of PANSS as possible surrogate for CGI. Turning to R^2_{indiv}, it ranges between 0.4904 and 0.5230, depending on the method of analysis, which is relatively low. To conclude, based on the investigators as unit of analysis, PANSS does not seem a promising surrogate for CGI.

19.5 Non-Gaussian Endpoints

As is clear from the formalism in Section 19.4, one needs the joint distribution of the random variables governing the surrogate and true endpoints. The easiest, though not the only, situation is where both are Gaussian random variables, but one also encounters binary (e.g., CD4+ counts over 500 mm^{-3}, tumor shrinkage), categorical (e.g., cholesterol levels <200 mg/dL, 200–299 mg/dL, 300+ mg/dL, tumor response as complete response, partial response, stable disease, progressive disease), censored continuous (e.g., time to undetectable viral load, time to cardiovascular death), longitudinal (e.g., CD4+ counts over time, blood pressure over time), and multivariate longitudinal (e.g., CD4+ and viral load over time jointly, various dimensions of quality of life over time) endpoints. The models used to validate a surrogate for a clinical endpoint will depend on the type of variables observed in the problem at hand. Table 19.2 shows some examples of potential surrogate endpoints in various diseases. In what follows, we will briefly discuss the settings of binary endpoints, failure-time endpoints, the combination of an ordinal and a survival endpoint, and longitudinal endpoints.

19.5.1 Binary Endpoints

Renard et al. (2002) have shown that extension to this situation is easily done using a latent variable formulation. That is, one posits the existence of a

TABLE 19.2

Examples of Possible Surrogate Endpoints in Various Diseases

Disease	Surrogate Endpoint	Type	Final Endpoint	Type
Resectable solid tumor	Time to recurrence	Censored	Survival	Censored
Advanced cancer	Tumor response	Binary	Time to progression	Censored
Osteoporosis	Bone mineral density	Longitudinal	Fracture	Binary
Cardiovascular disease	Ejection fraction	Continuous	Myocardial infraction	Binary
Hypertension	Blood pressure	Longitudinal	Coronary heart disease	Binary
Arrhythmia	Arrhythmic episodes	Longitudinal	Survival	Censored
ARMD	6-month visual acuity	Continuous	24-month visual acuity	Continuous
Glaucoma	Intraocular pressure	Continuous	Vision loss	Censored
Depression	Biomarkers	Multivariate	Depression scale	Continuous
HIV infection	CD4 counts + viral load	Multivariate	Progression to AIDS	Censored

Note: AIDS, acquired immune deficiency syndrome; ARMD, age-related macular degeneration; HIV, human immunodeficiency virus.

pair of continuously distributed latent variable responses $\left(\widetilde{S}_{ij}, \widetilde{T}_{ij}\right)$ that produce the actual values of (S_{ij}, T_{ij}). These unobserved variables are assumed to have a joint normal distribution and the realized values follow by double dichotomization. On the latent variable scale, we obtain a model similar to (19.12) and (19.13) and in the matrix (19.14), the variances are set equal to unity in order to ensure identifiability. This leads to the following model:

$$\begin{cases} \Phi^{-1}\left(P\left[S_{ij} = 1 | Z_{ij}, m_{S_i}, a_i, m_{T_i}, b_i\right]\right) = \mu_S + m_{S_i} + (\alpha + a_i) Z_{ij}, \\ \Phi^{-1}\left(P\left[T_{ij} = 1 | Z_{ij}, m_{S_i}, a_i, m_{T_i}, b_i\right]\right) = \mu_T + m_{T_i} + (\beta + b_i) Z_{ij}, \end{cases}$$

where Φ denotes the standard normal cumulative distribution function. Renard et al. (2002) used pseudo-likelihood methods to estimate the model parameters. Similar ideas have been used in the case one of the endpoints is continuous, with the other one binary or categorical (Burzykowski et al. 2005), (Chapter 6). The case of two binary outcomes has received further attention, encompassing flexible software implementation (Tilahun et al. 2008).

19.5.2 Two Failure-Time Endpoints

Assume now that S_{ij} and T_{ij} are failure-time endpoints. Model (19.12) and (19.13) is replaced by a model for two correlated failure-time random variables. Burzykowski et al. (2004) used copulas to this end (Clayton 1978; Hougaard 1986). One then assumes that the joint survivor function of (S_{ij}, T_{ij}) can be written as

$$F(s,t) = P\left(S_{ij} \geq s, T_{ij} \geq t\right) = C_\delta\left\{F_{sij}(s), F_{Tij}(t)\right\}, \quad s, t \geq 0, \tag{19.24}$$

where

(F_{sij}, F_{Tij}) denote marginal survivor functions

C_δ is a copula, that is, a distribution function on $[0,1]^2$ with $\delta \in R^1$

When the hazard functions are specified, estimates of the parameters for the joint model can be obtained using maximum likelihood. Shih and Louis (1995) discuss alternative estimation methods. The association parameter is generally hard to interpret. However, it can be shown (Genest and McKay 1986) that there is a link with Kendall's τ:

$$\tau = 4 \int_0^1 \int_0^1 C_\delta(u, v) C_\delta(du, dv) - 1,$$

providing an easy measure of surrogacy at the individual level. At the second stage, R^2_{trial} can be computed based on the pairs of treatment effects estimated at the first stage.

19.5.3 Ordinal Surrogate and a Survival Endpoint

Assume that T is a failure-time random variable and S is a categorical variable with K ordered categories. To propose validation measures, similar to those introduced in the previous section, Burzykowski et al. (2004) also used bivariate copulas, combining ideas of Molenberghs et al. (2001) and Burzykowski et al. (2004). One marginal distribution is a proportional odds logistic regression, while the other is a proportional hazards model. The Plackett copula (Dale 1986) was chosen to capture the association between both endpoints. The ensuing global odds ratio is relatively easy to interpret.

19.5.4 Methods for Combined Binary and Normally Distributed Endpoints

Statistical problems where various outcomes of a combined nature are observed are common, especially with normally distributed outcomes on the one hand and binary or categorical outcomes on the other. Emphasis may be on the determination of the entire joint distribution of both outcomes or on specific aspects, such as the association in general or correlation in particular between both outcomes. Burzykowski et al. (2005) review extensions of the meta-analytic approach, ranging over continuous, binary, ordinal, time-to-event, and longitudinally measured outcomes. Here, we focus on the combination of continuous and binary outcomes.

In this section, we start with a bivariate nonhierarchical setting, where the joint distribution can always be expressed as the product of a marginal distribution of one of the responses and the conditional distribution of the remaining response given the former one. The main problem with this approach is that no easy expressions for the association between both endpoints are available. Thus, we opt for a symmetric treatment of both endpoints. We focus on the case where the true endpoint is continuous and the surrogate is binary, the reverse case being entirely similar.

Generalized linear mixed models for endpoints of different data types are challenging (Molenberghs and Verbeke 2005). Hence, we concentrate on two-stage fixed-effects models. In the first stage, let $\widetilde{S}_{ij}$ be a latent variable, of which S_{ij} is the dichotomized version. A bivariate normal model for $\widetilde{S}_{ij}$ and T_{ij} is given by Molenberghs et al. (2001):

$$\widetilde{S}_{ij} = \mu_{Si} + \alpha_i Z_{ij} + \varepsilon_{Sij}, \tag{19.25}$$

$$T_{ij} = \mu_{Ti} + \beta_i Z_{ij} + \varepsilon_{Tij}, \tag{19.26}$$

where

μ_{Si} and μ_{Ti} are trial-specific intercepts

α_i and β_i are trial-specific effects of treatment Z_{ij} on the endpoints in trial i

ε_{Si} and ε_{Ti} are correlated error terms, assumed to be zero-mean normally distributed with covariance matrix

$$\Sigma = \begin{pmatrix} \frac{1}{(1-\rho^2)} & \frac{\rho\sigma}{\sqrt{(1-\rho^2)}} \\ & \sigma \end{pmatrix}, \tag{19.27}$$

where

σ is the variance of the continuous outcome

ρ is the correlation between both outcomes

The variance of $\widetilde{S}_{ij}$ is chosen for computational reasons. Using a probit formulation like Molenberghs et al. (2001) and owing to the replication at the trial level, we can impose a distribution on the trial-specific parameters. At the second stage, we assume

$$\begin{pmatrix} \mu_{Si} \\ \mu_{Ti} \\ \alpha_i \\ \beta_i \end{pmatrix} = \begin{pmatrix} \mu_S \\ \mu_T \\ \alpha \\ \beta \end{pmatrix} + \begin{pmatrix} m_{Si} \\ m_{Ti} \\ a_i \\ b_i \end{pmatrix}, \tag{19.28}$$

where the second term on the right hand of (19.28) is assumed to follow a zero-mean normal distribution with dispersion matrix (19.16). Measures to assess the quality of the surrogate both at the trial and individual level are then obtained. This case has received full attention in Assam et al. (2007).

19.5.5 Longitudinal Endpoints

Most of the earlier work focuses on univariate responses. Alonso et al. (2003) showed that going from a univariate setting to a multivariate framework represents new challenges. The R^2 measures proposed by Buyse et al. (2000) are no longer applicable. Alonso et al. (2003) based their calculations of surrogacy measures on a two-stage approach rather than a full random-effects approach. They assume that information from $i = 1, \ldots, N$ trials is available, in the ith of which, $j = 1, \ldots, n_i$ subjects are enrolled and they denote further the time at which subject j in trial i is measured as t_{ijk}. If T_{ijk} and S_{ijk} represent the associated true and surrogate endpoints, respectively, and Z_{ij} is a binary indicator variable for treatment, then along the ideas of Galecki (1994), they proposed the following joint model, at the first stage, for both responses

$$T_{ijk} = \mu_{Ti} + \beta_i Z_{ij} + g_{Tij}\left(t_{ijk}\right) + \varepsilon_{Tijk},$$
$$S_{ijk} = \mu_{Si} + \alpha_i Z_{ij} + g_{Sij}\left(t_{ijk}\right) + \varepsilon_{Sijk}, \tag{19.29}$$

where
μ_{Ti} and μ_{Si} are trial-specific intercepts
β_i and α_i are trial-specific effects of treatment Z_{ij} on the two endpoints
g_{Tij} and g_{Sij} are trial-subject-specific time functions that can include treatment-by-time interactions

They also assume that the vectors, collecting all information over time for patient j in trial i, $\widetilde{\varepsilon}_{Tij}$ and $\widetilde{\varepsilon}_{Sij}$ are correlated error terms, following a mean-zero multivariate normal distribution with covariance matrix

$$\Sigma_i = \begin{pmatrix} \Sigma_{TTi} & \Sigma_{TSi} \\ \Sigma'_{TSi} & \Sigma_{SSi} \end{pmatrix} = \begin{pmatrix} \sigma_{TTi} & \sigma_{TSi} \\ \sigma_{TSi} & \sigma_{SSi} \end{pmatrix} \otimes R_i. \tag{19.30}$$

Here, R_i is a correlation matrix for the repeated measurements.

If treatment effect can be assumed constant over time, then (19.19) can still be useful to evaluate surrogacy at the trial level. However, at the individual level, the situation is totally different, the R^2_{ind} is no longer applicable, and new concepts are needed.

Using multivariate ideas, Alonso et al. (2003) proposed the *variance reduction factor* (*VRF*) to capture individual-level surrogacy in this more elaborate setting. They quantified the relative reduction in the true endpoint variance after adjustment by the surrogate as

$$VRF_{ind} = \frac{\sum_i \{tr(\Sigma_{TTi}) - tr(\Sigma_{(T|S)i})\}}{\sum_i tr(\Sigma_{TTi})},$$ (19.31)

where $\Sigma_{(T|S)_i}$ denotes the conditional variance–covariance matrix of $\tilde{\varepsilon}_{T_{ij}}$ given $\tilde{\varepsilon}_{S_{ij}}$: $\Sigma_{(T|S)i} = \Sigma_{TTi} - \Sigma_{TSi}\Sigma_{SSi}^{-1}\Sigma'_{TSi}$. Here, Σ_{TTi} and Σ_{SSi} are the variance–covariance matrices associated with the true and surrogate endpoint, respectively, and Σ_{TSi} contains the covariances between the surrogate and the true endpoint. Alonso et al. (2003) showed that the VRF_{ind} ranges between zero and one, and that $VRF_{ind} = R^2_{ind}$ when the endpoints are measured only once.

An alternative proposal is

$$\theta_p = \sum_i \frac{1}{Np_i} tr\left\{(\Sigma_{TTi} - \Sigma_{(T|S)i})\Sigma_{TTi}^{-1}\right\}.$$ (19.32)

Structurally, both VRF and θ_p are similar, the difference being the reversal of summing the trace and calculating the ratio. In spite of this strong structural similarity, the VRF is not symmetric in S and T and it is only invariant with respect to linear orthogonal transformations, whereas θ_p is both symmetric and invariant with respect to the broader class of linear bijective transformations.

A common problem of all previous proposals is that they are strongly based on the normality assumption and extensions to nonnormal settings are difficult. To overcome this limitation, Alonso et al. (2005) introduced a new parameter, the so-called R^2_Λ, to evaluate surrogacy at the individual level when both responses are measured over time or in general when multivariate or repeated measures are available

$$R^2_\Lambda = \frac{1}{N}\sum_i(1 - \Lambda_i),$$ (19.33)

where $\Lambda_i = |\Sigma_i|/\{|\Sigma_{TTi}||\Sigma_{SSi}|\}$. This parameter not only allows the detection of more general patterns of association but can also be extended to more general settings than those defined by the normal distribution. They proved that R^2_Λ ranges between zero and one, and that in the cross-sectional case $R^2_\Lambda = R^2_{ind}$. These authors have shown that $R^2_\Lambda = 1$ whenever there is a deterministic relationship between two linear combinations of both endpoints, allowing the detection of strong associations in cases where the VRF or θ_p would fail in doing so.

19.6 Unified Approach

The longitudinal method of the previous section, while elegant, hinges upon normality of the outcome. First using the likelihood reduction factor (LRF) (Section 19.6.1) and then an information-theoretic approach (Section 19.6.2), extension and unification will be achieved.

19.6.1 Likelihood Reduction Factor

Estimating individual-level surrogacy, as the previous developments clearly show, has frequently been based on a variance–covariance matrix coming from the distribution of the residuals. However, if we move away from the normal distribution, it is not always clear how to quantify the association between both endpoints after adjusting for treatment and trial effect. To address this problem, Alonso et al. (2005) and Alonso and Molenberghs (2007) considered the following generalized linear models

$$g_T \{E(T_{ij})\} = \mu_{Ti} + \beta_i Z_{ij}, \tag{19.34}$$

$$g_T \{E(T_{ij}|S_{ij})\} = \theta_{0i} + \theta_{1i} Z_{ij} + \theta_{2i} S_{ij}, \tag{19.35}$$

where
g_T is an appropriate link function
μ_{Ti} are the trial-specific intercepts
β_i are trial-specific effects of treatment Z on the true endpoint in trial i

θ_{0i} and θ_{1i} are trial-specific intercepts and effects of treatment on the true endpoint when the surrogate endpoint is known. Note that (19.34) and (19.35) can be readily extended to incorporate more complex settings. Other extensions, such as nonlinearity between S_{ij} and $g_T \{E(T_{ij})\}$, are possible. We assume a linear relationship between S_{ij} and $g_T \{E(T_{ij})\}$ but consider extensions of (19.34) and (19.35) in the light of simplified modeling strategy, as presented by Tibaldi et al. (2003). They suggested several simplifications for the case of continuous true and surrogate endpoints. They have introduced the concept of three possible dimensions along which simplifications can be made: the trial, endpoint, and measurement error dimensions. Their ideas can be applied outside the original mixed model–based framework. We consider their trial and measurement error dimensions.

The trial dimension provides a choice between treating the trial-specific effects as fixed or random. The former is often chosen out of necessity, when the latter is too challenging. If the trial-specific effects are chosen fixed, then (19.34) and (19.35) are used to validate the surrogate endpoint. On the other hand, if the trial-specific effects are considered random, we extend (19.34) and (19.35) to appropriate generalized linear mixed-effects models:

$$g_T \left\{ E\left(T_{ij}\right) \right\} = \mu_T + m_{Ti} + \beta Z_{ij} + b_i Z_{ij}, \qquad (19.36)$$

$$g_T \left\{ E\left(T_{ij}|S_{ij}\right) \right\} = \theta_0 + c_{Ti} + \theta_1 Z_{ij} + a_i Z_{ij} + \theta_{2i} S_{ij}, \qquad (19.37)$$

where
- μ_T and β are a fixed intercept and treatment effect on the true endpoint
- m_{Ti} and b_i are a random intercept and treatment effects on the true endpoint
- θ_0 and θ_1 are a fixed intercept and treatment effect on the true endpoint when the surrogate is known
- c_{Ti} and a_i are a random intercept and treatment effects on the true endpoint when the surrogate is known

It is often the case in practice that different trials in meta-analysis have different sizes. Because univariate models are used to evaluate surrogacy in the information-theoretic approach, there is a need to adjust for the heterogeneity in information content between trial-specific contributions. This is the target of the choices along the so-called measurement error dimension. One way to account for a variable amount of information per trial is by weighting the contributions according to trial size, thus giving rise to a weighted linear regression models, particularly when estimating measures for trial-level surrogacy.

Let us turn to the so-called LRF. Observe that, in the case where the true endpoint is continuous and normally distributed, (19.34) and (19.35) reduce to normal regression models and (19.36) and (19.37) reduce to linear mixed models. On the other hand, when the true endpoint is binary, (19.34) and (19.35) reduce to logistic regression models. Alonso and Molenberghs (2007) used the LRF to evaluate individual-level surrogacy, which is obtained by

$$\text{LRF} = 1 - \frac{1}{N} \sum_i \exp\left(-\frac{G_i^2}{n_i}\right), \qquad (19.38)$$

where G_i^2 denotes the log-likelihood ratio test statistic to compare (19.34) and (19.35) or (19.36) and (19.37) within trial i. Alonso et al. (2005) established a number of properties for LRF, in particular, its ranging in the unit interval and, importantly, its reduction to R^2_{ind} in the cross-sectional case.

19.6.2 Information-Theoretic Unification

This proposal avoids the needs for a joint, hierarchical model, and allows for unification across different types of endpoints. The entropy of a random variable (Shannon 1948), a time-honored measure of randomness or uncertainty, is defined in the following way for the case of a discrete random variable Y, taking values $\{k_1, k_2, \ldots, k_m\}$, and with probability function $P(Y = k_i) = p_i$:

$$H(Y) = \sum_i p_i \log\left(\frac{1}{p_i}\right). \tag{19.39}$$

The differential entropy $h_d(X)$ of a continuous variable X with density $f_X(x)$ and support S_{f_X} equals

$$h_d(X) = -E[\log f_X(X)] = -\int_{S_{f_X}} f_X(x)\log f_X(x)\, dx. \tag{19.40}$$

The joint and conditional (differential) entropies are defined in an analogous fashion. Defining the information of a single event as $I(A) = \log p_A$, the entropy is $H(A) = -I(A)$. No information is gained from a totally certain event, $p_A \approx 1$, so $I(A) \approx 0$, while an improbable event is informative.

$H(Y)$ is the average uncertainty associated with P. Entropy is always nonnegative, satisfies $H(Y|X) \leq H(Y)$ for any pair of random variables, with equality holding under independence, and is invariant under a bijective transformation (Cover and Tomas 1991). Differential entropy enjoys some but not all properties of entropy: it can be infinitely large, negative, or positive, and is coordinate dependent. For a bijective transformation $Y = y(X)$, it follows that $h_d(Y) = h_d(X) - E_Y\left(\log\left|\frac{dx}{dy}(y)\right|\right)$.

We can now quantify the amount of uncertainty in Y, expected to be removed if the value of X was known, by $I(X, Y) = h_d(Y) - h_d(Y|X)$, the so-called *mutual information*. It is always nonnegative, zero if and only if X and Y are independent, symmetric, invariant under bijective transformations of X and Y, and $I(X, X) = h_d(X)$. The mutual information measures the information of X, shared by Y.

We will now introduce the entropy-power (Shannon 1948) for comparison of continuous random variables. Let X be a continuous n-dimensional random vector. The entropy-power of X is

$$EP(X) = \frac{1}{(2\pi e)^n} e^{2h(X)}. \tag{19.41}$$

The differential entropy of a continuous normal random variable is $h(X) = \frac{1}{2}\log(2\pi\sigma^2)$, a simple function of the variance and, on the natural logarithmic scale: $EP(X) = \sigma^2$. In general, $EP(X) \leq \mathrm{Var}(X)$ with equality if and only if X is normally distributed.

We can now define an information-theoretic measure of association (Schemper and Stare 1996):

$$R_h^2 = \frac{\text{EP}(Y) - \text{EP}(Y|X)}{\text{EP}(Y)}, \tag{19.42}$$

which ranges in the unit interval, equals zero if and only if (X, Y) are independent, is symmetric, is invariant under bijective transformation of X and Y, and, when $R_h^2 \to 1$ for continuous models, there is usually some degeneracy appearing in the distribution of (X, Y). There is a direct link between R_h^2 and the mutual information: $R_h^2 = 1 - e^{-2I(X,Y)}$. For Y discrete, it follows that $R_h^2 \leq 1 - e^{-2H(Y)}$, implying that R_h^2 then has an upper bound smaller than 1. We then redefine

$$\widetilde{R}_h^2 = \frac{R_h^2}{1 - e^{-2H(Y)}},$$

reaching 1 when both endpoints are deterministically related.

We can now redefine surrogacy, while preserving previous proposals as special cases. While we will focus on individual-level surrogacy, all results apply to the trial level too. Let $Y = T$ and $X = S$ be the true and surrogate endpoints, respectively. We consider S a good surrogate for T at the individual (trial) level, if a *large* amount of uncertainty about T (the treatment effect on T) is reduced when S (the treatment effect on S) is known. Equivalently, we term S a good surrogate for T at the individual level, if our lack of knowledge about the true endpoint is substantially reduced when the surrogate endpoint is known.

A meta-analytic framework, with N clinical trials, produces N_q different R_{hi}^2, and Alonso and coworkers proposed a meta-analytic R_h^2:

$$R_h^2 = \sum_{i=1}^{N_q} \alpha_i R_{hi}^2 = 1 - \sum_{i=1}^{N_q} \alpha_i e^{-2I_i(S_i, T_i)},$$

where $\alpha_i > 0$ for all i and $\sum_{i=1}^{N_q} \alpha_i = 1$. Different choices for α_i lead to different proposals, producing an uncountable family of parameters. This opens the additional issue of finding an *optimal* choice. In particular, for the cross-sectional normal–normal case, Alonso and Molenberghs (2007) have shown that $R_h^2 = R_{\text{ind}}^2$. The same holds for R_Λ^2 for the longitudinal case. Finally, when the true and surrogate endpoints have distributions in the exponential family, then LRF $\xrightarrow{P} R_h^2$ when the number of subjects per trial goes to infinity.

Alonso and Molenberghs (2007) developed asymptotic confidence intervals for R_h^2, based on the idea of Kent (1983), to build confidence intervals for $2I(T, S)$. Let $\hat{a} = 2n\hat{I}(T, S)$, where n is the number of patients. Define $\kappa_{1:\alpha}(a)$

and $\delta_{1:\alpha}(a)$ by $P\left(\chi_1^2(\kappa_{1:\alpha}(a)) \geq a\right) = \alpha$ and $P\left(\chi_1^2(\delta_{1:\delta}(a)) \leq a\right) = \alpha$. Here, χ_1^2 is a chi-squared random variable with 1 degree of freedom. If $P\left(\chi_1^2(0) \geq a\right) = \alpha$, then we set $\kappa_{1:\alpha}(a) = 0$. A conservative two-sided $1 - \alpha$ asymptotic confidence interval for R_h^2 is

$$\sum_i \alpha_i \left[n_i^{-1}\kappa_{1:\alpha}^i\left(\hat{a}\right), n_i^{-1}\delta_{1:\alpha}^i\left(\hat{a}\right)\right], \qquad (19.43)$$

where $1 - \alpha_i$ is the Bonferroni confidence level for the trial intervals (Alonso and Molenberghs 2007). This asymptotic interval has considerable computational advantage with respect to the bootstrap approach used by Alonso et al. (2005). Although it involves substantial mathematics, its implementation in practice is fairly straightforward and less computer-intensive than the meta-analytic approach. This is a direct consequence of the fact that the models used in the former are univariate models, which can be fitted using any standard regression software. However, the performance of this approach has not been studied in the mixed continuous and binary endpoint settings.

19.6.3 Fano's Inequality and the Theoretical Plausibility of Finding a Good Surrogate

Fano's inequality shows the relationship between entropy and prediction:

$$E\left[(T - g(S))^2\right] \geq EP(T)\left(1 - R_h^2\right) \qquad (19.44)$$

where $EP(T) = e^{2h(T)}/(2\pi e)$. Note that nothing has been assumed about the distribution of our responses and no specific form has been considered for the prediction function g. Also, (19.44) shows that the predictive quality strongly depends on the characteristics of the endpoint, specifically on its power-entropy. Fano's inequality states that the prediction error increases with $EP(T)$ and therefore, if our endpoint has a large power-entropy, then a surrogate should produce a large R_h^2 to have some predictive value. This means that, for some endpoints, the search for a good surrogate can be a dead end street: the larger the entropy of T, the more difficult it is to predict. Studying the power-entropy before trying to find a surrogate is therefore advisable.

19.6.4 Application to the Meta-Analysis of Five Clinical Trials in Schizophrenia

We will treat CGI as the true endpoint and PANSS as surrogate, although the reverse would be sensible, too. In practice, these endpoints are frequently dichotomized in a clinically meaningful way. Our binary true endpoint $T = CGId = 1$ for patients classified from *Very much improved* to *Improved*, and 0 otherwise. The binary surrogate $S = \text{PANSSd} = 1$ for patients with at least 20 points reduction versus baseline, and 0 otherwise. We will start from probit and Plackett-Dale models and compare results with the ones from the information-theoretic approach.

In line with Section 19.5.1, we formulate two continuous latent variables $\left(\widetilde{\text{CGI}}_{ij}, \widetilde{\text{PANSS}}_{ij}\right)$ assumed to follow a bivariate normal distribution. The following probit model can be fitted:

$$
\begin{pmatrix}
\tilde{\mu}_{ij}^{T} \\
\tilde{\mu}_{ij}^{S} \\
\ln(\sigma^2) \\
\ln\left(\dfrac{1+\tilde{\rho}}{1+\tilde{\rho}}\right)
\end{pmatrix}
=
\begin{pmatrix}
\tilde{\mu}_{T_i} + \tilde{\beta}_i Z_{ij} \\
\tilde{\mu}_{S_i} + \tilde{\alpha}_i Z_{ij} \\
c_{\sigma^2} \\
c_{\tilde{\rho}}
\end{pmatrix},
\tag{19.45}
$$

where $\tilde{\mu}_{ij}^{T} = E\left(\widetilde{\text{CGI}}_{ij}\right)$, $\tilde{\mu}_{ij}^{S} = E\left(\widetilde{\text{PANSS}}_{ij}\right)$, $\text{Var}\left(\widetilde{\text{CGI}}_{ij}\right) = 1$, $\sigma^2 = \text{Var}\left(\widetilde{\text{PANSS}}_{ij}\right)$ and $\tilde{\rho} = \text{corr}\left(\widetilde{\text{CGI}}_{ij}, \widetilde{\text{PANSS}}_{ij}\right)$ denotes the correlation between the true and surrogate endpoint latent variables. We can then use the estimated values of $(\tilde{\mu}_{S_i}, \tilde{\alpha}_i, \tilde{\beta}_i)$ to evaluate trial level surrogacy through the R^2_{trial}. At the individual level, $\tilde{\rho}^2$ is used to capture surrogacy.

Alternatively, the Dale (1986) formulation can be used, based on

$$
\begin{pmatrix}
\text{logit}\left(\pi_{ij}^{T}\right) \\
\text{logit}\left(\pi_{ij}^{S}\right) \\
\ln(\psi)
\end{pmatrix}
=
\begin{pmatrix}
\mu_{T_i} + \beta_i Z_{ij} \\
\mu_{S_i} + \alpha_i Z_{ij} \\
c_{\psi}
\end{pmatrix},
\tag{19.46}
$$

where $\pi_{ij}^{T} = E\left(\text{CGId}_{ij}\right)$, $\pi_{ij}^{S} = E\left(\text{PANSSd}_{ij}\right)$ and ψ is the global odds ratio associated to both endpoint. As before, the estimated values of $(\mu_{S_i}, \alpha_i, \beta_i)$ can be used to evaluate surrogacy at the trial level and the individual-level surrogacy is quantified using the global odds ratio.

In the information-theoretic approach, the following three models are fitted independently:

$$
\Phi\left(\pi_{ij}^{T}\right) = \mu_{T_i} + \beta_i Z_{ij},
\tag{19.47}
$$

$$
\Phi\left(\pi_{ij}^{T|S}\right) = \mu_{T_i}^{S} + \beta_i^{S} Z_{ij} + \gamma_{ij} S_{ij},
\tag{19.48}
$$

$$
\Phi\left(\pi_{ij}^{S}\right) = \mu_{S_i} + \alpha_i Z_{ij},
\tag{19.49}
$$

where $\pi_{ij}^{T} = E\left(\text{CGId}_{ij}\right)$, $\pi_{ij}^{T|S} = E\left(\text{CGId}_{ij}|\text{PANSSd}_{ij}\right)$, $\pi_{ij}^{S} = E\left(\text{PANSSd}_{ij}\right)$ and Φ denotes the cumulative standard normal distribution. At the trial level, the estimated values of $(\mu_{S_i}, \alpha_i, \beta_i)$ obtained from (19.47) and (19.49) can be used to calculate the R^2_{trial}, whereas at the individual level, we can quantify surrogacy using R^2_h. As it was stated before, the LRF is a consistent estimator of R^2_h; however, in principle, other estimators could be used as well. We will then quantify surrogacy at the individual level by $\hat{R}^2_h = 1 - \exp(-G^2/n)$, where G^2 is the loglikelihood ratio test to compare (19.47) with (19.48) and n denotes total number of patients. Furthermore, when applied to the binary–binary setting, Fanos's inequality takes the form

$$P(T \neq S) \geq \frac{1}{\log|\Psi|}\left[H(T) - 1 + \frac{1}{2}\ln\left(1 - R^2_h\right)\right],$$

where $\Psi = \{0, 1\}$ and $|\Psi|$ denotes the cardinality of Ψ. Here, again, Fano's inequality gives a lower bound for the probability of incorrect prediction.

Table 19.3 shows the results at the trial and individual level obtained with the different approaches described earlier. At the trial level, all the methods produced very similar values for the validation measure. In all cases, $R^2_{\text{trial}} \simeq 0.50$. It is also remarkable that the probit approach, in spite of being based on treatment effects defined at a latent level, produced a R^2_{trial} value similar to the ones obtained with the information-theoretic and Plackett-Dale approaches. However, as Alonso et al. (2003) showed, there is a linear relationship between the mean parameters defined at the latent level and the mean parameters of the model based on the observable endpoints and that could explain the agreement between the probit and the other two procedures. Therefore, at the trial level, we could conclude that knowing the treatment effect on the surrogate will reduce our uncertainty about the treatment effect on the true endpoint by 50%.

TABLE 19.3

Schizophrenia Study: Trial-Level and Individual-Level Validation Measures (95% Confidence Intervals)—Binary–Binary Case

Parameter	Estimate	95% CI
Trial-level R^2_{trial} measures		
1.1 Information-theoretic	0.49	(0.21, 0.81)
1.2 Probit	0.51	(0.18, 0.78)
1.3 Plackett-Dale	0.51	(0.21, 0.81)
Individual-level measures		
R^2_h	0.27	(0.24, 0.33)
$R^2_{h\,\text{max}}$	0.39	(0.35, 0.48)
Probit	0.67	(0.55, 0.76)
Plackett-Dale ψ	25.12	(14.66, 43.02)
Fano's lower bound	0.08	

At the individual level, the probit approach gives the strongest association between the surrogate and the true endpoint. Nevertheless, this value describes the association at an unobservable latent level, rendering its interpretation more awkward than with information theory, since it is not clear how this latent association could be relevant from a clinical point of view or how it could be translated into an association for the observable endpoints. The Plackett-Dale procedure quantifies surrogacy using a global odds ratio, making the comparison between this method and the others more difficult. Note that even though odds ratios are widely used in biomedical fields, the lack of an upper bound makes difficult their interpretation in this setting.

On the other hand, the value of the $R^2_{h\,max}$ illustrates that the surrogate can merely explain 39% of our uncertainty about the true endpoint, a relatively low value. Additionally, the lower bound for Fano's inequality clearly shows that using the value of PANSS to predict the outcome on CGI would be misleading in at least 8% of the cases. Even though this value is relatively low, it is only a lower bound and the real probability of error could be much larger.

At the trial level, the information-theoretic approach produces results similar to the ones from the conventional methods, but does so by means of models that are generally much easier to fit. At the individual level, the information-theoretic approach avoids the problem common with the probit model in that the correlation of the latter is formulated at the latent scale and therefore less relevant for practice. In addition, the information-theoretic measure ranges between 0 and 1, circumventing interpretational problems arising from using the unbounded Plackett-Dale based odds ratio.

19.7 Alternatives and Extensions

As a result of the aforementioned computational problems, several alternative strategies have been considered. For example, Shkedy and Torres Barbosa (2005) study in detail the use of Bayesian methodology and conclude that even relatively noninformative prior has a strongly beneficial impact on the algorithms' performance.

Cortinas et al. (2008) start from the information-theoretic approach, in the contexts of: (1) normally distributed endpoints; (2) a copula model for a categorical surrogate and a survival true endpoint; and (3) a joint modeling approach for longitudinal surrogate and true endpoints. Rather than fully relying on the methods described in Section 19.5, they use cross-validation to obtain adequate estimates of the trial-level surrogacy measure. Also, they explore the use of regression tree analysis, bagging regression analysis, random forests, and support vector machine methodology. They concluded that performance of such methods, in simulations and case studies, in terms of point and interval estimation, ranges from very good to excellent.

These are variations to the meta-analytic theme, as described here, in Burzykowski et al. (2005), and of which Daniels and Hughes (1997) is an early instance. There are a number of alternative paradigms. Frangakis and Rubin (2004) employ so-called principal stratification, still using the data from a single trial only. Drawing from the causality literature, Robins and Greenland (1992); Pearl (2001); and Taylor et al. (2005) use the direct/indirect-effect machinery.

It took two decades after the publication of Prentice's seminal paper until an attempt was made to review, classify, and study similarities and differences between the various paradigms (Joffe and Greene 2008). Joffe and Greene saw two important dimensions. First, some methods are based on a single trial while others use several trials, that is, meta-analysis. Second, some approaches are based on association, while others are based on causation. Because the meta-analytic framework described earlier is based on association and uses multiple trials, on the one hand, and because the causal framework initially used a single trial, on the other, the above dimensions got convoluted and it appeared that correlation/meta-analysis had to be a pair, just like causal/single trial. However, it is useful to disentangle the two dimensions and to keep in mind that proper evaluation of the relationship between the treatment effect on the surrogate and true endpoints is ideally based on meta-analysis. Joffe and Green state that the meta-analytic approach is essentially causal insofar as the treatment effects observed in all trials are in fact average causal effects. If a meta-analysis of several trials is not possible, then causal effects must be estimated for individual patients, which requires strong and unverifiable assumptions to be made. Recently, progress has been made regarding the relationship between the association and causal frameworks (Alonso et al. 2013). These authors consider a quadruple $Y_{ij} = \left[T_{ij} \left(Z_{ij} = 0 \right), T_{ij} \left(Z_{ij} = 1 \right), S_{ij} \left(Z_{ij} = 0 \right), S_{ij} \left(Z_{ij} = 1 \right) \right]'$, which is observable only if patient j in trial i would be assessed under both control and experimental treatment. Clearly, this is not possible and hence some of the outcomes in the quadruple are *counterfactual.* Counterfactuals are essential to the causal-inference framework, while this equation also carries a meta-analytic structure. Alonso et al. (2013) assume a multivariate normal for Y_{ij}, to derive insightful expressions. It is clear that both paradigms base their validation approach, upon causal effects of treatment. However, there is an important difference. While the causal inference line of thinking places emphasis on individual causal effects, in a meta-analytic approach, the focus is on the expected causal treatment effect. These authors show that, under broad circumstances, when a surrogate is considered acceptable from a meta-analytic perspective, at both the trial and individual level, then it would be good as well from a causal-inference angle. These authors also carefully show, in line with comments made earlier, that a surrogate, valid from a single-trial framework perspective, using individual causal effects, may not pass the test from a meta-analytic view-point, when heterogeneity from one trial to another is large and the causal association is low. Evidently, more work is needed, especially for endpoints of a different type, but at the same time it is comforting that, when based on multiple trials, the frameworks appear to show a good amount of agreement.

19.8 Prediction and Design Aspects

Until now, we have focused on quantifying surrogacy through a slate of measures, culminating in the information-theoretic ones. In practice, one may want to go beyond merely quantifying the strength of surrogacy, and further use a surrogate endpoint to predict the treatment effect on the true endpoint *without measuring the latter*. Put simply, the issue then is to obtain point and interval predictions for the treatment effect on the true endpoint based on the surrogate. This issue has been studied by Burzykowski and Buyse (2006) for the original meta-analytic approach for continuous endpoints and will be reviewed here.

The key motivation for validating a surrogate endpoint is the ability to predict the effect of treatment on the true endpoint based on the observed effect of treatment on the surrogate endpoint. It is essential, therefore, to explore the quality of prediction by (1) information obtained in the validation process based on trials $i = 1, \ldots, N$ and (2) the estimate of the effect of Z on S in a new trial $i = 0$. Fitting the mixed-effects model (19.12) and (19.13) to data from a meta-analysis provides estimates for the parameters and the variance components. Suppose then that a new trial $i = 0$ is considered for which data are available on the surrogate endpoint but not on the true endpoint. We can then fit the following linear model to the surrogate outcomes S_{0j}:

$$S_{0j} = \mu_{s0} + \alpha_0 Z_{0j} + \varepsilon_{s0j}. \tag{19.50}$$

We are interested in an estimate of the effect $\beta + b_0$ of Z on T, given the effect of Z on S. To this end, one can observe that $(\beta + b_0 | m_{s0}, a_0)$, where m_{s0} and a_0 are, respectively, the surrogate-specific random intercept and treatment effect in the new trial, follows a normal distribution with mean linear in μ_{s0}, μ_s, α_0, and α, and variance

$$\text{Var}\,(\beta + b_0 | m_{s0}, a_0) = \left(1 - R_{\text{trial}}^2\right) \text{Var}\,(b_0). \tag{19.51}$$

Here, Var (b_0) denotes the unconditional variance of the trial-specific random effect, related to the effect of Z on T (in the past or the new trials). The smaller the conditional variance (19.51), the higher the precision of the prediction, as captured by R_{trial}^2. Let us use ϑ to group the fixed-effects parameters and variance components related to the mixed-effects model (19.12) and (19.13), with $\widehat{\vartheta}$ denoting the corresponding estimates. Fitting the linear model (19.50) to data on the surrogate endpoint from the new trial provides estimates for m_{s0} and a_0. The prediction variance can be written as

$$\text{Var}\,(\beta + b_0 | \mu_{s0}, \alpha_0, \vartheta) \approx f\,\{\text{Var}\,(\widehat{\mu}_{s0}, \widehat{\alpha}_0)\} + f\{\text{Var}(\widehat{\vartheta})\} + \left(1 - R^2_{\text{trial}}\right) \text{Var}\,(b_0),$$

$$(19.52)$$

where $f\,\{\text{Var}\,(\widehat{\mu}_{s0}, \widehat{\alpha}_0)\}$ and $f\,\{\text{Var}\,(\widehat{\vartheta})\}$ are functions of the asymptotic variance–covariance matrices of $(\widehat{\mu}_{s0}, \widehat{\alpha}_0)^T$ and $\widehat{\vartheta}$, respectively. The third term on the right hand side of (19.52), which is equivalent to (19.51), describes the prediction's variability if μ_{s0}, α_0, and ϑ were known. The first two terms describe the contribution to the variability due to the use of the estimates of these parameters. It is useful to consider three scenarios.

Scenario 1. Estimation error in both the meta-analysis and the new trial: If the parameters of models (19.12) and (19.13) and (19.50) have to be estimated, as is the case in reality, the prediction variance is given by (19.52). From the equation, it is clear that in practice, the reduction of the variability of the estimation of $\beta + b_0$, related to the use of the information on m_{s0} and a_0, will always be smaller than that indicated by R^2_{trial}. The latter coefficient can thus be thought of as measuring the *potential* validity of a surrogate endpoint at the trial-level, assuming precise knowledge (or infinite numbers of trials and sample sizes per trial available for the estimation) of the parameters of models (19.12) and (19.13) and (19.50). See also Scenario 3.

Scenario 2. Estimation error only in the meta-analysis: This scenario is possible only in theory, as it would require an infinite sample size in the new trial. But it can provide information of practical interest, since, with an infinite sample size, the parameters of the single-trial regression model (19.50) would be known. Consequently, the first term on the right hand side of (19.52), $f\,\{\text{Var}\,(\widehat{\mu}_{s0}, \widehat{\alpha}_0)\}$, would vanish and (19.52) would reduce to

$$\text{Var}\,(\beta + b_0 | \mu_{s0}, \alpha_0, \vartheta) \approx f\{\text{Var}(\widehat{\vartheta})\} + \left(1 - R^2_{\text{trial}}\right) \text{Var}\,(b_0).$$

$$(19.53)$$

Expression (19.53) can thus be interpreted as indicating the minimum variance of the prediction of $\beta + b_0$, achievable in the actual application of the surrogate endpoint. In practice, the size of the meta-analytic data providing an estimate of ϑ will necessarily be finite and fixed. Consequently, the first term on the right hand side of (19.53) will always be present. Based on this observation, Gail et al. (2000) conclude that the use of surrogates validated through the meta-analytic approach will always be less efficient than the direct use of the true endpoint. Of course, even so, a surrogate can be of great use in terms of reduced sample size, reduce trial length, gain in number of life years, etc.

Scenario 3. No estimation error: If the parameters of the mixed-effects model (19.12) and (19.13) and the single-trial regression model (19.50) were known, the prediction variance for $\beta + b_0$ would only contain the last term on the right hand side of (19.52). Thus, the variance would be reduced to (19.51), which is clearly linked with (19.44). While this situation is, strictly speaking, of theoretical relevance only, as it would require infinite numbers of trials and sample sizes per trial available for the estimation in the meta-analysis and in the new trial, it provides important insight.

Based on these scenarios, one can argue that in a particular application, the size of the minimum variance (19.53) is of importance. The reason is that (19.53) is associated with the minimum width of the prediction interval for $\beta + b_0$ that might be approached in a particular application by letting the sample size for the new trial increase toward infinity. This minimum width will be responsible for the loss of efficiency related to the use of the surrogate, pointed out in Gail et al. (2000). It would thus be important to quantify the loss of efficiency, since it may be counterbalanced by a shortening of trial duration. One might consider using the ratio of (19.53) to $\mathrm{Var}(b_0)$, the unconditional variance of $\beta + b_0$. However, Burzykowski and Buyse (2006) considered another way of expressing this information, which should be more meaningful clinically.

19.8.1 Surrogate Threshold Effect

We will outline the proposal made by Burzykowski and Buyse (2006) and first focus on the case where the surrogate and true endpoints are jointly normally distributed. Assume that the prediction of $\beta + b_0$ can be made independently of μ_{s0}. Under this assumption, the conditional mean of $\beta + b_0$ is a simple linear function of α_0, the treatment effect on the surrogate, while the conditional variance can be written as

$$\mathrm{Var}\left(\beta + b_0 | \alpha_0, \vartheta\right) = \mathrm{Var}\left(b_0\right)\left(1 - R_{\mathrm{trial(r)}}^2\right). \tag{19.54}$$

The coefficient of determination $R_{\mathrm{trial(r)}}^2$ in (19.54) is simply the square of the correlation coefficient of trial-specific random effects b_i and a_i. If ϑ were known and α_0 could be observed without measurement error (i.e., assuming an infinite sample size for the new trial), the prediction variance would equal (19.54). In practice, an estimate $\widehat{\vartheta}$ is used and then prediction variance (19.53) ought to be applied:

$$\mathrm{Var}\left(\beta + b_0 | \alpha_0, \vartheta\right) \approx f\left\{\mathrm{Var}\left(\widehat{\vartheta}\right)\right\} + \left(1 - R_{\mathrm{trial(r)}}^2\right)\mathrm{Var}\left(b_0\right). \tag{19.55}$$

Since in linear mixed models, the maximum likelihood estimates of the covariance parameters are asymptotically independent of the fixed effects parameters (Verbeke and Molenberghs 2000), one can show that the prediction variance (19.55) can be expressed approximately as a quadratic function of α_0.

Consider a $(1 - \gamma)100\%$ prediction interval for $\beta + b_0$:

$$E\,(\beta + b_0|\alpha_0, \vartheta) \pm z_{1-\frac{\gamma}{2}}\sqrt{\mathrm{Var}\,(\beta + b_0|\alpha_0, \vartheta)}, \tag{19.56}$$

where $z_{1-\gamma/2}$ is the $(1-\gamma/2)$ quantile of the standard normal distribution. The limits of the interval (19.56) are functions of α_0. Define the lower and upper prediction limit functions of α_0 as

$$l(\alpha_0), u(\alpha_0) \equiv E(\beta + b_0|\alpha_0, \vartheta) \pm z_{1-\frac{\gamma}{2}}\sqrt{\mathrm{Var}(\beta + b_0|\alpha_0, \vartheta)}. \tag{19.57}$$

One might then compute a value of α_0 such that

$$l(\alpha_0) = 0. \tag{19.58}$$

Depending on the setting, one could also consider the upper limit $u\,(\alpha_0)$. We will call this value the *surrogate threshold effect* (STE). Its magnitude depends on the variance of the prediction. The larger the variance, the larger the absolute value of STE. From a clinical point of view, a large value of STE points to the need of observing a large treatment effect on the surrogate endpoint in order to conclude a nonzero effect on the true endpoint. In such a case, the use of the surrogate would not be reasonable, even if the surrogate were *potentially* valid, that is, with $R^2_{\mathrm{trial}(r)} \simeq 1$. The STE can thus provide additional important information about the usefulness of the surrogate in a particular application.

Note that the interval (19.56) and the prediction limit function $l\,(\alpha_0)$ can be constructed using the variances given by (19.54) or (19.55). Consequently, one might get two versions of STE. The version obtained from using (19.54) will be denoted by $\mathrm{STE}_{\infty,\infty}$. The infinity signs indicate that the measure assumes the knowledge of both of ϑ as well as of α_0, achievable only with an infinite number of infinite-sample-size trials in the meta-analytic data and an infinite sample size for the new trial. In practice, $\mathrm{STE}_{\infty,\infty}$ will be computed using estimates. A large value of $\mathrm{STE}_{\infty,\infty}$ would point to the need of observing a large treatment effect on the surrogate endpoint even if there were no estimation error present. In this case, one would question even the *potential* validity of the surrogate.

If the variance (19.55) is used to define $l(\alpha_0)$, we will denote the STE by $\mathrm{STE}_{N,\infty}$, with N indicating the need for the estimation of ϑ. $\mathrm{STE}_{N,\infty}$ captures the *practical* validity of the surrogate, which accounts for the need of estimating parameters of model (19.12) and (19.13). It is possible that a surrogate might seem to be *potentially valid* (low $\mathrm{STE}_{\infty,\infty}$ value), but might not be valid *practically* (large $\mathrm{STE}_{N,\infty}$ value), owing to the loss of precision resulting from estimation of the mixed-effects model parameters. The roots of (19.58) can be obtained by solving a quadratic equation. The number of solutions of the equation depends on the parameter configuration in $l(\alpha_0)$ (Burzykowski and Buyse 2006).

$STE_{\infty,\infty}$ and $STE_{N,\infty}$ can address concerns about the usefulness of the meta-analytic approach, expressed by Gail et al. (2000). They noted that, even for a valid surrogate, the variance of the prediction of treatment effect on the true endpoint cannot be reduced to 0, even in the absence of any estimation error. $STE_{N,\infty}$ can be used to quantify this loss of efficiency.

Interestingly, the STE can be expressed in terms of treatment effect on the surrogate necessary to be observed to predict a significant treatment effect on the true endpoint. In a practical application, one would seek a value of STE (preferably, $STE_{N,\infty}$) well within the range of treatment effects on surrogates observed in previous clinical trials, as close as possible to the (weighted) mean effect.

STE and its estimation have been developed under the mixed-effects model (19.12) and (19.13), but Burzykowski and Buyse (2006) also derived the STE when, perhaps for numerical convenience, the two-stage approach of Section 19.4.2 is used. Furthermore, STE can be computed for any type of surrogate. To this aim, one merely needs to use an appropriate joint model for surrogate and true endpoints, capable of providing the required treatment effect. Burzykowski and Buyse (2006) presented time-to-event applications.

19.9 Concluding Remarks

Over the years, a variety of surrogate marker evaluation strategies have been proposed, cast within a meta-analytic framework. With an increasing range of endpoint types considered, such as continuous, binary, time-to-event, and longitudinal endpoints, also the scatter of types of measures proposed has increased. Some of these measures are difficult to calculate from fully specified hierarchical models, which has sparked off the formulation of simplified strategies. We reviewed the ensuing divergence of proposals, which then has triggered efforts of convergence, eventually leading to the information-theoretic approach, which is both general and simple to implement. These developments have been illustrated using data from clinical trials in schizophrenia.

While quantifying surrogacy is important, so is prediction of the treatment effect in a new trial based on the surrogate. Work done in this area has been reviewed, with emphasis on the so-called STE and the sources of variability involved in the prediction process. A connection with the information-theoretic approach is pointed out.

Even though more work is called for, we believe the information-theoretic approach and the STE are promising paths toward effective assessment and use of surrogate endpoints in practice. Software implementations for methodology described here and beyond are available from www.ibiostat.be.

A key issue is whether a surrogate is still valid if, in a new trial, the same surrogate and true endpoints, but a different drug is envisaged. This is the so-called *class* question. It is usually argued that a surrogate could still be used if the new drug belongs to the same class of drugs as the ones in the evaluation exercise. Of course, this in itself is rather subjective and clinical expertise is necessary to meaningfully delineate a drug class.

Acknowledgment

The authors gratefully acknowledge support from IAP research Network P7/06 of the Belgian Government (Belgian Science Policy).

References

Alonso, A., Geys, H., Molenberghs, G., and Vangeneugden, T. (2003). Valida-tion of surrogate markers in multiple randomized clinical trials with repeated measurements. *Biometrical Journal*, 45:931–945.

Alonso, A. and Molenberghs, G. (2007). Surrogate marker evaluation from an infor-mation theoretic perspective. *Biometrics*, 63:180–186.

Alonso, A., Molenberghs, G., Geys, H., and Buyse, M. (2005). A unifying approach for surrogate marker validation based on Prentice's criteria. *Statistics in Medicine*, 25:205–211.

Alonso, A., Van der Elst, W., Molenberghs, G., Buyse, M., and Burzykowski, T. (2013). On the relationship between the causal-inference and meta-analytic paradigms for the validation of surrogate endpoints. Technical report.

Assam, P., Tilahun, A., Alonso, A., and Molenberghs, G. (2007). Information-theory based surrogate marker evaluation from several randomized clinical trials with continuous true and binary surrogate endpoints. *Clinical Trials*, 4:587–597.

Begg, C. and Leung, D. (2000). On the use of surrogate endpoints in randomized trials. *Journal of the Royal Statistical Society, Series A*, 163:26–27.

Burzykowski, T. and Buyse, M. (2006). Surrogate threshold effect: An alternative mea-sure for meta-analytic surrogate endpoint validation. *Pharmaceutical Statistics*, 5:173–186.

Burzykowski, T., Molenberghs, G., and Buyse, M. (2004). The validation of surrogate endpoints using data from randomized clinical trials: A case-study in advanced colorectal cancer. *Journal of the Royal Statistical Society, Series A*, 167:103–124.

Burzykowski, T., Molenberghs, G., and Buyse, M. (2005). *The Evaluation of Surrogate Endpoints*. Springer, New York.

Buyse, M. and Molenberghs, G. (1998). Criteria for the validation of surrogate end-points in randomized experiments. *Biometrics*, 54:1014–1029.

Buyse, M., Molenberghs, G., Burzykowski, T., Renard, D., and Geys, H. (2000). The validation of surrogate endpoints in meta-analyses of randomized experiments. *Biostatistics*, 1:49–68.

Choi, S., Lagakos, S., Schooley, R., and Volberding, P. (1993). CD4+ lymphocytes are an incomplete surrogate marker for clinical progression in persons with asymptomatic HIV infection taking zidovudine. *Annals of Internal Medicine*, 118:674–680.

Clayton, D. (1978). A model for association in bivariate life tables and its application in epidemiological studies of familial tendency in chronic disease incidence. *Biometrika*, 65:141–151.

Cortinas Abrahantes, J., Molenberghs, G., Burzykowski, T., Shkedy, Z., and Renard, D. (2004). Choice of units of analysis and modeling strategies in multi-level hierarchical models. *Computational Statistics and Data Analysis*, 47:537–563.

Cover, T. and Tomas, J. (1991). *Elements of Information Theory*. John Wiley & Sons, New York.

Dale, J. (1986). Global cross ratio models for bivariate, discrete, ordered responses. *Biometrics*, 42:909–917.

Daniels, M. and Hughes, M. (1997). Meta-analysis for the evaluation of potential surrogate markers. *Statistics in Medicine*, 16:1965–1982.

DeGruttola, V., Fleming, T., Lin, D., and Coombs, R. (1997). Validating surrogate markers—Are we being naive? *Journal of Infectious Diseases*, 175:237–246.

DeGruttola, V. and Tu, X. (1994). Modelling progression of CD-4 lymphocyte count and its relationship to survival time. *Biometrics*, 50:1003–1014.

Dunn, N. and Mann, R. (1999). Prescription-event and other forms of epidemiological monitoring of side-effects in the UK. *Clinical and Experimental Allergy*, 29:217–239.

Ellenberg, S. and Hamilton, J. (1989). Surrogate endpoints in clinical trials: Cancer. *Statistics in Medicine*, 8:405–413.

Ferentz, A. (2002). Integrating pharmacogenomics into drug development. *Pharmacogenomics*, 3:453–467.

Flandre, P. and Saidi, Y. (1999). Letter to the editor: Estimating the proportion of treatment effect explained by a surrogate marker. *Statistics in Medicine*, 18:107–115.

Fleming, T. (1994). Surrogate markers in AIDS and cancer trials. *Statistics in Medicine*, 13:1423–1435.

Fleming, T. and DeMets, D. (1996). Surrogate end points in clinical trials: Are we being misled? *Annals of Internal Medicine*, 125:605–613.

Frangakis, C. and Rubin, D. (2004). Principal stratification in causal inference. *Biometrics*, 58:21–29.

Freedman, L., Graubard, B., and Schatzkin, A. (1992). Statistical validation of intermediate endpoints for chronic diseases. *Statistics in Medicine*, 11:167–178.

Gail, M., Pfeiffer, R., van Houwelingen, H., and Carroll, R. (2000). On meta-analytic assessment of surrogate outcomes. *Biostatistics*, 1:231–246.

Galecki, A. (1994). General class of covariance structures for two or more repeated factors in longitudinal data analysis. *Communications in Statistics: Theory and Methods*, 23:3105–3119.

Genest, C. and McKay, J. (1986). The joy of copulas: Bivariate distributions with uniform marginals. *American Statistician*, 40:280–283.

Group, B. D. W. (2001). Biomarkers and surrogate endpoints: Preferred definitions and conceptual framework. *Clinical Pharmacological Therapy*, 69:89–95.

Heise, C., Sampson-Johannes, A., Williams, A., McCormick, F., Von Hoff, D., and Kirn, D. (1997). Onyx-015, an e1b gene-attenuated adenovirus, causes tumor-specific cytolysis and antitumoral efficacy that can be augmented by standard chemo-therapeutic agents. *Nature Medicine*, 3:639–645.

Hougaard, P. (1986). Survival models for heterogeneous populations derived from stable distributions. *Biometrika*, 73:387–396.

Investigators, C. A. S. T. C. (1989). Preliminary report: Effect of encainide and flecainide on mortality in a randomized trial of arrhythmia suppression after myocardial infraction. *New England Journal of Medicine*, 321:406–412.

Joffe, M. and Greene, T. (2008). Related causal frameworks for surrogate outcomes. *Biometrics*, 64:1–10.

Jones, T. (2001). Call for a new approach to the process of clinical trials and drug registration. *British Medical Journal*, 322:920–923.

Kay, S., Fiszbein, A., and Opler, L. (1987). The positive and negative syndrome scale (PANSS) for schizophrenia. *Schizophrenia Bulletin*, 13:261–276.

Kay, S., Opler, L., and Lindenmayer, J. (1988). Reliability and validity of the positive and negative syndrome scale for schizophrenics. *Psychiatric Research*, 23:99–110.

Kent, J. (1983). Information gain and a general measure of correlation. *Biometrika*, 70:163–173.

Lagakos, S. and Hoth, D. (1992). Surrogate markers in AIDS: Where are we? where are we going? *Annals of Internal Medicine*, 116:599–601.

Lesko, L. and Atkinson, A. (2001). Use of biomarkers and surrogate endpoints in drug development and regulatory decision making: Criteria, validation, strategies. *Annual Review of Pharmacological Toxicology*, 41:347–366.

Lin, D., Fleming, T., and DeGruttola, V. (1997). Estimating the proportion of treatment effect explained by a surrogate marker. *Statistics in Medicine*, 16:1515–1527.

Molenberghs, G., Burzykowski, T., Alonso, A., Assam, P., Tilahun, A., and Buyse, M. (2010). A unified framework for the evaluation of surrogate endpoints in clinical trials. *Statistical Methods in Medical Research*, 19:205–236.

Molenberghs, G., Geys, H., and Buyse, M. (2001). Evaluation of surrogate end-points in randomized experiments with mixed discrete and continuous outcomes. *Statistics in Medicine*, 20:3023–3038.

Molenberghs, G. and Verbeke, G. (2005). *Models for Discrete Longitudinal Data*. Springer, New York.

Pearl, J. (2001). *Causality: Models, Reasoning, and Inference*. Cambridge University Press, Cambridge, U.K.

Prentice, R. (1989). Surrogate endpoints in clinical trials: Definitions and operational criteria. *Statistics in Medicine*, 8:431–440.

Renard, D., Geys, H., Molenberghs, G., Burzykowski, T., and Buyse, M. (2002). Validation of surrogate endpoints in multiple randomized clinical trials with discrete outcomes. *Biometrical Journal*, 44:1–15.

Robins, J. and Greenland, S. (1992). Identifiability and exchangeability for direct and indirect effects. *Epidemiology*, 3:143–155.

Schatzkin, A. and Gail, M. (2002). The promise and peril of surrogate end points in cancer research. *Nature Reviews Cancer*, 2:19–27.

Schemper, M. and Stare, J. (1996). Explained variation in survival analysis. *Statistics in Medicine*, 15:1999–2012.

Shannon, C. (1948). A mathematical theory of communication. *Bell System Technical Journal*, 27:379–423 and 623–656.

Shih, J. and Louis, T. (1995). Inferences on association parameter in copula models for bivariate survival data. *Biometrics*, 51:1384–1399.

Taylor, J., Wang, Y., and Thiébaut, R. (2005). Counterfactual links to the proportion of treatment effect explained by a surrogate marker. *Biometrics*, 61:1102–1111.

The International Conference on Harmonisation of Technical Requirements for Registration of Pharmaceuticals for Human Use, (ICH) (1998). ICH harmonised tripartite guideline. Statistical principles for clinical trials. *Federal Register* 63:179:49583.

Tibaldi, F., Cortiñas Abrahantes, J., Molenberghs, G., Renard, D., Burzykowski, T., Buyse, M., Parmar, M., Stijnen, T., and Wolfinger, R. (2003). Simplified hierarchical linear models for the evaluation of surrogate endpoints. *Journal of Statistical Computation and Simulation*, 73:643–658.

Tilahun, A., Assam, P., Alonso, A., and Molenberghs, G. (2008). Information theory-based surrogate marker evaluation from several randomized clinical trials with binary endpoints, using SAS. *Journal of Biopharmaceutical Statistics*, 18:326–341.

Verbeke, G. and Molenberghs, G. (2000). *Linear Mixed Models for Longitudinal Data*. Springer, New York.

Volberding, P., Lagakos, S., Koch, M. A. et al. (1990). Zidovudine in asymptomatic human immunodeficiency virus infection:a controlled trial in persons with fewer than 500 cd4-positive cells per cubic millimeter. *New England Journal of Medicine*, 322(14):941–949.

Index

Printed and bound by CPI Group (UK) Ltd, Croydon, CR0 4YY

24/10/2024

01778301-0017